Discrete-Event System Simulation
FOURTH EDITION

Jerry Banks
Independent Consultant

John S. Carson II
Brooks Automation

Barry L. Nelson
Northwestern University

David M. Nicol
University of Illinois, Urbana-Champaign

PRENTICE-HALL INTERNATIONAL SERIES IN INDUSTRIAL AND SYSTEMS ENGINEERING
W. J. Fabrycky and J. H. Mize, editors

PEARSON
Prentice
Hall

Upper Saddle River, New Jersey 07458

Library of Congress Cataloging-in-Publication Data

Discrete-event system simulation / Jerry Banks... [et al].–4th ed.
   p. cm.
  ISBN 0-13-144679-7
   1. Simulation methods. I. Banks, Jerry.

T57.62.D53 2004
003'.83–dc22

2004053354

Vice President and Editorial Director, ECS: *Marcia J. Horton*
Vice President and Director of Production and Manufacturing, ESM: *David W. Riccardi*
Executive Managing Editor: *Vince O'Brien*
Managing Editor: *David A. George*
Production Editors: *Daniel Sandin and Wendy Kopf*
Director of Creative Services: *Paul Belfanti*
Art Director: *Jayne Conte*
Cover Designer: *Bruce Kenselaar*
Art Editor: *Greg Dulles*
Manufacturing Buyer: *Lisa McDowell*
Senior Marketing Manager: *Holly Stark*

© 2005, 2001, 1999 by Pearson Education, Inc.
Pearson Prentice Hall
Pearson Education, Inc.
Upper Saddle River, NJ 07458

Pearson Prentice Hall™ is a trademark of Pearson Education, Inc.

Mathematica and the Mathematica logo are registered trademarks of Wolfram Research, Inc., 100 Trade Center Drive, Champaign, IL 61820-7237.

The author and publisher of this book have used their best efforts in preparing this book. These efforts include the development, research, and testing of the theories and programs to determine their effectiveness. The author and publisher make no warranty of any kind, expressed or implied, with regard to these programs or the documentation contained in this book. The author and publisher shall not be liable in any event for incidental or consequential damages in connection with, or arising out of, the furnishing, performance, or use of these programs.

Printed in the United States of America
10 9 8 7 6 5 4 3 2 1

ISBN 0-13-144679-7

Pearson Education Ltd., *London*
Pearson Education Australia Pty. Ltd., *Sydney*
Pearson Education Singapore, Pte. Ltd.
Pearson Education North Asia Ltd., *Hong Kong*
Pearson Education Canada, Inc., *Toronto*
Pearson Educación de Mexico, S.A. de C.V.
Pearson Education—Japan, *Tokyo*
Pearson Education Malaysia, Pte. Ltd.
Pearson Education, Inc., *Upper Saddle River, New Jersey*

Contents

Preface

The objective of the text is to provide a basic treatment of all of the important aspects of discrete-event simulation, with particular emphasis on applications in manufacturing, services, and computing. The fourth edition, like earlier editions, is meant for an upper-level-undergraduate or master's-level introduction to simulation or for a second course with applications. We have updated the material extensively, revised some chapters completely, and added a new chapter on the simulation of computer networks. The associated website www.bcnn.net is now closely integrated with the text, and all students and instructors should visit the site.

Chapter 1, *Introduction to Simulation*, has been generally updated, and every example in Chapter 2, *Simulation Examples*, has an Excel spreadsheet solution on the website. Exercises have been prepared that require downloading these spreadsheet solutions and using them. To reflect the continuing evolution of simulation software, Chapter 3, *General Principles*, has been modernized to include properties and operations of current simulation languages; in Chapter 4, *Simulation Software*, simulation in Java replaces C++. We also have maintained an up-to-date discussion of the features of currently available simulation software. Simulation software changes so rapidly, however, that we point to the websites of all of the software vendors mentioned in the text.

Chapter 5, *Statistical Models in Simulation*, incorporates some additional models: the beta and negative binomial distributions and the nonstationary Poisson process. These are backed up by new material on simulating (Chapter 8, *Random-Variate Generation*) and fitting (Chapter 9, *Input Modeling*) the models. For clarity, Chapter 8 has been substantially reorganized. In Chapter 7, *Random-Number Generation*, we have deemphasized statistical testing of random-number generators since the period length of modern generators has become so long that sampling-based tests are no longer feasible.

Chapter 10, *Verification and Validation*, replaces hypothesis testing by a confidence-interval approach for input–output validation.

The core chapters on the analysis of simulation output are Chapter 11, *Output Analysis for a Single Model*, and Chapter 12, *Comparison and Evaluation of Alternative System Designs*. Chapter 11 has been significantly reorganized, and there is new material on prediction intervals and on estimating probabilities and quantiles from only summary statistics. Chapter 12 contains a new procedure for screening a large number of system designs to extract a smaller group of the best.

Chapter 13, *Simulation of Manufacturing and Material-Handling Systems*, adds an extended example and analysis of a small manufacturing system.

Chapter 14, *Simulation of Computer Systems*, replaces the discussion of C++ simulation tools for computer simulation with one focused on Java in general and on the SSFNet simulator in particular. Chapter 15, *Simulation of Computer Networks*, is new. The website has examples (in Java) of simulations discussed in this Chapter and provides extensive links to supporting material.

Discrete-Event System Simulation can serve as a textbook in the following types of courses:

1. An introductory simulation course in engineering, computer science, or management (Chapters 1–9 and selected parts of Chapters 10–12 when no companion language text is used; if a companion language text is used, skip Chapter 4, and use the application Chapters 13, 14, and 15, as appropriate);
2. A second course in simulation (all of Chapters 10–12, a companion language text, and an outside project; add Chapter 13, 14, or 15, as appropriate)

We gratefully acknowledge the cheerful aid of Gamze Tokol and Dave Goldsman in converting some of the chapters to LaTeX for those coauthors who do not speak the language, and the assistance of Feng Yang, who checked the references and exercises in many chapters.

JERRY BANKS
JOHN S. CARSON II
BARRY L. NELSON
DAVID M. NICOL

About the Authors

Jerry Banks retired in 1999 as a professor in the School of Industrial and Systems Engineering, Georgia Institute of Technology, after which he worked as senior simulation technology advisor for Brooks Automation; he is currently an independent consultant. He is the author, coauthor, editor, or coeditor of eleven books, one set of proceedings, several chapters in texts, and numerous technical papers. He is the editor of the *Handbook of Simulation*, published in 1998 by John Wiley, which won the award for Excellence in Engineering Handbooks from the Professional Scholarly Publishing Division of the Association of American Publishers, Inc. He is also author or coauthor of *Getting Started with AutoMod*, Second Edition, *Introduction to SIMAN V and CINEMA V*, *Getting Started with GPSS/H*, Second Edition, *Forecasting and Management of Technology* and *Principles of Quality Control*. He was a founding partner in the simulation-consulting firm Carson/Banks & Associates, Inc., which was purchased by AutoSimulations, Inc. (now part of Brooks Automation). He is a full member of many technical societies, among them the Institute of Industrial Engineers (IIE); he served eight years as that organization's representative to the Board of the Winter Simulation Conference, including two years as board chair. He is the recipient of the INFORMS College on Simulation Distinguished Service Award for 1999 and was named a fellow of IIE in 2002.

John S. Carson II is the consulting technical manager for the AutoMod Group at Brooks Automation. He has over 28 years experience in simulation in a wide range of application areas, including manufacturing, distribution, warehousing and material handling, transportation and rapid transit systems, port operations (container terminals and bulk handling), and health-care systems. Currently, he is involved in the design of next-generation simulation products and in the development of tools to speed up model development for semi-conductor manufacturing, distribution centers, container terminals and other areas of special interest. He co-founded and managed an independent simulation services company for 8 years, has been an independent simulation consultant, and has taught at the Georgia Institute of Technology, the University of Florida, and the University of Wisconsin-Madison.

Barry L. Nelson is the James N. and Margie M. Krebs Professor in the Department of Industrial Engineering and Management Sciences at Northwestern University and is director of the Master of Engineering Management Program there. His research centers on the design and analysis of computer-simulation experiments on models of stochastic systems, concentrating on multivariate input modeling and output analysis and on optimization via simulation. He has published numerous papers and two books. He has served as the simulation area editor of *Operations Research* and as president of the INFORMS (then TIMS) College on Simulation, and he has held many positions for the annual Winter Simulation Conference, including program chair in 1997 and board member currently.

David M. Nicol is professor of electrical and computer engineering at the University of Illinois at Urbana-Champaign. He is a long-time contributor in the field of parallel and distributed discrete-event simulations, having written one of the early Ph.D. dissertations on the topic. He has also worked in parallel algorithms, algorithms for mapping workload in parallel architectures, performance analysis, and reliability modeling and analysis. His research contributions extend to 150 articles in leading computer-science journals and conferences. His research is driven largely by problems encountered in industry and government—he has worked closely with researchers at NASA, IBM, AT&T, Bellcore, Motorola, and the Los Alamos, Sandia, and Oak Ridge National Laboratories. His current interests lie in modeling and simulation of very large systems, particularly communications and other infrastructure, with applications in evaluating system security. From 1997 to 2003 he was the editor-in-chief of the *ACM Transactions on Modeling and Computer Simulation*. Professor Nicol is a Fellow of the IEEE.

Part I

Introduction to Discrete-Event System Simulation

1

Introduction to Simulation

A *simulation* is the imitation of the operation of a real-world process or system over time. Whether done by hand or on a computer, simulation involves the generation of an artificial history of a system and the observation of that artificial history to draw inferences concerning the operating characteristics of the real system.

The behavior of a system as it evolves over time is studied by developing a simulation *model*. This model usually takes the form of a set of assumptions concerning the operation of the system. These assumptions are expressed in mathematical, logical, and symbolic relationships between the *entities*, or objects of interest, of the system. Once developed and validated, a model can be used to investigate a wide variety of "what if" questions about the real-world system. Potential changes to the system can first be simulated, in order to predict their impact on system performance. Simulation can also be used to study systems in the design stage, before such systems are built. Thus, simulation modeling can be used both as an analysis tool for predicting the effect of changes to existing systems and as a design tool to predict the performance of new systems under varying sets of circumstances.

In some instances, a model can be developed which is simple enough to be "solved" by mathematical methods. Such solutions might be found by the use of differential calculus, probability theory, algebraic methods, or other mathematical techniques. The solution usually consists of one or more numerical parameters, which are called measures of performance of the system. However, many real-world systems are so complex that models of these systems are virtually impossible to solve mathematically. In these instances, numerical, computer-based simulation can be used to imitate the behavior of the system over time. From the simulation, data are collected as if a real system were being observed. This simulation-generated data is used to estimate the measures of performance of the system.

This book provides an introductory treatment of the concepts and methods of one form of simulation modeling—discrete-event simulation modeling. The first chapter initially discusses when to use simulation, its advantages and disadvantages, and actual areas of its application. Then the concepts of system and model are explored. Finally, an outline is given of the steps in building and using a simulation model of a system.

1.1 WHEN SIMULATION IS THE APPROPRIATE TOOL

The availability of special-purpose simulation languages, of massive computing capabilities at a decreasing cost per operation, and of advances in simulation methodologies have made simulation one of the most widely used and accepted tools in operations research and systems analysis. Circumstances under which simulation is the appropriate tool to use have been discussed by many authors, from Naylor *et al.* [1966] to Shannon [1998]. Simulation can be used for the following purposes:

1. Simulation enables the study of, and experimentation with, the internal interactions of a complex system or of a subsystem within a complex system.
2. Informational, organizational, and environmental changes can be simulated, and the effect of these alterations on the model's behavior can be observed.
3. The knowledge gained during the designing of a simulation model could be of great value toward suggesting improvement in the system under investigation.
4. Changing simulation inputs and observing the resulting outputs can produce valuable insight into which variables are the most important and into how variables interact.
5. Simulation can be used as a pedagogical device to reinforce analytic solution methodologies.
6. Simulation can be used to experiment with new designs or policies before implementation, so as to prepare for what might happen.
7. Simulation can be used to verify analytic solutions.
8. Simulating different capabilities for a machine can help determine the requirements on it.
9. Simulation models designed for training make learning possible without the cost and disruption of on-the-job instruction.
10. Animation shows a system in simulated operation so that the plan can be visualized.
11. The modern system (factory, wafer fabrication plant, service organization, etc.) is so complex that its internal interactions can be treated only through simulation.

1.2 WHEN SIMULATION IS NOT APPROPRIATE

This section is based on an article by Banks and Gibson [1997], who gave ten rules for evaluating when simulation is not appropriate. The first rule indicates that simulation should not be used when the problem can be solved by common sense. An example is given of an automobile tag facility serving customers who arrive randomly at an average rate of 100/hour and are served at a mean rate of 12/hour. To determine the minimum number of servers needed, simulation is not necessary. Just compute $100/12 = 8.33$ indicating that nine or more servers are needed.

The second rule says that simulation should not be used if the problem can be solved analytically. For example, under certain conditions, the average waiting time in the example above can be found from curves that were developed by Hillier and Lieberman [2002].

The next rule says that simulation should not be used if it is easier to perform direct experiments. An example of a fast-food drive-in restaurant is given where it was less expensive to stage a person taking orders using a hand-held terminal and voice communication to determine the effect of adding another order station on customer waiting time.

The fourth rule says not to use simulation if the costs exceed the savings. There are many steps in completing a simulation, as will be discussed in Section 1.11, and these must be done thoroughly. If a simulation study costs $20,000 and the savings might be $10,000, simulation would not be appropriate.

Rules five and six indicate that simulation should not be performed if the resources or time are not available. If the simulation is estimated to cost $20,000 and there is only $10,000 available, the suggestion is not to venture into a simulation study. Similarly, if a decision in needed in two weeks and a simulation will take a month, the simulation study is not advised.

Simulation takes data, sometimes lots of data. If no data is available, not even estimates, simulation is not advised. The next rule concerns the ability to verify and validate the model. If there is not enough time or if the personnel are not available, simulation is not appropriate.

If managers have unreasonable expectations, if they ask for too much too soon, or if the power of simulation is overestimated, simulation might not be appropriate.

Last, if system behavior is too complex or can't be defined, simulation is not appropriate. Human behavior is sometimes extremely complex to model.

1.3 ADVANTAGES AND DISADVANTAGES OF SIMULATION

Simulation is intuitively appealing to a client because it mimics what happens in a real system or what is perceived for a system that is in the design stage. The output data from a simulation should directly correspond to the outputs that could be recorded from the real system. Additionally, it is possible to develop a simulation model of a system without dubious assumptions (such as the same statistical distribution for every random variable) of mathematically solvable models. For these and other reasons, simulation is frequently the technique of choice in problem solving.

In contrast to optimization models, simulation models are "run" rather than solved. Given a particular set of input and model characteristics, the model is run and the simulated behavior is observed. This process of changing inputs and model characteristics results in a set of scenarios that are evaluated. A good solution, either in the analysis of an existing system or in the design of a new system, is then recommended for implementation.

Simulation has many advantages, but some disadvantages. These are listed by Pegden, Shannon, and Sadowski [1995]. Some advantages are these:

1. New policies, operating procedures, decision rules, information flows, organizational procedures, and so on can be explored without disrupting ongoing operations of the real system.

2. New hardware designs, physical layouts, transportation systems, and so on can be tested without committing resources for their acquisition.
3. Hypotheses about how or why certain phenomena occur can be tested for feasibility.
4. Time can be compressed or expanded to allow for a speed-up or slow-down of the phenomena under investigation.
5. Insight can be obtained about the interaction of variables.
6. Insight can be obtained about the importance of variables to the performance of the system.
7. Bottleneck analysis can be performed to discover where work in process, information, materials, and so on are being delayed excessively.
8. A simulation study can help in understanding how the system operates rather than how individuals think the system operates.
9. "What if" questions can be answered. This is particularly useful in the design of new systems.

Some disadvantages are these:

1. Model building requires special training. It is an art that is learned over time and through experience. Furthermore, if two models are constructed by different competent individuals, they might have similarities, but it is highly unlikely that they will be the same.
2. Simulation results can be difficult to interpret. Most simulation outputs are essentially random variables (they are usually based on random inputs), so it can be hard to distinguish whether an observation is a result of system interrelationships or of randomness.
3. Simulation modeling and analysis can be time consuming and expensive. Skimping on resources for modeling and analysis could result in a simulation model or analysis that is not sufficient to the task.
4. Simulation is used in some cases when an analytical solution is possible, or even preferable, as was discussed in Section 1.2. This might be particularly true in the simulation of some waiting lines where closed-form queueing models are available.

In defense of simulation, these four disadvantages, respectively, can be offset as follows:

1. Vendors of simulation software have been actively developing packages that contain models that need only input data for their operation. Such models have the generic tag "simulator" or "template."
2. Many simulation software vendors have developed output-analysis capabilities within their packages for performing very thorough analysis.
3. Simulation can be performed faster today than yesterday and will be even faster tomorrow, because of advances in hardware that permit rapid running of scenarios and because of advances in many simulation packages. For example, some simulation software contains constructs for modeling material handling that uses such transporters as fork-lift trucks, conveyors, and automated guided vehicles.
4. Closed-form models are not able to analyze most of the complex systems that are encountered in practice. In many years of consulting practice by two of the authors, not one problem was encountered that could have been solved by a closed-form solution.

1.4 AREAS OF APPLICATION

The applications of simulation are vast. The Winter Simulation Conference (WSC) is an excellent way to learn more about the latest in simulation applications and theory. There are also numerous tutorials at both the beginning and the advanced levels. WSC is sponsored by six technical societies and the National Institute of Standards and Technology (NIST). The technical societies are American Statistical Association (ASA), Association for Computing Machinery/Special Interest Group on Simulation (ACM/SIGSIM), Institute of Electrical and Electronics Engineers: Computer Society (IEEE/CS), Institute of Electrical and Electronics Engineers: Systems, Man and Cybernetics Society (IEEE/SMCS), Institute of Industrial Engineers (IIE), Institute for Operations Research and the Management Sciences: College on Simulation (INFORMS/CS) and The Society for Computer Simulation (SCS). Note that IEEE is represented by two bodies. Information about the upcoming WSC can be obtained from www.wintersim.org. WSC programs with full papers are available from www.informs-cs.org/wscpapers.html. Some presentations, by area, from a recent WSC are listed next:

Manufacturing Applications

Dynamic modeling of continuous manufacturing systems, using analogies to electrical systems

Benchmarking of a stochastic production planning model in a simulation test bed

Paint line color change reduction in automobile assembly

Modeling for quality and productivity in steel cord manufacturing

Shared resource capacity analysis in biotech manufacturing

Neutral information model for simulating machine shop operations

Semiconductor Manufacturing

Constant time interval production planning with application to work-in-process control

Accelerating products under due-date oriented dispatching rules

Design framework for automated material handling systems in 300-mm wafer fabrication factories

Making optimal design decisions for next-generation dispensing tools

Application of cluster tool modeling in a 300-mm wafer fabrication factory

Resident-entity based simulation of batch chamber tools in 300-mm semiconductor manufacturing

Construction Engineering and Project Management

Impact of multitasking and merge bias on procurement of complex equipment

Application of lean concepts and simulation for drainage operations maintenance crews

Building a virtual shop model for steel fabrication

Simulation of the residential lumber supply chain

Military Applications

Frequency-based design for terminating simulations: A peace-enforcement example

A multibased framework for supporting military-based interactive simulations in 3D environments

Specifying the behavior of computer-generated forces without programming

Fidelity and validity: Issues of human behavioral representation

Assessing technology effects on human performance through trade-space development and evaluation

Impact of an automatic logistics system on the sortie-generation process

Research plan development for modeling and simulation of military operations in urban terrain

Logistics, Supply Chain, and Distribution Applications

Inventory analysis in a server–computer manufacturing environment

Comparison of bottleneck detection methods for AGV systems

Semiconductor supply-network simulation

Analysis of international departure passenger flows in an airport terminal

Application of discrete simulation techniques to liquid natural gas supply chains

Online simulation of pedestrian flow in public buildings

Transportation Modes and Traffic

Simulating aircraft-delay absorption

Runway schedule determination by simulation optimization

Simulation of freeway merging and diverging behavior

Modeling ambulance service of the Austrian Red Cross

Simulation modeling in support of emergency firefighting in Norfolk

Modeling ship arrivals in ports

Optimization of a barge transportation system for petroleum delivery

Iterative optimization and simulation of barge traffic on an inland waterway

Business Process Simulation

Agent-based modeling and simulation of store performance for personalized pricing

Visualization of probabilistic business models

Modeling and simulation of a telephone call center

Using simulation to approximate subgradients of convex performance measures in service systems

Simulation's role in baggage screening at airports

Human-fatigue risk simulations in continuous operations

Optimization of a telecommunications billing system

Segmenting the customer base for maximum returns

Health Care

Modeling front office and patient care in ambulatory health care practices

Evaluation of hospital operations between the emergency department and a medical telemetry unit

Estimating maximum capacity in an emergency room

Reducing the length of stay in an emergency department

Simulating six-sigma improvement ideas for a hospital emergency department

A simulation–integer-linear-programming-based tool for scheduling emergency room staff

Some general trends in simulation applications are as follows: At present, simulation for risk analysis is growing, including in such areas as insurance, options pricing, and portfolio analysis. Another growing area is call-center analysis, which is not amenable to queuing models because of its complexity. Simulation of large-scale systems such as the internet backbone, wireless networks, and supply chains are growing as hardware and software increase their capability to handle extremely large numbers of entities in a reasonable time.

Lastly, simulation models of automated material handling systems (AMHS) are being used as test beds for the development and functional testing of control-system software. Called an emulation, the simulation model is connected in real time to the control-system software or to a software emulator; it is used to provide the same responses to a control system as the real AMHS does (for example, a box blocking or clearing a photo eye, or a command to start picking an order). Software development can begin much earlier during AMHS installation and commissioning, to reduce the time spent in the field on trying to debug control software while attempting to ramp up a new system or continue running an existing one. Models have been driven by control systems at various levels—from high-level supervisory systems, such as warehouse management systems (WMS) or AGV dispatching systems, to such low-level control as programmable logic controllers (PLCs) controlling merges on a conveyor system.

1.5 SYSTEMS AND SYSTEM ENVIRONMENT

To model a system, it is necessary to understand the concept of a system and the system boundary. A *system* is defined as a group of objects that are joined together in some regular interaction or interdependence toward the accomplishment of some purpose. An example is a production system manufacturing automobiles. The machines, component parts, and workers operate jointly along an assembly line to produce a high-quality vehicle.

A system is often affected by changes occurring outside the system. Such changes are said to occur in the *system environment* [Gordon, 1978]. In modeling systems, it is necessary to decide on the *boundary* between the system and its environment. This decision may depend on the purpose of the study.

In the case of the factory system, for example, the factors controlling the arrival of orders may be considered to be outside the influence of the factory and therefore part of the environment. However, if the effect of supply on demand is to be considered, there will be a relationship between factory output and arrival of orders, and this relationship must be considered an activity of the system. Similarly, in the case of a bank system, there could be a limit on the maximum interest rate that can be paid. For the study of a single bank, this would be regarded as a constraint imposed by the environment. In a study of the effects of monetary laws on the banking industry, however, the setting of the limit would be an activity of the system. [Gordon, 1978]

1.6 COMPONENTS OF A SYSTEM

In order to understand and analyze a system, a number of terms need to be defined. An *entity* is an object of interest in the system. An *attribute* is a property of an entity. An *activity* represents a time period of specified length. If a bank is being studied, customers might be one of the entities, the balance in their checking accounts might be an attribute, and making deposits might be an activity.

The collection of entities that compose a system for one study might only be a subset of the overall system for another study [Law and Kelton, 2000]. For example, if the aforementioned bank is being studied to determine the number of tellers needed to provide for paying and

Table 1.1 Examples of Systems and Components

System	Entities	Attributes	Activities	Events	State Variables
Banking	Customers	Checking-account balance	Making deposits	Arrival; departure	Number of busy tellers; number of customers waiting
Rapid rail	Riders	Origination; destination	Traveling	Arrival at station; arrival at destination	Number of riders waiting at each station; number of riders in transit
Production	Machines	Speed; capacity; breakdown rate	Welding; stamping	Breakdown	Status of machines (busy, idle, or down)
Communications	Messages	Length; destination	Transmitting	Arrival at destination	Number waiting to be transmitted
Inventory	Warehouse	Capacity	Withdrawing	Demand	Levels of inventory; backlogged demands

receiving, the system can be defined as that portion of the bank consisting of the regular tellers and the customers waiting in line. If the purpose of the study is expanded to determine the number of special tellers needed (to prepare cashier's checks, to sell traveler's checks, etc.), the definition of the system must be expanded.

The *state* of a system is defined to be that collection of variables necessary to describe the system at any time, relative to the objectives of the study. In the study of a bank, possible state variables are the number of busy tellers, the number of customers waiting in line or being served, and the arrival time of the next customer. An *event* is defined as an instantaneous occurrence that might change the state of the system. The term *endogenous* is used to describe activities and events occurring within a system, and the term *exogenous* is used to describe activities and events in the environment that affect the system. In the bank study, the arrival of a customer is an exogenous event, and the completion of service of a customer is an endogenous event.

Table 1.1 lists examples of entities, attributes, activities, events, and state variables for several systems. Only a partial listing of the system components is shown. A complete list cannot be developed unless the purpose of the study is known. Depending on the purpose, various aspects of the system will be of interest, and then the listing of components can be completed.

1.7 DISCRETE AND CONTINUOUS SYSTEMS

Systems can be categorized as discrete or continuous. "Few systems in practice are wholly discrete or continuous, but since one type of change predominates for most systems, it will usually be possible to classify a system as being either discrete or continuous" [Law and Kelton, 2000]. A *discrete system* is one in which the state variable(s) change only at a discrete set of points in time. The bank is an example of a discrete system: The state variable, the number of customers in the bank, changes only when a customer arrives or when the service provided a customer is completed. Figure 1.1 shows how the number of customers changes only at discrete points in time.

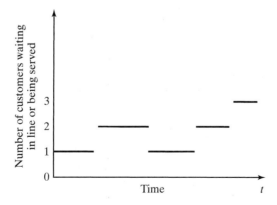

Figure 1.1   Discrete-system state variable.

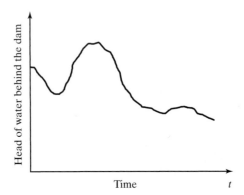

Figure 1.2   Continuous-system state variable.

A *continuous system* is one in which the state variable(s) change continuously over time. An example is the head of water behind a dam. During and for some time after a rain storm, water flows into the lake behind the dam. Water is drawn from the dam for flood control and to make electricity. Evaporation also decreases the water level. Figure 1.2 shows how the state variable *head of water behind the dam* changes for this continuous system.

1.8 MODEL OF A SYSTEM

Sometimes it is of interest to study a system to understand the relationships between its components or to predict how the system will operate under a new policy. To study the system, it is sometimes possible to experiment with the system itself. However, this is not always possible. A new system might not yet exist; it could be in only hypothetical form or at the design stage. Even if the system exists, it might be impractical to experiment with it. For example, it might not be wise or possible to double the unemployment rate to discover the effect of employment on inflation. In the case of a bank, reducing the numbers of tellers to study the effect on the length of waiting lines might infuriate the customers so greatly that they move their accounts to a competitor. Consequently, studies of systems are often accomplished with a model of a system.

We had a consulting job for the simulation of a redesigned port in western Australia. At \$200 millions for a loading/unloading berth, it's not advisable to invest that amount only to find that the berth is inadequate for the task.

A *model* is defined as a representation of a system for the purpose of studying the system. For most studies, it is only necessary to consider those aspects of the system that affect the problem under investigation. These aspects are represented in a model of the system; the model, by definition, is a simplification of the system. On the other hand, the model should be sufficiently detailed to permit valid conclusions to be drawn about the real system. Different models of the same system could be required as the purpose of investigation changes.

Just as the components of a system were entities, attributes, and activities, models are represented similarly. However, the model contains only those components that are relevant to the study. The components of a model are discussed more extensively in Chapter 3.

1.9 TYPES OF MODELS

Models can be classified as being mathematical or physical. A mathematical model uses symbolic notation and mathematical equations to represent a system. A simulation model is a particular type of mathematical model of a system.

Simulation models may be further classified as being static or dynamic, deterministic or stochastic, and discrete or continuous. A *static* simulation model, sometimes called a Monte Carlo simulation, represents a system at a particular point in time. *Dynamic* simulation models represent systems as they change over time. The simulation of a bank from 9:00 A.M. to 4:00 P.M. is an example of a dynamic simulation.

Simulation models that contain no random variables are classified as *deterministic*. Deterministic models have a known set of inputs, which will result in a unique set of outputs. Deterministic arrivals would occur at a dentist's office if all patients arrived at the scheduled appointment time. A *stochastic* simulation model has one or more random variables as inputs. Random inputs lead to random outputs. Since the outputs are random, they can be considered only as estimates of the true characteristics of a model. The simulation of a bank would usually involve random interarrival times and random service times. Thus, in a stochastic simulation, the output measures—the average number of people waiting, the average waiting time of a customer—must be treated as statistical estimates of the true characteristics of the system.

Discrete and continuous systems were defined in Section 1.7. Discrete and continuous models are defined in an analogous manner. However, a discrete simulation model is not always used to model a discrete system, nor is a continuous simulation model always used to model a continuous system. Tanks and pipes are modeled discretely by some software vendors, even though we know that fluid flow is continuous. In addition, simulation models may be mixed, both discrete and continuous. The choice of whether to use a discrete or continuous (or both discrete and continuous) simulation model is a function of the characteristics of the system and the objective of the study. Thus, a communication channel could be modeled discretely if the characteristics and movement of each message were deemed important. Conversely, if the flow of messages in aggregate over the channel were of importance, modeling the system via continuous simulation could be more appropriate. The models considered in this text are discrete, dynamic, and stochastic.

1.10 DISCRETE-EVENT SYSTEM SIMULATION

This is a textbook about discrete-event system simulation. Discrete-event systems simulation is the modeling of systems in which the state variable changes only at a discrete set of points in time. The simulation models are analyzed by numerical methods rather than by analytical methods. *Analytical* methods employ the deductive reasoning of mathematics to "solve" the model. For example, differential calculus can be used to compute the minimum-cost policy for some inventory models. *Numerical* methods employ computational procedures to "solve" mathematical models. In the case of simulation models, which employ numerical methods, models are "run" rather than solved—that is, an artificial history of the system is generated from the model assumptions, and observations are collected to be analyzed and to estimate the true system performance measures. Real-world simulation models are rather large, and the amount of data stored and manipulated is vast, so such runs are usually conducted with the aid of a computer. However, much insight can be obtained by simulating small models manually.

In summary, this textbook is about discrete-event system simulation in which the models of interest are analyzed numerically, usually with the aid of a computer.

1.11 STEPS IN A SIMULATION STUDY

Figure 1.3 shows a set of steps to guide a model builder in a thorough and sound simulation study. Similar figures and discussion of steps can be found in other sources [Shannon, 1975; Gordon, 1978; Law and Kelton, 2000]. The number beside each symbol in Figure 1.3 refers to the more detailed discussion in the text. The steps in a simulation study are as follows:

Problem formulation.   Every study should begin with a statement of the problem. If the statement is provided by the policymakers, or those that have the problem, the analyst must ensure that the problem being described is clearly understood. If a problem statement is being developed by the analyst, it is important that the policymakers understand and agree with the formulation. Although not shown in Figure 1.3, there are occasions where the problem must be reformulated as the study progresses. In many instances, policymakers and analysts are aware that there is a problem long before the nature of the problem is known.

Setting of objectives and overall project plan.   The objectives indicate the questions to be answered by simulation. At this point, a determination should be made concerning whether simulation is the appropriate methodology for the problem as formulated and objectives as stated. Assuming that it is decided that simulation is appropriate, the overall project plan should include a statement of the alternative systems to be considered and of a method for evaluating the effectiveness of these alternatives. It should also include the plans for the study in terms of the number of people involved, the cost of the study, and the number of days required to accomplish each phase of the work, along with the results expected at the end of each stage.

Model conceptualization.   The construction of a model of a system is probably as much art as science. Pritsker [1998] provides a lengthy discussion of this step. "Although it is not possible to provide a set of instructions that will lead to building successful and appropriate models in every instance, there are some general guidelines that can be followed" [Morris, 1967]. The art of modeling is enhanced by an ability to abstract the essential features of a problem, to select and modify basic assumptions that characterize the system, and then to enrich and elaborate the model until a useful approximation results. Thus, it is best to start with a simple model and build toward greater complexity. However, the model complexity need not exceed that required to accomplish the purposes for which the model is intended. Violation of this principle will only add to model-building and computer expenses. It is not necessary to have a one-to-one mapping between the model and the real system. Only the essence of the real system is needed.

It is advisable to involve the model user in model conceptualization. Involving the model user will both enhance the quality of the resulting model and increase the confidence of the model user in the application of the model. (Chapter 2 describes a number of simulation models. Chapter 6 describes queueing models that can be solved analytically. However, only experience with real systems—versus textbook problems—can "teach" the art of model building.)

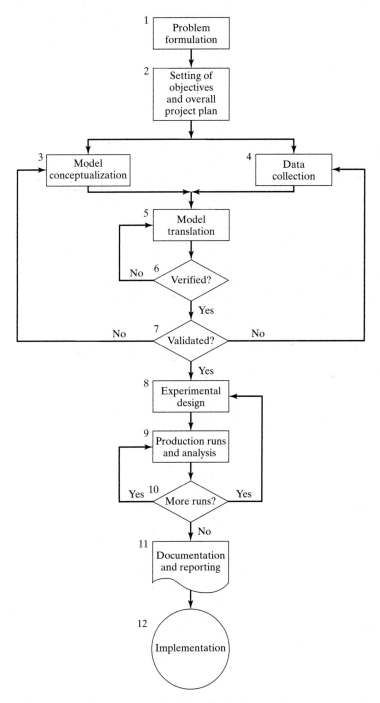

Figure 1.3 Steps in a simulation study.

Data collection. There is a constant interplay between the construction of the model and the collection of the needed input data [Shannon, 1975]. As the complexity of the model changes, the required data elements can also change. Also, since data collection takes such a large portion of the total time required to perform a simulation, it is necessary to begin it as early as possible, usually together with the early stages of model building.

The objectives of the study dictate, in a large way, the kind of data to be collected. In the study of a bank, for example, if the desire is to learn about the length of waiting lines as the number of tellers change, the types of data needed would be the distributions of interarrival times (at different times of the day), the service-time distributions for the tellers, and historic distributions on the lengths of waiting lines under varying conditions. This last type of data will be used to validate the simulation model. (Chapter 9 discusses data collection and data analysis; Chapter 5 discusses statistical distributions that occur frequently in simulation modeling. See also an excellent discussion by Henderson [2003].)

Model translation. Most real-world systems result in models that require a great deal of information storage and computation, so the model must be entered into a computer-recognizable format. We use the term "program" even though it is possible to accomplish the desired result in many instances with little or no actual coding. The modeler must decide whether to program the model in a simulation language, such as GPSS/H (discussed in Chapter 4), or to use special-purpose simulation software. For manufacturing and material handling, Chapter 4 discusses Arena®, AutoMod™, Extend™, Flexsim, MicroSaint, ProModel®, Quest®, SIMUL8®, and WITNESS™. Simulation languages are powerful and flexible. However, if the problem is amenable to solution with the simulation software, the model development time is greatly reduced. Furthermore, most of the simulation-software packages have added features that enhance their flexibility, although the amount of flexibility varies greatly.

Verified? Verification pertains to the computer program prepared for the simulation model. Is the computer program performing properly? With complex models, it is difficult, if not impossible, to translate a model successfully in its entirety without a good deal of debugging; if the input parameters and logical structure of the model are correctly represented in the computer, verification has been completed. For the most part, common sense is used in completing this step. (Chapter 10 discusses verification of simulation models, and Balci [2003] also discusses this topic.)

Validated? Validation usually is achieved through the calibration of the model, an iterative process of comparing the model against actual system behavior and using the discrepancies between the two, and the insights gained, to improve the model. This process is repeated until model accuracy is judged acceptable. In the example of a bank previously mentioned, data was collected concerning the length of waiting lines under current conditions. Does the simulation model replicate this system measure? This is one means of validation. (Chapter 10 discusses the validation of simulation models, and Balci [2003] also discusses this topic.)

Experimental design. The alternatives that are to be simulated must be determined. Often, the decision concerning which alternatives to simulate will be a function of runs that have been completed and analyzed. For each system design that is simulated, decisions need to be made concerning the length of the initialization period, the length of simulation runs, and the number of replications to be made of each run. (Chapters 11 and 12 discuss issues associated with the experimental design, and Kleijnen [1998] discusses this topic extensively.)

Production runs and analysis. Production runs, and their subsequent analysis, are used to estimate measures of performance for the system designs that are being simulated. (Chapters 11 and 12 discuss the analysis of simulation experiments, and Chapter 4 discusses software to aid in this step, including AutoStat (in AutoMod), OptQuest (in several pieces of simulation software), SimRunner (in ProModel), and WITNESS Optimizer (in WITNESS).

More Runs? Given the analysis of runs that have been completed, the analyst determines whether additional runs are needed and what design those additional experiments should follow.

Documentation and reporting. There are two types of documentation: program and progress. Program documentation is necessary for numerous reasons. If the program is going to be used again by the same or different analysts, it could be necessary to understand how the program operates. This will create confidence in the program, so that model users and policy-makers can make decisions based on the analysis. Also, if the program is to be modified by the same or a different analyst, this step can be greatly facilitated by adequate documentation. One experience with an inadequately documented program is usually enough to convince an analyst of the necessity of this important step. Another reason for documenting a program is so that model users can change parameters at will in an effort to learn the relationships between input parameters and output measures of performance or to discover the input parameters that "optimize" some output measure of performance.

Musselman [1998] discusses progress reports that provide the important, written history of a simulation project. Project reports give a chronology of work done and decisions made. This can prove to be of great value in keeping the project on course.

Musselman suggests frequent reports (monthly, at least) so that even those not involved in the day-to-day operation can keep abreast. The awareness of these others can often enhance the successful completion of the project by surfacing misunderstandings early, when the problem can be solved easily. Musselman also suggests maintaining a project log providing a comprehensive record of accomplishments, change requests, key decisions, and other items of importance.

On the reporting side, Musselman suggests frequent deliverables. These may or may not be the results of major accomplishments. His maxim is that "it is better to work with many intermediate milestones than with one absolute deadline." Possibilities prior to the final report include a model specification, prototype demonstrations, animations, training results, intermediate analyses, program documentation, progress reports, and presentations. He suggests that these deliverables should be timed judiciously over the life of the project.

The result of all the analysis should be reported clearly and concisely in a final report. This will enable the model users (now, the decision makers) to review the final formulation, the alternative systems that were addressed, the criterion by which the alternatives were compared, the results of the experiments, and the recommended solution to the problem. Furthermore, if decisions have to be justified at a higher level, the final report should provide a vehicle of certification for the model user/decision maker and add to the credibility of the model and of the model-building process.

Implementation. The success of the implementation phase depends on how well the previous 11 steps have been performed. It is also contingent upon how thoroughly the analyst has involved the ultimate model user during the entire simulation process. If the model user has been thoroughly involved during the model-building process and if the model user understands the nature of the model and its outputs, the likelihood of a vigorous implementation is enhanced [Pritsker, 1995]. Conversely, if the model and its underlying assumptions have not been properly communicated, implementation will probably suffer, regardless of the simulation model's validity.

The simulation-model building process shown in Figure 1.3 can be broken down into four phases. The first phase, consisting of steps 1 (Problem Formulation) and 2 (Setting of Objective and Overall Design), is a period of discovery or orientation. The initial statement of the problem is usually quite "fuzzy," the initial objectives will usually have to be reset, and the original project plan will usually have to be fine-tuned. These recalibrations and clarifications could occur in this phase, or perhaps will occur after or during another phase (i.e., the analyst might have to restart the process).

The second phase is related to model building and data collection and includes steps 3 (Model Conceptualization), 4 (Data Collection), 5 (Model Translation), 6 (Verification), and 7 (Validation). A continuing interplay is required among the steps. Exclusion of the model user during this phase can have dire implications at the time of implementation.

The third phase concerns the running of the model. It involves steps 8 (Experimental Design), 9 (Production Runs and Analysis), and 10 (Additional Runs). This phase must have a thoroughly conceived plan for experimenting with the simulation model. A discrete-event stochastic simulation is in fact a statistical experiment. The output variables are estimates that contain random error, and therefore a proper statistical analysis is required. Such a philosophy is in contrast to that of the analyst who makes a single run and draws an inference from that single data point.

The fourth phase, implementation, involves steps 11 (Documentation and Reporting) and 12 (Implementation). Successful implementation depends on continual involvement of the model user and on the successful completion of every step in the process. Perhaps the most crucial point in the entire process is step 7 (Validation), because an invalid model is going to lead to erroneous results, which, if implemented, could be dangerous, costly, or both.

REFERENCES

BALCI, O. [2003], "Verification, Validation, and Certification of Modeling and Simulation Applications," in *Proceedings of the Winter Simulation Conference*, Ed., S. Chick, P. J. Sánchez, D. Ferrin, and D. J. Morrice, New Orleans, LA, Dec. 7–10, pp. 150–158.

BANKS, J., AND R. R. GIBSON [1997], "Don't Simulate When: 10 Rules for Determining when Simulation Is Not Appropriate," *IIE Solutions*, September.

GORDON, G. [1978], *System Simulation*, 2d ed., Prentice-Hall, Englewood Cliffs, NJ.

HENDERSON, S. G. [2003], "Input Model Uncertainty: Why Do We Care and What Should We Do About It?" in *Proceedings of the Winter Simulation Conference*, Ed., S. Chick, P. J. Sánchez, D. Ferrin, and D. J. Morrice, New Orleans, LA, Dec. 7–10, pp. 90–100.

HILLIER, F. S., AND G. J. LIEBERMAN [2002], *Introduction to Operations Research*, 7th ed., McGraw–Hill, New York.

KLEIJNEN, J. P. C. [1998], "Experimental Design for Sensitivity Analysis, Optimization, and Validation of Simulation Models," in *Handbook of Simulation*, Ed., Jerry Banks, John Wiley, New York.

LAW, A. M., AND W. D. KELTON [2000], *Simulation Modeling and Analysis*, 3d ed., McGraw–Hill, New York.

MORRIS, W. T. [1967], "On the Art of Modeling," *Management Science*, Vol. 13, No. 12.

MUSSELMAN, K. J. [1998], "Guidelines for Success," in *Handbook of Simulation*, Ed., Jerry Banks, John Wiley, New York.

NAYLOR, T. H., J. L. BALINTFY, D. S. BURDICK, AND K. CHU [1966], *Computer Simulation Techniques*, Wiley, New York.

PEGDEN, C. D., R. E. SHANNON, AND R. P. SADOWSKI [1995], *Introduction to Simulation Using SIMAN*, 2d ed., McGraw–Hill, New York.

PRITSKER, A. A. B. [1995], *Introduction to Simulation and SLAM II*, 4th ed., Wiley & Sons, New York.

PRITSKER, A. A. B. [1998], "Principles of Simulation Modeling," in *Handbook of Simulation*, Ed., Jerry Banks, John Wiley, New York.

SHANNON, R. E. [1975], *Systems Simulation: The Art and Science*, Prentice-Hall, Englewood Cliffs, NJ.

SHANNON, R. E. [1998], "Introduction to the Art and Science of Simulation," in *Proceedings of the Winter Simulation Conference*, Eds., D. J. Medeiros, E. F. Watson, J. S. Carson, and M. S. Manivannan, Washington, DC, Dec. 13–16, pp. 7–14.

EXERCISES

1. Name several entities, attributes, activities, events, and state variables for the following systems:

    (a) A small-appliance repair shop
    (b) A cafeteria
    (c) A grocery store
    (d) A laundromat
    (e) A fast-food restaurant
    (f) A hospital emergency room
    (g) A taxicab company with 10 taxis
    (h) An automobile assembly line

2. Consider the simulation process shown in Figure 1.3.

    (a) Reduce the steps by at least two by combining similar activities. Give your rationale.
    (b) Increase the steps by at least two by separating current steps or enlarging on existing steps. Give your rationale.

3. A simulation of a major traffic intersection is to be conducted, with the objective of improving the current traffic flow. Provide three iterations, in increasing order of complexity, of steps 1 and 2 in the simulation process of Figure 1.3.

4. A simulation is to be conducted of cooking a spaghetti dinner to discover at what time a person should start in order to have the meal on the table by 7:00 P.M. Read a recipe for preparing a spaghetti dinner (or ask a friend or relative for the recipe). As best you can, trace what you understand to be needed, in the data-collection phase of the simulation process of Figure 1.3, in order to perform a simulation in which the model includes each step in the recipe. What are the events, activities, and state variables in this system?

5. What are the events and activities associated with the operation of your checkbook?

6. Read an article on the application of simulation related to your major area of study or interest, in the current WSC Proceedings, and prepare a report on how the author accomplishes the steps given in Figure 1.3.

7. Get a copy of a recent WSC Proceedings and report on the different applications discussed in an area of interest to you.

8. Get a copy of a recent WSC Proceedings and report on the most unusual application that you can find.

9. Go to the Winter Simulation Conference website at: www.wintersim.org and address the following:

   (a) What advanced tutorials were offered at the previous WSC or are planned at the next WSC?
   (b) Where and when will the next WSC be held?

10. Go to the Winter Simulation Conference website at www.wintersim.org and address the following:

    (a) When was the largest (in attendance) WSC, and how many attended?
    (b) In what calendar year, from the beginning of WSC, was there no Conference?
    (c) What was the largest expanse of time, from the beginning of WSC, between occurrences of the Conference?
    (d) Beginning with the 25th WSC, can you discern a pattern for the location of the Conference?

11. Using your favorite search engine, search the web for 'discrete event simulation output analysis' and prepare a report discussing what you find.

12. Using your favorite search engine, search the web for 'supply chain simulation' and prepare a report discussing what you find.

13. Using your favorite search engine, search the web for 'web based simulation' and prepare a report discussing what you find.

2

Simulation Examples

This chapter presents several examples of simulations that can be performed by devising a simulation table either manually or with a spreadsheet. The simulation table provides a systematic method for tracking system state over time. These examples provide insight into the methodology of discrete-system simulation and the descriptive statistics used for predicting system performance.

The simulations in this chapter are carried out by following three steps:

1. Determine the characteristics of each of the inputs to the simulation. Quite often, these are modeled as probability distributions, either continuous or discrete.
2. Construct a simulation table. Each simulation table is different, for each is developed for the problem at hand. An example of a simulation table is shown in Table 2.1. In this example, there are p inputs, $x_{ij}, j = 1, 2, \ldots, p$, and one response, y_i, for each of repetitions (or, trials) $i = 1, 2, \ldots, n$. Initialize the table by filling in the data for repetition 1.
3. For each repetition i, generate a value for each of the p inputs, and evaluate the function, calculating a value of the response y_i. The input values may be computed by sampling values from the distributions chosen in step 1. A response typically depends on the inputs and one or more previous responses.

This chapter gives a number of simulation examples in queueing, inventory, reliability, and network analysis. The two queueing examples provide a single-server and two-server system, respectively. (Chapter 6 provides more insight into queueing models.) The first of the inventory examples involves a problem that has a closed-form solution; thus, the simulation solution can be compared to the mathematical solution. The second inventory example pertains to the classic order-level model.

Table 2.1   Simulation Table

	Inputs						Response
Repetitions	x_{i1}	x_{i2}	$\cdots$	x_{ij}	$\cdots$	x_{ip}	y_i
1							
2							
3							
.							
.							
.							
n							

Next, there is an example that introduces the concept of random normal numbers and a model for the simulation of lead-time demand. The examples conclude with the analysis of a network.

2.1 SIMULATION OF QUEUEING SYSTEMS

A queueing system is described by its calling population, the nature of the arrivals, the service mechanism, the system capacity, and the queueing discipline. These attributes of a queueing system are described in detail in Chapter 6. A simple single-channel queueing system is portrayed in Figure 2.1.

In the single-channel queue, the calling population is infinite; that is, if a unit leaves the calling population and joins the waiting line or enters service, there is no change in the arrival rate of other units that could need service. Arrivals for service occur one at a time in a random fashion; once they join the waiting line, they are eventually served. In addition, service times are of some random length according to a probability distribution which does not change over time. The system capacity has no limit, meaning that any number of units can wait in line. Finally, units are served in the order of their arrival (often called FIFO: first in, first out) by a single server or channel.

Arrivals and services are defined by the distribution of the time between arrivals and the distribution of service times, respectively. For any simple single- or multichannel queue, the overall effective arrival rate must be less than the total service rate, or the waiting line will grow without bound. When queues grow without bound, they are termed "explosive" or unstable. (In some re-entrant queueing networks in which units return a number of times to the same server

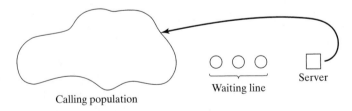

Calling population   Waiting line   Server

Figure 2.1   Queueing system.

before finally exiting from the system, the condition that arrival rate be less than service rate might not guarantee stability. See Harrison and Nguyen [1995] for more explanation. Interestingly, this type of instability was noticed first, not in theory, but in actual manufacturing in semiconductor manufacturing plants.) More complex situations can occur—for example, arrival rates that are greater than service rates for short periods of time, or networks of queues with routing. However, this chapter sticks to the most basic queues.

Prior to our introducing several simulations of queueing systems, it is necessary to understand the concepts of system state, events, and simulation clock. (These concepts are studied systematically in Chapter 3.) The state of the system is the number of units in the system and the status of the server, busy or idle. An event is a set of circumstances that causes an instantaneous change in the state of the system. In a single-channel queueing system, there are only two possible events that can affect the state of the system. They are the entry of a unit into the system (the arrival event) and the completion of service on a unit (the departure event). The queueing system includes the server, the unit being serviced (if one is being serviced), and the units in the queue (if any are waiting). The simulation clock is used to track simulated time.

If a unit has just completed service, the simulation proceeds in the manner shown in the flow diagram of Figure 2.2. Note that the server has only two possible states: It is either busy or idle.

The arrival event occurs when a unit enters the system. The flow diagram for the arrival event is shown in Figure 2.3. The unit will find the server either idle or busy; therefore, either the unit begins service immediately, or it enters the queue for the server. The unit follows the course of action shown in Figure 2.4. If the server is busy, the unit enters the queue. If the server is idle and the queue is empty, the unit begins service. It is not possible for the server to be idle while the queue is nonempty.

After the completion of a service, the server either will become idle or will remain busy with the next unit. The relationship of these two outcomes to the status of the queue is shown in Figure 2.5. If the queue is not empty, another unit will enter the server and it will be busy. If the queue is empty, the server will be idle after a service is completed. These two possibilities are shown as the shaded portions of Figure 2.5. It is impossible for the server to become busy if the queue is empty when a service is completed. Similarly, it is impossible for the server to be idle after a service is completed when the queue is not empty.

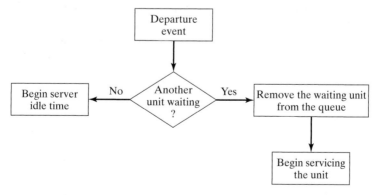

Figure 2.2 *Service just completed* flow diagram.

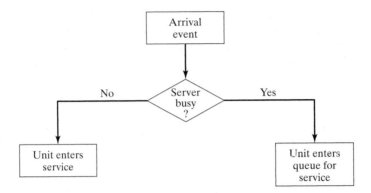

Figure 2.3   *Unit entering system* flow diagram.

		Queue status	
		Not empty	Empty
Server status	Busy	Enter queue	Enter queue
	Idle	Impossible	Enter service

Figure 2.4   Potential unit actions upon arrival.

		Queue status	
		Not empty	Empty
Server outcomes	Busy	//////////	Impossible
	Idle	Impossible	//////////

Figure 2.5   Server outcomes after the completion of service.

Now, how can the events described above occur in simulated time? Simulations of queueing systems generally require the maintenance of an event list for determining what happens next. The event list tracks the future times at which the different types of events occur. Simulations using event lists are described in Chapter 3. This chapter simplifies the simulation by tracking each unit explicitly. Simulation clock times for arrivals and departures are computed in a simulation table customized for each problem. In simulation, events usually occur at random times, the randomness imitating uncertainty in real life. For example, it is not known with certainty when the next customer will arrive at a grocery checkout counter, or how long the bank teller will take to complete a transaction. In these cases, a statistical model of the data is developed either from data collected and analyzed or from subjective estimates and assumptions.

The randomness needed to imitate real life is made possible through the use of "random numbers." Random numbers are distributed uniformly and independently on the interval (0, 1). Random digits are uniformly distributed on the set $\{0, 1, 2, \dots, 9\}$. Random digits can be used to form random numbers by selecting the proper number of digits for each random number and placing a decimal point to the left of the value selected. The proper number of digits is dictated by the accuracy of the data being used for input purposes. If the input distribution has values with two decimal places, two digits are taken from a random digits table (such as Table A.1) and the decimal point is placed to the left to form a random number.

Random numbers also can be generated in simulation packages and in spreadsheets (such as Excel). For example, Excel has a macro function called RAND() that returns a "random" number between 0 and 1. When numbers are generated by using a procedure, they are often referred to as pseudo-random numbers. Because the procedure is fully known, it is always possible to predict the sequence of numbers that will be generated prior to the simulation. The most commonly used methods for generating random numbers are discussed in Chapter 7.

In a single-channel queueing simulation, interarrival times and service times are generated from the distributions of these random variables. The examples that follow show how such times are generated. For simplicity, assume that the times between arrivals were generated by rolling a die five times and recording the up face. Table 2.2 contains a set of five interarrival times generated in this manner. These five interarrival times are used to compute the arrival times of six customers at the queueing system.

The first customer is assumed to arrive at clock time 0. This starts the clock in operation. The second customer arrives two time units later, at clock time 2. The third customer arrives four time units later, at clock time 6; and so on.

The second time of interest is the service time. Table 2.3 contains service times generated at random from a distribution of service times. The only possible service times are one, two, three, and four time units. Assuming that all four values are equally likely to occur, these values could

Table 2.2 Interarrival and Clock Times

Customer	Interarrival Time	Arrival Time on Clock
1	—	0
2	2	2
3	4	6
4	1	7
5	2	9
6	6	15

Table 2.3 Service Times

Customer	Service Time
1	2
2	1
3	3
4	2
5	1
6	4

Table 2.4   Simulation Table Emphasizing Clock Times

A	B	C	D	E
	Arrival	*Time Service*	*Service*	*Time Service*
Customer	*Time*	*Begins*	*Time*	*Ends*
Number	*(Clock)*	*(Clock)*	*(Duration)*	*(Clock)*
1	0	0	2	2
2	2	2	1	3
3	6	6	3	9
4	7	9	2	11
5	9	11	1	12
6	15	15	4	19

Table 2.5   Chronological Ordering of Events

	Customer	*Clock*
Event Type	*Number*	*Time*
Arrival	1	0
Departure	1	2
Arrival	2	2
Departure	2	3
Arrival	3	6
Arrival	4	7
Departure	3	9
Arrival	5	9
Departure	4	11
Departure	5	12
Arrival	6	15
Departure	6	19

have been generated by placing the numbers one through four on chips and drawing the chips from a hat with replacement, being sure to record the numbers selected. Now, the interarrival times and service times must be meshed to simulate the single-channel queueing system. As is shown in Table 2.4, the first customer arrives at clock time 0 and immediately begins service, which requires two minutes. Service is completed at clock time 2. The second customer arrives at clock time 2 and is finished at clock time 3. Note that the fourth customer arrived at clock time 7, but service could not begin until clock time 9. This occurred because customer 3 did not finish service until clock time 9.

Table 2.4 was designed specifically for a single-channel queue that serves customers on a first-in–first-out (FIFO) basis. It keeps track of the clock time at which each event occurs. The second column of Table 2.4 records the clock time of each arrival event, while the last column records the clock time of each departure event. The occurrence of the two types of events in chronological order is shown in Table 2.5 and Figure 2.6.

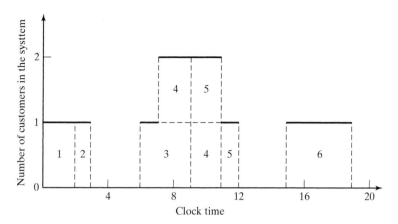

Figure 2.6   Number of customers in the system.

It should be noted that Table 2.5 is ordered by clock time, in which case the events may or may not be ordered by customer number. The chronological ordering of events is the basis of the approach to discrete-event simulation described in Chapter 3.

Figure 2.6 depicts the number of customers in the system at the various clock times. It is a visual image of the event listing of Table 2.5. Customer 1 is in the system from clock time 0 to clock time 2. Customer 2 arrives at clock time 2 and departs at clock time 3. No customers are in the system from clock time 3 to clock time 6. During some time periods, two customers are in the system, such as at clock time 8, when customers 3 and 4 are both in the system. Also, there are times when events occur simultaneously, such as at clock time 9, when customer 5 arrives and customer 3 departs.

Example 2.1 follows the logic described above while keeping track of a number of attributes of the system. Example 2.2 is concerned with a two-channel queueing system. The flow diagrams for a multichannel queueing system are slightly different from those for a single-channel system. The development and interpretation of these flow diagrams is left as an exercise for the reader.

Example 2.1:   Single-Channel Queue

A small grocery store has only one checkout counter. Customers arrive at this checkout counter at random times that are from 1 to 8 minutes apart. Each possible value of interarrival time has the same probability of occurrence, as shown in Table 2.6. The service times vary from 1 to 6 minutes, with the probabilities shown in Table 2.7. The problem is to analyze the system by simulating the arrival and service of 100 customers.

In actuality, 100 customers is too small a sample size to draw any reliable conclusions. The accuracy of the results is enhanced by increasing the sample size, as is discussed in Chapter 11. However, the purpose of the exercise is to demonstrate how simple simulations can be carried out in a table, either manually or with a spreadsheet, not to recommend changes in the grocery store. A second issue, discussed thoroughly in Chapter 11, is that of initial conditions. A simulation of a grocery store that starts with an empty system is not realistic unless the intention is to model the system from startup or to model until steady-state operation is reached. Here, to keep calculations simple, starting conditions and concerns are overlooked.

Table 2.6   Distribution of Time Between Arrivals

Time between Arrivals (Minutes)	Probability	Cumulative Probability	Random Digit Assignment
1	0.125	0.125	001–125
2	0.125	0.250	126–250
3	0.125	0.375	251–375
4	0.125	0.500	376–500
5	0.125	0.625	501–625
6	0.125	0.750	626–750
7	0.125	0.875	751–875
8	0.125	1.000	876–000

Table 2.7   Service-Time Distribution

Service Time (Minutes)	Probability	Cumulative Probability	Random Digit Assignment
1	0.10	0.10	01–10
2	0.20	0.30	11–30
3	0.30	0.60	31–60
4	0.25	0.85	61–85
5	0.10	0.95	86–95
6	0.05	1.00	96–00

A set of uniformly distributed random numbers is needed to generate the arrivals at the checkout counter. Such random numbers have the following properties:

1. The set of random numbers is uniformly distributed between 0 and 1.
2. Successive random numbers are independent.

With tabular simulations, random digits such as those found in Table A.1 in the Appendix can be converted to random numbers. Random digits are converted to random numbers by placing a decimal point appropriately. Since the probabilities in Table 2.6 are accurate to 3 significant digits, three-place random numbers will suffice. It is necessary to list 99 random numbers to generate the times between arrivals. Why only 99 numbers? The first arrival is assumed to occur at time 0, so only 99 more arrivals need to be generated to end up with 100 customers. Similarly, for Table 2.7, two-place random numbers will suffice.

The rightmost two columns of Tables 2.6 and 2.7 are used to generate random arrivals and random service times. The third column in each table contains the cumulative probability for the distribution. The rightmost column contains the random digit assignment. In Table 2.6, the first random digit assignment is 001–125. There are 1000 three-digit values possible (001 through 000). The probability of a time-between-arrival of 1 minute is 0.125, so 125 of the 1000 random digit values are assigned to such an occurrence. Times between arrival for 99 customers are generated by listing 99 three-digit values from Table A.1 and comparing them to the random digit assignment of Table 2.6.

For manual simulations, it is good practice to start at a random position in the random digit table and proceed in a systematic direction, never re-using the same stream of digits in a given problem. If the same pattern is used repeatedly, bias could result from the same pattern's being generated.

The time-between-arrival determination is shown in Table 2.8. Note that the first random digits are 064. To obtain the corresponding time between arrivals, enter the fourth column of Table 2.6 and read 1 minute from the first column of the table. Alternatively, we see that 0.064 is between the cumulative probabilities 0.001 and 0.125, again resulting in 1 minute as the generated time.

Service times for the first 18 and the 100th customers are shown in Table 2.9. These service times were generated via the methodology described above, together with the aid of Table 2.7. (The entire table can be generated by using the Excel spreadsheet for Example 2.1 at www.bcnn.net.)

Table 2.8 Time-Between-Arrival Determination

Customer	Random Digits	Time between Arrivals (Minutes)	Customer	Random Digits	Time between Arrivals (Minutes)
1	—	—	11	413	4
2	064	1	12	462	4
3	112	1	13	843	7
4	678	6	14	738	6
5	289	3	15	359	3
6	871	7	16	888	8
7	583	5	17	902	8
8	139	2	18	212	2
9	423	4	⋮	⋮	⋮
10	039	1	100	538	5

Table 2.9 Service Times Generated

Customer	Random Digits	Service Time (Minutes)	Customer	Random Digits	Service Time (Minutes)
1	84	4	11	94	5
2	18	2	12	32	3
3	87	5	13	79	4
4	81	4	14	92	5
5	06	1	15	46	3
6	91	5	16	21	2
7	79	4	17	73	4
8	09	1	18	55	3
9	64	4	⋮	⋮	⋮
10	38	3	100	26	2

The first customer's service time is 4 minutes, because the random digits 84 fall in the bracket 61–85—or, alternatively, because the derived random number 0.84 falls between the cumulative probabilities 0.61 and 0.85.

The essence of a manual simulation is the simulation table. These tables are designed for the problem at hand, with columns added to answer the questions posed. The simulation table for the single-channel queue, shown in Table 2.10, is an extension of the type of table already seen in Table 2.4. The first step is to initialize the table by filling in cells for the first customer. The first customer is assumed to arrive at time 0. Service begins immediately and finishes at time 4. The customer was in the system for 4 minutes. After the first customer, subsequent rows in the table are based on the random numbers for interarrival time, service time, and the completion time of the previous customer. For example, the second customer arrives at time 1. But service could not begin until time 4; the server (checkout person) was busy until that time. The second customer waited in the queue for three minutes. The second customer was in the system for 5 minutes. Skip down to the fifth customer. Service ends at time 16, but the sixth customer does not arrive until time 18, at which time service began. The server (checkout person) was idle for two minutes. This process continues for all 100 customers. The rightmost two columns have been added to collect statistical measures of performance, such as each customer's time in system and the server's idle time (if any) since the previous customer departed. In order to compute summary statistics, totals are formed as shown for service times, time customers spend in the system, idle time of the server, and time the customers wait in the queue.

Some of the findings from the simulation in Table 2.10 are as follows:

1. The average waiting time for a customer is 1.74 minutes. This is computed in the following manner:

$$\text{Average waiting time (minutes)} = \frac{\text{total time customers wait in queue (minutes)}}{\text{total numbers of customers}}$$

$$= \frac{174}{100} = 1.74 \text{ minutes}$$

2. The probability that a customer has to wait in the queue is 0.46. This is computed in the following manner:

$$\text{Probability (wait)} = \frac{\text{numbers of customers who wait}}{\text{total number of customers}}$$

$$= \frac{46}{100} = 0.46$$

3. The proportion of idle time of the server is 0.24. This is computed in the following manner:

$$\text{Probability of idle server} = \frac{\text{total idle time of server (minutes)}}{\text{total run time of simulation (minutes)}}$$

$$= \frac{101}{418} = 0.24$$

The probability of the server's being busy is the complement of 0.24, namely, 0.76.

Table 2.10 Simulation Table for Single-Channel Queuing Problem

Customer	Interarrival Time (Minutes)	Arrival Time	Service Time (Minutes)	Time Service Begins	Waiting Time in Queue (Minutes)	Time Service Ends	Time Customer Spends in System (Minutes)	Idle Time of Server (Minutes)
		Clock		*Clock*		*Clock*		
1		0	4	0	0	4	4	
2	1	1	2	4	3	6	5	0
3	1	2	5	6	4	11	9	0
4	6	8	4	11	3	15	7	0
5	3	11	1	15	4	16	5	0
6	7	18	5	18	0	23	5	2
7	5	23	4	23	0	27	4	0
8	2	25	1	27	2	28	3	0
9	4	29	4	29	0	33	4	1
10	1	30	3	33	3	36	6	0
11	4	34	5	36	2	41	7	0
12	4	38	3	41	3	44	6	0
13	7	45	4	45	0	49	4	1
14	6	51	5	51	0	56	5	2
15	3	54	3	56	2	59	5	0
16	8	62	2	62	0	64	2	3
17	8	70	4	70	0	74	4	6
18	2	72	3	74	2	77	5	0
19	7	79	1	79	0	80	1	2
20	4	83	2	83	0	85	2	3
⋯	⋯	⋯	⋯	⋯	⋯	⋯	⋯	⋯
100	5	415	2	416	1	418	3	0
Total	415		317		174		491	101

31

4. The average service time is 3.17 minutes. This is computed in the following manner:

$$\begin{array}{c}\text{Average service time} \\ \text{(minutes)}\end{array} = \frac{\text{total service time (minutes)}}{\text{total number of customers}}$$

$$= \frac{317}{100} = 3.17 \text{ minutes}$$

This result can be compared with the expected service time by finding the mean of the service-time distribution, using the equation

$$E(S) = \sum_{s=0}^{\infty} sp(s)$$

Applying the expected-value equation to the distribution in Table 2.7 gives

Expected service time =

$$1(0.10) + 2(0.20) + 3(0.30) + 4(0.25) + 5(0.10) + 6(0.05) = 3.2 \text{ minutes}$$

The expected service time is slightly higher than the average service time in the simulation. The longer the simulation, the closer the average will be to $E(S)$.

5. The average time between arrivals is 4.19 minutes. This is computed in the following manner:

$$\begin{array}{c}\text{Average time between} \\ \text{arrivals (minutes)}\end{array} = \frac{\begin{array}{c}\text{sum of all times} \\ \text{between arrivals (minutes)}\end{array}}{\text{number of arrivals} - 1}$$

$$= \frac{415}{99} = 4.19 \text{ minutes}$$

One is subtracted from the denominator because the first arrival is assumed to occur at time 0. This result can be compared to the expected time between arrivals by finding the mean of the discrete uniform distribution whose endpoints are $a = 1$ and $b = 8$. The mean is given by

$$E(A) = \frac{a+b}{2} = \frac{1+8}{2} = 4.5 \text{ minutes}$$

The expected time between arrivals is slightly higher than the average. However, as the simulation becomes longer, the average value of the time between arrivals should approach the theoretical mean, $E(A)$.

6. The average waiting time of those who wait is 3.22 minutes. This is computed in the following manner:

$$\begin{array}{c}\text{Average waiting time of} \\ \text{those who wait} \\ \text{(minutes)}\end{array} = \frac{\text{total time customers wait in queue (minutes)}}{\text{total number of customers that wait}}$$

$$= \frac{174}{54} = 3.22 \text{ minutes}$$

7. The average time a customer spends in the system is 4.91 minutes. This can be found in two ways. First, the computation can be achieved by the following relationship:

$$\begin{array}{c}\text{Average time customer}\\ \text{spends in the system}\\ \text{(minutes)}\end{array} = \dfrac{\begin{array}{c}\text{total time customers spend in the}\\ \text{system (minutes)}\end{array}}{\text{total number of customers}}$$

$$= \frac{491}{100} = 4.91 \text{ minutes}$$

The second way of computing this same result is to realize that the following relationship must hold:

$$\begin{array}{c}\text{Average time}\\ \text{customer spends}\\ \text{in the system}\\ \text{(minutes)}\end{array} = \begin{array}{c}\text{average time}\\ \text{customer spends}\\ \text{waiting in the}\\ \text{queue (minutes)}\end{array} + \begin{array}{c}\text{average time}\\ \text{customer spends}\\ \text{in service}\\ \text{(minutes)}\end{array}$$

From findings 1 and 4, this results in

$$\text{Average time customer spends in the system} = 1.74 + 3.17 = 4.91 \text{ minutes}$$

A decision maker would be interested in results of this type, but a longer simulation would increase the accuracy of the findings. However, some tentative inferences can be drawn at this point. About half of the customers have to wait; however, the average waiting time is not excessive. The server does not have an undue amount of idle time. More reliable statements about the results would depend on balancing the cost of waiting against the cost of additional servers.

Excel spreadsheets have been constructed for each of the examples in this chapter. The spreadsheets can be found at www.bcnn.net. The spreadsheets have a common format. The first sheet is One-Trial. The second sheet is Experiment. The third sheet is entitled Explain. Here, the logic in the spreadsheet is discussed, and questions pertaining to that logic are asked of the reader. Use the default seed '12345' to reproduce the One-Trial output shown in the examples in the text, and use the appropriate number of trials (or replications) to reproduce the Experiment shown in the text, again using the default seed '12345'.

Exercises relating to the spreadsheets have been prepared also. These are the last set of exercises at the end of this chapter. The first set of exercises is for manual simulation.

The spreadsheets allow for many entities to flow through the system. (In Example 2.1, the entities are customers.) For instance, the spreadsheet for Example 2.1 has 100 customers going through the system, and the number of trials can vary from one to 400. Let's say that 200 trials are selected. Then, 200 trials of the simulation, each of 100 customers, will be conducted.

For Example 2.1, the frequency of waiting time in queue for the first trial of 100 customers is shown in Figure 2.7. (Note: In all histograms in the remainder of this chapter, the upper limit of the bin is indicated on the legend on the x-axis, even if the legend is shown centered within the bin.) As mentioned previously, 46% of them did not have to wait, and 42% waited less than four minutes (but more than zero minutes).

Figure 2.7   Frequency of waiting time in queue.

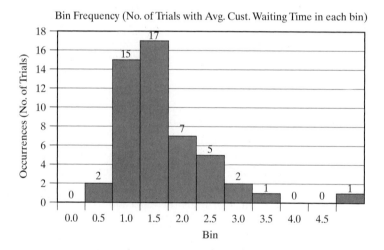

Figure 2.8   Frequency distribution of average waiting times.

From the Experiment sheet of the Excel spreadsheet, the average waiting time over 50 trials was 1.50 minutes. Figure 2.8 shows a histogram of the 50 average waiting times for the 50 trials. The overall average (1.50 minutes) is just to the right of the two most popular bins.

The exercises ask that you experiment with this spreadsheet. But, you can also experiment on your own to discover the effect of randomness and of the input data. For example, what if you run 400 trials instead of 50? Does the shape of the distribution in Figure 2.8 change? What if you run 25 trials instead of 50? How much does the shape change as you generate new trials?

Example 2.2:   The Able–Baker Call Center Problem _____

This example illustrates the simulation procedure when there is more than one service channel. Consider a computer technical support center where personnel take calls and provide service. The time between calls ranges from 1 to 4 minutes, with distribution as shown in Table 2.11. There are two technical support people—Able and Baker. Able is more experienced and can provide service faster than Baker. The distributions of their service times are shown in Tables 2.12 and 2.13. Times are usually a continuous measure. But, in this and other time-based examples in this chapter, we make them discrete for ease of explanation.

The simulation proceeds in a manner similar to Example 2.1, except that it is more complex because of the two servers. A simplifying rule is that Able gets the call if both technical support people are idle. Able has seniority. (The solution would be different if the decision were made at random or by any other rule.)

The problem is to find how well the current arrangement is working. To estimate the system measures of performance, a simulation of the first 100 callers is made. A simulation with more callers would yield more reliable results, but, for purposes of this illustration, a 100-caller simulation has been selected.

Table 2.11   Interarrival Distribution of Calls for Technical Support

Time between Arrivals (Minutes)	Probability	Cumulative Probability	Random-Digit Assignment
1	0.25	0.25	01–25
2	0.40	0.65	26–65
3	0.20	0.85	66–85
4	0.15	1.00	86–00

Table 2.12   Service Distribution of Able

Service Time (Minutes)	Probability	Cumulative Probability	Random-Digit Assignment
2	0.30	0.30	01–30
3	0.28	0.58	31–58
4	0.25	0.83	59–83
5	0.17	1.00	84–00

Table 2.13   Service Distribution of Baker

Service Time (Minutes)	Probability	Cumulative Probability	Random-Digit Assignment
3	0.35	0.35	01–35
4	0.25	0.60	36–60
5	0.20	0.80	61–80
6	0.20	1.00	81–00

The simulation proceeds in accordance with the following set of steps:

Step 1. For Caller k, generate an interarrival time A_k. Add it to the previous arrival time T_{k-1} to get the arrival time of Caller k as $T_k = T_{k-1} + A_k$.

Step 2. If Able is idle, Caller k begins service with Able at the current time T_{now}.

Able's service completion time, $T_{fin,A}$ is given by $T_{fin,A} = T_{now} + T_{svc,A}$ where $T_{svc,A}$ is the service time generated from Able's Service Time Distribution.

Caller k's time in system, T_{sys}, is given by $T_{sys} = T_{fin,A} - T_k$.

Because Able was idle, Caller k's delay, T_{wait}, is given by $T_{wait} = 0$.

If Able is busy, but Baker is idle, Caller k begins service with Baker at the current time T_{now}. Baker's service completion time, $T_{fin,B}$ is given by $T_{fin,B} = T_{now} + T_{svc,B}$ where $T_{svc,B}$ is the service time generated from Baker's Service Time Distribution.

Caller k's time in system, T_{sys}, is given by $T_{sys} = T_{fin,B} - T_k$.

Because Baker was idle, Caller k's delay, T_{wait}, is given by $T_{wait} = 0$.

Step 3. If Able and Baker are both busy, then calculate the time at which the first one becomes available, as follows:

$$T_{beg} = min(T_{fin,A}, T_{fin,B}).$$

Caller k begins service at T_{beg}. When service for Caller k begins, set $T_{now} = T_{beg}$.

Then compute $T_{fin,A}$ or $T_{fin,B}$ as in Step 2.

Caller k's time in system, T_{sys}, is given by $T_{sys} = T_{fin,A} - T_k$ or $T_{sys} = T_{fin,B} - T_k$, as appropriate.

The preceding steps have been implemented in an Excel spreadsheet that is available on the website www.bcnn.net. The reader is strongly encouraged to examine the Excel spreadsheet with particular emphasis on how the cell values are calculated. Also, there are exercises at the end of this chapter that ask the reader to run a variety of experiments using the spreadsheet.

A portion of the output in the Excel spreadsheet is given in Table 2.14 to clarify the steps previously listed. Caller 1 arrives at clock time 0 to get the simulation started. Able is idle, so Caller 1 begins service with Able at clock time 0. The service time, 2 minutes, is generated from information given in Table 2.12 by following the procedure in Example 2.1. Thus, Caller 1 completes service at clock time 2 minutes and was not delayed.

An interarrival time of 2 minutes is generated from Table 2.11 by following the procedure in Example 2.1. So, the arrival of Caller 2 is at clock time 2 minutes. Able is idle at the time, having just completed service on Caller 1, so Caller 2 is served by Able.

Now, skip down to Caller 4, serviced by Able from clock time 8 minutes to clock time 12 minutes. Note that Caller 5 arrives at clock time 9 minutes. Because Able is busy with Caller 4 at that time, but Baker is available, Baker services Caller 5, completing service at clock time 12 minutes.

Table 2.14 Simulation Table for Call-Center Example

Caller Number	Interarrival Time (Minutes)	Arrival Time (Clock)	When Able Available (Clock)	When Baker Available (Clock)	Server Chosen	Service Time (Minutes)	Time Service Begins (Clock)	Able's Service Completion Time (Clock)	Baker's Service Completion Time (Clock)	Caller Delay (Minutes)	Time in System (Minutes)
1		0	0	0	Able	2	0	2		0	2
2	2	2	2	0	Able	2	2	4		0	2
3	4	6	4	0	Able	2	6	8		0	2
4	2	8	8	0	Able	4	8	12		0	4
5	1	9	12	0	Baker	3	9		12	0	3
⋮	⋮	⋮	⋮	⋮	⋮	⋮	⋮	⋮	⋮	⋮	⋮
100	1	219	221	219	Baker	4	219		223	0	4
Total										211	564

37

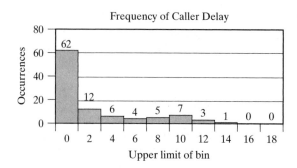

Figure 2.9 Frequency of caller delay for first trial.

For the trial under discussion (the simulation of 100 callers), the frequency diagram shown in Figure 2.9 (from the Excel spreadsheet in www.bcnn.net) shows that 62 of 100 (62%) of the callers had no delay, 12% had a delay of one or two minutes, and so on.

The distribution in Figure 2.9 is skewed to the right. Push 'Generate New Trial' repeatedly and notice that this is usually what happens, but not always. 'Generate New Trial' recalculates the spreadsheet for one trial (100 callers). One trial does not provide sufficient information on which to form a conclusion, but this is one of the great advantages to having the spreadsheet—the effect of variability is quite evident.

In Table 2.14, it is seen that the total customer delay was 211 minutes, or about 2.1 minutes per caller. It is also seen in Table 2.14 that the total time in system was 564 minutes, or about 5.6 minutes per caller.

In the second sheet of the spreadsheet, we run an experiment with 400 trials (each trial consisting of the simulation of 100 callers) to generate Figure 2.10. It is seen that about 19% of the average delays (78 of 400) are longer than one minute. Only 2.75% (11 of 400) are longer than 2 minutes.

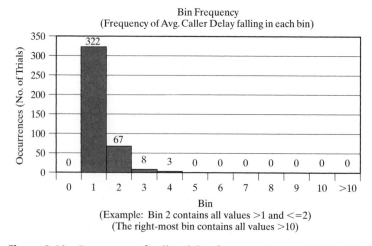

Figure 2.10 Frequency of caller delay for experiment of 400 trials.

In summary, one server cannot handle all the callers, and three servers would probably be more than are necessary. Adding an additional server would surely reduce the waiting time to nearly zero; however, the cost of waiting would have to be quite high to justify an additional server.

2.2 SIMULATION OF INVENTORY SYSTEMS

An important class of simulation problems involves inventory systems. A simple inventory system is shown in Figure 2.11. This inventory system has a periodic review of length N, at which time the inventory level is checked. An order is made to bring the inventory up to the level M. At the end of the first review period, an order quantity, Q_1, is placed. In this inventory system, the lead time (i.e., the length of time between the placement and receipt of an order) is zero. Demands are not usually known with certainty, so the order quantities are probabilistic. Demand is shown as being uniform over the time period in Figure 2.11. In actuality, demands are not usually uniform and do fluctuate over time. One possibility is that demands all occur at the beginning of the cycle. Another is that the lead time is random of some positive length.

Notice that, in the second cycle, the amount in inventory drops below zero, indicating a shortage. In Figure 2.11, these units are backordered; when the order arrives, the demand for the backordered items is satisfied first. To avoid shortages, a buffer, or safety, stock would need to be carried.

Carrying stock in inventory has an associated cost attributed to the interest paid on the funds borrowed to buy the items (this also could be considered as the loss from not having the funds available for other investment purposes). Other costs can be placed in the carrying or holding cost column: renting of storage space, hiring of guards, and so on. An alternative to carrying high inventory is to make more frequent reviews and, consequently, more frequent purchases or replenishments. This has an associated cost: the ordering cost. Also, there is a cost in being short. Customers could get angry, with a subsequent loss of good will. Larger inventories decrease the

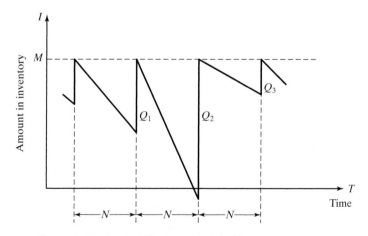

Figure 2.11 Probabilistic order-level inventory system.

Table 2.15   Distribution of Newspapers Demanded Per Day

	Demand Probability Distribution		
Demand	*Good*	*Fair*	*Poor*
40	0.03	0.10	0.44
50	0.05	0.18	0.22
60	0.15	0.40	0.16
70	0.20	0.20	0.12
80	0.35	0.08	0.06
90	0.15	0.04	0.00
100	0.07	0.00	0.00

possibilities of shortages. These costs must be traded off in order to minimize the total cost of an inventory system.

The total cost (or total profit) of an inventory system is the measure of performance. This can be affected by the policy alternatives. For example, in Figure 2.11, the decision maker can control the maximum inventory level, M, and the length of the cycle, N. What effect does changing N have on the various costs?

In an (M, N) inventory system, the events that may occur are the demand for items in the inventory, the review of the inventory position, and the receipt of an order at the end of each review period. When the lead time is zero, as in Figure 2.11, the last two events occur simultaneously.

In the following example for deciding how many newspapers to buy, only a single time period of specified length is relevant, and only a single procurement is made. Inventory remaining at the end of the single time period is sold for scrap or discarded. A wide variety of real-world problems are of this form, including the stocking of spare parts, perishable items, style goods, and special seasonal items [Hadley and Whitin, 1963].

Example 2.3:   The News Dealer's Problem

A classical inventory problem concerns the purchase and sale of newspapers. The newsstand buys the papers for 33 cents each and sells them for 50 cents each. Newspapers not sold at the end of the day are sold as scrap for 5 cents each. Newspapers can be purchased in bundles of 10. Thus, the newsstand can buy 50, 60, and so on. There are three types of newsdays: "good"; "fair"; and "poor"; they have the probabilities 0.35, 0.45, and 0.20, respectively. The distribution of newspapers demanded on each of these days is given in Table 2.15. The problem is to compute the optimal number of papers the newsstand should purchase. This will be accomplished by simulating demands for 20 days and recording profits from sales each day.

The profits are given by the following relationship:

$$\text{Profit} = \begin{pmatrix} \text{revenue} \\ \text{from sales} \end{pmatrix} - \begin{pmatrix} \text{cost of} \\ \text{newspapers} \end{pmatrix} - \begin{pmatrix} \text{lost profit from} \\ \text{excess demand} \end{pmatrix} + \begin{pmatrix} \text{salvage from sale} \\ \text{of scrap papers} \end{pmatrix}$$

From the problem statement, the revenue from sales is 50 cents for each paper sold. The cost of newspapers is 33 cents for each paper purchased. The lost profit from excess demand is 17 cents for each paper demanded that could not be provided. Such a shortage cost is somewhat

Table 2.16   Random Digit Assignment for Type of Newsday

Type of Newsday	Probability	Cumulative Probability	Random Digit Assignment
Good	0.35	0.35	01–35
Fair	0.45	0.80	36–80
Poor	0.20	1.00	81–00

Table 2.17   Random Digit Assignments for Newspapers Demanded

Demand	Cumulative Distribution			Random Digit Assignment		
	Good	Fair	Poor	Good	Fair	Poor
40	0.03	0.10	0.44	01–03	01–10	01–44
50	0.08	0.28	0.66	04–08	11–28	45–66
60	0.23	0.68	0.82	09–23	29–68	67–82
70	0.43	0.88	0.94	24–43	69–88	83–94
80	0.78	0.96	1.00	44–78	89–96	95–00
90	0.93	1.00	1.00	79–93	97–00	
100	1.00	1.00	1.00	94–00		

controversial, but makes the problem much more interesting. The salvage value of scrap papers is 5 cents each.

Tables 2.16 and 2.17 provide the random digit assignments for the types of newsdays and the demands for those newsdays. To solve this problem by simulation requires setting a policy of buying a certain number of papers each day, then simulating the demands for papers over the 20-day time period to determine the total profit. The policy (number of newspapers purchased) is changed to other values and the simulation repeated until the best value is found.

The simulation table for the decision to purchase 70 newspapers is shown in Table 2.18.

On day 1, the demand is for 80 newspapers, but only 70 newspapers are available. The revenue from the sale of 70 newspapers is $35.00. The lost profit for the excess demand of 10 newspapers is $1.70. The profit for the first day is computed as follows:

$$\text{Profit} = \$35.00 - \$23.10 - \$1.70 + 0 = \$10.20$$

On the fourth day, the demand is less than the supply. The revenue from sales of 50 newspapers is $25.00. Twenty newspapers are sold for scrap at $0.05 each yielding $1.00. The daily profit is determined as follows:

$$\text{Profit} = \$25.00 - \$23.10 - 0 + \$1.00 = \$2.90$$

The profit for the 20-day period is the sum of the daily profits, $131.00. It can also be computed from the totals for the 20 days of the simulation as follows:

$$\text{Total profit} = \$600.00 - \$462.00 - \$17.00 + \$10.00 = \$131.00$$

Table 2.18   Simulation Table for Purchase of 70 Newspapers

Day	Random Digits for Type of Newsday	Type of Newsday	Random Digits for Demand	Demand	Revenue from Sales	Lost Profit from Excess Demand	Salvage from Sale of Scrap	Daily Profit
1	58	Fair	93	80	$35.00	$1.70	–	$10.20
2	17	Good	63	80	35.00	1.70	–	10.20
3	21	Good	31	70	35.00	–	–	11.90
4	45	Fair	19	50	25.00	–	1.00	2.90
5	43	Fair	91	80	35.00	1.70	–	10.20
6	36	Fair	75	70	35.00	–	–	11.90
7	27	Good	84	90	35.00	3.40	–	8.50
8	73	Fair	37	60	30.00	–	0.50	7.40
9	86	Poor	23	40	20.00	–	1.50	- 1.60
10	19	Good	02	40	20.00	–	1.50	-1.60
11	93	Poor	53	50	25.00	–	1.00	2.90
12	45	Fair	96	80	35.00	1.70	–	10.20
13	47	Fair	33	60	30.00	–	0.50	7.40
14	30	Good	86	90	35.00	3.40	–	8.50
15	12	Good	16	60	30.00	–	0.50	7.40
16	41	Fair	07	40	20.00	–	1.50	- 1.60
17	65	Fair	64	60	30.00	–	0.50	7.40
18	57	Fair	94	80	35.00	1.70	–	10.20
19	18	Good	55	80	35.00	1.70	–	10.20
20	98	Poor	13	40	20.00	–	1.50	- 1.60
					$600.00	$17.00	$10.00	$131.00

where the cost of newspapers for 20 days is $(20 \times \$0.33 \times 70) = \$462.00$. In general, because the results of one day are independent of previous days, inventory problems of this type are easier than queueing problems when solved in a spreadsheet such as is shown in www.bcnn.net and discussed shortly.

Figure 2.12 shows the result of 400 trials, each of twenty days, with a policy of purchasing 70 newspapers per day. For these trials, the average total (20-day) profit was $137.61. The minimum 20-day profit was $64.70 and the maximum was $186.10. Figure 2.12 shows that only 45 of the 400 trials resulted in a total 20-day profit of more than $160.

The manual solution shown in Table 2.18 had a profit of $131.00. This one 20-day is not far from the average over the 400 trials, $137.61; but the result for one 20-day simulation could have been the minimum value or the maximum value. Such an occurrence demonstrates the usefulness of conducting many trials.

On the One Trial sheet, look at the Daily Profit that results when clicking the button 'Generate New Trial'. The results vary quite a bit both in the histogram called 'Frequency of Daily Profit' (showing what happened on each of the 20 days) and in the total profits for those 20 days. The histograms are almost like snowflakes, in that no two are alike! The first two histograms generated are shown in Figure 2.13.

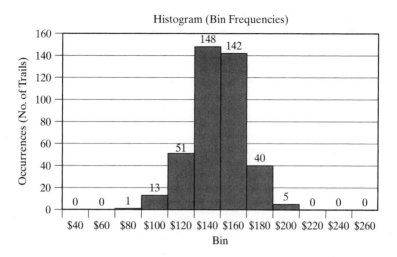

Figure 2.12   Frequency of total (20-day) profits with purchasing of 70 newspapers per day.

Figure 2.13   First two histograms of daily profit.

Example 2.4:   Simulation of an Order-Up-To-Level Inventory System _____

Consider a situation in which a company sells refrigerators. The system they use for maintaining inventory is to review the situation after a fixed number of days (say N) and make a decision about what is to be done. The policy is to order up to a level (the *order up to* level—say, M), using the following relationship:

Order quantity = (Order-up-to level) − (Ending inventory) + (Shortage quantity)

Let's say that the order-up-to level (M) is 11 and the ending inventory is three. Further, let's say that the review period (N) is five days. Thus, on the fifth day of the cycle, 8 refrigerators will be ordered from the supplier. If there is a shortage of two refrigerators on the fifth day, then 13 refrigerators will be ordered. (There can't be both ending inventory and shortages at the same time.) If there were a shortage of three refrigerators, the first three received would be provided to the customers when the order arrived. That's called "making up backorders." The *lost sales* case occurs when customer demand is lost if the inventory is not available.

The number of refrigerators ordered each day is randomly distributed as shown in Table 2.19. Another source of randomness is the number of days after the order is placed with the supplier before arrival, or *lead time*. The distribution of lead time is shown in Table 2.20. Assume that the orders are placed at the end of the day. If the lead time is zero, the order from the supplier will arrive the next morning, and the refrigerators will be available for distribution that next day. If the lead time is one day, the order from the supplier arrives the second morning after, and will be available for distribution that day.

The simulation has been started with the inventory level at 3 refrigerators and an order for 8 refrigerators to arrive in 2 days' time. The simulation table is shown in Table 2.21.

Following the simulation table for several selected days indicates how the process operates. The order for 8 refrigerators is available on the morning of the third day of the first cycle, raising

Table 2.19   Random Digit Assignments for Daily Demand

Demand	Probability	Cumulative Probability	Random Digit Assignment
0	0.10	0.10	01–10
1	0.25	0.35	11–35
2	0.35	0.70	36–70
3	0.21	0.91	71–91
4	0.09	1.00	92–00

Table 2.20   Random Digit Assignments for Lead Time

Lead Time (Days)	Probability	Cumulative Probability	Random Digit Assignment
1	0.6	0.6	1–6
2	0.3	0.9	7–9
3	0.1	1.0	0

Table 2.21 Simulation Tables for (M, N) Inventory System

Day	Cycle	Day within Cycle	Beginning Inventory	Random Digits for Demand	Demand	Ending Inventory	Shortage Quantity	Order Quantity	Random Digits for Demand	Lead Time (days)	Days until Order Arrives
1	1	1	3	26	1	2	0	—	—	—	1
2	1	2	2	68	2	0	0	—	—	—	—
3	1	3	8	33	1	7	0	—	—	—	—
4	1	4	7	39	2	5	0	—	—	—	—
5	1	5	5	86	3	2	0	9	8	2	2
6	2	1	2	18	1	1	0	—	—	—	1
7	2	2	1	64	2	0	1	—	—	—	—
8	2	3	9	79	3	5	0	—	—	—	—
9	2	4	5	55	2	3	0	—	—	—	—
10	2	5	3	74	3	0	0	11	7	2	2
11	3	1	0	21	1	0	1	—	—	—	1
12	3	2	0	43	2	0	3	—	—	—	—
13	3	3	11	49	2	6	0	—	—	—	—
14	3	4	6	90	3	3	0	—	—	—	—
15	3	5	3	35	1	2	0	9	2	1	1
16	4	1	2	08	0	2	0	—	—	—	—
17	4	2	11	98	4	7	0	—	—	—	—
18	4	3	7	61	2	5	0	—	—	—	—
19	4	4	5	85	3	2	0	—	—	—	—
20	4	5	2	81	3	0	1	12	3	1	1
21	5	1	0	53	2	0	3	—	—	—	—
22	5	2	12	15	1	8	0	—	—	—	—
23	5	3	8	94	4	4	0	—	—	—	—
24	5	4	4	19	1	3	0	—	—	—	—
25	5	5	3	44	2	1	0	10	1	1	1
Total						68	9				
Average					2.04	2.72	0.36				

the inventory level from zero refrigerators to 8 refrigerators. Demands during the remainder of the first cycle reduced the ending inventory level to 2 refrigerators on the fifth day. Thus, an order for 9 refrigerators was placed. The lead time for this order was 2 days. The order for 9 refrigerators was added to inventory on the morning of day 3 of cycle 2.

Notice that the beginning inventory on the fifth day of the fourth cycle was 2. An order for 3 refrigerators on that day led to a shortage condition. One refrigerator was backordered on that day. Twelve refrigerators were ordered $(11 + 1)$, and they had a lead time of one day. On the next day, the demand was two, so additional shortages resulted.

At the beginning of the next day, the order had arrived. Three refrigerators were used to make up the backorders and there was a demand for one refrigerator, so the ending inventory was 8.

From five cycles of simulation, the average ending inventory is approximately 2.72 (68/25) units. On 5 of 25 days, a shortage condition existed.

In this example, there cannot be more than one order outstanding from the supplier at any time, but there are situations where lead times are so long that the relationship shown so far needs to be modified to the following:

Order quantity = (Order-up-to level) − (Ending inventory) − (On order)
　　　　　　　　 + (Shortage quantity)

This relationship makes sure that extra ordering doesn't occur. To make an estimate of the mean refrigerators in ending inventory by using simulation, many trials would have to be simulated. The Excel spreadsheet solution in www.bcnn.net offers an opportunity to perform such a simulation.

The Excel spreadsheet allows for numerous changes in the input. The policy can be changed (i.e., the values of M and N). The distribution of daily demand and lead time can be changed within the limits of the demand and lead time—that is, demand can be 0, 1, 2, 3, or 4 refrigerators per day and lead times can be 1, 2, or 3 days. Clicking on Generate New Trial will recalculate the spreadsheet. Looking at one figure after another in the One Trial sheet, with the values as given in the problem statement above shows that there is no consistent result.

However, setting the number of trials to 100 in the Experiment sheet and recalculating the spreadsheet produces little variability in the average inventory. It's usually in the range from 2.69 to 3.01, leaving the values the same as in the problem definition above. Nor is there much change in the distribution of the average ending inventory, as shown in Figure 2.14.

2.3 OTHER EXAMPLES OF SIMULATION

This section includes examples of the simulation of a reliability problem, a bombing mission, the generation of the lead-time demand distribution when given the distributions of demand and lead time, and an activity network.

Example 2.5:　A Reliability Problem

A milling machine has three different bearings that fail in service. The distribution of the life of each bearing is identical, as shown in Table 2.22. When a bearing fails, the mill stops, a repairperson is called, and a new bearing is installed. The delay time of the repairperson's arriving at the milling machine is also a random variable having the distribution given in Table 2.23.

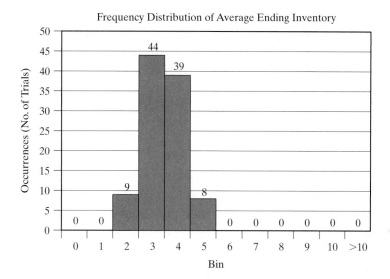

Figure 2.14 Average ending inventory for 100 trials (each 25 days).

Table 2.22 Bearing-Life Distribution

Bearing Life (Hours)	Probability	Cumulative Probability	Random Digit Assignment
1000	0.10	0.10	01–10
1100	0.13	0.23	11–23
1200	0.25	0.48	24–48
1300	0.13	0.61	49–61
1400	0.09	0.70	62–70
1500	0.12	0.82	71–82
1600	0.02	0.84	83–84
1700	0.06	0.90	85–90
1800	0.05	0.95	91–95
1900	0.05	1.00	96–00

Downtime for the mill is estimated at $10 per minute. The direct on-site cost of the repairperson is $30 per hour. It takes 20 minutes to change one bearing, 30 minutes to change two bearings, and 40 minutes to change three bearings. A proposal has been made to replace all three bearings whenever a bearing fails. Management needs an evaluation of the proposal. The total cost per 10,000 bearing-hours will be used as the measure of performance.

Table 2.24 represents a simulation of 15 bearing changes under the current method of operation. Note that there are instances where more than one bearing fails at the same time. This is unlikely to occur in practice and is due to using a rather coarse grid of 100 hours for bearing life. It will be assumed in this example that the times are never exactly the same and thus no more than one

Table 2.23   Delay-Time Distribution

Delay Time (Minutes)	Probability	Cumulative Probability	Random Digit Assignment
5	0.6	0.6	1–6
10	0.3	0.9	7–9
15	0.1	1.0	0

Table 2.24   Bearing Replacement under Current Method

	Bearing 1				Bearing 2				Bearing 3		
RD[a]	Life (Hours)	RD	Delay (Minutes)	RD	Life (Hours)	RD	Delay (Minutes)	RD	Life (Hours)	RD	Delay (Minutes)
1	67	1,400	7	10	71	1,500	8	10	18	1,100	6	5
2	55	1,300	3	5	21	1,100	3	5	17	1,100	2	5
3	98	1,900	1	5	79	1,500	3	5	65	1,400	2	5
4	76	1,500	6	5	88	1,700	1	5	03	1,000	9	10
5	53	1,300	4	5	93	1,800	0	15	54	1,300	8	10
6	69	1,400	8	10	77	1,500	6	5	17	1,100	3	5
7	80	1,500	5	5	08	1,000	9	10	19	1,100	6	5
8	93	1,800	7	10	21	1,100	8	10	09	1,000	7	10
9	35	1,200	0	15	13	1,100	3	5	61	1,300	1	5
10	02	1,000	5	5	03	1,100	2	5	84	1,600	0	15
11	99	1,900	9	10	14	1,000	1	5	11	1,100	5	5
12	65	1,400	4	5	05	1,000	0	15	25	1,200	2	5
13	53	1,300	7	10	29	1,200	2	5	86	1,700	8	10
14	87	1,700	1	5	07	1,000	4	5	65	1,400	3	5
15	90	1,700	2	5	20	1,100	3	5	44	1,200	4	5
Total				110				110				105

[a]RD, random digits.

bearing is changed at any breakdown. The cost of the current system is estimated as follows:

Cost of bearing = 45 bearings × $32/bearing = $1,440
Cost of delay time = (110 + 110 + 105) minutes × $10/minute = $3,250
Cost of downtime during repair = 45 bearings × 20 minutes/bearing × $10 minute
= $9,000
Cost of repairpersons = 45 bearings × 20 minutes/bearing × $30/60 minutes
= $450
Total cost = $1,440 + $3,250 + $9,000 + $450 = $14,140

The total life of the bearings is (22,300 + 18,700 + 18,600) 59,600 hours. Therefore, the total cost per 10,000 bearing-hours is ($14,140/5.96) = $2,372.

Table 2.25 is a simulation of the proposed method. Note that the random digits are not shown. For the first set of bearings, the earliest failure is at 1,000 hours. All three bearings are replaced

Table 2.25   Bearing Replacement under Proposed Method

	Bearing 1 Life (Hours)	Bearing 2 Life (Hours)	Bearing 3 Life (Hours)	First Failure (Hours)	Delay (Minutes)
1	1,700	1,100	1,000	1,000	10
2	1,000	1,800	1,200	1,000	5
3	1,500	1,700	1,300	1,300	5
4	1,300	1,100	1,800	1,100	5
5	1,200	1,100	1,300	1,100	5
6	1,000	1,200	1,200	1,000	10
7	1,500	1,700	1,200	1,200	5
8	1,300	1,700	1,000	1,000	10
9	1,800	1,200	1,100	1,100	15
10	1,300	1,300	1,100	1,100	5
11	1,400	1,300	1,900	1,300	10
12	1,500	1,300	1,400	1,300	5
13	1,500	1,800	1,200	1,200	10
14	1,000	1,900	1,400	1,000	5
15	1,300	1,700	1,700	1,300	5
Total					110

at that time, even though the remaining bearings had more life in them. For example, Bearing 1 would have lasted 700 additional hours.

The cost of the proposed system is estimated as follows:

Cost of bearings = 45 bearings × $32/bearing = $1,440
Cost of delay time = 110 minutes × $10/minute = $1,100
Cost of downtime during repairs = 15 sets × 40 minutes/set × $10/minute
$$= \$6,000$$
Cost of repairpersons = 15 sets × 40 minutes/set × $30/60 minutes
$$= \$300$$
Total cost = $1,440 + $1,100 + $6,000 + $300 = $8,840

The total life of the bearings is $(17,000 \times 3)$ 51,000 hours. Therefore, the total cost per 10,000 bearing-hours is ($8,840/5.1) = $1,733.

The new policy generates a savings of $634 per 10,000 hours of bearing-life. If the machine runs continuously, the savings are about $556 per year.

There are two Excel spreadsheet models for Example 2.5 at www.bcnn.net. These are Example 2.5C (the current system) and Example 2.5P (the proposed system). Much flexibility is offered with respect to the inputs on these models. The user can change the distribution of bearing life (making sure that the cumulative probability is exactly 1.00). The distribution of delay time can be changed (again, making sure that the cumulative probability is exactly 1.00). Also, the parameters of the problem can be changed (bearing cost per unit, and so forth). As in other spreadsheet models, the number of trials can be varied from 1 to 400. Finally, in the Experiment sheet, the endpoints of the bins can be changed for observing the frequency of total cost for 10,000 hours of bearing life.

Example 2.6: Random Normal Numbers

Consider a bomber attempting to destroy an ammunition depot, as shown in Figure 2.15. (This bomber has conventional rather than laser-guided weapons.) If a bomb falls anywhere on the target, a hit is scored; otherwise, the bomb is a miss. The bomber flies in the horizontal direction and carries 10 bombs. The aiming point is (0,0). The point of impact is assumed to be normally distributed around the aiming point with a standard deviation of 400 meters in the direction of flight and 200 meters in the perpendicular direction. The problem is to simulate the operation and make statements about the number of bombs on target.

Recall that the standardized normal variate, Z, having mean 0 and standard deviation 1, is distributed as

$$Z = \frac{X - \mu}{\sigma}$$

where X is a normal random variable, μ is the mean of the distribution of X, and σ is the standard deviation of X. Then,

$$X = Z\sigma_x$$

$$Y = Z\sigma_Y$$

where (X, Y) are the simulated coordinates of the bomb after it has fallen. With $\sigma_x = 400$ and $\sigma_Y = 200$, we have

$$X = 400Z_i$$

$$Y = 200Z_j$$

Figure 2.15   Ammunition depot.

Table 2.26   Simulated Bombing Run

Bomb	RNN_x	X Coordinate $(400\ RNN_x)$	RNN_y	Y Coordinate $(200\ RNN_y)$	Result[a]
1	2.2296	891.8	−0.1932	−38.6	Miss
2	−2.0035	−801.4	1.3034	260.7	Miss
3	−3.1432	−1257.3	0.3286	65.7	Miss
4	−0.7968	−318.7	−1.1417	−228.3	Miss
5	1.0741	429.6	0.7612	152.2	Hit
6	0.1265	50.6	−0.3098	−62.0	Hit
7	0.0611	24.5	−1.1066	−221.3	Hit
8	1.2182	487.3	0.2487	49.7	Hit
9	−0.8026	−321.0	−1.0098	−202.0	Miss
10	0.7324	−293.0	0.2552	51.0	Hit

[a]Total: 5 hits, 5 misses

The i and j subscripts have been added to indicate that the values of Z should be different. What are these Z values and where can they be found? The values of Z are random normal numbers. These can be generated from uniformly distributed random numbers, as will be discussed in Chapter 7. A small sample of random normal numbers is given in Table A.2. The table of random normal numbers is used in the same way as the table of random numbers: that is, start at a random place in the table and proceed in a systematic direction, avoiding overlap.

An example of one bomber's run will indicate how the simulation is performed. Table 2.26 shows the results of a simulated run. The random normal numbers in Table 2.26 are shown to four-decimal-place accuracy.

The mnemonic RNN_x stands for "random normal number to compute the x coordinate" and corresponds to Z_i. The first random normal number used was 2.2296, generating the x-coordinate $400(2.2296) = 891.8$. The random normal number to generate the y-coordinate was −0.1932, resulting in the y-coordinate −38.6. Taken together, $(891.8, -38.6)$ is a miss, for it is off the target. As shown in Table 2.26, there were 5 hits and 5 misses.

The spreadsheet for Example 2.6 at www.bcnn.net makes it possible to conduct lots of experiments. Note that the target shown in the One Trial sheet is flexible. It can be changed by editing the (X, Y) coordinates, shown in green, so long as a convex shape is maintained. With the standard deviation in the x-direction set at 400 meters and the standard deviation in the y-direction set at 200 meters, and with the shape of the target unchanged, an experiment was run with 400 trials (each trial being 10 bombs). A result is shown in Figure 2.16.

Notice that the results range from two hits to ten hits. The average is 6.72 hits. If only one mission (trial) is run, a very misleading result could occur, but Figure 2.16 provides useful descriptive information. For instance, 44% $[(175/400) \times 100\%]$ of the bombing runs there are six or fewer hits. In about 71% of the cases, $[(283/400) \times 100\%]$ there were six, seven, or eight hits.

Example 2.7:   Lead-Time Demand

Lead-time demand occurs in an inventory system when the lead time is not instantaneous. The lead time is the time from placement of an order until the order is received. Assume that lead time

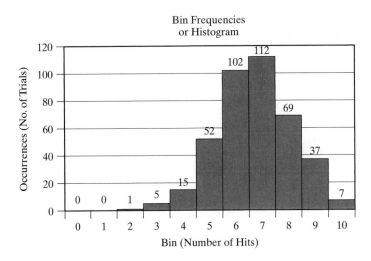

Figure 2.16   Results of 400 trials of the bombing mission.

is a random variable. During the lead time, demands also occur at random. Lead-time demand is thus a random variable defined as the sum of the demands over the lead time, or $\sum_{i=0}^{T} D_i$, where i is the time period of the lead time, $i = 0, 1, 2, \ldots$; D_i is the demand during the ith time period; and T is the lead time. The distribution of lead-time demand is found by simulating many cycles of lead time and building a histogram based on the results.

A firm sells bulk rolls of newsprint. The daily demand is given by the following probability distribution:

Daily Demand (Rolls)	3	4	5	6
Probability	0.20	0.35	0.30	0.15

The lead time is the number of days from placing an order until the firm receives the order from the supplier. In this instance, lead time is a random variable given by the following distribution:

Lead Time (Days)	1	2	3
Probability	0.36	0.42	0.22

Table 2.27 shows the random digit assignment for demand and Table 2.28 does the same for lead time. The incomplete simulation table is shown in Table 2.29. The random digits for the first cycle were 57. This generates a lead time of 2 days. Thus, two pairs of random digits must be generated for the daily demand. The first of these pairs is 11, which leads to the demand 3. This

Table 2.27 Random Digit Assignment for Demand

Daily Demand	Probability	Cumulative Probability	Random Digit Assignment
3	0.20	0.20	01–20
4	0.35	0.55	21–55
5	0.30	0.85	56–85
6	0.15	1.00	86–00

Table 2.28 Random Digit Assignment for Lead Time

Lead Time (Days)	Probability	Cumulative Probability	Random Digit Assignment
1	0.36	0.36	01–36
2	0.42	0.78	37–78
3	0.22	1.00	79–00

Table 2.29 Simulation Table for Lead-Time Demand

Cycle	Random Digits for Lead Time	Lead Time (Days)	Random Digits for Demand	Demand	Lead-Time Demand
1	57	2	11	3	
			64	5	8
2	33	1	37	4	4
3	46	2	13	3	
			80	5	8
4	91	3	27	4	
			66	5	
			47	4	13
.	.	.	.	.	.
.	.	.	.	.	.
.	.	.	.	.	.

is followed by the demand 5. The lead-time demand for the first cycle is 8. After many cycles are simulated, a histogram is generated. Click on "Generate New Trial" repeatedly to see the effect of randomness on a 20-cycle trial.

Although the probabilities for each value of lead-time demand can be generated in this case, simulation can also be used to sample from one or more distributions. The resulting lead-time distribution of lead-time demand might be like that in Figure 2.17. This result was obtained from simulating 20 cycles of lead-time demand, using the Excel spreadsheet found at www.bcnn.net.

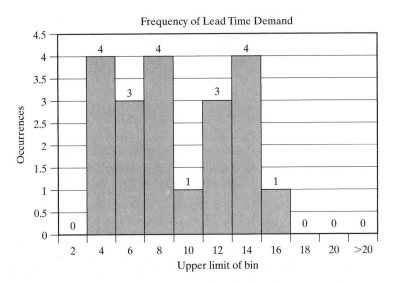

Figure 2.17 Frequency of lead-time demand.

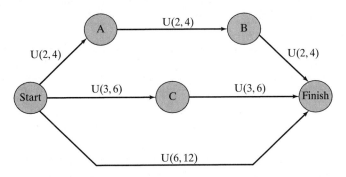

Figure 2.18 Activity network.

Suppose you have a project that requires the completion of a number of activities. Some activities must be carried out sequentially; others can be done in parallel. The project can be represented by a network of activities, as shown in Figure 2.18. There are three paths through the network, each path representing a sequence of activities that must be completed in order. The activities on two different paths can be carried out in parallel.

In the activity network in Figure 2.18, the arcs represent activities and the nodes represent the start or end of an activity. The time to complete all activities on a path is the sum of the activity times along the path. To complete the entire project, all activities must be completed; therefore, project completion time is the maximum over all path completion times.

The topmost path is along the path Start → A→ B → Finish. The middle path is along the path Start → C → Finish. The bottom path is given by Start → Finish.

Example 2.8:   Project Simulation
Here's a concrete example using Figure 2.18. Let's say that three friends wanted to cook bacon, eggs, and toast for breakfast for some weekend visitors. Each friend was going to prepare one of the three items. The activities might be as follows:

Top path:	Start	$\rightarrow$	A	Crack eggs
	A	$\rightarrow$	B	Scramble eggs
	B	$\rightarrow$	Finish	Cook eggs
Middle path:	Start	$\rightarrow$	C	Make toast
	C	$\rightarrow$	Finish	Butter toast
Bottom path:	Start	$\rightarrow$	Finish	Fry the bacon

Let's say that the times to accomplish each of the activities in preparing this breakfast are variable and can be represented by a uniform distribution between a lower and upper limit, as shown in Figure 2.18. You want to know something about the preparation time so that you can tell the visitors to be at the breakfast table on time. Perhaps you want to estimate the probability of preparing breakfast within a specified amount of time.

The activity times are shown on the arcs of the activity network. For example, the activity time from Start $\rightarrow$ A (i.e., crack the eggs) is assumed to be uniformly distributed between 2 and 4 minutes. That means that all times between 2 and 4 are equally likely to occur. The expected value, or mean time, for this activity is the midpoint, three minutes.

Applying that logic, the expected value along the topmost path is nine minutes, which is determined by adding the three expected values $(3 + 3 + 3)$. The shortest possible completion time, which is determined by adding the smallest values, is six minutes $(2 + 2 + 2)$. The largest possible time along the top path is twelve minutes $(4 + 4 + 4)$.

Similarly, the expected value through the middle path is nine minutes, while the smallest and largest times are six and twelve minutes, respectively. The bottom path has the same expected value and the same extreme values.

The time that the project will be completed is the maximum time through any of the paths. (Thinking again about the preparation of this breakfast, the time that everything will be ready is when the eggs, the toast and the bacon are ready.) But, since activity times are assumed to have some random variability, the times through the paths are not constant.

Pritsker [1995] showed how such a project could be analyzed with a simulation of independent replications of the activity times. For a uniform distribution, a simulated activity time is given by

Simulated Activity Time = Lower limit
$$+ \text{ (Upper limit } - \text{ Lower limit) } * \text{ Random number}$$

With a table of random numbers, the time for each simulated activity can be computed manually. For example, for activity Start $\rightarrow$ A, if the random number is 0.7943, the simulated activity time is $2 + (4 - 2) * 0.7943 = 3.5886$, or 3.59 minutes.

The Experiment worksheet in the Excel workbook for Example 2.8 found at www.bcnn.net allows from 1 to 400 trials and computes the average, median, minimum and maximum values.

With 400 trials and using the default seed, the results are as follows:

Mean 10.12 minutes
Minimum 6.85 minutes
Maximum 12.00 minutes

The so-called critical path is the path that takes the longest time; that is, its time is the project completion time. For each of the 400 trials, the experiment determines which path was critical, with these results:

Top path 30.00% of the trials
Middle path 31.25%
Bottom path 38.75%

We conclude that the chance of the bacon being the last item ready is 38.75%. Is this a fluke? Why aren't the paths each represented about 1/3 of the time? The answer to this question is left for a later exercise.

Lastly, the project completion times were placed in a frequency chart. These differ each time that the spreadsheet is recalculated, but, in any large number of trials, the basic shape of the frequency chart (or histogram) will remain roughly the same. Starting from the default seed, the resulting frequency chart for project completion time is shown in Figure 2.19. Inferences that could be drawn include the following:

13.5% of the time (54 of 400), the breakfast will be ready in 9 minutes or less.
20.5% of the time (82 of 400), it will take from 11 to 12 minutes.

Figure 2.19   Bin frequencies for completion times.

2.4 SUMMARY

This chapter introduced simulation concepts by means of examples, to illustrate general areas of application and to motivate the remaining chapters. In the next chapter, we give a more systematic presentation of the basic concepts.

Ad-hoc simulation tables were used in completing each example. Events in the tables were generated by using uniformly distributed random numbers and, in one case, random normal numbers. The examples illustrate the need for working out the characteristics of the input data, generating random variables from the input models, and analyzing the resulting response. The queueing examples, especially the two-channel queue, illustrate some of the complex dependencies that can occur—in this example, between subsequent customers visiting the queue. Because of these complexities, the ad-hoc simulation table approach fails, or becomes unbearably complex, even with relatively simple networks of queues. For this and other reasons, a more systematic methodology, such as the event scheduling approach described in Chapter 3, is needed. These subjects are treated in more detail in the remaining chapters of the text.

Examples are drawn principally from queueing and inventory systems, because a large number of simulations concern problems in these areas. Additional examples are given in the areas of reliability, static simulation, the generation of a random sample from an unknown distribution, and project management. All of the examples are modeled using Excel. These are available at www.bcnn.net.

REFERENCES

HADLEY G., AND T. M. WHITIN [1963], *Analysis of Inventory Systems*, Prentice-Hall, Englewood Cliffs, NJ.

HARRISON, J.M., AND V. NGUYEN [1995], "Some Badly Behaved Closed Queueing Networks", *Proceedings of IMA Workshop on Stochastic Networks*, eds. F. Kelly and R. J. Williams.

PRITSKER, A. A. B. [1995], *Introduction to Simulation and SLAM II*, 4th ed., John Wiley, New York.

SEILA, A., V. CERIC, AND P. TADIKAMALLA [2003], *Applied Simulation Modeling*, Duxbury, Belmont, CA.

EXERCISES

Manual Exercises

Most of these exercises could also be solved in Excel. For hints on implementation in Excel, study the spreadsheet solutions at www.bcnn.net and read the "Explain" worksheet, as appropriate for the problem being solved. Another suggestion for solving these exercises in Excel is to take advantage of the VB functions in the spreadsheet solutions in www.bcnn.net. Start with a supplied example, delete the existing inputs and simulation table in the One Trial worksheet, and use what remains. On the Experiment worksheet, change the response in the cell just below the word "Link."

1. Consider the following continuously operating job shop. Interarrival times of jobs are distributed as follows:

Time Between Arrivals (Hours)	Probability
0	.23
1	.37
2	.28
3	.12

Processing times for jobs are normally distributed, with mean 50 minutes and standard deviation 8 minutes. Construct a simulation table, and perform a simulation for 10 new customers. Assume that, when the simulation begins, there is one job being processed (scheduled to be completed in 25 minutes) and there is one job with a 50-minute processing time in the queue.

(a) What was the average time in the queue for the 10 new jobs?
(b) What was the average processing time of the 10 new jobs?
(c) What was the maximum time in the system for the 10 new jobs?

2. A baker is trying to figure out how many dozens of bagels to bake each day. The probability distribution of the number of bagel customers is as follows:

Number of Customers/Day	8	10	12	14
Probability	0.35	0.30	0.25	0.10

Customers order 1, 2, 3, or 4 dozen bagels according to the following probability distribution.

Number of Dozen Ordered/Customer	1	2	3	4
Probability	0.4	0.3	0.2	0.1

Bagels sell for $8.40 per dozen. They cost $5.80 per dozen to make. All bagels not sold at the end of the day are sold at half-price to a local grocery store. Based on 5 days of simulation, how many dozen (to the nearest 5 dozen) bagels should be baked each day?

3. Develop and interpret flow diagrams analogous to Figures 2.2 and 2.3 for a queueing system with i channels.

4. Smalltown Taxi operates one vehicle during the 9:00 A.M. to 5:00 P.M. period. Currently, consideration is being given to the addition of a second vehicle to the fleet. The demand for taxis follows the distribution shown:

Time Between Calls (Minutes)	15	20	25	30	35
Probability		0.14	0.22	0.43	0.17	0.04

The distribution of time to complete a service is as follows:

Service Time (Minutes)	5	15	25	35	45
Probability		0.12	0.35	0.43	0.06	0.04

Simulate five individual days of operation of the current system and of the system with an additional taxicab. Compare the two systems with respect to the waiting times of the customers and any other measures that might shed light on the situation.

5. The random variables X, Y, and Z are distributed as follows:

$$X \sim N(\mu = 100, \sigma^2 = 100)$$
$$Y \sim N(\mu = 300, \sigma^2 = 225)$$
$$Z \sim N(\mu = 40, \sigma^2 = 64)$$

Simulate 50 values of the random variable

$$W = \frac{X + Y}{Z}$$

Prepare a histogram of the resulting values, using class intervals of width equal to 3.

6. Given A, B, and C, which are uncorrelated random variables. Variable A is normally distributed with $\mu = 100$ and $\sigma^2 = 400$. Variable B is discrete uniformly distributed with probability distribution given by $p(b) = 1/5$ with $b = 0, 1, 2, 3$ and 4. Variable C is distributed in accordance with the following table.

Value of C	Probability
10	.10
20	.25
30	.50
40	.15

Use simulation to estimate the mean of a new variable D, that is defined as

$$D = (A - 25B)/(2C)$$

Use a sample of size 10.

7. Estimate, by simulation, the average number of lost sales per week for an inventory system that functions as follows:

   (a) Whenever the inventory level falls to or below 10 units, an order is placed. Only one order can be outstanding at a time.
   (b) The size of each order is equal to $20 - I$, where I is the inventory level when the order is placed.
   (c) If a demand occurs during a period when the inventory level is zero, the sale is lost.
   (d) Daily demand is normally distributed, with a mean of 5 units and a standard deviation of 1.5 units. (Round off demands to the closest integer during the simulation, and, if a negative value results, give it a demand of zero.)
   (e) Lead time is distributed uniformly between zero and 5 days—integers only.
   (f) The simulation will start with 18 units in inventory.
   (g) For simplicity, assume that orders are placed at the close of the business day and received after the lead time has occurred. Thus, if lead time is one day, the order is available for distribution on the morning of the second day of business following the placement of the order.
   (h) Let the simulation run for 5 weeks.

8. An elevator in a manufacturing plant carries exactly 400 kilograms of material. There are three kinds of material, and these are in boxes, that arrive for a ride on the elevator. These materials and their distributions of time between arrivals are as follows:

Material	Weight (kilograms)	Interarrival Time (Minutes)
A	200	5 ± 2 (uniform)
B	100	6 (constant)
C	50	$P(2) = 0.33$
		$P(3) = 0.67$

   It takes the elevator 1 minute to go up to the second floor, 2 minutes to unload, and 1 minute to return to the first floor. The elevator does not leave the first floor unless it has a full load. Simulate 1 hour of operation of the system. What is the average transit time for a box of material A (time from its arrival until it is unloaded)? What is the average waiting time for a box of material B? How many boxes of material C made the trip in 1 hour?

9. The random variables X and Y are distributed as follows:
$$X \sim 10 \pm 10 \text{ (uniform)}$$
$$Y \sim 10 \pm 8 \text{ (uniform)}$$

   (a) Simulate 200 values of the random variable
$$Z = XY$$
   Prepare a histogram of the resulting values. What is the range of Z and what is its average value?
   (b) Same as (a), except that
$$Z = X/Y$$

10. Consider the assembly of two steel plates, each plate having a hole drilled in its center. The plates are to be joined by a pin. The plates are aligned for assembly relative to the bottom left corner (0,0). The hole placement is centered at (3,2) on each plate. The standard deviation in each direction is 0.0045. The hole diameter is normally distributed, with a mean of 0.3 and a standard deviation of 0.005. The pin diameter is also distributed normally, with a mean of 0.29 and a standard deviation of 0.004. What fraction of pins will go through the assembled plates? Base your answer on a simulation of 50 observations.

Hint: Clearance = $\text{Min}(h_1, h_2) - [(x_1 - x_2)^2 + (y_1 - y_2)^2]^{.5} - p$, where

h_i = hole diameter, i = plate 1, 2
p = pin diameter
x_i = distance to center of plate hole, horizontal direction, $i = 1, 2$
y_i = distance to center of plate hole, vertical direction, $i = 1, 2$

11. In the exercise above, the pin will wobble if it is too loose. Wobbling occurs if $\text{Min}(h_1, h_2) - p \geq 0.006$. What fraction of the assemblies wobble? (Conditional probability—i.e., given that the pins go through)

12. Three points are chosen at random on the circumference of a circle. Estimate the probability that they all lie on the same semicircle, by Monte Carlo sampling methods. Perform 5 replications.

13. Two theorems from statistics are as follows:

Theorem 1 Let $Z_1, Z_2, \ldots, Z_k$ be normally and independently distributed random variables with mean $\mu = 0$ and variance $\sigma^2 = 1$. Then the random variable

$$\chi^2 = Z_1^2 + Z_2^2 + \cdots + Z_k^2$$

is said to follow the chi-squared distribution with k degrees of freedom, abbreviated χ_k^2.

Theorem 2 Let $Z \sim N(0, 1)$ and V be a chi-square random variable with k degrees of freedom. If Z and V are independent, then the random variable

$$T = \frac{Z}{\sqrt{V/k}}$$

is said to follow the t distribution with k degrees of freedom, abbreviated t_k.

Generate a t-distributed random variate with 3 degrees of freedom. Use the following random values in the order shown:

random digits	random normal numbers
6729	1.06
1837	−0.72
2572	0.28
8134	−0.18
5251	−0.63

14. A bank has one drive-in teller and room for one additional customer to wait. Customers arriving when the queue is full park and go inside the bank to transact business. The times between arrivals and the service-time distributions follow:

Time Between Arrivals (Minutes)	Probability	Service Time (Minutes)	Probability
0	0.09	1	0.20
1	0.17	2	0.40
2	0.27	3	0.28
3	0.20	4	0.12
4	0.15		
5	0.12		

Simulate the operation of the drive-in teller for 10 new customers. The first of the 10 new customers arrives at a time determined at random. Start the simulation with one customer being served, leaving at time 3, and one in the queue. How many customers went into the bank to transact business?

15. The number of acres covered by Cedar Bog Lake is being estimated. The lake is shown next, with the scale shown in feet:

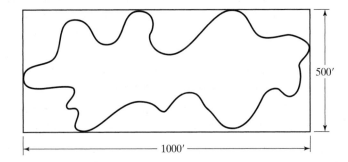

Use simulation to estimate the size of the lake. (One acre = 43,560 square feet).

16. In Example 2.2, assume that the average delay with another server, Charlie, is virtually zero. But another server costs $20/hour. If the average caller delay costs $200 per hour, at what point is it advisable to add another server?

17. In Example 2.8, why isn't each path represented about 1/3 of the time?

18. In Example 2.7, the only way to have a lead-time demand of 3 or 4 is to have lead time of one day and a demand of 3 or 4. The probability of that happening is $0.36(0.20 + 0.35) = 0.198$.

Using this logic, what is the probability of having lead-time demand equal 11 or 12? (Use computation, not simulation.)

19. In Example 2.7, what is the probability of having a lead-time demand of 18? (Use computation, not simulation.)

Spreadsheet Exercises

20. In Example 2.1, let the arrival distribution be uniformly distributed between 1 and 10 minutes (integers only). Run the experiment for 50 trials. Is there any difference between the bin frequencies shown and those of Figure 2.8?

21. In Example 2.1, let the service-time distribution be changed to the following:

Service Time (Minutes)	1	2	3	4	5	6
Probability		0.05	0.10	0.20	0.30	0.25	0.10

Run the experiment for 50 trials. Is there any difference between the average waiting time here and that in Example 2.1 before the change was made? If there is a difference, to what do you attribute it?

22. Run the experiment in Example 2.1 for 25, 50, 100, 200, and 400 trials. (All trials are with different sets of random numbers.) What are the differences in the minimum and maximum values of the average waiting times? If there are differences, how do you explain them?

23. In Example 2.2, run 10 experiments, each with 100 trials. What was the fraction of trials in each experiment such that the callers had an average delay of three minutes or less? For the 1000 trials just simulated, what fraction of the callers had an average delay of two minutes or less?

24. In Example 2.2, conduct an experiment of 400 trials. Explain the large spread between the minimum average delay and the maximum average delay.

25. In Example 2.2, run 10 experiments with 50 trials each and 10 experiments with 400 trials each. Explain the differences in output that you observe, if any, between the two experiments.

26. In Example 2.2, modify the spreadsheet so that the number of calls taken by Able and Baker can be displayed. What fraction did each take in an experiment of 400 trials?

27. In Example 2.3, determine the best policy for ordering newspapers based on experiments of 50 trials.

28. Run 10 experiments, each with 400 trials, in Example 2.3. What is the largest difference that you got between the maximum and minimum daily profit?

29. Use the spreadsheet for Example 2.4. For 100 trials, what is the average ending inventory if $M = 11$ and $N = 4, 5, 6$?

30. Use the spreadsheet for Example 2.4. For 100 trials, what is the average ending inventory if $N = 5$ and $M = 10, 11, 12$?

31. At what bearing cost in Example 2.5 is the total cost per 10,000 hours virtually the same for the current and proposed systems? Base your estimate on 10 experiments of the current system and of the proposed system, each experiment consisting of 400 trials.

32. Change the cell widths on the experiment in the Excel spreadsheet for Example 2.5 (current or proposed) to a width of $50 beginning below the minimum value (for example, if the minimum value is $1528.46, let the first cell begin at $1500). What is the advantage of doing this?

33. Using the spreadsheet for Example 2.5 (proposed), run 10 experiments of 40 trials each, and record the range (maximum value − minimum value) of the results. Next, compute the average range. Then, do the same as before, except use 400 trials in each experiment. If there is a difference, how do you explain it?

34. Set $\sigma_x = 600$ meters and $\sigma_y = 300$ meters in the spreadsheet for Example 2.6. Leave the target intact. Conduct a simulation with 200 trials. What was the average number of hits?

35. Set $\sigma_x = 300$ meters and $\sigma_y = 600$ meters in the spreadsheet for Example 2.6. Leave the target intact. Conduct a simulation with 200 trials. What was the average number of hits?

36. Set $\sigma_x = 300$ meters and $\sigma_y = 150$ meters in the spreadsheet for Example 2.6. Leave the target intact. Conduct a simulation with 200 trials. What was the average number of hits? Explain the difference between this exercise and the previous exercise.

37. Set $\sigma_x = 2\sigma_y$ meters in the spreadsheet for Example 2.6. Leave the target intact. What is the value of σ_x if the average number of hits is to be about 6.0, based on one experiment of 400 trials.

38. In Example 2.7, suppose you wanted a better estimate of the distribution of lead-time demand: Would you (1) increase the number of cycles on the"One Trial" worksheet? or (2) increase the number of trials on the "Experiment" worksheet? Explain your answer.

39. In Example 2.7, suppose you wanted a better estimate of the mean lead-time demand? What would you do?

40. In Example 2.7, let the demand probabilities be equal for the possible values 3, 4, 5, and 6. Run an experiment with 400 trials. Compare the average lead-time demand from using the original input data and from using the new input data.

41. In Example 2.7, let the lead-time probabilities be equal for the possible values 1, 2, and 3. Run an experiment with 400 trials. Compare the average lead-time demand from using the original input data and from using the new input data.

42. In Example 2.8, recalculate the spreadsheet 20 times, each with 400 trials. Record the number of times that the middle path was the longest. What is your best guess of the true mean of the fraction of time that the middle path is taken?

43. In the above exercise, what is the smallest value and the largest value encountered for the number of times that the middle path was selected? What if you had conducted one simulation,

gotten the smallest (or largest) value, and reported that as the result? How could you argue yourself out of that jam?

44. In Example 2.8, suppose the third pathway (the bottom one in Figure 2.18) is changed so that it consists of six U(1,2) activities. Modify the spreadsheet to accommodate this. Which is the most frequently occurring path now? What insight does this exercise provide?

45. Using Excel, generate 1000 values in a column, using the formula
$= -10*LN(RAND())$.

    (a) Compute descriptive statistics about the data in that column, including the minimum value, the maximum value, the mean, the median, and the standard deviation.
    (b) Tabulate the values into 10 bins, each of width equal to five: the first bin beginning at 0, and the eleventh bin for overflow (if any).
    (c) Does the histogram resemble any distribution with which you are familiar? If so, what is its name?

    Hint: Use FREQUENCY in Excel to form bins.

46. Using Excel, generate 12 columns, each with 250 values, using the formula
=RAND().

    In cell M1, place the formula
    =SUM(A1:L1)-6 and copy it to the 249 cells below M1 in column M.

    (a) Compute descriptive statistics about the data in that column, including minimum value, maximum value, mean, median, and standard deviation.
    (b) Tabulate the values with 9 bins: the first bin will include all values less than or equal to −3.5; the next six bins are of width one; the last bin will include all values greater than 3.5.
    (c) Does the histogram resemble any distribution with which you are familiar? If so, what is its name?

    Hint 1: Use FREQUENCY in Excel to form bins.

    Hint 2: The values in Column M can be used instead of those in Table A.2.

47. Using Excel, generate a table that can be used instead of Table A.1, using the formula
=INT(RAND()*100000).

48. Use Excel to estimate the value of π, as in the equation for the area of a circle,

$$A = \pi r^2.$$

    Hint: Consider a quarter circle with unit radius inside a unit square.

49. A college student is making financial plans for the next year. Expenses include the following known costs:

    Tuition $8,400
    Dormitory $5,400

There are also a lot of uncertain expenses, with ranges as follows:

Meals	$900–$1,350
Entertainment	$600–$1,200
Transportation	$200–$600
Books	$400–$800

The student foresees income to pay for next year. The only certain amounts are as follows:

State scholarship	$3,000
Parental contribution	$4,000

Some of the income is variable, with ranges as follows:

Waiting tables	$3,000–$5,000
Library aide	$2,000–$3,000

The student predicts that a loan will be needed. Use Excel (1,000 trials) to forecast the probability distribution for the size of the loan. Treat any amounts that are variable as uniformly distributed over the range indicated.

50. The 22-stop Bally 1090 slot machine had the following number of symbols on each reel:

Symbols	Reel 1	Reel 2	Reel 3
Cherries	2	5	2
Oranges	2	3	7
Plums	5	1	10
Bells	10	2	1
Bars	2	10	1
7's	1	1	1

Payoffs were as follows:

Combination	Payoff	Combination	Payoff
Left cherry	2	Three plums	20
Right cherry	2	Plum plum bar	14
Any two cherries	5	Bar plum plum	14
Three cherries	20	Three bells	20
Three oranges	20	Bell bell bar	18
Orange orange bar	10	Bar bell bell	18
Bar orange orange	10	Three bars	50
		Three 7's	100

Using Excel, simulate 1,000 plays of this slot machine, each for $1.

How much was the gain or loss?

3

General Principles

This chapter develops a common framework for the modeling of complex systems by using discrete-event simulation. It covers the basic building blocks of all discrete-event simulation models: entities and attributes, activities, and events. In discrete-event simulation, a system is modeled in terms of its state at each point in time; of the entities that pass through the system and the entities that represent system resources; and of the activities and events that cause the system state to change. Discrete-event models are appropriate for those systems for which changes in system state occur only at discrete points in time.

The simulation languages and software (collectively called simulation packages) described in Chapter 4 are fundamentally packages for discrete-event simulation. A few of the packages also include the capability to model continuous variables in a purely continuous simulation or a mixed discrete–continuous model. The discussion in this chapter focuses on the discrete-event concepts and methodologies. The discussion in Chapter 4 focuses more on the capabilities of the individual packages and on some of their higher-level constructs.

This chapter introduces and explains the fundamental concepts and methodologies underlying all discrete-event simulation packages. These concepts and methodologies are not tied to any particular package. Many of the packages use different terminology from that used here, and most have a number of higher-level constructs designed to make modeling simpler and more straightforward for their application domain. For example, this chapter discusses the fundamental abstract concept of an entity, but Chapter 4 discusses more concrete realizations of entities, such as machines, conveyors, and vehicles that are built into some of the packages to facilitate modeling in the manufacturing, material handling, or other domains.

Applications of simulation in specific contexts are discussed in Part Five of this text. Topics covered include the simulation of manufacturing and material handling systems in Chapter 13, the simulation of computer systems in Chapter 14, and the simulation of communications systems in Chapter 15.

Section 3.1 covers the general principles and concepts of discrete-event simulation, the event scheduling/time advance algorithm, and the three prevalent world views: event scheduling, process interaction, and activity scanning. Section 3.2 introduces some of the notions of list processing, one of the more important methodologies used in discrete-event simulation software. Chapter 4 covers the implementation of the concepts in a number of the more widely used simulation packages.

3.1 CONCEPTS IN DISCRETE-EVENT SIMULATION

The concept of a system and a model of a system were discussed briefly in Chapter 1. This chapter deals exclusively with dynamic, stochastic systems (i.e., involving time and containing random elements) that change in a discrete manner. This section expands on these concepts and proposes a framework for the development of a discrete-event model of a system. The major concepts are briefly defined and then illustrated by examples:

System A collection of entities (e.g., people and machines) that interact together over time to accomplish one or more goals.

Model An abstract representation of a system, usually containing structural, logical, or mathematical relationships that describe a system in terms of state, entities and their attributes, sets, processes, events, activities, and delays.

System state A collection of variables that contain all the information necessary to describe the system at any time.

Entity Any object or component in the system that requires explicit representation in the model (e.g., a server, a customer, a machine).

Attributes The properties of a given entity (e.g., the priority of a waiting customer, the routing of a job through a job shop).

List A collection of (permanently or temporarily) associated entities, ordered in some logical fashion (such as all customers currently in a waiting line, ordered by "first come, first served," or by priority).

Event An instantaneous occurrence that changes the state of a system (such as an arrival of a new customer).

Event notice A record of an event to occur at the current or some future time, along with any associated data necessary to execute the event; at a minimum, the record includes the event type and the event time.

Event list A list of event notices for future events, ordered by time of occurrence; also known as the future event list (FEL).

Activity A duration of time of specified length (e.g., a service time or interarrival time), which is known when it begins (although it may be defined in terms of a statistical distribution).

Delay A duration of time of unspecified indefinite length, which is not known until it ends (e.g., a customer's delay in a last-in–first-out waiting line which, when it begins, depends on future arrivals).

Clock A variable representing simulated time, called CLOCK in the examples to follow.

Different simulation packages use different terminology for the same or similar concepts—for example, lists are sometimes called sets, queues, or chains. Sets or lists are used to hold both entities and event notices. The entities on a list are always ordered by some rule, such as first-in–first-out or last-in–first-out, or are ranked by some entity attribute, such as priority or due date. The future event list is always ranked by the event time recorded in the event notice. Section 3.2 discusses a number of methods for handling lists and introduces some of the methodologies for efficient processing of ordered sets or lists.

An activity typically represents a service time, an interarrival time, or any other processing time whose duration has been characterized and defined by the modeler. An activity's duration may be specified in a number of ways:

1. Deterministic—for example, always exactly 5 minutes;
2. Statistical—for example, as a random draw from among 2, 5, 7 with equal probabilities;
3. A function depending on system variables and/or entity attributes—for example, loading time for an iron ore ship as a function of the ship's allowed cargo weight and the loading rate in tons per hour.

However it is characterized, the duration of an activity is computable from its specification at the instant it begins. Its duration is not affected by the occurrence of other events (unless, as is allowed by some simulation packages, the model contains logic to cancel an event). To keep track of activities and their expected completion time, at the simulated instant that an activity duration begins, an event notice is created having an event time equal to the activity's completion time. For example, if the current simulated time is CLOCK = 100 minutes and an inspection time of exactly 5 minutes is just beginning, then an event notice is created that specifies the type of event (an end-of-inspection event), and the event time ($100 + 5 = 105$ minutes).

In contrast to an activity, a delay's duration is not specified by the modeler ahead of time, but rather is determined by system conditions. Quite often, a delay's duration is measured and is one of the desired outputs of a model run. Typically, a delay ends when some set of logical conditions becomes true or one or more other events occur. For example, a customer's delay in a waiting line may be dependent on the number and duration of service of other customers ahead in line as well as the availability of servers and equipment.

A delay is sometimes called a *conditional wait*, an activity an *unconditional wait*. The completion of an activity is an event, often called a *primary event*, that is managed by placing an event notice on the FEL. In contrast, delays are managed by placing the associated entity on another list, perhaps representing a waiting line, until such time as system conditions permit the processing of the entity. The completion of a delay is sometimes called a conditional or secondary event, but such events are not represented by event notices, nor do they appear on the FEL.

The systems considered here are dynamic, that is, changing over time. Therefore, system state, entity attributes and the number of active entities, the contents of sets, and the activities and delays currently in progress are all functions of time and are constantly changing over time. Time itself is represented by a variable called CLOCK.

Example 3.1:   Call Center, Revisited

Consider the Able–Baker call center system of Example 2.2. A discrete-event model has the following components:

System state

$L_Q(t)$, the number of callers waiting to be served at time t;

$L_A(t)$, 0 or 1 to indicate Able as being idle or busy at time t;

$L_B(t)$, 0 or 1 to indicate Baker as being idle or busy at time t.

Entities   Neither the callers nor the servers need to be explicitly represented, except in terms of the state variables, unless certain caller averages are desired (compare Examples 3.4 and 3.5).

Events

Arrival event;

Service completion by Able;

Service completion by Baker.

Activities

Interarrival time, defined in Table 2.11;

Service time by Able, defined in Table 2.12;

Service time by Baker, defined in Table 2.13.

Delay   A caller's wait in queue until Able or Baker becomes free.

The definition of the model components provides a static description of the model. In addition, a description of the dynamic relationships and interactions between the components is also needed. Some questions that need answers include:

1. How does each event affect system state, entity attributes, and set contents?
2. How are activities defined (i.e., deterministic, probabilistic, or some other mathematical equation)? What event marks the beginning or end of each activity? Can the activity begin regardless of system state, or is its beginning conditioned on the system being in a certain state? (For example, a machining "activity" cannot begin unless the machine is idle, not broken, and not in maintenance.)
3. Which events trigger the beginning (and end) of each type of delay? Under what conditions does a delay begin or end?
4. What is the system state at time 0? What events should be generated at time 0 to "prime" the model—that is, to get the simulation started?

A discrete-event simulation is the modeling over time of a system all of whose state changes occur at discrete points in time—those points when an event occurs. A discrete-event simulation (hereafter called a simulation) proceeds by producing a sequence of system snapshots (or system images) that represent the evolution of the system through time. A given snapshot at a given time (CLOCK = t) includes not only the system state at time t, but also a list (the FEL) of all activities currently in progress and when each such activity will end, the status of all entities and current membership of all sets, plus the current values of cumulative statistics and counters that will be used to calculate summary statistics at the end of the simulation. A prototype system snapshot is shown in Figure 3.1. (Not all models will contain every element exhibited in Figure 3.1. Further illustrations are provided in the examples in this chapter.)

Clock	System state	Entities and attributes	Set 1	Set 2	. . .	Future event list, FEL	Cumulative statistics and counters
t	$(x, y, z, \ldots)$					$(3, t_1)-$ Type 3 event to occur at time t_1 $(1, t_2)-$ Type 1 event to occur at time t_2	
	. . .					. . .	. . .

Figure 3.1   Prototype system snapshot at simulation time t.

3.1.1 The Event Scheduling/Time Advance Algorithm

The mechanism for advancing simulation time and guaranteeing that all events occur in correct chronological order is based on the future event list (FEL). This list contains all event notices for events that have been scheduled to occur at a future time. Scheduling a future event means that, at the instant an activity begins, its duration is computed or drawn as a sample from a statistical distribution; and that the end-activity event, together with its event time, is placed on the future event list. In the real world, most future events are not scheduled but merely happen—such as random breakdowns or random arrivals. In the model, such random events are represented by the end of some activity, which in turn is represented by a statistical distribution.

At any given time t, the FEL contains all previously scheduled future events and their associated event times (called $t_1, t_2, \ldots$ in Figure 3.1). The FEL is ordered by event time, meaning that the events are arranged chronologically—that is, the event times satisfy

$$t < t_1 \leq t_2 \leq t_3 \leq \cdots \leq t_n.$$

Time t is the value of CLOCK, the current value of simulated time. The event associated with time t_1 is called the imminent event; that is, it is the next event that will occur. After the system snapshot at simulation time CLOCK $= t$ has been updated, the CLOCK is advanced to simulation time CLOCK $= t_1$, the imminent event notice is removed from the FEL, and the event is executed. Execution of the imminent event means that a new system snapshot for time t_1 is created, one based on the old snapshot at time t and the nature of the imminent event. At time t_1, new future events may or might not be generated, but if any are, they are scheduled by creating event notices and putting them into their proper position on the FEL. After the new system snapshot for time t_1 has been updated, the clock is advanced to the time of the new imminent event and that event is executed. This process repeats until the simulation is over. The sequence of actions that a simulator (or simulation language) must perform to advance the clock and build a new system snapshot is called the *event-scheduling/time-advance algorithm*, whose steps are listed in Figure 3.2 (and explained thereafter).

Old system snapshot at time t

CLOCK	System state	· · ·	Future event list	· · ·
t	$(5, 1, 6)$		$(3, t_1)$ — Type 3 event to occur at time t_1 $(1, t_2)$ — Type 1 event to occur at time t_2 $(1, t_3)$ — Type 1 event to occur at time t_3 . . . . $(2, t_n)$ — Type 2 event to occur at time t_n	

Event-scheduling/time-advance algorithm

Step 1. *Remove the event notice for the imminent event*
    (event 3, time t_1) from FEL.
Step 2. *Advance CLOCK to imminent event time*
    (i.e., advance CLOCK from t to t_1).
Step 3. *Execute imminent event: update system state,*
    change entity attributes, and set membership as needed.
Step 4. *Generate future events (if necessary) and*
    place their event notices on FEL, ranked by event time.
    (Example: Event 4 to occur at time t^*, where $t_2 < t^* < t_3$.)
Step 5. *Update cumulative statistics and counters.*

New system snapshot at time t_1

CLOCK	System state	· · ·	Future event list	· · ·
t_1	$(5, 1, 5)$		$(1, t_2)$ — Type 1 event to occur at time t_2 $(4, t^*)$ — Type 4 event to occur at time t^* $(1, t_3)$ — Type 1 event to occur at time t_3 . . . . $(2, t_n)$ — Type 2 event to occur at time t_n	

Figure 3.2   Advancing simulation time and updating system image.

The length and contents of the FEL are constantly changing as the simulation progresses, and thus its efficient management in a computerized simulation will have a major impact on the efficiency of the computer program representing the model. The management of a list is called *list processing*. The major list processing operations performed on a FEL are removal of the imminent event, addition of a new event to the list, and occasionally removal of some event (called cancellation of an event). As the imminent event is usually at the top of the list, its removal is as efficient as possible. Addition of a new event (and cancellation of an old event) requires a search of the list. The efficiency of this search depends on the logical organization of the list and on how the search is conducted. In addition to the FEL, all the sets in a model are maintained in some logical order, and the operations of addition and removal of entities from the set also require efficient list-processing techniques. A brief introduction to list processing in simulation is given in Section 3.2.

The removal and addition of events from the FEL is illustrated in Figure 3.2. Event 3 with event time t_1 represents, say, a service completion event at server 3. Since it is the imminent event at time t, it is removed from the FEL in step 1 (Figure 3.2) of the event-scheduling/time-advance algorithm. When event 4 (say, an arrival event) with event time t^* is generated at step 4, one possible way to determine its correct position on the FEL is to conduct a top-down search:

If $t^* < t_2$,   place event 4 at the top of the FEL.

If $t_2 \leq t^* < t_3$,   place event 4 second on the list.

If $t_3 \leq t^* < t_4$,   place event 4 third on the list.

$\vdots$

If $t_n \leq t^*$,   place event 4 last on the list.

(In Figure 3.2, it was assumed that t^* was between t_2 and t_3.) Another way is to conduct a bottom-up search. The least efficient way to maintain the FEL is to leave it as an unordered list (additions placed arbitrarily at the top or bottom), which would require at step 1 of Figure 3.2 a complete search of the list for the imminent event before each clock advance. (The imminent event is the event on the FEL with the lowest event time.)

The system snapshot at time 0 is defined by the initial conditions and the generation of the so-called exogenous events. The specified initial conditions define the system state at time 0. For example, in Figure 3.2, if $t = 0$, then the state (5, 1, 6) might represent the initial number of customers at three different points in the system. An exogenous event is a happening "outside the system" that impinges on the system. An important example is an arrival to a queueing system. At time 0, the first arrival event is generated and is scheduled on the FEL (meaning that its event notice is placed on the FEL). The interarrival time is an example of an activity. When the clock eventually is advanced to the time of this first arrival, a second arrival event is generated. First, an interarrival time is generated, a^*; it is added to the current time, CLOCK $= t$; the resulting (future) event time, $t + a^* = t^*$, is used to position the new arrival event notice on the FEL. This method of generating an external arrival stream is called bootstrapping; it provides one example of how future events are generated in step 4 of the event-scheduling/time-advance algorithm. Bootstrapping is illustrated in Figure 3.3. The first three interarrival times generated are 3.7, 0.4, and 3.3 time units. The end of an interarrival interval is an example of a primary event.

A second example of how future events are generated (step 4 of Figure 3.2) is provided by a service completion event in a queueing simulation. When one customer completes service, at current time CLOCK $= t$, if the next customer is present, then a new service time, s^*, will be generated for the next customer. The next service completion event will be scheduled to occur at future time $t^* = t + s^*$, by placing onto the FEL a new event notice, of type *service completion*, with event time t^*. In addition, a service-completion event will be generated and scheduled at the time of an arrival event provided that, upon arrival, there is at least one idle server in the server group. A service time is an example of an activity. Beginning service is a conditional event, because its occurrence is triggered only on the condition that a customer is present and a server is free. Service completion is an example of a primary event. Note that a conditional event, such as beginning service, is triggered by a primary event's occurring and by certain conditions prevailing in the system. Only primary events appear on the FEL.

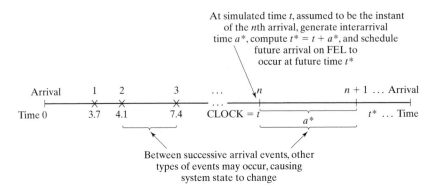

Figure 3.3   Generation of an external arrival stream by bootstrapping.

A third important example is the alternate generation of runtimes and downtimes for a machine subject to breakdowns. At time 0, the first runtime will be generated, and an end-of-runtime event will be scheduled. Whenever an end-of-runtime event occurs, a downtime will be generated, and an end-of-downtime event will be scheduled on the FEL. When the CLOCK is eventually advanced to the time of this end-of-downtime event, a runtime is generated, and an end-of-runtime event is scheduled on the FEL. In this way, runtimes and downtimes continually alternate throughout the simulation. A runtime and a downtime are examples of activities, and *end of runtime* and *end of downtime* are primary events.

Every simulation must have a stopping event, here called E, which defines how long the simulation will run. There are generally two ways to stop a simulation:

1. At time 0, schedule a stop simulation event at a specified future time T_E. Thus, before simulating, it is known that the simulation will run over the time interval $[0, T_E]$. Example: Simulate a job shop for $T_E = 40$ hours.
2. Run length T_E is determined by the simulation itself. Generally, T_E is the time of occurrence of some specified event E. Examples: T_E is the time of the 100th service completion at a certain service center. T_E is the time of breakdown of a complex system. T_E is the time of disengagement or total kill (whichever occurs first) in a combat simulation. T_E is the time at which a distribution center ships the last carton in a day's orders.

In case 2, T_E is not known ahead of time. Indeed, it could be one of the statistics of primary interest to be produced by the simulation.

3.1.2 World Views

When using a simulation package or even when doing a manual simulation, a modeler adopts a world view or orientation for developing a model. The most prevalent world views are the event-scheduling world view, as discussed in the previous section, the process-interaction world view, and the activity-scanning world view. Even if a particular package does not directly support one or more of the world views, understanding the different approaches could suggest alternative ways to model a given system.

To summarize the previous discussion, when using the event-scheduling approach, a simulation analyst concentrates on events and their effect on system state. This world view will be illustrated by the manual simulations of Section 3.1.3 and the Java simulation in Chapter 4.

When using a package that supports the process-interaction approach, a simulation analyst thinks in terms of processes. The analyst defines the simulation model in terms of entities or objects and their life cycle as they flow through the system, demanding resources and queueing to wait for resources. More precisely, a process is the life cycle of one entity. This life cycle consists of various events and activities. Some activities might require the use of one or more resources whose capacities are limited. These and other constraints cause processes to interact, the simplest example being an entity forced to wait in a queue (on a list) because the resource it needs is busy with another entity. The process interaction approach is popular because it has intuitive appeal and because the simulation packages that implement it allow an analyst to describe the process flow in terms of high-level block or network constructs, while the interaction among processes is handled automatically.

In more precise terms, a process is a time-sequenced list of events, activities and delays, including demands for resources, that define the life cycle of one entity as it moves through a system. An example of a "customer process" is shown in Figure 3.4. In this figure, we see the interaction between two customer processes as customer $n + 1$ is delayed until the previous customer's "end service event" occurs. Usually, many processes are active simultaneously in a model, and the interaction among processes could be quite complex.

Underlying the implementation of the process interaction approach in a simulation package, but usually hidden from a modeler's view, events are being scheduled on a future event list and entities are being placed onto lists whenever they face delays, causing one process to temporarily suspend its execution while other processes proceed. It is important that the modeler have a basic understanding of the concepts and, for the simulation package being used, a detailed understanding of the built-in but hidden rules of operation. Schriber and Brunner [2003] provide understanding in this area.

Both the event-scheduling and the process-interaction approach use a variable time advance—that is, when all events and system state changes have occurred at one instant of simulated time,

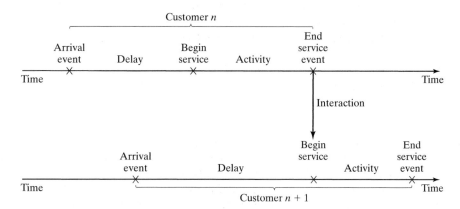

Figure 3.4 Two interacting customer processes in a single-server queue.

the simulation clock is advanced to the time of the next imminent event on the FEL. The activity-scanning approach, in contrast, uses a fixed time increment and a rule-based approach to decide whether any activities can begin at each point in simulated time.

With the activity-scanning approach, a modeler concentrates on the activities of a model and those conditions, simple or complex, that allow an activity to begin. At each clock advance, the conditions for each activity are checked, and, if the conditions are true, then the corresponding activity begins. Proponents claim that the activity-scanning approach is simple in concept and leads to modular models that are more easily maintained, understood, and modified by other analysts at later times. They admit, however, that the repeated scanning to discover whether an activity can begin results in slow runtime on computers. Thus, the pure activity-scanning approach has been modified (and made conceptually somewhat more complex) by what is called the three-phase approach, which combines some of the features of event scheduling with activity scanning to allow for variable time advance and the avoidance of scanning when it is not necessary, but keeps the main advantages of the activity-scanning approach.

In the three-phase approach, events are considered to be activities of duration zero time units. With this definition, activities are divided into two categories, which are called B and C.

B activities   activities bound to occur; all primary events and unconditional activities.

C activities   activities or events that are conditional upon certain conditions being true.

The B-type activities and events can be scheduled ahead of time, just as in the event-scheduling approach. This allows variable time advance. The FEL contains only B-type events. Scanning to learn whether any C-type activities can begin or C-type events occur happens only at the end of each time advance, after all B-type events have completed. In summary, with the three-phase approach, the simulation proceeds with repeated execution of the 3 phases until it is completed:

Phase A   Remove the imminent event from the FEL and advance the clock to its event time. Remove from the FEL any other events that have the same event time.

Phase B   Execute all B-type events that were removed from the FEL. (This could free a number of resources or otherwise change system state.)

Phase C   Scan the conditions that trigger each C-type activity and activate any whose conditions are met. Rescan until no additional C-type activities can begin and no events occur.

The three-phase approach improves the execution efficiency of the activity-scanning method. In addition, proponents claim that the activity scanning and three-phase approaches are particularly good at handling complex resource problems in which various combinations of resources are needed to accomplish different tasks. These approaches guarantee that all resources being freed at a given simulated time will all be freed before any available resources are reallocated to new tasks.

Example 3.2:   Call Center, Back Again

The events and activities were identified in Example 3.1. Under the three-phase approach, the conditions for beginning each activity in Phase C are as follows:

Activity	Condition
Service time by Able	A caller is in queue and Able is idle
Service time by Baker	A caller is in queue, Baker is idle and Able is busy

Using the process-interaction approach, we view the model from the viewpoint of a caller and its "life cycle." Considering a life cycle as beginning upon arrival, a customer process is pictured in Figure 3.4.

In summary, as will be illustrated in Chapter 4, the process-interaction approach has been adopted by the simulation packages most popular in the USA. On the other hand, a number of activity-scanning packages are popular in the UK and Europe. Some of the packages allow portions of a model to be event-scheduling based, if that orientation is convenient, mixed with the process-interaction approach. Finally, some of the packages are based on a flow chart, block diagram or network structure, which upon closer examination turns out to be a specific implementation of the process-interaction concept.

3.1.3 Manual Simulation Using Event Scheduling

In the conducting of an event-scheduling simulation, a simulation table is used to record the successive system snapshots as time advances.

Example 3.3:   Single-Channel Queue

Reconsider the grocery store with one checkout counter that was simulated in Example 2.1 by an ad hoc method. The system consists of those customers in the waiting line plus the one (if any) checking out. A stopping time of 60 minutes is set for this example. The model has the following components:

System state $(LQ(t), LS(t))$, where $LQ(t)$ is the number of customers in the waiting line, and $LS(t)$ is the number being served (0 or 1) at time t.

Entities   The server and customers are not explicitly modeled, except in terms of the state variables.

Events
Arrival (A);
Departure (D);
Stopping event (E), scheduled to occur at time 60.

Event notices
(A, t), representing an arrival event to occur at future time t;
(D, t), representing a customer departure at future time t;
$(E, 60)$, representing the simulation stop event at future time 60.

Activities
Interarrival time, defined in Table 2.6;
Service time, defined in Table 2.7.

Delay   Customer time spent in waiting line.

The event notices are written as (event type, event time). In this model, the FEL will always contain either two or three event notices. The effect of the arrival and departure events was first shown in Figures 2.2 and 2.3 and is shown in more detail in Figures 3.5 and 3.6.

The simulation table for the checkout counter is given in Table 3.1. The reader should cover all system snapshots except one, starting with the first, and attempt to construct the next snapshot

Figure 3.5   Execution of the arrival event.

from the previous one and the event logic in Figures 3.5 and 3.6. The interarrival times and service times will be identical to those used in Table 2.10:

Interarrival Times	1	1	6	3	7	5	2	4	1···
Service Times		4	2	5	4	1	5	4	1	4···

Initial conditions are that the first customer arrive at time 0 and begin service. This is reflected in Table 3.1 by the system snapshot at time zero (CLOCK = 0), with $LQ(0) = 0$, $LS(0) = 1$, and both a departure event and arrival event on the FEL. Also, the simulation is scheduled to stop at time 60. Only two statistics, server utilization and maximum queue length, will be collected. Server utilization is defined by total server busy time (B) divided by total time (T_E). Total busy time, B, and maximum queue length, MQ, will be accumulated as the simulation progresses. A column headed "comments" is included to aid the reader. (a^* and s^* are the generated interarrival and service times, respectively.)

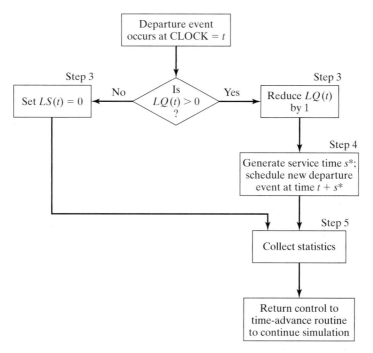

Figure 3.6   Execution of the departure event.

As soon as the system snapshot at time CLOCK $= 0$ is complete, the simulation begins. At time 0, the imminent event is $(A, 1)$. The CLOCK is advanced to time 1, and $(A, 1)$ is removed from the FEL. Because $LS(t) = 1$ for $0 \leq t \leq 1$ (i.e., the server was busy for 1 minute), the cumulative busy time is increased from $B = 0$ to $B = 1$. By the event logic in Figure 3.6, set $LS(1) = 1$ (the server becomes busy). The FEL is left with three future events, $(A, 2)$, $(D, 4)$, and $(E, 60)$. The simulation CLOCK is next advanced to time 2, and an arrival event is executed. The interpretation of the remainder of Table 3.1 is left to the reader.

The simulation in Table 3.1 covers the time interval [0,23]. At simulated time 23, the system empties, but the next arrival also occurs at time 23. The server was busy for 21 of the 23 time units simulated, and the maximum queue length was two. This simulation is, of course, too short to draw any reliable conclusions. Exercise 1 asks the reader to continue the simulation and to compare the results with those in Example 2.1. Note that the simulation table gives the system state at all times, not just the listed times. For example, from time 11 to time 15, there is one customer in service and one in the waiting line.

When an event-scheduling algorithm is computerized, only one snapshot (the current one or partially updated one) is kept in computer memory. With the idea of implementing event scheduling in Java or some other general-purpose language, the following rule should be followed. A new snapshot can be derived only from the previous snapshot, newly generated random variables, and the event logic (Figures 3.5 and 3.6). Past snapshots should be ignored for advancing of the clock. The current snapshot must contain all information necessary to continue the simulation.

Table 3.1   Simulation Table for Checkout Counter (Example 3.3)

	System State				Cumulative Statistics	
Clock	LQ(t)	LS(t)	Future Event List	Comment	B	MQ
0	0	1	(A, 1) (D, 4) (E, 60)	First A occurs ($a^* = 1$) Schedule next A ($s^* = 4$) Schedule first D	0	0
1	1	1	(A, 2) (D, 4) (E, 60)	Second A occurs: (A, 1) ($a^* = 1$) Schedule next A (Customer delayed)	1	1
2	2	1	(D, 4) (A, 8) (E, 60)	Third A occurs: (A, 2) ($a^* = 6$) Schedule next A (Two customers delayed)	2	2
4	1	1	(D, 6) (A, 8) (E, 60)	First D occurs: (D, 4) ($s^* = 2$) Schedule next D (Customer delayed)	4	2
6	0	1	(A, 8) (D, 11) (E, 60)	Second D occurs: (D, 6) ($s^* = 5$) Schedule next D	6	2
8	1	1	(D, 11) (A, 11) (E, 60)	Fourth A occurs: (A, 8) ($a^* = 3$ Schedule next A (Customer delayed)	8	2
11	1	1	(D, 15) (A, 18) (E, 60)	Fifth A occurs: (A, 11) ($a^* = 7$) Schedule next A Third D occurs: (D, 11) ($s^* = 4$) Schedule next D Customer delayed	11	2
15	0	1	(D, 16) (A, 18) (E, 60)	Fourth D occurs: (D, 15) ($s^* = 1$) Schedule next D	15	2
16	0	0	(A, 18) (E, 60)	Fifth D occurs: (D, 16)	16	2
18	0	1	(D, 23) (A, 23) (E, 60)	Sixth A occurs ($a^* = 5$) Schedule next A ($s^* = 5$) Schedule next D	16	2
23	0	1	(A, 25) (D, 27) (E, 60)	Seventh A occurs: (A, 23) ($a^* = 2$) Schedule next Arrival Sixth D occurs: (D, 23)	21	2

Example 3.4:   The Checkout-Counter Simulation, Continued _____

Suppose that, in the simulation of the checkout counter in Example 3.3, the simulation analyst desires to estimate mean response time and mean proportion of customers who spend 5 or more minutes in the system. A response time is the length of time a customer spends in the system. In order to estimate these customer averages, it is necessary to expand the model in Example 3.3 to represent the individual customers explicitly. In addition, to be able to compute an individual customer's response time when that customer departs, it will be necessary to know that customer's arrival time. Therefore, a customer entity with arrival time as an attribute will be added to the list of model components in Example 3.3. These customer entities will be stored in a list to be called "CHECKOUT LINE"; they will be called $C1, C2, C3, \ldots$ Finally, the event notices on the FEL will be expanded to indicate which customer is affected. For example, $(D, 4, C1)$ means that customer $C1$ will depart at time 4. The additional model components are the following:

Entities (Ci, t), representing customer Ci who arrived at time t
Event notices
   (A, t, Ci), the arrival of customer Ci at future time t
   (D, t, Cj), the departure of customer Cj at future time t
Set "CHECKOUT LINE," the set of all customers currently at the checkout counter (being served or waiting to be served), ordered by time of arrival

Three new cumulative statistics will be collected: S, the sum of customer response times for all customers who have departed by the current time; F, the total number of customers who spend 5 or more minutes at the checkout counter; and N_D, the total number of departures up to the current simulation time. These three cumulative statistics will be updated whenever the departure event occurs; the logic for collecting these statistics would be incorporated into step 5 of the departure event in Figure 3.6.

The simulation table for Example 3.4 is shown in Table 3.2. The same data for interarrival and service times will be used again; so Table 3.2 essentially repeats Table 3.1, except that the new components are included (and the comment column has been deleted). These new components are needed for the computation of the cumulative statistics S, F, and N_D. For example, at time 4, a departure event occurs for customer $C1$. The customer entity $C1$ is removed from the list called "CHECKOUT LINE"; the attribute "time of arrival" is noted to be 0, so the response time for this customer was 4 minutes. Hence, S is incremented by 4 minutes. N_D is incremented by one customer, but F is not incremented, for the time in system was less than five minutes. Similarly, at time 23, when the departure event $(D, 23, C6)$ is being executed, the response time for customer $C6$ is computed by

$$\text{Response time} = \text{CLOCK TIME–attribute "time of arrival"}$$
$$= \quad 23 \quad - \quad 18$$
$$= 5 \text{ minutes}$$

Then S is incremented by 5 minutes, and F and N_D by one customer.

For a simulation run length of 23 minutes, the average response time was $S/N_D = 35/6 = 5.83$ minutes, and the observed proportion of customers who spent 5 or more minutes in the system was $F/N_D = 0.83$. Again, this simulation was far too short to regard these estimates as having any degree of accuracy. The purpose of Example 3.4, however, was to illustrate the notion that, in many simulation models, the information desired from the simulation (such as the statistics S/N_D and F/N_D) to some extent dictates the structure of the model.

Example 3.5: The Dump-Truck Problem

Six dump trucks are used to haul coal from the entrance of a small mine to the railroad. Figure 3.7 provides a schematic of the dump-truck operation. Each truck is loaded by one of two loaders. After a loading, the truck immediately moves to the scale, to be weighed as soon as possible. Both

Table 3.2   Simulation Table for Example 3.4

Clock	System State			CHECKOUT LINE	Future Event List	Cumulative Statistics		
	LQ(t)	LS(t)				S	N_D	F
0	0	1	(C1,0)		(A,1,C2) (D,4, C1) (E,60)	0	0	0
1	1	1	(C1,0) (C2,1)		(A,2,C3) (D,4,C1) (E,60)	0	0	0
2	2	1	(C1,0) (C2,1) (C3,2)		(D,4,C1) (A,8,C4) (E,60)	0	0	0
4	1	1	(C2,1) (C3,2)		(D,6,C2) (A,8,C4) (E, 60)	4	1	0
6	0	1	(C3,2)		(A,8,C4) (D,11,C3) (E,60)	9	2	1
8	1	1	(C3,2) (C4,8)		(D,11,C3) (A,11,C5) (E,60)	9	2	1
11	1	1	(C4,8) (C5,11)		(D,15,C4) (A,18,C6) (E,60)	18	3	2
15	0	1	(C5,11)		(D,16,C5) (A,18,C6) (E,60)	25	4	3
16	0	0			(A,18,C6) (E,60)	30	5	4
18	0	1	(C6,18)		(D,23,C6) (A,23,C7) (E,60)	30	5	4
23	0	1	(C7,23)		(A,25,C8) (D,27,C7) (E,60)	35	6	5

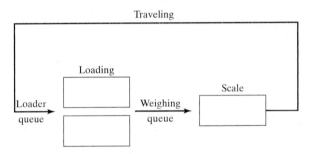

Figure 3.7   Dump truck problem.

the loaders and the scale have a first-come–first-served waiting line (or queue) for trucks. Travel time from a loader to the scale is considered negligible. After being weighed, a truck begins a travel time (during which time the truck unloads) and then afterward returns to the loader queue. The distributions of loading time, weighing time, and travel time are given in Tables 3.3, 3.4, and 3.5, respectively, together with the random digit assignment for generating these variables by using random digits from Table A.1. The purpose of the simulation is to estimate the loader and scale utilizations (percentage of time busy). The model has the following components:

System state $[LQ(t), L(t), WQ(t), W(t)]$, where
  $LQ(t)$ = number of trucks in loader queue
  $L(t)$ = number of trucks (0, 1, or 2) being loaded
  $WQ(t)$ = number of trucks in weigh queue
  $W(t)$ = number of trucks (0 or 1) being weighed, all at simulation time t
Event notices
  (ALQ, t, DTi), dump truck i arrives at loader queue (ALQ) at time t
  (EL, t, DTi), dump truck i ends loading (EL) at time t
  (EW, t, DTi), dump truck i ends weighing (EW) at time t

Table 3.3 Distribution of Loading Time for the Dump Trucks

Loading Time	Probability	Cumulative Probability	Random Digit Assignment
5	0.30	0.30	1−3
10	0.50	0.80	4−8
15	0.20	1.00	9−0

Table 3.4 Distribution of Weighing Time for the Dump Trucks

Weighing Time	Probability	Cumulative Probability	Random Digit Assignment
12	0.70	0.70	1−7
16	0.30	1.00	8−0

Table 3.5 Distribution of Travel Time for the Dump Trucks

Travel Time	Probability	Cumulative Probability	Random Digit Assignment
40	0.40	0.40	1−4
60	0.30	0.70	5−7
80	0.20	0.90	8−9
100	0.10	1.00	0

Entities The six dump trucks ($DT1, \ldots, DT6$)

Lists

Loader queue, all trucks waiting to begin loading, ordered on a first-come–first-served basis

Weigh queue, all trucks waiting to be weighed, ordered on a first-come–first-served basis

Activities Loading time, weighing time, and travel time

Delays Delay at loader queue, and delay at scale

The simulation table is given in Table 3.6. It has been assumed that five of the trucks are at the loaders and one is at the scale at time 0. The activity times are taken from the following list as needed:

Loading Time	10	5	5	10	15	10	10
Weighing Time	12	12	12	16	12	16	
Travel Time	60	100	40	40	80		

Table 3.6 Simulation Table for Dump-Truck Operation (Example 3.5)

Clock t	System State LQ(t)	System State L(t)	System State WQ(t)	System State W(t)	Lists Loader Queue	Lists Weigh Queue	Future Event List	Cumulative Statistics B_L	Cumulative Statistics B_S
0	3	2	0	1	DT4 DT5 DT6		(EL, 5, DT3) (EL, 10, DT2) (EW, 12, DT1)	0	0
5	2	2	1	1	DT5 DT6	DT3	(EL, 10, DT2) (EL, 5 + 5, DT4) (EW, 12, DT1)	10	5
10	1	2	2	1	DT6	DT3 DT2	(EL, 10, DT4) (EW, 12, DT1) (EL, 10 + 10, DT5)	20	10
10	0	2	3	1		DT3 DT2 DT4	(EW, 12, DT1) (EL, 20, DT5) (EL, 10 + 15, DT6)	20	10
12	0	2	2	1		DT2 DT4	(EL, 20, DT5) (EW, 12 + 12, DT3) (EL, 25, DT6) (ALQ, 12 + 60, DT1)	24	12
20	0	1	3	1		DT2 DT4 DT5	(EW, 24, DT3) (EL, 25, DT6) (ALQ, 72, DT1)	40	20
24	0	1	2	1		DT4 DT5	(EL, 25, DT6) (EW, 24 + 12, DT2) (ALQ, 72, DT1) (ALQ, 24 + 100, DT3)	44	24
25	0	0	3	1		DT4 DT5 DT6	(EW, 36, DT2) (ALQ, 72, DT1) (ALQ, 124, DT3)	45	25
36	0	0	2	1		DT5 DT6	(EW, 36 + 16, DT4) (ALQ, 72, DT1) (ALQ, 36 + 40, DT2) (ALQ, 124, DT3)	45	36
52	0	0	1	1		DT6	(EW, 52 + 12, DT5) (ALQ, 72, DT1) (ALQ, 76, DT2) (ALQ, 52 + 40, DT4) (ALQ, 124, DT3)	45	52
64	0	0	0	1			(ALQ, 72, DT1) (ALQ, 76, DT2) (EW, 64 + 16, DT6) (ALQ, 92, DT4) (ALQ, 124, DT3) (ALQ, 64 + 80, DT5)	45	64
72	0	1	0	1			(ALQ, 76, DT2) (EW, 80, DT6) (EL, 72 + 10, DT1) (ALQ, 92, DT4) (ALQ, 124, DT3) (ALQ, 144, DT5)	45	72
76	0	2	0	1			(EW, 80, DT6) (EL, 82, DT1) (EL, 76 + 10, DT2) (ALQ, 92, DT4) (ALQ, 124, DT3) (ALQ, 144, DT5)	49	76

When an end-loading (EL) event occurs, say for truck j at time t, other events might be triggered. If the scale is idle [$W(t) = 0$], truck j begins weighing and an end-weighing event (EW) is scheduled on the FEL; otherwise, truck j joins the weigh queue. If, at this time, there is another truck waiting for a loader, it will be removed from the loader queue and will begin loading by the scheduling of an *end-loading* event (EL) on the FEL. Both this logic for the occurrence of the end-loading event and the appropriate logic for the other two events, should be incorporated into an event diagram, as in Figures 3.5 and 3.6 of Example 3.3. The construction of these event-logic diagrams is left as an exercise for the reader (Exercise 2).

As an aid to the reader, in Table 3.6, whenever a new event is scheduled, its event time is written as "$t + $ (activity time)." For example, at time 0, the imminent event is an EL event with event time 5. The clock is advanced to time $t = 5$, dump truck 3 joins the weigh queue (because the scale is occupied), and truck 4 begins to load. Thus, an EL event is scheduled for truck 4 at future time 10, computed by (present time) + (loading time) = $5 + 5 = 10$.

In order to estimate the loader and scale utilizations, two cumulative statistics are maintained:

$$B_L = \text{ total busy time of both loaders from time 0 to time } t$$
$$B_S = \text{ total busy time of the scale from time 0 to time } t$$

Both loaders are busy from time 0 to time 20, so $B_L = 40$ at time $t = 20$—but, from time 20 to time 24, only one loader is busy; thus, B_L increases by only 4 minutes over the time interval [20, 24]. Similarly, from time 25 to time 36, both loaders are idle ($L(25) = 0$), so B_L does not change. For the relatively short simulation in Table 3.6, the utilizations are estimated as follows:

$$\text{Average loader utilization} = \frac{49/2}{76} = 0.32$$

$$\text{Average scale utilization} = \frac{76}{76} = 1.00$$

These estimates cannot be regarded as accurate estimates of the long-run "steady-state" utilizations of the loader and scale; a considerably longer simulation would be needed to reduce the effect of the assumed conditions at time 0 (five of the six trucks at the loaders) and to realize accurate estimates. On the other hand, if the analyst were interested in the so-called transient behavior of the system over a short period of time (say 1 or 2 hours), given the specified initial conditions, then the results in Table 3.6 can be considered representative (or constituting one sample) of that transient behavior. Additional samples can be obtained by conducting additional simulations, each one having the same initial conditions but using a different stream of random digits to generate the activity times.

Table 3.6, the simulation table for the dump-truck operation, could have been simplified somewhat by not explicitly modeling the dump trucks as entities—that is, the event notices could be written as (EL, t), and so on, and the state variables used to keep track merely of the number of trucks in each part of the system, not which trucks were involved. With this representation, the same utilization statistics could be collected. On the other hand, if mean "system" response time, or proportion of trucks spending more than 30 minutes in the "system," were being estimated, where "system" refers to the loader queue and loaders and the weigh queue and scale, then dump truck entities (DTi), together with an attribute equal to arrival time at the loader queue, would be indispensable. Whenever a truck left the scale, that truck's response time could be computed as current simulation time (t) minus the arrival-time attribute. This new response time would be used to update the cumulative statistics: S = total response time of all trucks that have been through the "system" and F = number of truck response times that have been greater than 30 minutes. This example again illustrates the notion that, to some extent, the complexity of the model depends on the performance measures being estimated.

Example 3.6:   The Dump-Truck Problem Revisited

The events and activities were identified in Example 3.5. Under the activity-scanning approach, the conditions for beginning each activity are as follows:

Activity	Condition
Loading time	Truck is at front of loader queue, and at least one loader is idle.
Weighing time	Truck is at front of weigh queue, and weigh scale is idle.
Travel time	Truck has just completed a weighing.

Using the process-interaction approach, we view the model from the viewpoint of one dump truck and its "life cycle." Considering a life cycle as beginning at the loader queue, we can picture a dump-truck process as in Figure 3.8.

3.2 LIST PROCESSING

List processing deals with methods for handling lists of entities and the future event list. Simulation packages provide, both explicitly for an analyst's use and covertly in the simulation mechanism behind the language, facilities for an analyst or the model itself to use lists and to perform the basic operations on lists.

Section 3.2.1 describes the basic properties and operations performed on lists. Section 3.2.2 discusses the use of arrays for processing lists and the use of array indices to create linked lists, arrays being a simpler mechanism for describing the basic operations than the more general dynamically allocated linked lists discussed in section 3.2.3. Finally, section 3.2.4 briefly introduces some of the more advanced techniques for managing lists.

The purpose of this discussion of list processing is not to prepare the reader to implement lists and their processing in a general-purpose language such as FORTRAN, C++, or Java, but rather to increase the reader's understanding of lists and of their underlying concepts and operations.

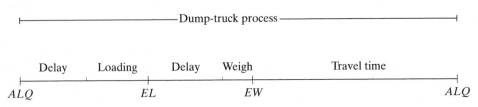

Figure 3.8   The dump-truck process.

3.2.1 Lists: Basic Properties and Operations

As has previously been discussed, lists are a set of ordered or ranked records. In simulation, each record represents one entity or one event notice.

Lists are ranked, so they have a *top* or *head* (the first item on the list); some way to traverse the list (to find the second, third, etc. items on the list); and a *bottom* or *tail* (the last item on the list). A head pointer is a variable that points to or indicates the record at the top of the list. Some implementations of lists also have a tail pointer that points to the bottom item on the list.

For purposes of discussion, an entity, along with its attributes or an event notice, will be referred to as a *record*. An entity identifier and its attributes are *fields* in the entity record; the event type, event time, and any other event-related data are fields in the event-notice record. Each record on a list will also have a field that holds a "next pointer" that points to the next record on the list, providing a way to traverse the list. Some lists also require a "previous pointer," to allow for traversing the list from bottom to top.

For either type of list, the main activities in list processing are adding a record to a list and removing a record from a list. More specifically, the main operations on a list are the following:

1. removing a record from the top of the list;
2. removing a record from any location on the list;
3. adding an entity record to the top or bottom of the list;
4. adding a record at an arbitrary position in the list, one specified by the ranking rule.

The first and third operations, removing or adding a record to the top or bottom of the list, can be carried out in minimal time by adjusting two record pointers and the head or tail pointer, the other two operations require at least a partial search through the list. Making these two operations efficient is the goal of list-processing techniques.

In the event-scheduling approach, when time is advanced and the imminent event is due to be executed, the removal operation takes place first—namely, the event at the top of the FEL is removed from the FEL. If an arbitrary event is being canceled, or an entity is removed from a list based on some of its attributes (say, for example, its priority and due date) to begin an activity, then the second removal operation is performed. When an entity joins the back of a first-in–first-out queue implemented as a list, then the third operation, adding an entity to the bottom of a list, is performed. Finally, if a queue has the ranking rule *earliest due date first*, then, upon arrival at the queue, an entity must be added to the list not at the top or bottom, but at the position determined by the due-date ranking rule.

For simulation on a computer, whether via a general-purpose language (such as FORTRAN, C++, or Java) or via a simulation package, each entity record and event notice is stored in a physical location in computer memory. There are two basic possibilities: (a) All records are stored in arrays. Arrays hold successive records in contiguous locations in computer memory. They therefore can be referenced by an array index that can be thought of as a row number in a matrix. (b) All entities and event notices are represented by structures (as in C) or classes (as in Java), allocated from RAM memory as needed, and tracked by pointers to a record or structure.

Most simulation packages use dynamically allocated records and pointers to keep track of lists of items, as arrays are conceptually simpler, so the concept of linked lists is first explained through arrays and array indices in Section 3.2.2 and then applied to dynamically allocated records and pointers in Section 3.2.3.

3.2.2 Using Arrays for List Processing

The array method of list storage is typical of FORTRAN, but it may be used in other procedural languages. Most versions of FORTRAN do not have actual record-type data structures, but a record may be implemented as a row in a 2-dimensional array or as a number of parallel arrays. For convenience, we use the notation R(i) to refer to the i-th record in the array, however it may be stored in the language being used. Most modern simulation packages do not use arrays for list storage, but rather use dynamically allocated records—that is, records that are created upon first being needed and destroyed when they are no longer needed.

Arrays are advantageous in that any specified record, say the i-th, can be retrieved quickly without searching, merely by referencing R(i). Arrays are disadvantaged when items are added to the middle of a list or the list must be rearranged. In addition, arrays typically have a fixed size, determined at compile time or upon initial allocation when a program first begins to execute. In simulation, the maximum number of records for any list could be difficult (or impossible) to predict, and the current number of them in a list may vary widely over the course of the simulation run. Worse yet, most simulations require more than one list; if they are kept in separate arrays, each would have to be dimensioned to the largest the list would ever be, potentially using excessive amounts of computer memory.

In the use of arrays for storing lists, there are two basic methods for keeping track of the ranking of records in a list. One method is to store the first record in R(1), the second in R(2), and so on, and the last in R(*tailptr*), where *tailptr* is used to refer to the last item in the list. Although simple in concept and easy to understand, this method will be extremely inefficient for all except the shortest lists, those of less than five or so records, for adding an item, for example, in position 41 in a list of 100 items, will require that the last 60 records be physically moved down one array position to make space for the new record. Even if the list were a first-in–first-out list, removing the top item from the list would be inefficient, as all remaining items would have to be physically moved up one position in the array. The physical rearrangement method of managing lists will not be discussed further. What is needed is a method to track and rearrange the logical ordering of items in a list without having to move the records physically in computer memory.

In the second method, a variable called a head pointer, with name *headptr*, points to the record at the top of the list. For example, if the record in position R(11) were the record at the top of the list, then *headptr* would have the value 11. In addition, each record has a field that stores the index or pointer of the next record in the list. For convenience, let R(i, next) represent the next index field.

Example 3.7:   A List for the Dump Trucks at the Weigh Queue _____

In Example 3.5, the dump-truck problem, suppose that a waiting line of three dump trucks occurred at the weigh queue, specifically, DT3, DT2, and DT4, in that order, at exactly CLOCK time 10 in Table 3.6. Suppose further that the model is tracking one attribute of each dump truck: its arrival time at the weigh queue, updated each time it arrives. Finally, suppose that the entities are stored in records in an array dimensioned from 1 to 6, one record for each dump truck. Each entity is represented by a record with 3 fields: the first is an entity identifier; the second is the arrival time at the weigh queue; the last is a pointer field to "point to" the next record, if any, in the list representing the weigh queue, as follows:

$$[DTi, \text{ arrival time at weigh queue, next index}]$$

Before its first arrival at the weigh queue, and before being added to the weigh queue list, a dump truck's second and third fields are meaningless. At time 0, the records would be initialized as follows:

$$R(1) = [\text{ DT1, 0.0, 0}]$$
$$R(2) = [\text{ DT2, 0.0, 0}]$$
$$\vdots$$
$$R(6) = [\text{ DT6, 0.0, 0}]$$

Then, at CLOCK time 10 in the simulation in Table 3.6, the list of entities in the weigh queue would be defined by

$$\text{headptr} = 3$$
$$R(1) = [\text{ DT1, 0.0, 0}]$$
$$R(2) = [\text{ DT2, 10.0, 4}]$$
$$R(3) = [\text{ DT3, 5.0, 2}]$$
$$R(4) = [\text{ DT4, 10.0, 0}]$$
$$R(5) = [\text{ DT5, 0.0, 0}]$$
$$R(6) = [\text{ DT6, 0.0, 0}]$$
$$\text{tailptr} = 4$$

To traverse the list, start with the head pointer, go to that record, retrieve that record's next pointer, and proceed, to create the list in its logical order—for example,

$$\text{headptr} = 3$$
$$R(3) = [\text{ DT3, 5.0, 2}]$$
$$R(2) = [\text{ DT2, 10.0, 4}]$$
$$R(4) = [\text{ DT4, 10.0, 0}]$$

The zero entry for next pointer in $R(4)$, as well as tailptr $= 4$, indicates that DT4 is at the end of the list.

Using next pointers for a first-in–first-out list, such as the weigh queue in this example, makes the operations of adding and removing entity records, as dump trucks join and leave the weigh queue, particularly simple. At CLOCK time 12, dump truck DT3 begins weighing and thus leaves the weigh queue. To remove the DT3 entity record from the top of the list, update the head pointer by setting it equal to the next pointer value of the record at the top of the list:

$$\text{headptr} = R(\text{headptr, next})$$

In this example, we get

$$\text{headptr} = R(3, \text{ next}) = 2$$

meaning that dump truck DT2 in $R(2)$ is now at the top of the list.

Similarly, at CLOCK time 20, dump truck DT5 arrives at the weigh queue and joins the rear of the queue. To add the DT5 entity record to the bottom of the list, the following steps are taken:

R(tailptr, next) = 5 (update the next pointer field of the previously last item)
tailptr = 5   (update the value of the tail pointer)

This approach becomes slightly more complex when a list is a ranked list, such as the future event list, or an entity list ranked by an entity attribute. For ranked lists, to add or remove an item anywhere except to the head or tail of the list, searching is usually required. See Example 3.8.

Note that, in the dump-truck problem, the loader queue could also be implemented as a list using the same six records and the same array, because each dump-truck entity will be on at most one of the two lists and, while loading, weighing or traveling, a dump truck will be on neither list.

3.2.3 Using Dynamic Allocation and Linked Lists

In procedural languages, such as C++ and Java, and in most simulation languages, entity records are dynamically created when an entity is created and event notice records are dynamically created whenever an event is scheduled on the future event list. The languages themselves, or the operating systems on which they are running, maintain a linked list of free chunks of computer memory and allocate a chunk of desired size upon request to running programs. (Another use of linked lists!) When an entity "dies," that is, exits from the simulated system, and also after an event occurs and the event notice is no longer needed, the corresponding records are freed, making that chunk of computer memory available for later reuse; the language or operating system adds the chunk to the list of free memory.

In this text, we are not concerned with the details of allocating and freeing computer memory, so we will assume that the necessary operations occur as needed. With dynamic allocation, a record is referenced by a pointer instead of by an array index. When a record is allocated in C++ or Java, the allocation routine returns a pointer to the allocated record, which must be stored in a variable or a field of another record for later use. A pointer to a record can be thought of as the physical or logical address in computer memory of the record.

In our example, we will use a notation for records identical to that in the previous section (3.2.2):

Entities: [ID, attributes, next pointer]
Event notices: [event type, event time, other data, next pointer]

but we will not reference them by the array notation R(i) as before, because it would be misleading. If for some reason we wanted the third item on the list, we would have to traverse the list, counting items until we reached the third record. Unlike in arrays, there is no way to retrieve the i-th record in a linked list directly, because the actual records could be stored at any arbitrary location in computer memory and are not stored contiguously, as arrays are.

Example 3.8:   The Future Event List and the Dump-Truck Problem _____
Beginning from Table 3.6, event notices in the dump-truck problem of Example 3.5 are expanded to include a pointer to the next event notice on the future event list and can be represented by

[event type, event time, DTi, nextptr],

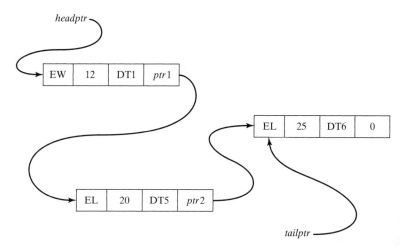

Figure 3.9 The dump-truck future event list as a linked list.

—for example,

$$[EL, 10, DT3, nextptr]$$

where EL is the end-loading event to occur at future time 10 for dump truck DT3, and the field *nextptr* points to the next record on the FEL. Keep in mind that the records may be stored anywhere in computer memory, and in particular are not necessarily stored contiguously. Figure 3.9 represents the future event list at CLOCK time 10, taken from Table 3.6. The fourth field in each record is a pointer value to the next record in the future event list.

The C++ and Java languages, and other general-purpose languages, use different notation for referencing data from pointer variables. For discussion purposes, if R is a pointer to a record, then

$$R{\rightarrow}eventtype, \ R{\rightarrow}eventtime, \ R{\rightarrow}next$$

are the event type, the event time and the next record for the event notice that R points to. For example, if R is set equal to the head pointer for the FEL at CLOCK time 10, then

$$R{\rightarrow}eventtype = EW$$
$$R{\rightarrow}eventtime = 12$$
$$R{\rightarrow}next \text{ is the pointer for the second event notice on the FEL,}$$

so that

$$R{\rightarrow}next{\rightarrow}eventtype = EL$$
$$R{\rightarrow}next{\rightarrow}eventtime = 20$$
$$R{\rightarrow}next{\rightarrow}next \text{ is the pointer to the third event notice on the FEL}$$

If one of the pointer fields is zero (or null), then that record is the last item in the list, and the pointer variable *tailptr* points to that last record, as depicted in Figure 3.9.

What we have described are called singly-linked lists, because there is a one-way linkage from the head of the list to its tail. The tail pointer is kept mostly for convenience and efficiency, especially for lists for which items are added at the bottom of the list. Of course, a tail pointer is not strictly necessary, because the last item can always be found by traversing the list, but it does make some operations more efficient.

For some purposes, it is desirable to traverse or search a list by starting at the tail in addition to the head. For such purposes, a doubly-linked list can be used. Records on a doubly-linked list have two pointer fields, one for the next record and one for the previous record. Good references that discuss arrays, singly- and doubly-linked lists, and searching and traversing of lists are Cormen, *et al.* [2001] and Sedgewick [1998].

3.2.4 Advanced Techniques

Many of the modern simulation packages use techniques and representations of lists that are more efficient than searching through a doubly-linked list. Most of these topics are too advanced for this text. The purpose of this section is to introduce some of the more advanced ideas briefly.

One idea to speed up processing or doubly-linked lists is to use a middle pointer in addition to a head and tail pointer. With special techniques, the *mid* pointer will always point to the approximate middle of the list. Then, when a new record is being added to the list, the algorithm first examines the middle record to decide whether to begin searching at the head of the list or the middle of the list. Theoretically, except for some overhead due to maintenance of the mid pointer, this technique should cut search times in half. A few advanced techniques use one or more mid pointers, depending on the length of the list.

Some advanced algorithms use list representations other than a doubly-linked list, such as heaps or trees. These topics are beyond the scope of this text. Good references are Cormen, *et al.* [2001] and Sedgewick [1998].

3.3 SUMMARY

This chapter introduced the major concepts and building blocks in simulation, the most important being entities and attributes, events, and activities. The three major world views—event scheduling, process interaction, activity scanning—were discussed. Finally, to gain an understanding of one of the most important underlying methodologies, Section 3.2 introduced some of the basic notions of list processing.

The next chapter will provide a survey of some of the most widely used and popular simulation packages, most of which either use exclusively or allow the process-interaction approach to simulation.

REFERENCES

CORMEN, T. H., C. E. LEISEROON, AND R. L. RIVEST [2001], *Introduction to Algorithms*, 2nd ed., McGraw–Hill, New York.

SCHRIBER, T. J., AND D. T. BRUNNER [2003], "Inside Simulation Software: How it Works and why it Matters," *Proceedings of the 2003 Winter Simulation Conference*, S. Chick, P. J. Sánchez, D. Ferrin, and D. J. Morrice, eds., New Orleans., LA, Dec. 7–10, pp. 113–123.
SEDGEWICK, R. [1998], *Algorithms in C++*, 3d ed., Addison–Wesley, Reading, MA.

EXERCISES

Instructions to the reader: For most exercises, the reader should first construct a model by explicitly defining the following:

1. system state;

2. system entities and their attributes;

3. sets, and the entities that may be put into the sets;

4. events and activities; event notices;

5. variables needed to collect cumulative statistics.

Second, the reader should either (1) develop the event logic (as in Figures 3.5 and 3.6 for Example 3.3) in preparation for using the event-scheduling approach, or (2) develop the system processes (as in Figure 3.4) in preparation for using the process-interaction approach.

Most problems contain activities that are uniformly distributed over an interval $[a, b]$. When conducting a manual simulation, assume that $a, a + 1, a + 2, \ldots , b$ are the only possible values; that is, the activity time is a *discrete* random variable. The discreteness assumption will simplify the manual simulation.

1. (a) Using the event-scheduling approach, continue the (manual) checkout-counter simulation in Example 3.3, Table 3.1. Use the same interarrival and service times that were previously generated and used in Example 2.1. When the last interarrival time is used, continue the simulation until time 60 using the data in Tables 2.8 and 2.9.

    (b) Do exercise 1(a) again, adding the model components necessary to estimate mean response time and proportion of customers who spend 5 or more minutes in the system. (*Hint:* See Example 3.4, Table 3.2.)

    (c) Comment on the relative merits of manual versus computerized simulations.

2. Construct the event-logic diagrams for the dump-truck problem, Example 3.5.

3. In the dump-truck problem of Example 3.5, it is desired to estimate mean response time and the proportion of response times which are greater than 30 minutes. A response time for a truck begins when that truck arrives at the loader queue and ends when the truck finishes weighing. Add the model components and cumulative statistics needed to estimate these two measures of system performance. Simulate for 8 hours.

4. Prepare a table in the manner of Table 3.2, until the CLOCK reaches time 15, using the interarrival and service times given below in the order shown. The stopping event will be at time 30.

   Interarrival times 1 5 6 3 8

   Service times 3 5 4 1 5

5. Continue Table 3.2 until caller C7 departs.

6. The data for Table 3.2 are changed to the following:

   Interarrival times 4 5 2 8 3 7

   Service times 5 3 4 6 2 7

   Prepare a table in the manner of Table 3.2 with a stopping event at time 25.

7. Redo Example 2.2 (the Able–Baker call center problem) by a manual simulation, using the event-scheduling approach.

8. Redo Example 2.4 (the (M, N) inventory system) by a manual simulation, using the event-scheduling approach.

9. Redo Example 2.5 (the bearing-replacement problem) by a manual simulation, using the event-scheduling approach.

10. Rework Example 3.5, but using the following data:

    Loading times 10 5 10 10 5 10 5

    Weigh times 12 16 12 12 16 12 12

    Travel times 40 60 40 80 100 40

4

Simulation Software

In this chapter, we first discuss the history of simulation software. Simulation software has a history that is just reaching middle age. We base this history on our collective experience, articles written by Professor Richard Nance, and panel discussions at the annual Winter Simulation Conference.

Next, we discuss features and attributes of simulation software. If you were about to purchase simulation software, what would concern you? Would it be the cost, the ease of learning, the ease of use, or would it be the power to model the kind of system with which you are concerned? Or would it be the animation capabilities? Following the discussion of features, we discuss other issues and concerns related to the selection of simulation software.

Software used to develop simulation models can be divided into three categories. First, there are the general-purpose programming languages, such as C, C++, and Java. Second, there are simulation programming languages, examples being GPSS/H™, SIMAN V® and SLAM II®. Third, there are the simulation environments. This category includes many products that are distinguished one way or another (by, for example, cost, application area, or type of animation), but have common characteristics, such as a graphical user interface and an environment that supports all (or most) aspects of a simulation study. Many simulation environments contain a simulation programming language, but some take a graphical approach similar to process-flow diagramming.

In the first category, we discuss simulation in Java. Java is a general-purpose programming language that was not specifically designed for use in simulation. Java was chosen since it is widely used and widely available. Today very few people writing discrete-event simulation models are using programming languages alone; however, in certain application areas, some people are using packages based on Java or on another general-purpose language. Understanding how to develop a model in a general-purpose language helps to understand how the basic concepts and algorithms discussed in Chapter 3 are implemented.

In the second category, we discuss GPSS/H, a highly structured process-interaction simulation language. GPSS was designed for relatively easy simulation of queuing systems, such as in job shops, but it has been used to simulate systems of great complexity. It was first introduced by IBM; today, there are various implementations of GPSS, GPSS/H being one of the most widely used.

In the third category, we have selected a number of simulation software packages for discussion. There are many simulation packages currently available; we have selected a few that have survived and thriven for a number of years, to represent different approaches for model-building.

One of the important components of a simulation environment is the output analyzer, which is used to conduct experimentation and assist with analyses. To illustrate the range of desirable characteristics, we look at four tools incorporated into some of the simulation environments. Typically these statistical analysis tools compute summary statistics, confidence intervals, and other statistical measures. Some support warmup determination, design of experiments, and sensitivity analyses. Many packages now offer optimization techniques based on genetic algorithms, evolutionary strategies, tabu search, scatter search, and other recently developed heuristic methods. In addition to the support for statistical analysis and optimization, the simulation environments offer data management, scenario definition, and run management. Data management offers support for managing all the input and output data associated with the analyses.

4.1 HISTORY OF SIMULATION SOFTWARE

Our discussion of the history of simulation software is based on Nance [1995], who breaks the years from 1955 through 1986 into five periods. Additional historical information is taken from a panel discussion at the 1992 Winter Simulation Conference entitled "Perspectives of the Founding Fathers" [Wilson, 1992], during which eight early users of simulation presented their historical perspectives. We add a sixth and most recent period:

1955–60	The Period of Search
1961–65	The Advent
1966–70	The Formative Period
1971–78	The Expansion Period
1979–86	The Period of Consolidation and Regeneration
1987–?	The Period of Integrated Environments

The following subsections provide a brief presentation of this history. As indicated in [Nance, 1995], there were at least 137 simulation programming languages reported as of 1981, and many more since then. This brief history is far from all-inclusive. The languages and packages we mention have stood the test of time by surviving to the present day or were the historical forerunner of a package in present use.

4.1.1 The Period of Search (1955–60)

In the early years, simulation was conducted in FORTRAN or other general-purpose programming language, without the support of simulation-specific routines. In the first period (1955–1960), much effort was expended on the search for unifying concepts and the development of reusable routines to facilitate simulation. The General Simulation Program of K.D. Tocher and D.G. Owen [Tocher, 1960] is considered the first "language effort." Tocher identified and developed routines that could be reused in subsequent simulation projects.

4.1.2 The Advent (1961–65)

The forerunners of the simulation programming languages (SPLs) in use today appeared in the period 1961–65. As Harold Hixson said in [Wilson, 1992], "in the beginning there were FORTRAN, ALGOL, and GPSS"—that is, there were the FORTRAN-based packages (such as SIMSCRIPT and GASP), the ALGOL descendent SIMULA, and GPSS.

The first process interaction SPL, GPSS was developed by Geoffrey Gordon at IBM and appeared in about 1961. Gordon developed GPSS (General Purpose Simulation System) for quick simulations of communications and computer systems, but its ease of use quickly spread its popularity to other application areas. GPSS is based on a block-diagram representation (similar to a process-flow diagram) and is suited for queuing models of all kinds. As reported by Reitman in [Wilson, 1992], as early as 1965 GPSS was connected to an interactive display terminal that could interrupt and display intermediate results, a foreshadow of the interactive simulations of today, but far too expensive at the time to gain widespread use.

Harry Markowitz (later to receive a Nobel Prize for his work in portfolio theory) provided the major conceptual guidance for SIMSCRIPT, first appearing in 1963. The RAND Corporation developed the language under sponsorship of the U.S. Air Force. SIMSCRIPT originally was heavily influenced by FORTRAN, but, in later versions, its developers broke from its FORTRAN base and created its own SPL. The initial versions were based on event scheduling.

Phillip J. Kiviat of the Applied Research Laboratory of the United States Steel Corporation began the development of GASP (General Activity Simulation Program) in 1961. Originally, it was based on the general-purpose programming language ALGOL, but later a decision was made to base it on FORTRAN. GASP, like GPSS, used flowchart symbols familiar to engineers. It was not a language proper, but rather a collection of FORTRAN routines to facilitate simulation in FORTRAN.

Numerous other SPLs were developed during this time period. Notably, they included SIMULA, an extension of ALGOL, developed in Norway and widely used in Europe, and The Control and Simulation Language (CSL), which took an activity-scanning approach.

4.1.3 The Formative Period (1966–70)

During this period, concepts were reviewed and refined to promote a more consistent representation of each language's worldview. The major SPLs matured and gained wider usage.

Rapid hardware advancements and user demands forced some languages, notably GPSS, to undergo major revisions. GPSS/360, with its extensions to earlier versions of GPSS, emerged for the IBM 360 computer. Its popularity motivated at least six other hardware vendors and other groups to produce their own implementation of GPSS or a look-alike.

SIMSCRIPT II represented a major advancement in SPLs. In its free-form English-like language and "forgiving" compiler, an attempt was made to give the user major consideration in the language design.

ECSL, a descendent of CSL, was developed and became popular in the UK. In Europe, SIMULA added the concept of classes and inheritance, thus becoming a precursor of the modern object-oriented programming languages.

4.1.4 The Expansion Period (1971–78)

Major advances in GPSS during this period came from outside IBM. Julian Reitman of Norden Systems headed the development of GPSS/NORDEN, a pioneering effort that offered an interactive, visual online environment. James O. Henriksen of Wolverine Software developed GPSS/H, released in 1977 for IBM mainframes, later for mini-computers and the PC. It was notable for being compiled and reportedly 5 to 30 times faster than standard GPSS. With the addition of new features, including an interactive debugger, it has become the principal version of GPSS in use today.

Alan Pritsker at Purdue made major changes to GASP, with GASP IV appearing in 1974. It incorporated state events in addition to time events, thus adding support for the activity-scanning worldview in addition to the event-scheduling worldview.

Efforts were made during this period to attempt to simplify the modeling process. Using SIMULA, an attempt was made to develop a system definition from a high-level user perspective that could be translated automatically into an executable model. Similar efforts included interactive program generators, the "Programming by Questionnaire," and natural-language interfaces, together with automatic mappings to the language of choice. As did earlier over-optimistic beliefs in automatic programming, these efforts ran into severe limitations in the generality of what could be modeled—that is, they ran into the unavoidable complexity of real-world systems. Nevertheless, efforts to simplify simulation modeling continue, with the most success seen in simulation systems designed for application to narrow domains.

4.1.5 Consolidation and Regeneration (1979–86)

The fifth period saw the beginnings of SPLs written for, or adapted to, desktop computers and the microcomputer. During this period, the predominant SPLs extended their implementation to many computers and microprocessors while maintaining their basic structure.

Two major descendants of GASP appeared: SLAM II and SIMAN. The Simulation Language for Alternative Modeling (SLAM), produced by Pritsker and Associates, Inc., sought to provide multiple modeling perspectives and combined modeling capabilities [Pritsker and Pegden, 1979]—that is, it had an event-scheduling perspective based on GASP, a network worldview (a variant of the process-interaction perspective), and a continuous component. With SLAM, you could select one worldview or use a mix of all three.

SIMAN (SIMulation ANalysis) possessed a general modeling capability found in SPLs such as GASP IV, but also had a block-diagram component similar in some respects to that in SLAM and GPSS. C. Dennis Pegden developed SIMAN as a one-person faculty project over a period of about two years; he later founded Systems Modeling Corporation to market SIMAN. SIMAN was the first major simulation language executable on the IBM PC and designed to run under MS-DOS constraints. Similar to GASP, SIMAN allowed an event-scheduling approach by programming in FORTRAN with a supplied collection of FORTRAN routines, a block-diagram approach (another variant of the process-interaction worldview) analogous in some ways to that of GPSS and SLAM, and a continuous component.

4.1.6 Integrated Environments (1987–Present)

The most recent period is notable by the growth of SPLs on the personal computer and the emergence of simulation environments with graphical user interfaces, animation, and other visualization tools. Many of these environments also contain input-data analyzers and output analyzers. Some packages attempt to simplify the modeling process by the use of process-flow or block diagramming and of "fill-in-the-blank" windows that avoid the need to learn programming syntax. Animation ranges from schematic-like representations to 2-D and 3-D scale drawings.

Recent advancements have been made in web-based simulation. Much discussion has taken place concerning a role for simulation in supply-chain management. The combination of simulation and emulation shows promise.

Information about various software packages is given in Section 4.7 , including the websites of the vendors. A view of current developments in simulation software is available from these websites.

4.2 SELECTION OF SIMULATION SOFTWARE

This chapter includes a brief introduction to a number of simulation-software packages. Every two years, OR/MS Today publishes a simulation-software survey [Swain, 2003]. The 2003 issue had 48 products, including simulation support packages such as input-data analyzers.

There are many features that are relevant when selecting simulation software [Banks, 1996]. Some of these features are shown, along with a brief description, in Tables 4.1 to 4.5. We offer the following advice when evaluating and selecting simulation software:

1. Do not focus on a single issue, such as ease of use. Consider the accuracy and level of detail obtainable, ease of learning, vendor support, and applicability to your applications.
2. Execution speed is important. Do not think exclusively in terms of experimental runs that take place at night and over the weekend. Speed affects development time. During debugging, an analyst might have to wait for the model to run up to the point in simulated time where an error occurs many times before the error is identified.
3. Beware of advertising claims and demonstrations. Many advertisements exploit positive features of the software only. Similarly, the demonstrations solve the test problem very well, but perhaps not your problem.
4. Ask the vendor to solve a small version of your problem.
5. Beware of "checklists" with "yes" and "no" as the entries. For example, many packages claim to have a conveyor entity. However, implementations have considerable variation

Table 4.1 Model-Building Features

Feature	Description
Modeling worldview	Process interaction, event perspectives, and continuous modeling, depending on needs
Input-data analysis capability	Estimate empirical or statistical distributions from raw data
Graphical model-building	Process-flow, block-diagram, or network approach
Conditional routing	Route entities based on prescribed conditions or attributes
Simulation programming	Capability to add procedural logic through a high-level powerful simulation language
Syntax	Easily understood, consistent, unambiguous, English-like
Input flexibility	Accepts data from external files, data bases, spreadsheets, or interactively
Modeling conciseness	Powerful actions, blocks, or nodes
Randomness	Random-variate generators for all common distributions, e.g., exponential triangular uniform normal
Specialized components and templates	Material handling: vehicles, conveyors, bridge cranes, AS/RS, etc. Handling of liquids and bulk materials Communication systems Computer systems Call centers Others
User-built custom objects	Reusable objects, templates and submodels
Continuous flow	Tanks and pipes or bulk conveyors
Interface with general-programming language	Link code in C, C++, Java, or other general-programming language

Table 4.2 Runtime Environment

Feature	Description
Execution speed	Many runs needed for scenarios and replications. Impacts development as well as experimentation
Model size; number of variables and attributes	Should be no built-in limits
Interactive debugger	Monitor the simulation in detail as it progresses. Ability to break, trap, run until, step; to display status, attributes and variables; etc.
Model status and statistics	Display at any time during simulation
Runtime license	Ability to change parameters and run a model (but not to change logic or build a new model)

Table 4.3 Animation and Layout Features

Feature	Description
Type of animation	True to scale or iconic (such as process-flow diagram)
Import drawing and object files	From CAD (vector formats) drawings or icons (bit-mapped or raster graphics)
Dimension	2-D, 2-D with perspective, 3-D
Movement	Motion of entities or status indicators
Quality of motion	Smooth or jerky
Libraries of common objects	Extensive predrawn graphics
Navigation	Panning, zooming, rotation
Views	User defined, named
Display step	Control of animation speed
Selectable objects	Dynamic status and statistics displayed upon selection
Hardware requirements	Standard or special video card, RAM requirements

and level of fidelity. Implementation and capability are what is important. As a second example, most packages offer a runtime license, but these vary considerably in price and features.

6. Simulation users ask whether the simulation model can link to and use code or routines written in external languages such as C, C++, or Java. This is a good feature, especially when the external routines already exist and are suitable for the purpose at hand. However, the more important question is whether the simulation package and language are sufficiently powerful to avoid having to write logic in any external language.

7. There may be a significant trade-off between the graphical model-building environments and ones based on a simulation language. Graphical model-building removes the learning curve due to language syntax, but it does not remove the need for procedural logic in most real-world models and the debugging to get it right. Beware of "no programming required"

Table 4.4 Output Features

Feature	Description
Scenario manager	Create user-defined scenarios to simulate
Run manager	Make all runs (scenarios and replications) and save results for future analyses
Warmup capability	For steady-state analysis
Independent replications	Using a different set of random numbers
Optimization	Genetic algorithms, tabu search, etc.
Standardized reports	Summary reports including averages, counts, minimum and maximum, etc.
Customized reports	Tailored presentations for managers
Statistical analysis	Confidence intervals, designed experiments, etc.
Business graphics	Bar charts, pie charts, time lines, etc.
Costing module	Activity-based costing included
File export	Input to spreadsheet or database for custom processing and analysis
Database maintenance	Store output in an organized manner

Table 4.5 Vendor Support and Product Documentation

Feature	Description
Training	Regularly scheduled classes of high quality
Documentation	Quality, completeness, online
Help system	General or context-sensitive
Tutorials	For learning the package or specific features
Support	Telephone, e-mail, web
Upgrades, maintenance	Regularity of new versions and maintenance releases that address customer needs
Track record	Stability, history, customer relations

unless either the package is a near-perfect fit to your problem domain or programming (customized procedural logic) is possible with the supplied blocks, nodes, or process-flow diagram—in which case "no programming required" refers to syntax only and not the development of procedural logic.

4.3 AN EXAMPLE SIMULATION

Example 4.1: The Checkout Counter: A Typical Single-Server Queue _____

The system, a grocery checkout counter, is modeled as a single-server queue. The simulation will run until 1000 customers have been served. In addition, assume that the interarrival times of customers are exponentially distributed with a mean of 4.5 minutes, and that the service times are (approximately) normally distributed with a mean of 3.2 minutes and a standard deviation of

Table 4.6   Definitions of Variables, Functions, and Subroutines in the Java Model of the Single-Server Queue

Variables	*Description*
System state	
QueueLength	Number of customers enqueued (but not in service) at current simulated time
NumberInService	Number being served at current simulated time
Entity attributes and sets	
Customers	FCFS Queue of customers in system
Future event list	
FutureEventList	Priority-ordered list of pending events
Activity durations	
MeanInterArrivalTime	The interarrival time between the previous customer's arrival and the next arrival
MeanServiceTime	The service time of the most recent customer to begin service
Input parameters	
MeanInterarrivalTime	Mean interarrival time (4.5 minutes)
MeanServiceTime	Mean service time (3.2 minutes)
SIGMA	Standard deviation of service time (0.6 minute)
TotalCustomers	The stopping criterion—number of customers to be served (1000)
Simulation variables	
Clock	The current value of simulated time
Statistical Accumulators	
LastEventTime	Time of occurrence of the last event
TotalBusy	Total busy time of server (so far)
MaxQueueLength	Maximum length of waiting line (so far)
SumResponseTime	Sum of customer response times for all customers who have departed (so far)
NumberOfDepartures	Number of departures (so far)
LongService	Number of customers who spent 4 or more minutes at the checkout counter (so far)
Summary statistics	
RHO =BusyTime/Clock	Proportion of time server is busy (here the value of Clock is the final value of simulated time)
AVGR	Average response time (equal to SumResponseTime/TotalCustomers)
PC4	Proportion of customers who spent 4 or more minutes at the checkout counter

(*continued overleaf*)

Table 4.6   (*continued*)

Functions	Description
`exponential(mu)`	Function to generate samples from an exponential distribution with mean `mu`
`normal(xmu,SIGMA)`	Function to generate samples from a normal distribution with mean `xmu` and standard deviation `SIGMA`

Methods	Description
`Initialization`	Initialization method
`ProcessArrival`	Event method that executes the arrival event
`ProcessDeparture`	Event method that executes the departure event
`ReportGeneration`	Report generator

0.6 minute. (The approximation is that service times are always positive.) When the cashier is busy, a queue forms with no customers turned away. This example was simulated manually in Examples 3.3 and 3.4 by using the event-scheduling point of view. The model contains two events: the arrival and departure events. Figures 3.5 and 3.6 provide the event logic.

The following three sections illustrate the simulation of this single-server queue in Java, GPSS/H, and SSF. Although this example is much simpler than models that arise in the study of complex systems, its simulation contains the essential components of all discrete-event simulations.

4.4 SIMULATION IN JAVA

Java is a widely used programming language that has been used extensively in simulation. It does not, however, provide any facilities directly aimed at aiding the simulation analyst, who therefore must program all details of the event-scheduling/time-advance algorithm, the statistics-gathering capability, the generation of samples from specified probability distributions, and the report generator. However, the runtime library does provide a random-number generator. Unlike with FORTRAN or C, the object-orientedness of Java does support modular construction of large models. For the most part, the special-purpose simulation languages hide the details of event scheduling, whereas in Java all the details must be explicitly programmed. However, to a certain extent, simulation libraries such as SSF (Cowie 1999) alleviate the development burden by providing access to standardized simulation functionality and by hiding low-level scheduling minutiae.

There are many online resources for learning Java; we assume a prior working knowledge of the language. Any discrete-event simulation model written in Java contains the components discussed in Section 4.3: system state, entities and attributes, sets, events, activities and delays, and the components listed shortly. To facilitate development and debugging, it is best to organize the Java model in a modular fashion by using methods. The following components are common to almost all models written in Java:

Clock A variable defining simulated time

Initialization method A method to define the system state at time 0

Min-time event method A method that identifies the imminent event, that is, the element of the future event list (`FutureEventList`) that has the smallest time-stamp

Event methods For each event type, a method to update system state (and cumulative statistics) when that event occurs

Random-variate generators Methods to generate samples from desired probability distributions

Main program To maintain overall control of the event-scheduling algorithm

Report generator A method that computes summary statistics from cumulative statistics and prints a report at the end of the simulation

The overall structure of a Java simulation program is shown in Figure 4.1. This flowchart is an expansion of the event-scheduling/time-advance algorithm outlined in Figure 3.2. (The steps mentioned in Figure 4.1 refer to the five steps in Figure 3.2.)

The simulation begins by setting the simulation `Clock` to zero, initializing cumulative statistics to zero, generating any initial events (there will always be at least one) and placing them on the `FutureEventList`, and defining the system state at time 0. The simulation program then cycles, repeatedly passing the current least-time event to the appropriate event methods until the simulation is over. At each step, after finding the imminent event but before calling the event method, the simulation `Clock` is advanced to the time of the imminent event. (Recall that, during the simulated time between the occurrence of two successive events, the system state and entity attributes do not change in value. Indeed, this is the definition of discrete-event simulation: The system state changes only when an event occurs.) Next, the appropriate event method is called to execute the imminent event, update cumulative statistics, and generate future events (to be placed on the `FutureEventList`). Executing the imminent event means that the system state, entity attributes, and set membership are changed to reflect the fact that the event has occurred. Notice that all actions in an event method take place at one instant of simulated time. The value of the variable `Clock` does not change in an event method. If the simulation is not over, control passes again to the time-advance method, then to the appropriate event method, and so on. When the simulation is over, control passes to the report generator, which computes the desired summary statistics from the collected cumulative statistics and prints a report.

The efficiency of a simulation model in terms of computer runtime is determined to a large extent by the techniques used to manipulate the `FutureEventList` and other sets. As discussed earlier in Section 4.3, removal of the imminent event and addition of a new event are the two main operations performed on the `FutureEventList`. Java includes general, efficient data structures for searching and priority lists; it is usual to build a customized interface to these to suit the application. In the example to follow, we use customized interfaces to implement the event list and the list of waiting customers. The underlying priority-queue organization is efficient, in the sense of having access costs that grow only in the logarithm of the number of elements in the list.

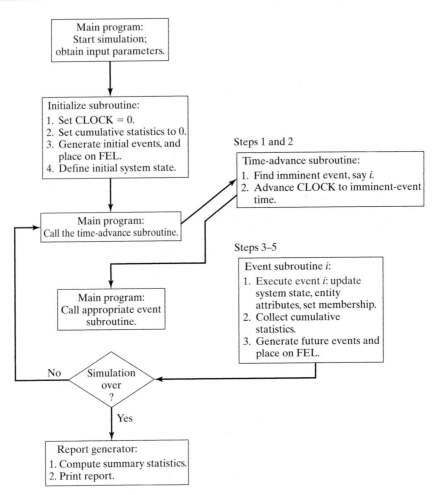

Figure 4.1   Overall structure of an event-scheduling simulation program.

Example 4.2:   Single-Server Queue Simulation in Java

The grocery checkout counter, defined in detail in Example 4.1, is now simulated by using Java. A version of this example was simulated manually in Examples 3.3 and 3.4, where the system state, entities and attributes, sets, events, activities, and delays were analyzed and defined.

Class `Event` represents an event. It stores a code for the event type (`arrival` or `departure`), and the event time-stamp. It has associated methods (functions) for creating an event and accessing its data. It also has an associated method **compareTo**, which compares the event with another (passed as an argument) and reports whether the first event should be considered to be smaller, equal, or larger than the argument event. The methods for this model and the flow of control are shown in Figure 4.2, which is an adaptation of Figure 4.1 for this particular problem. Table 4.6 lists the variables used for system state, entity attributes and sets, activity durations, and

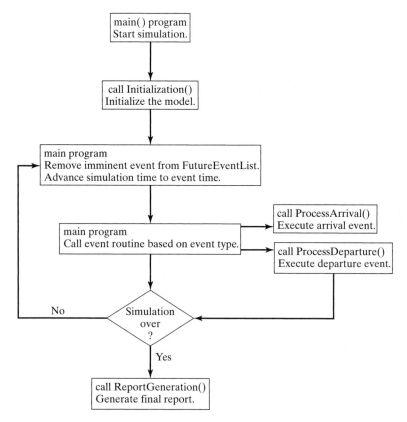

Figure 4.2   Overall structure of Java simulation of a single-server queue.

cumulative and summary statistics; the functions used to generate samples from the exponential and normal distributions; and all the other methods needed.

The entry point of the program and the location of the control logic is through class Sim, shown in Figure 4.3. Variables of classes EventList and Queue are declared. As these classes are all useful for programs other than Sim, their declarations are given in other files, per Java rules. A variable of the Java built-in class Random is also declared; instances of this class provided random-number streams. The main method controls the overall flow of the event-scheduling/time-advance algorithm.

The main program method first gives values to variables describing model parameters; it creates instances of the random-number generator, event list, and customer queue; and then it calls method Initialization to initialize other variables, such as the statistics-gathering variables. Control then enters a loop which is exited only after TotalCustomers customers have received service. Inside the loop, a copy of the imminent event is obtained by calling the getMin method of the priority queue, and then that event is removed from the event list by a call to dequeue. The global simulation time Clock is set to the time-stamp contained in the imminent event, and then either ProcessArrival or ProcessDeparture is called, depending on the type of the

```
class Sim {

// Class Sim variables
public static double Clock, MeanInterArrivalTime, MeanServiceTime,
        SIGMA, LastEventTime, TotalBusy, MaxQueueLength, SumResponseTime;
public static long  NumberOfCustomers, QueueLength, NumberInService,
        TotalCustomers, NumberOfDepartures, LongService;

public final static int arrival = 1;
public final static int departure = 2;

public static EventList FutureEventList;
public static Queue Customers;
public static Random stream;

public static void main(String argv[]) {

  MeanInterArrivalTime = 4.5; MeanServiceTime = 3.2;
  SIGMA                = 0.6; TotalCustomers  = 1000;
  long seed            =   Long.parseLong(argv[0]);
  stream = new Random(seed);            // initialize rng stream
  FutureEventList = new EventList();
  Customers = new Queue();

  Initialization();

  // Loop until first "TotalCustomers" have departed
  while(NumberOfDepartures < TotalCustomers ) {
    Event evt = (Event)FutureEventList.getMin();  // get imminent event
    FutureEventList.dequeue();                     // be rid of it
    Clock = evt.get_time();                        // advance in time
    if( evt.get_type() == arrival ) ProcessArrival(evt);
    else  ProcessDeparture(evt);
    }
  ReportGeneration();
 }
```

Figure 4.3 Java main program for the single-server queue simulation.

event. When the simulation is finally over, a call is made to method `ReportGeneration` to create and print out the final report.

A listing for the `Sim` class method `Initialization` is given in Figure 4.4. The simulation clock, system state, and other variables are initialized. Note that the first arrival event is created by generating a local `Event` variable whose constructor accepts the event's type and time. The event time-stamp is generated randomly by a call to `Sim` class method `exponential` and is passed to the random-number stream to use with the mean of the exponential distribution from which to sample. The event is inserted into the future event list by calling method `enqueue`. This logic assumes that the system is empty at simulated time `Clock = 0`, so that no departure can be scheduled. It is straightforward to modify the code to accommodate alternative starting conditions by adding events to `FutureEventList` and `Customers` as needed.

```
public static void Initialization()   {
  Clock = 0.0;
  QueueLength = 0;
  NumberInService = 0;
  LastEventTime = 0.0;
  TotalBusy = 0 ;
  MaxQueueLength = 0;
  SumResponseTime = 0;
  NumberOfDepartures = 0;
  LongService = 0;

  // create first arrival event
  Event evt =
   new Event(arrival, exponential( stream, MeanInterArrivalTime));
  FutureEventList.enqueue( evt );
 }
```

Figure 4.4   Java initialization method for the single-server queue simulation.

```
public static void ProcessArrival(Event evt) {
  Customers.enqueue(evt);
  QueueLength++;
  // if the server is idle, fetch the event, do statistics
  // and put into service
  if( NumberInService == 0) ScheduleDeparture();
  else TotalBusy += (Clock - LastEventTime);  // server is busy

  // adjust max queue length statistics
  if (MaxQueueLength < QueueLength) MaxQueueLength = QueueLength;

  // schedule the next arrival
  Event next_arrival =
   new Event(arrival, Clock+exponential(stream,MeanInterArrivalTime));
  FutureEventList.enqueue( next_arrival );
  LastEventTime = Clock;
 }
```

Figure 4.5   Java arrival event method for the single-server queue simulation.

Figure 4.5 gives a listing of Sim class method ProcessArrival, which is called to process each arrival event. The basic logic of the arrival event for a single-server queue was given in Figure 3.5 (where LQ corresponds to QueueLength and LS corresponds to NumberInService). First, the new arrival is added to the queue Customers of customers in the system. Next, if the server is idle (NumberInService == 0) then the new customer is to go immediately into service, so Sim class method ScheduleDeparture is called to do that scheduling. An arrival to an idle queue does not update the cumulative statistics, except possibly the maximum queue length. An arrival to a busy queue does *not* cause the scheduling of a departure, but does increase the total busy time by the amount of simulation time between the current event and the one immediately preceding it (because, if the server is busy now, it had to have had at least one customer in service by the end of processing the previous event). In either case, a new arrival is responsible for scheduling the next arrival, one random interarrival time into the future. An arrival event is created with simulation time equal to the current `Clock` value plus an exponential increment, that event is inserted into the future event list, the variable `LastEventTime` recording the time of the last event processed is set to the current time, and control is returned to the `main` method of class `Sim`.

Sim class method `ProcessDeparture`, which executes the departure event, is listed in Figure 4.6, as is method `ScheduleDeparture`. A flowchart for the logic of the departure event was given in Figure 3.6. After removing the event from the queue of all customers, the number in service is examined. If there are customers waiting, then the departure of the next one to enter service is scheduled. Then, cumulative statistics recording the sum of all response times, sum of busy time, number of customers who used more than 4 minutes of service time, and number of departures are updated. (Note that the maximum queue length cannot change in value when a departure occurs.) Notice that customers are removed from `Customers` in FIFO order; hence, the response time `response` of the departing customer can be computed by subtracting the arrival time of the job leaving service (obtained from the copy of the arrival event removed from the `Customers` queue) from the current simulation time. After the incrementing of the total number of departures and the saving of the time of this event, control is returned to the main program.

Figure 4.6 also gives the logic of method `ScheduleDeparture`, called by both `Process-Arrival` and `ProcessDeparture` to put the next customer into service. The `Sim` class method

```
public static void ScheduleDeparture() {
  double ServiceTime;
  // get the job at the head of the queue
  while (( ServiceTime = normal(stream,MeanServiceTime, SIGMA)) < 0 );
  Event depart = new Event(departure,Clock+ServiceTime);
  FutureEventList.enqueue( depart );
  NumberInService = 1;
  QueueLength--;
}

public static void ProcessDeparture(Event e) {
  // get the customer description
  Event finished = (Event) Customers.dequeue();
  // if there are customers in the queue then schedule
  // the departure of the next one
  if( QueueLength > 0 ) ScheduleDeparture();
  else NumberInService = 0;
  // measure the response time and add to the sum
  double response = (Clock - finished.get_time());
  SumResponseTime += response;
  if( response > 4.0 ) LongService++; // record long service
  TotalBusy += (Clock - LastEventTime );
  NumberOfDepartures++;
  LastEventTime = Clock;
}
```

Figure 4.6   Java departure event method for the single-server queue simulation.

normal, which generates normally distributed service times, is called until it produces a nonnegative sample. A new event with type departure is created, with event time equal to the current simulation time plus the service time just sampled. That event is pushed onto FutureEventList, the number in service is set to one, and the number waiting (QueueLength) is decremented to reflect the fact that the customer entering service is waiting no longer.

The report generator, Sim class method ReportGeneration, is listed in Figure 4.7. The summary statistics, RHO, AVGR, and PC4, are computed by the formulas in Table 4.6; then the input parameters are printed, followed by the summary statistics. It is a good idea to print the input parameters at the end of the simulation, in order to verify that their values are correct and that these values have not been inadvertently changed.

Figure 4.8 provides a listing of Sim class methods exponential and normal, used to generate random variates. Both of these functions call method nextDouble, which is defined for the built-in Java Random class generates a random number uniformly distributed on the (0,1) interval. We use Random here for simplicity of explanation; superior random-number generators can be built by hand, as described in Chapter 7. The techniques for generating exponentially

```
public static void ReportGeneration() {
double RHO    = TotalBusy/Clock;
double AVGR   = SumResponseTime/TotalCustomers;
double PC4    = ((double)LongService)/TotalCustomers;


System.out.print( "SINGLE SERVER QUEUE SIMULATION ");
System.out.println( "- GROCERY STORE CHECKOUT COUNTER ");
System.out.println( "\tMEAN INTERARRIVAL TIME                    "
   + MeanInterArrivalTime );
System.out.println( "\tMEAN SERVICE TIME                         "
   + MeanServiceTime );
System.out.println( "\tSTANDARD DEVIATION OF SERVICE TIMES       "
   + SIGMA );
System.out.println( "\tNUMBER OF CUSTOMERS SERVED                "
   + TotalCustomers );
System.out.println();
System.out.println( "\tSERVER UTILIZATION                        "
   + RHO );
System.out.println( "\tMAXIMUM LINE LENGTH                       "
   + MaxQueueLength );
System.out.println( "\tAVERAGE RESPONSE TIME                     "
   + AVGR + "  MINUTES" );
System.out.println( "\tPROPORTION WHO SPEND FOUR ");
System.out.println( "\t MINUTES OR MORE IN SYSTEM                "
   + PC4 );
System.out.println( "\tSIMULATION RUNLENGTH                      "
   + Clock + " MINUTES" );
System.out.println( "\tNUMBER OF DEPARTURES                      "
   + TotalCustomers );
 }
```

Figure 4.7   Java report generator for the single-server queue simulation.

```
public static double exponential(Random rng, double mean) {
 return -mean*Math.log( rng.nextDouble() );
}

public static double SaveNormal;
public static int  NumNormals = 0;
public static final double  PI = 3.1415927 ;

public static double normal(Random rng, double mean, double sigma) {
        double ReturnNormal;
        // should we generate two normals?
        if(NumNormals == 0 ) {
          double r1 = rng.nextDouble();
          double r2 = rng.nextDouble();
          ReturnNormal = Math.sqrt(-2*Math.log(r1))*Math.cos(2*PI*r2);
          SaveNormal   = Math.sqrt(-2*Math.log(r1))*Math.sin(2*PI*r2);
          NumNormals = 1;
        } else {
          NumNormals = 0;
          ReturnNormal = SaveNormal;
        }
        return ReturnNormal*sigma + mean ;
 }
```

Figure 4.8 Random-variate generators for the single-server queue simulation.

and normally distributed random variates, discussed in Chapter 8, are based on first generating a $U(0,1)$ random number. For further explanation, the reader is referred to Chapters 7 and 8.

The output from the grocery-checkout-counter simulation is shown in Figure 4.9. It should be emphasized that the output statistics are estimates that contain random error. The values shown are influenced by the particular random numbers that happened to have been used, by the initial conditions at time 0, and by the run length (in this case, 1000 departures). Methods for estimating the standard error of such estimates are discussed in Chapter 11.

In some simulations, it is desired to stop the simulation after a fixed length of time, say TE = 12 hours = 720 minutes. In this case, an additional event type, stop event, is defined and is scheduled to occur by scheduling a stop event as part of simulation initialization. When the stopping event does occur, the cumulative statistics will be updated and the report generator called. The main program and method Initialization will require minor changes. Exercise 1 asks the reader to make these changes. Exercise 2 considers the additional change that any customer at the checkout counter at simulated time Clock = TE should be allowed to exit the store, but no new arrivals are allowed after time TE.

4.5 SIMULATION IN GPSS

GPSS is a highly structured, special-purpose simulation programming language based on the process-interaction approach and oriented toward queueing systems. A block diagram provides

```
SINGLE SERVER QUEUE SIMULATION - GROCERY STORE CHECKOUT COUNTER
        MEAN INTERARRIVAL TIME                 4.5
        MEAN SERVICE TIME                      3.2
        STANDARD DEVIATION OF SERVICE TIMES    0.6
        NUMBER OF CUSTOMERS SERVED             1000

        SERVER UTILIZATION                     0.671
        MAXIMUM LINE LENGTH                    9.0
        AVERAGE RESPONSE TIME                  6.375 MINUTES
        PROPORTION WHO SPEND FOUR
         MINUTES OR MORE IN SYSTEM             0.604
        SIMULATION RUNLENGTH                   4728.936 MINUTES
        NUMBER OF DEPARTURES                   1000
```

Figure 4.9   Output from the Java single-server queue simulation.

a convenient way to describe the system being simulated. There are over 40 standard blocks in GPSS. Entities called transactions may be viewed as flowing through the block diagram. Blocks represent events, delays, and other actions that affect transaction flow. Thus, GPSS can be used to model any situation where transactions (entities, customers, units of traffic) are flowing through a system (e.g., a network of queues, with the queues preceding scarce resources). The block diagram is converted to block statements, control statements are added, and the result is a GPSS model.

The first version of GPSS was released by IBM in 1961. It was the first process-interaction simulation language and became popular; it has been implemented anew and improved by many parties since 1961, with GPSS/H being the most widely used version in use today. Example 4.3 is based on GPSS/H.

GPSS/H is a product of Wolverine Software Corporation, Annandale, VA (Banks, Carson, and Sy, 1995; Henriksen, 1999). It is a flexible, yet powerful tool for simulation. Unlike the original IBM implementation, GPSS/H includes built-in file and screen I/O, use of an arithmetic expression as a block operand, an interactive debugger, faster execution, expanded control statements, ordinary variables and arrays, a floating-point clock, built-in math functions, and built-in random-variate generators.

The animator for GPSS/H is Proof Animation™, another product of Wolverine Software Corporation (Henriksen, 1999). Proof Animation provides a 2-D animation, usually based on a scale drawing. It can run in postprocessed mode (after the simulation has finished running) or concurrently. In postprocessed mode, the animation is driven by two files: the layout file for the static background, and a trace file that contains commands to make objects move and produce other dynamic events. It can work with any simulation package that can write the ASCII trace file. Alternately, it can run concurrently with the simulation by sending the trace file commands as messages, or it can be controlled directly by using its DLL (dynamic link library) version.

Example 4.3:   Single-Server Queue Simulation in GPSS/H _____

Figure 4.10 exhibits the block diagram and Figure 4.11 the GPSS program for the grocery-store checkout-counter model described in Example 4.2. Note that the program (Figure 4.11) is a translation of the block diagram together with additional definition and control statements.

In Figure 4.10, the GENERATE block represents the arrival event, with the interarrival times specified by RVEXPO(1,&*IAT*). RVEXPO stands for "random variable, exponentially distributed,"

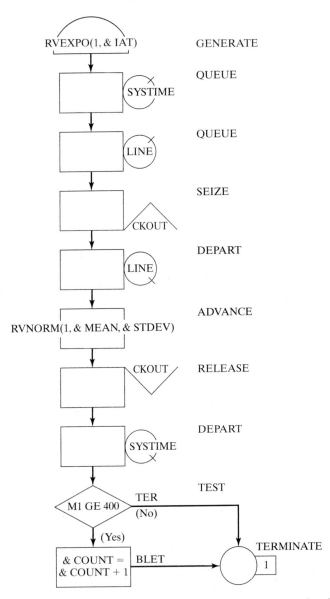

Figure 4.10 GPSS block diagram for the single-server queue simulation.

the 1 indicates the random-number stream to use, and *&IAT* indicates that the mean time for the exponential distribution comes from a so-called ampervariable *&IAT*. Ampervariable names begin with the "&" character; Wolverine added ampervariables to GPSS because the original IBM implementation had limited support for ordinary global variables, with no user freedom for naming them. (In the discussion that follows, all nonreserved words are shown in italics.)

```
        SIMULATE
*
*       Define Ampervariables
*
        INTEGER         &LIMIT
        REAL            &IAT,&MEAN,&STDEV,&COUNT
        LET             &IAT=4.5
        LET             &MEAN=3.2
        LET             &STDEV=.6
        LET             &LIMIT=1000
*
*       Write Input Data to File
*
        PUTPIC          FILE=OUT,LINES=5,(&IAT,&MEAN,&STDEV,&LIMIT)
        Mean interarrival time              **.**   minutes
        Mean service time                   **.**   minutes
        Standard deviation of service time  **.**   minutes
        Number of customers to be served    *****


*
*       GPSS/H Block Section
*
        GENERATE        RVEXPO(1,&IAT)  Exponential arrivals
        QUEUE           SYSTIME         Begin response time data collection
        QUEUE           LINE            Customer joins waiting line
        SEIZE           CHECKOUT        Begin checkout at cash register
        DEPART          LINE            Customer starting service leaves queue
        ADVANCE         RVNORM(1,&MEAN,&STDEV)  Customer's service time
        RELEASE         CHECKOUT        Customer leaves checkout area
        DEPART          SYSTIME         End response time data collection
        TEST GE         M1,4,TER        Is response time GE 4 minutes?
        BLET            &COUNT=&COUNT+1 If so, add 1 to counter
TER     TERMINATE       1
*
        START           &LIMIT          Simulate for required number
*
*       Write Customized Output Data to File
*
        PUTPIC          FILE=OUT,LINES=7,(FR(CHECKOUT)/1000,QM(LINE),_
        QT(SYSTIME),&COUNT/N(TER),AC1,N(TER))
        Server utilization                  .***
        Maximum line length                 **
        Average response time               **.**     minutes
        Proportion who spend four minutes   .***
             or more in the system
        Simulation runlength                ****.** minutes
        Number of departures                ****
*
        END
```

Figure 4.11   GPSS/H program for the single-server queue simulation.

The next block is a QUEUE with a queue named *SYSTIME*. It should be noted that the QUEUE block is not needed for queues or waiting lines to form in GPSS. The true purpose of the QUEUE block is to work in conjunction with the DEPART block to collect data on queues or any other subsystem. In Example 4.3, we want to measure the system response time—that is, the time a transaction spends in the system. Placing a QUEUE block at the point that transactions enter the system and placing the counterpart of the QUEUE block, the DEPART block, at the point that the transactions complete their processing causes the response times to be collected automatically. The purpose of the DEPART block is to signal the end of data collection for an individual transaction. The QUEUE and DEPART block combination is not necessary for queues to be modeled, but rather is used for statistical data collection.

The next QUEUE block (with name *LINE*) begins data collection for the waiting line before the cashier. The customers may or may not have to wait for the cashier. Upon arrival to an idle checkout counter, or after advancing to the head of the waiting line, a customer captures the cashier, as represented by the SEIZE block with the resource named *CHECKOUT*. Once the transaction representing a customer captures the cashier represented by the resource CHECKOUT, the data collection for the waiting-line statistics ends, as represented by the DEPART block for the queue named LINE. The transaction's service time at the cashier is represented by an ADVANCE block. RVNORM indicates "random variable, normally distributed." Again, random-number stream 1 is being used, the mean time for the normal distribution is given by ampervariable *&MEAN*, and its standard deviation is given by ampervariable *&STDEV*. Next, the customer gives up the use of the facility *CHECKOUT* with a RELEASE block. The end of the data collection for response times is indicated by the DEPART block for the queue *SYSTIME*.

Next, there is a TEST block that checks to see whether the time in the system, M1, is greater than or equal to 4 minutes. (Note that M1 is a reserved word in GPSS/H; it automatically tracks transaction total time in system.) In GPSS/H, the maxim is "if true, pass through." Thus, if the customer has been in the system four minutes or longer, the next BLET block (for block LET) adds one to the counter *&COUNT*. If not true, the escape route is to the block labeled *TER*. That label appears before the TERMINATE block whose purpose is the removal of the transaction from the system. The TERMINATE block has a value "1" indicating that one more transaction is added toward the limiting value (or "transactions to go").

The control statements in this example are all of those lines in Figure 4.11 that precede or follow the block section. (There are eleven blocks in the model from the GENERATE block to the TERMINATE block.) The control statements that begin with an "*" are comments, some of which are used for spacing purposes. The control statement SIMULATE tells GPSS/H to conduct a simulation; if it is omitted, GPSS/H compiles the model and checks for errors only. The ampervariables are defined as integer or real by control statements INTEGER and REAL. It seems that the ampervariable *&COUNT* should be defined as an integer; however, it will be divided later by a real value. If it is integer, the result of an integer divided by a real value is truncation, and that is not desired in this case. The four assignment statements (LET) provide data for the simulation. These four values could have been placed directly in the program; however, the preferred practice is to place them in ampervariables at the top of the program so that changes can be made more easily or the model can be modified to read them from a data file.

To ensure that the model data is correct, and for the purpose of managing different scenarios simulated, it is good practice to echo the input data. This is accomplished with a PUTPIC (for

```
Mean interarrival time                    4.50   minutes
Mean service time                         3.20   minutes
Standard deviation of service time        0.60   minutes
Number of customers to be served          1000

Server utilization                        0.676
Maximum line length                       7
Average response time                     6.33     minutes
Proportion who spend four minutes         0.646
     or more in the system
Simulation runlength                      4767.27 minutes
Number of departures                      1000
```

Figure 4.12   Customized GPSS/H output report for the single-server queue simulation.

"put picture") control statement. The five lines following PUTPIC provide formatting information, with the asterisks being markers (called picture formatting) in which the values of the four ampervariables replace the asterisks when PUTPIC is executed. Thus, "**.**" indicates a value that may have two digits following the decimal point and up to two before it.

The START control statement controls simulation execution. It starts the simulation, sets up a "termination-to-go" counter with initial value its operand (*&LIMIT*), and controls the length of the simulation.

After the simulation completes, a second PUTPIC control statement is used to write the desired output data to the same file *OUT*. The printed statistics are all gathered automatically by GPSS. The first output in the parenthesized list is the server utilization. FR*(CHECKOUT)*/1000 indicates that the fractional utilization of the facility *CHECKOUT* is printed. Because FR*(CHECKOUT)* is in parts per thousand, the denominator is provided to compute fractional utilization. QM*(LINE)* is the maximum value in the queue *LINE* during the simulation. QT*(SYSTIME)* is the average time in the queue *SYSTIME*. *&COUNT*/N*(TER)* is the number of customers who had a response time of four or more minutes divided by the number of customers that went through the block with label *TER*, or N*(TER)*. AC1 is the clock time, whose last value gives the length of the simulation.

The contents of the custom output file *OUT* are shown in Figure 4.12. The standard GPSS/H output file is displayed in Figure 4.13. Although much of the same data shown in the file *OUT* can be found in the standard GPSS/H output, the custom file is more compact and uses the language of the problem rather than GPSS jargon. There are many other reasons that customized output files are useful. For example, if 50 replications of the model are to be made and the lowest, highest, and average value of a response are desired, this can be accomplished by using control statements, with the results in a very compact form, rather than extracting the desired values from 50 standard output files.

4.6 SIMULATION IN SSF

The Scalable Simulation Framework (SSF) is an Application Program Interface (API) that describes a set of capabilities for object-oriented, process-view simulation. The API is sparse and

```
RELATIVE CLOCK: 4767.2740   ABSOLUTE CLOCK: 4767.2740


BLOCK CURRENT        TOTAL BLOCK CURRENT    TOTAL
1                    1003 TER               1000
2                    1003
3             3      1003
4                    1000
5                    1000
6                    1000
7                    1000
8                    1000
9                    1000
10                    646
```

FACILITY	--AVG-UTIL-DURING--			ENTRIES	AVERAGE	CURRENT	PERCENT	SEIZING	PREEMPTING
	TOTAL	AVAIL	UNAVL		TIME/XACT	STATUS	AVAIL	XACT	XACT
	TIME	TIME	TIME						
CHECKOUT	0.676			1000	3.224	AVAIL			

QUEUE	MAXIMUM CONTENTS	AVERAGE CONTENTS	TOTAL ENTRIES	ZERO ENTRIES	PERCENT ZEROS	AVERAGE TIME/UNIT	$AVERAGE TIME/UNIT	QTABLE NUMBER	CURRENT CONTENTS
SYSTIME	8	1.331	1003	0		6.325	6.235		3
LINE	7	0.655	1003	334	33.3	3.111	4.665		3

RANDOM STREAM	ANTITHETIC VARIATES	INITIAL POSITION	CURRENT POSITION	SAMPLE COUNT	CHI-SQUARE UNIFORMITY
1	OFF	100000	103004	3004	0.83

Figure 4.13 Standard GPSS/H output report for the single-server queue simulation.

was designed to allow implementations to achieve high performance (e.g. on parallel computers). SSF APIs exist for both C++ and in Java, and implementations exist in both languages. SSF has a wide user base—particularly in network simulation by using the add-on framework SSFNet (www.ssfnet.org). Our chapter on network simulation uses SSFNet.

The SSF API defines five base classes. process is a class that implements threads of control; the action method of a derived class contains the execution body of the thread. The Entity class is used to describe simulation objects. It contains state variables, processes, and communication endpoints. The inChannel and outChannel classes are communication endpoints. The Event class defines messages sent between entities. One model entity communicates with another by "writing" an Event into an outChannel; at some later time, it is available at one or more inChannels. A process that expects input on an inChannel can suspend, waiting for an event on it. These points, and others, will be elaborated upon as we work through an SSF implementation of the single-server queue.

Source code given in Figure 4.14 expresses the logic of arrival generation in SSF for the single-server queue example. The example is built on two SSF processes. One of these generates jobs and adds them to the system; the other services the enqueued jobs. Class SSQueue is a class that contains the whole simulation experiment. It uses the auxiliary classes Random (for random-number generation) and Queue (to implement FIFO queueing of general objects). SSQueue

```
// SSF MODEL OF JOB ARRIVAL PROCESS
class SSQueue extends Entity {

  private static Random rng;
  public static final double MeanServiceTime = 3.2;
  public static final double SIGMA = 0.6;
  public static final double MeanInterarrivalTime = 4.5;
  public static final long ticksPerUnitTime = 1000000000;
  public long generated=0;
  public Queue Waiting;
  outChannel out;
  inChannel in;

  public static long TotalCustomers=0, MaxQueueLength=0, TotalServiceTime=0;
  public static long LongResponse=0, SumResponseTime=0, jobStart;

  class arrival {
   long id,  arrival_time;
   public arrival(long num, long a) { id=num; arrival_time = a; }
  }

class Arrivals extends process {
   private Random rng;
   private SSQueue owner;
   public Arrivals (SSQueue _owner, long seed) {
       super(_owner); owner = _owner;
       rng = new Random(seed);
   }
   public boolean isSimple() { return true; }
   public void action() {
      if ( generated++ > 0 ) {
       // put a new Customer on the queue with the present arrival time
       int Size = owner.Waiting.numElements();
       owner.Waiting.enqueue( new arrival(generated, now()));
       if( Size == 0) owner.out.write( new Event() );   // signal start of burst
      }
      waitFor(owner.d2t( owner.exponential(rng, owner.MeanInterarrivalTime)) );
  }
 }
}
```

Figure 4.14   SSF Model of Job-Arrival Process.

defines experimental constants ("public static final" types) and contains SSF communication endpoints out and in, through which the two processes communicate. SSQueue also defines an inner class arrival, which stores the identity and arrival time of each job.

Class Arrivals is an SSF process. Its constructor stores the identity of the entity that owns it, and creates a random-number generator that is initialized with the seed passed to it. For all but the initial call, method action generates and enqueues a new arrival, then blocks (via SSF method waitFor) for an inter-arrival time; on the first call, it by-passes the job-generation step and blocks for an initial interarrival time. The call to waitFor highlights details needing explanation. An SSQueue object calls the Arrival constructor and is saved as the "owner." This class contains an auxiliary method exponential, which samples an exponential random variable with specified mean by using a specified random-number stream. It also contains methods d2t and t2d that translate between a discrete "tick"-based integer clock and a double-precision floating-point representation. In the waitFor call, we use the same code seen earlier to sample the exponential in double-precision format, but then use d2t to convert it into the simulator's integer clock format. The specific conversion factor is listed as a SSQueue constant, 10^9 ticks per unit time.

SSF interprocess communication is used sparingly in this example. Because service is nonpreemptive, when a job's service completes, the process providing service can examine the list of waiting customers (in variable owner.Waiting) to see whether it needs to give service to another customer. Thus, the only time the server process needs to be told that there is a job waiting is when a job arrives to an empty system. This is reflected in Arrivals.action by use of its owner's out channel.

A last point of interest is that Arrivals is, in SSF terminology, a "simple" process. This means that every statement in action that might suspend the process would be the last statement executed under normal execution semantics. The Arrivals class tells SSF that it is simple by overriding a default method isSimple to return the value true, rather than the default value (false). The key reason for using simple processes is performance—they require that no state be saved, only the condition under which the process ought to be reanimated. And, when it is reanimated, it starts executing at the first line of action.

Figure 4.15 illustrates the code for the Server process. Like process Arrival, its constructor is called by an instance of SSQueue and is given the identity of that instance and a random-number seed. Like Arrival, it is a simple process. It maintains state variable in_service to remember the specifics of a job in service and state variable service_time to remember the value of the service time sampled for the job in service. When the SSF kernel calls action, either a job has completed service, or the Arrival process has just signaled Server though the inChannel. We distinguish the cases by looking at variable in_service, which will be nonnull if a job is in service, just now completed. In this case, some statistics are updated. After this task is done, a test is made for customers waiting for service. The first waiting customer is dequeued from the waiting list and is copied into the in_service variable; the process then samples a service time and suspends through a waitFor call. If no customer was waiting, the process suspends on a waitOn statement until an event from the Arrival process awakens it.

SSF bridges the gap between models developed in pure Java and models developed in languages specifically designed for simulation. It provides the flexibility offered by a general-programming language, yet has essential support for simulation.

4.7 SIMULATION SOFTWARE

All the simulation packages described in later subsections run on a PC under Microsoft Windows 2000 or XP. Although in terms of specifics the packages all differ, generally they have many things in common.

```
// SSF MODEL OF SINGLE SERVER QUEUE : ACCEPTING JOBS
class Server extends process {
 private Random rng;
 private SSQueue owner ;
 private arrival in_service;
 private long service_time;

 public Server(SSQueue _owner, long seed) {
     super(_owner);
     owner = _owner;
     rng = new Random(seed);
 }
 public boolean isSimple() { return true; }

 public void action() {
   // executes due to being idle and getting a job, or by service time expiration.
   // if there is a job awaiting service, take it out of the queue
   // sample a service time, do statistics, and wait for the service epoch

   // if in_service is not null, we entered because of a job completion
   if( in_service != null ) {
       owner.TotalServiceTime += service_time;
       long in_system = (now() -in_service.arrival_time);
       owner.SumResponseTime  += in_system;
       if( owner.t2d(in_system) > 4.0 ) owner.LongResponse++;
       in_service = null;
       if( owner.MaxQueueLength < owner.Waiting.numElements() + 1 )
           owner.MaxQueueLength = owner.Waiting.numElements() + 1;
       owner.TotalCustomers++;
   }
   if( owner.Waiting.numElements() > 0 ) {
       in_service = (arrival)owner.Waiting.dequeue();
       service_time = -1;
       while ( service_time < 0.0 )
         service_time = owner.d2t(owner.normal( rng, owner.MeanServiceTime, owner.SIGMA));
                                                           // model service time
       waitFor( service_time );
   } else {
       waitOn( owner.in );  // we await a wake-up call
   }
 }
}
```

Figure 4.15   SSF Model of Single-Server Queue : Server.

Common characteristics include a graphical user interface, animation, and automatically collected outputs to measure system performance. In virtually all packages, simulation results may be displayed in tabular or graphical form in standard reports and interactively while running a simulation. Outputs from different scenarios can be compared graphically or in tabular form. Most provide statistical analyses that include confidence intervals for performance measures and comparisons, plus a variety of other analysis methods. Some of the statistical-analysis modules are described in Section 4.8.

All the packages described here take the process-interaction worldview. A few also allow event-scheduling models and mixed discrete–continuous models. For animation, some emphasize scale drawings in 2-D or 3-D; others emphasize iconic-type animations based on schematic drawings or process-flow diagrams. A few offer both scale drawing and schematic-type animations. Almost all offer dynamic business graphing in the form of time lines, bar charts, and pie charts.

In addition to the information contained in this chapter, the websites given below can be investigated:

Arena
`www.arenasimulation.com/`
AutoMod
`www.automod.com`
Delmia/QUEST
`www.delmia.com` and `www.3ds.com`
Extend
`www.imaginethatinc.com/`
Flexsim
`www.flexsim.com/`
Micro Saint
`www.maad.com`
ProModel
`www.promodel.com/`
SIMUL8
`www.simul8.com/`
WITNESS
`www.witness-for-simulation.com/`

4.7.1 Arena

Arena Basic, Standard, and Professional Editions are offered by Systems Modeling Corporation [Bapat and Sturrock, 2003]. Arena can be used for simulating discrete and continuous systems. A recent addition to the Arena family of products is OptQuest for Arena, an optimization software package (discussed in Section 4.8.2.)

The Arena Basic Edition is targeted at modeling business processes and other systems in support of high-level analysis needs. It represents process dynamics in a hierarchical flowchart and stores system information in data spreadsheets. It has built-in activity-based costing and is closely integrated with the flowcharting software Visio.

The Arena Standard Edition is designed for more detailed models of discrete and continuous systems. First released in 1993, Arena employs an object-based design for entirely graphical model development. Simulation models are built from graphical objects called modules to define system logic and such physical components as machines, operators, and clerks. Modules are represented by icons plus associated data entered in a dialog window. These icons are connected to represent entity flow. Modules are organized into collections called templates. The Arena template is the core collection of modules providing general-purpose features for modeling all types of applications. In addition to standard features, such as resources, queues, process logic, and system data, the Arena template includes modules focused on specific aspects of manufacturing and material-handling systems. Arena SE can also be used to model combined discrete/continuous systems, such as pharmaceutical and chemical production, through its built-in continuous-modeling capabilities.

The Arena Professional Edition enhances Arena SE with the capability to craft custom simulation objects that mirror components of the real system, including terminology, process logic, data, performance metrics, and animation. The Arena family also includes products designed specifically to model call centers and high-speed production lines, namely Arena Contact Center and Arena Packaging.

At the heart of Arena is the SIMAN simulation language. For animating simulation models, Arena's core modeling constructs are accompanied by standard graphics for showing queues, resource status, and entity flow. Arena's 2-D animations are created by using Arena's built-in drawing tools and by incorporating clip art, AutoCAD, Visio, and other graphics.

Arena's Input Analyzer automates the process of selecting the proper distribution and its parameters for representing existing data, such as process and interarrival times. The Output Analyzer and Process Analyzer (discussed in Section 4.8.2) automate comparison of different design alternatives.

4.7.2 AutoMod

The AutoMod Product Suite is offered by Brooks Automation [Rohrer, 2003]. It includes the AutoMod simulation package, AutoStat for experimentation and analysis, and AutoView for making AVI movies of the built-in 3-D animation. The main focus of the AutoMod simulation product is manufacturing and material-handling systems. AutoMod's strength is in detailed, large models used for planning, operational decision support, and control-systems testing.

AutoMod has built-in templates for most common material-handling systems, including vehicle systems, conveyors, automated storage and retrieval systems, bridge cranes, power and free conveyors, and kinematics for robotics. With its Tanks and Pipes module, it also supports continuous modeling of fluid and bulk-material flow.

The pathmover vehicle system can be used to model lift trucks, humans walking or pushing carts, automated guided vehicles, trucks, and cars. All the movement templates are based on a 3-D scale drawing (drawn or imported from CAD as 2-D or 3-D). All the components of a template are highly parameterized. For example, the conveyor template contains conveyor sections, stations for load induction or removal, motors, and photo-eyes. Sections are defined by length, width, speed, acceleration, and type (accumulating or nonaccumulating), plus other specialized parameters. Photo-eyes have blocked and cleared timeouts that facilitate modeling of detailed conveyor logic.

In addition to the material-handling templates, AutoMod contains a full simulation programming language. Its 3-D animation can be viewed from any angle or perspective in real time. The user can freely zoom, pan, or rotate the 3-D world.

An AutoMod model consists of one or more systems. A system can be either a process system, in which flow and control logic are defined, or a movement system based on one of the material-handling templates. A model may contain any number of systems, which can be saved and reused as objects in other models. Processes can contain complex logic to control the flow of either manufacturing materials or control messages, to contend for resources, or to wait for user-specified times. Loads can move between processes with or without using movement systems.

In the AutoMod worldview, loads (products, parts, etc.) move from process to process and compete for resources (equipment, operators, vehicles, and queues). The load is the active entity, executing action statements in each process. To move between processes, loads may use a conveyor or vehicle in a movement system.

AutoStat, described in Section 4.8.2, works with AutoMod models to provide a complete environment for the user to define scenarios, conduct experimentation, and perform analyses. It offers optimization based on an evolutionary strategies algorithm.

4.7.3 Extend

The Extend family of products is offered by Imagine That, Inc. [Krahl, 2003]. Extend OR, Industry, and Suite are used for simulating discrete and mixed discrete–continuous systems; Extend CP is for continuous modeling only. Extend combines a block-diagram approach to model-building with a development environment for creating new blocks.

Each Extend block has an icon and encapsulates code, parameters, user interface, animation, and online help. Extend includes a large set of elemental blocks; libraries of blocks for specific application areas, such as manufacturing, business processes, and high-speed processes, are also available. Third-party developers have created Extend libraries for vertical market applications, including supply-chain dynamics, reliability engineering, and pulp and paper processing.

Models are built by placing and connecting blocks and entering the parameters on the block's dialog window. Elemental blocks in Extend include Generator, Queue, Activity, Resource Pool, and Exit. The active entities, called items in Extend, are created at Generator blocks and move from block to block by way of item connectors. Separate value connectors allow the attachment of a calculation to a block parameter or the retrieval of statistical information for reporting purposes. Input parameters can be changed interactively during a model run and can come from external sources. Outputs are displayed dynamically and in graphical and tabular format. The Industry and Suite products also provide an embedded database for centralized information management.

Extend provides iconic process-flow animation of the block diagram. For scaled 2-D animation, Proof Animation [Henriksen, 2002] from Wolverine Software is included in the Suite product. Collections of blocks representing a submodel, such as a subassembly line or functional process, can be grouped into a hierarchical block on the model worksheet; hierarchical blocks can also be stored in a library for reuse. Parameters from the submodel can be grouped and displayed at the level of the hierarchical block for access to model I/O. Extend supports the Microsoft component object model (COM/ActiveX), open database connectivity (ODBC), and Internet data exchange. Activity-based costing, statistical analysis of output data with confidence intervals, and the Evolutionary Optimizer are included.

For creating new blocks, Extend comes with a compiled C-like programming environment. The message-based language includes simulation-specific functions and supports custom interface development. Extend has an open architecture; in most cases, the source code for blocks is available for custom development. The architecture also supports linking to and using code and routines written in external languages.

4.7.4 Flexsim

Flexsim simulation software is developed and owned by Flexsim Software Products, Inc. of Orem, Utah (Nordgren, 2003). Flexsim is a discrete-event, object-oriented simulator developed in C++, using Open GL technology. Animation can be shown in tree view, 2-D, 3-D, and virtual reality. All views can be shown concurrently during the model development of run phase. It integrates Microsoft's Visual C++ IDE and compiler within a graphical 3-D click-and-drag simulation environment.

Flexsim software is used to build models that behave like the actual physical or conceptual systems they represent. A simulation model of any flow system or process can be created in Flexsim by using drag-and-drop model-building objects.

Flexsim is used to improve production efficiencies and reduce operating costs through simulation, experimentation, and optimization of dynamic flow systems. Engineers and managers use Flexsim to evaluate plant capacity, balance packaging and manufacturing lines, manage bottlenecks, solve work-in-process problems, justify capital expenditures, plan equipment maintenance schedules, establish proper inventory levels, improve order-picking systems, and optimize production rates. Flexsim allows end users to introduce and simulate new conditions for the model and to analyze their effects and results in order to find ways to improve the system being studied. By using Flexsim, efficiencies—increased throughput and decreased costs—can be identified, tested, and proven prior to implementing them in the actual system. The results of each simulation can be analyzed graphically through 3-D animation and through statistical reports and graphs, which are all also useful in communicating a model's purpose and results to both technical and nontechnical audiences.

4.7.5 Micro Saint

Micro Saint is offered by Micro Analysis and Design, Inc. [Bloechle and Schunk, 2003]. Micro Saint is a general-purpose, discrete-event, network simulation-software package for building models that simulate real-life processes. With Micro Saint models, users can gain useful information about processes that might be too expensive or time-consuming to test in the real world.

Micro Saint does not use the terminology or graphic representations of a specific industry. A Micro Saint model can be built for any process that can be represented by a flowchart diagram. The terms that are used are defined by the user. In addition, the icons and background for the ActionView animation and the flowcharting symbols are customizable. Micro Saint provides two views of the simulation model. The network diagram view shows the process flowchart in action, and ActionView provides a realistic 2-D picture of the process.

Micro Saint supports the development of models of various complexity to match the user's needs. Simple, functional models can be built by drawing a network diagram and filling in the task-timing information. More complex models can also be built that include dynamically changing variables, probabilistic and tactical branching logic, sorted queues, conditional task execution, animation, optimization, and extensive data collection.

A separate module (called COM Services) is available that enables Micro Saint to exchange data with other software applications and makes it easy to customize the model. In addition, OptQuest optimization is included with Micro Saint and is designed to automatically search for and find optimal or near-optimal solutions to the model.

4.7.6 ProModel

ProModel is offered by PROMODEL Corporation [Harrell, 2003]. It is a simulation and animation tool designed to model manufacturing systems. The company also offers MedModel for healthcare systems and ServiceModel for service systems. ProModel offers 2-D animation with an optional

3-D like perspective view. ProModel's animation is generated automatically as the model is developed.

ProModel has manufacturing-oriented modeling elements and rule-based decision logic. Some systems can be modeled by selecting from ProModel's set of highly parameterized modeling elements. In addition, its simulation programming language provides for modeling special situations not covered by the built-in choices.

The modeling elements in ProModel are parts (entities), locations, resources, path networks, routing and processing logic, and arrivals. Parts arrive and follow the routing and processing logic from location to location. Resources are used to represent people, tools, or vehicles that transport parts between locations, perform an operation on a part at a location, or perform maintenance on a location or other resource that is down. Resources may travel on path networks with given speeds, accelerations, and pickup and setdown travel times. The routing and processing element allows user-defined procedural logic in ProModel's simulation-programming language.

ProModel includes logic for automatically generating cost data associated with a process. Costs can be added for location usage, resources, and entities.

ProModel comes complete with an output viewer, allowing for straightforward data presentation and useful graphics and charts, such as state diagrams.

ProModel's runtime interface allows a user to define multiple scenarios for experimentation. SimRunner (discussed in Section 4.8.2) adds the capability to perform an optimization. It is based on an evolutionary-strategy algorithm, a variant of the genetic algorithm approach. The OptQuest Optimizer (OptQuest for ProModel) is available as an add-on product.

4.7.7 QUEST

QUEST® is offered by Delmia Corp. QUEST (Queuing Event Simulation Tool) is a manufacturing-oriented simulation package. QUEST combines an object-based, true 3-D simulation environment with a graphical user interface and material-flow modules for modeling labor, conveyors, automated guided vehicles, kinematic devices, cranes, fluids, power and free conveyors, and automated storage and retrieval systems. QUEST models incorporate 2-D and 3-D CAD geometry to create a virtual factory environment.

Delmia also offers a number of workcell simulators, including IGRIP® for robotic simulation and programming and ERGO™ for ergonomic analyses. Robots and human-based workcells that are simulated in IGRIP and ERGO can be imported into QUEST models both visually and numerically.

Delmia provides even further integration with QUEST and other manufacturing technologies through PROCESS ENGINEER™, Delmia's process-planning environment. The Manufacturing Hub infrastructure behind this software consists of an object-oriented database for storing Product, Process, and Resource objects that are configuration-managed and effectivity-controlled. A QUEST model is automatically created from the information stored in the database, and the resulting model can be linked to the database for automatic update purposes. QUEST can be used to introduce and update resource-specific information and model output results into the Manufacturing Hub for use in other products.

A QUEST model consists of elements from a number of element classes. Built-in element classes include AGVs and transporters, subresources, buffers, conveyors, power and free systems, labor, machines, parts, container parts, and processes. Each element has associated geometric data and parameters that define its behavior. Parts may have a route and control rules to govern part flow. Commonly needed behavior logic is selected from comprehensive menus, many parameter-driven.

For unique problems, Delmia's QUEST Simulation Control Language (SCL) can be used. This structured simulation-programming language provides distributed processing with access to all system variables. SCL allows expert users to define custom behaviors and to gain control over the simulation.

Delmia QUEST's open architecture allows the advanced user to perform batch simulation runs to automatically collect and tabulate data by using the Batch Control Language (BCL). Replications and parameter optimization are controlled with batch command files or by the OptQuest optimization software, as described in Section 4.8.2.

Output is available both numerically (with the statistical reporting mechanisms) and visually (with a resulting virtual factory-like animation). Statistical output results are available internally through the graphical user interface or externally through HTML and can be customized by using XML or QUEST's own BCL. Digital movies can be created from the animation, or a read-only encrypted version of the model can be authored for viewing and experimentation in QUEST Express™, a "lite" version of QUEST.

4.7.8 SIMUL8

SIMUL8 is provided by SIMUL8 Corporation and was first introduced in 1995. In SIMUL8, models are created by drawing the flow of work with the computer mouse, using a series of icons and arrows to represent the resources and queues in the system. Default values are provided for all properties of the icons, so that the animation can be viewed very early in the modeling process. Drilling down in property boxes opens up progressively more detailed properties. The main focus of SIMUL8 is service industries where people are processing transactions.

Like some other packages, SIMUL8 has the concepts of "Templates" and "Components." Templates, or prebuilt simulations, focus on particular recurring decision types that can be quickly parameterized to fit a specific company issue. Components are user-defined icons that can be reused and shared across a company's simulations. This reduces the time to build simulations, standardizes how some situation are handled across a corporation, and often removes much of the data-collection phase of a simulation study.

SIMUL8 Corporation's approach to business is different from most of the other packages here in that they claim to be aiming to spread simulation very widely across businesses, rather than concentrate it in the hands of dedicated and highly trained simulation professionals. This means they have very different pricing and support policies, but it also means the software has to contain features that watch how the product is being used and provide assistance if some potentially invalid analysis is conducted.

SIMUL8 saves its simulation model and data in XML format so that it will be easy to transfer it to and from other applications. It provides some nonsimulation features that make it possible for the model-builder to create custom user interfaces in spreadsheet, dialog, or wizard form. SIMUL8 has a VBA interface and supports ActiveX/COM so that external applications can build and control SIMUL8 simulations.

The product is available in two levels, Standard and Professional. The two levels provide the same simulation features, but Professional adds 3-D, "Virtual Reality" views of the simulation, and database links to corporate databases and has certain features that are likely to be useful only to full-time simulation modelers. SIMUL8 Professional comes with a license to distribute simulations with a free SIMUL8 Viewer.

4.7.9 WITNESS

WITNESS is offered by the Lanner Group and has separate versions for manufacturing and service industries. It contains many elements for discrete-part manufacturing and also contains elements for continuous processing, such as the flow of fluids through processors, tanks, and pipes.

WITNESS models are based on template elements. These may be customized and combined into module elements and templates for reuse. The standard machine elements can be single, batch, production, assembly, multistation, or multicycle. Other discrete modeling elements include multiple types of conveyor, tracks, vehicles, labor, and carriers. The behavior of each element is described on a tabbed detail form in the WITNESS user interface.

The models are displayed in a 2-D layout animation with multiple windows and display layers; there are optional process-flow displays and element-routing overlays. Models can be changed at any point in a model run and saved at any run point for future reload.

Optional WITNESS modules include WITNESS VR, an integrated virtual reality 3-D view of the working model, where there is full mouse control of the camera flight and position. Options exist, too, for postprocessed VR with multiscreen projection and various headset technologies. Other WITNESS modules include links to CAD systems, a model documentor, and the WITNESS Optimizer outlined in the section below.

WITNESS has object-model and ActiveX control for simulation embedding and includes direct data links to Microsoft Excel, MINITAB, and any OLEDB database source. XML data format saves offer additional linkage functionality.

4.8 EXPERIMENTATION AND STATISTICAL-ANALYSIS TOOLS

4.8.1 Common Features

Virtually all simulation packages offer various degrees of support for statistical analysis of simulation outputs. In recent years, many packages have added optimization as one of the analysis tools. To support analysis, most packages provide scenario definition, run-management capabilities, and data export to spreadsheets and other external applications.

Optimization is used to find a "near-optimal" solution. The user must define an objective or fitness function, usually a cost or cost-like function that incorporates the trade-off between additional throughput and additional resources. Until recently, the methods available for optimizing a system had difficulty coping with the random and nonlinear nature of most simulation outputs. Advances in the field of metaheuristics have offered new approaches to simulation optimization, ones based on artificial intelligence, neural networks, genetic algorithms, evolutionary strategies, tabu search, and scatter search.

4.8.2 Products

This section briefly discusses Arena's Output and Process Analyzer, AutoStat for AutoMod, OptQuest (which is used in a number of simulation products), and SimRunner for ProModel.

Arena's Output and Process Analyzer

Arena comes with the Output Analyzer and Process Analyzer. In addition, Arena uses OptQuest for optimization.

The Output Analyzer provides confidence intervals, comparison of multiple systems, and warm-up determination to reduce initial condition biases. It creates various plots, charts, and histograms, smoothes responses, and does correlation analysis. To compute accurate confidence intervals, it does internal batching (both within and across replications, with no user intervention) and data truncation to provide stationary, independent, and normally distributed data sets.

The Process Analyzer adds sophisticated scenario-management capabilities to Arena for comprehensive design of experiments. It allows a user to define scenarios, make the desired runs, and analyze the results. It allows an arbitrary number of controls and responses. Responses can be added after runs have been completed. It will rank scenarios by any response and provide summaries and statistical measures of the responses. A user can view 2-D and 3-D charts of response values across either replications or scenarios.

AutoStat

AutoStat is the run manager and statistical-analysis product in the AutoMod product family [Rohrer, 2003]. AutoStat provides a number of analyses, including warm-up determination for steady-state analysis, absolute and comparison confidence intervals, design of experiments, sensitivity analysis, and optimization via an evolutionary strategy. The evolutionary-strategies algorithm used by AutoStat is well suited to finding a near-optimal solution without getting trapped at a local optimum.

With AutoStat, an end user can define any number of scenarios by defining factors and their range of values. Factors include single parameters, such as resource capacity or vehicle speed; single cells in a data file; and complete data files. By allowing a data file to be a factor, a user can experiment with, for example, alternate production schedules, customer orders for different days, different labor schedules, or any other numerical inputs typically specified in a data file. Any standard or custom output can be designated as a response. For each defined response, AutoStat computes descriptive statistics (average, standard deviation, minimum, and maximum) and confidence intervals. New responses can be defined after runs are made, because AutoStat archives and compresses the standard and custom outputs from all runs. Various charts and plots are available to provide graphical comparisons.

AutoStat supports correlated sampling (see Chapter 12) using common random numbers. This sampling technique minimizes variation between paired samples, giving a better indication of the true effects of model changes.

AutoStat is capable of distributing simulation runs across a local area network and pulling back all results to the user's machine. Support for multiple machines and CPU's gives users the ability to make many more runs of the simulation than would otherwise be possible, by using idle machines during off hours. This is especially useful in multifactor analysis and optimization, both of which could require large numbers of runs. AutoStat also has a diagnostics capability that automatically detects "unusual" runs, where the definition of "unusual" is user-definable.

AutoStat also works with two other products from AutoSimulations: the AutoMod Simulator, a spreadsheet-based job-shop simulator; and AutoSched AP, a rule-based simulation package for finite-capacity scheduling in the semiconductor industry.

OptQuest

OptQuest® was developed by Dr. Fred Glover of the University of Colorado, cofounder of OptTek Systems, Inc. [April *et al.*, 2003].

OptQuest is based on a combination of methods: scatter search, tabu search, linear/integer programming, and neural networks. Scatter search is a population-based approach where existing solutions are combined to create new solutions. Tabu search is then superimposed to prohibit the search from reinvestigating previous solutions, and neural networks screen out solutions likely to be poor. The combination of methods allows the search process to escape local optimality in the quest for the best solution.

Some of the differences between OptTek's methods and other methods include

- the ability to avoid being trapped in locally optimal solutions to problems that contain nonlinearities (which commonly are present in real-world problems);
- the ability to handle nonlinear and discontinuous relationships that are not specifiable by the kinds of equations and formulas that are used in standard mathematical programming formulations;
- the ability to solve problems that involve uncertainties, such as those arising from uncertain supplies, demands, prices, costs, flow rates, and queuing rates.

SimRunner

SimRunner was developed by PROMODEL Corporation out of the simulation-optimization research of Royce Bowden, Mississippi State University [Harrell *et al.*, 2003]. It is available for ProModel, MedModel, and ServiceModel.

SimRunner uses genetic algorithms and evolution strategies, which are variants of evolutionary algorithms. Evolutionary algorithms are population-based direct-search techniques. A user first specifies input factors (integer or real-valued decision variables) composed of ProModel macros and then specifies an objective function composed of simulation-output responses. SimRunner manipulates the input factors within boundaries specified by the user seeking to minimize, to maximize, or to achieve a user-specified target value for the objective function. The optimization-output report includes a confidence interval on the mean value of the objective function for each solution evaluated over the course of the optimization and displays 3-D plots of the simulation's output-response surface for the solutions evaluated. In addition to the multivariable optimization module, SimRunner has a utility for helping users estimate the end of the warm-up phase (initialization bias) of a steady-state simulation and the number of replications needed to obtain an estimate of the objective function's mean value to within a specified percentage error and confidence level.

REFERENCES

APRIL, J., F. GLOVER, J. P. KELLY, AND M. LAGUNA [2003], "Practical Introduction to Simulation Optimization," *Proceedings of the 2003 Winter Simulation Conference*, S. Chick, P. J. Sánchez, D. Ferrin, and D. J. Morrice, eds., New Orleans, LA, Dec. 7–10, pp. 71–78.

BANKS, J., J. S. CARSON, AND J. N. SY [1995], *Getting Started with GPSS/H*, 2d ed., Wolverine Software Corporation, Annandale, VA.

BANKS, J. [1996], "Interpreting Software Checklists," *OR/MS Today*, June.

BAPAT, V. AND D. STURROCK [2003]. "The Arena Product Family: Enterprise Modeling Solutions," *Proceedings of the 2003 Winter Simulation Conference*, S. Chick, P. J. Sánchez, D. Ferrin, and D. J. Morrice, eds., New Orleans, LA, Dec. 7–10, pp. 210–217.

BLOECHLE, W. K., AND D. SCHUNK [2003], "Micro Saint Sharp Simulation Software," *Proceedings of the 2003 Winter Simulation Conference*, S. Chick, P. J. Sánchez, D. Ferrin, and D. J. Morrice, eds., New Orleans, LA, Dec. 7–10, pp. 182–187.

COWIE, J. [1999], "Scalable Simulation Framework API Reference Manual," www.ssfnet.org/SSFdocs/ssfapiManual.pdf.

CRAIN, R. C., AND J. O. HENRIKSEN [1999], "Simulation Using GPSS/H," *Proceedings of the 1999 Winter Simulation Conference*, P. A. Farrington, H. B. Nembhard, D. T. Sturrock, G. W. Evans, eds., Phoenix, AZ, Dec. 5–8, pp. 182–187.

HARRELL, C. R., B. K. GHOSH, AND R. BOWDEN [2003], *Simulation Using ProModel*, 2d ed., New York: McGraw–Hill.

HARRELL, C. R., AND R. N. PRICE [2003], "Simulation Modeling Using ProModel," *Proceedings of the 2003 Winter Simulation Conference*, S. Chick, P. J. Sánchez, D. Ferrin, and D. J. Morrice, eds., New Orleans, LA, Dec. 7–10, pp. 175–181.

HENRIKSEN, J. O. [1999], "General-Purpose Concurrent and Post-Processed Animation with Proof," *Proceedings of the 1999 Winter Simulation Conference*, P. A. Farrington, H. B. Nembhard, D. T. Sturrock, G. W. Evans, eds., Phoenix, AZ, Dec. 5–8, pp. 176–181.

KRAHL, D. [2003], "Extend: An Interactive Simulation Environment," *Proceedings of the 2003 Winter Simulation Conference*, S. Chick, P. J. Sánchez, D. Ferrin, and D. J. Morrice, eds., New Orleans, LA, Dec. 7–10, pp. 188–196.

MEHTA, A., AND I. RAWLS [1999], "Business Solutions Using Witness," *Proceedings of the 1999 Winter Simulation Conference*, P. A. Farrington, H. B. Nembhard, D. T. Sturrock, G. W. Evans, eds., Phoenix, AZ, Dec. 5–8, pp. 230–233.

NANCE, R. E. [1995], "Simulation Programming Languages: An Abridged History," Proceedings of the 1995 Winter Simulation Conference, X. Alexopoulos, K. Kang, W. R. Lilegdon, and D. Goldsman, eds., Arlington, VA, Dec. 13–16, pp. 1307–1313.

NORDGREN, W. B. [2003], "Flexsim Simulation Environment," *Proceedings of the 2003 Winter Simulation Conference*, S. Chick, P. J. Sánchez, D. Ferrin, and D. J. Morrice, eds., New Orleans, LA, Dec. 7–10, pp. 197–200.

PRITSKER, A. A. B., AND C. D. PEGDEN [1979], *Introduction to Simulation and SLAM*, John Wiley, New York.

ROHRER, M. W. [2003], "Maximizing Simulation ROI with AutoMod," *Proceedings of the 2003 Winter Simulation Conference*, S. Chick, P. J. Sánchez, D. Ferrin, and D. J. Morrice, eds., New Orleans., LA, Dec. 7–10, pp. 201–209.

SWAIN, J. J. [2003], "Simulation Reloaded: Sixth Biennial Survey of Discrete-Event Software Tools," *OR/MS Today*, August, Vol. 30, No. 4, pp. 46–57.

TOCHER, D. D., AND D. G. OWEN [1960], "The Automatic Programming of Simulations," *Proceedings of the Second International Conference on Operational Research*, J. Banbury and J. Maitland, eds., pp. 50–68.

WILSON, J. R., *et al.* [1992], "The Winter Simulation Conference: Perspectives of the Founding Fathers," *Proceedings of the 1992 Winter Simulation Conference*, J. Swain, D. Goldsman, R. C. Crain, and J. R. Wilson, eds., Arlington, VA, Dec. 13–16, pp.37–62.

EXERCISES

For the exercises below, reader should code the model in a general-purpose language (such as C, C++, or Java), a special-purpose simulation language (such as GPSS/H), or any desired simulation package.

Most problems contain activities that are uniformly distributed over an interval $[a, b]$. Assume that all values between a and b are possible; that is, the activity time is a *continuous* random variable.

The uniform distribution is denoted by $U(a, b)$, where a and b are the endpoints of the interval, or by $m \pm h$, where m is the mean and h is the "spread" of the distribution. These four parameters are related by the equations

$$m = \frac{a+b}{2} \quad h = \frac{b-a}{2}$$

$$a = m - h \quad b = m + h$$

Some of the uniform-random-variate generators available require specification of a and b; others require m and h.

Some problems have activities that are assumed to be normally distributed, as denoted by $N(\mu, \sigma^2)$, where μ is the mean and σ^2 the variance. (Since activity times are nonnegative, the normal distribution is appropriate only if $\mu \geq k\sigma$, where k is at least 4 and preferably 5 or larger. If a negative value is generated, it is discarded.) Other problems use the exponential distribution with some rate λ or mean $1/\lambda$. Chapter 5 reviews these distributions; Chapter 8 covers the generation of random variates having these distributions. All of the languages have a facility to easily generate samples from these distributions. For C, C++, or Java simulations, the student may use the functions given in Section 4.4 for generating samples from the normal and exponential distributions.

1. Make the necessary modifications to the Java model of the checkout counter (Example 4.2) so that the simulation will run for exactly 60 hours.

2. In addition to the change in Exercise 1, also assume that any customers still at the counter at time 60 hours will be served, but no arrivals after time 60 hours are allowed. Make the necessary changes to the Java code and run the model.

3. Implement the changes in Exercises 1 and 2 in any of the simulation packages.

4. Ambulances are dispatched at a rate of one every 15 ± 10 minutes in a large metropolitan area. Fifteen percent of the calls are false alarms, which require 12 ± 2 minutes to complete. All other calls can be one of two kinds. The first kind are classified as serious. They constitute 15% of the non-false alarm calls and take 25 ± 5 minutes to complete. The remaining calls

take 20 ± 10 minutes to complete. Assume that there are a very large number of available ambulances, and that any number can be on call at any time. Simulate the system until 500 calls are completed.

5. In Exercise 4, estimate the mean time that an ambulance takes to complete a call.

6. (a) In Exercise 4, suppose that there is only one ambulance available. Any calls that arrive while the ambulance is out must wait. Can one ambulance handle the work load? (b) Simulate with x ambulances, where $x = 1, 2, 3$, or 4, and compare the alternatives on the basis of length of time a call must wait, percentage of calls that must wait, and percentage of time the ambulance is out on call.

7. Passengers arrive at the security screening area at Chattahoochee Airport according to a time given by N(20, 3) seconds. At the first point, the boarding pass and ID are checked by one of two people in a time that is distributed N(12, 1) seconds. (Passengers always pick the shortest line when there is an option.) The next step is the X-ray area which takes a time that is N(15, 2) seconds; there are two lanes open at all times. Some 15% of the people have to be rechecked for a time that N(100, 10) seconds. The number of recheckers needed is to be determined. Simulate this system for eight hours with one and two recheckers.

8. A superhighway connects one large metropolitan area to another. A vehicle leaves the first city every 20 ± 15 seconds. Twenty percent of the vehicles have 1 passenger, 30% of the vehicles have 2 passengers, 10% have 3 passengers, and 10% have 4 passengers. The remaining 30% of the vehicles are buses, which carry 40 people. It takes 60 ± 10 minutes for a vehicle to travel between the two metropolitan areas. How long does it take for 5000 people to arrive in the second city?

9. People arrive at a meat counter at a rate of one every 25 ± 10 seconds. There are two sections: one for beef and one for pork. People want goods from them in the following proportion: beef only, 50%; pork only, 30%; beef and pork, 20%. It takes 45 ± 20 seconds for a butcher to serve one customer for one order. All customers place one order, except "beef and pork" customers place two orders. Assume that there are enough butchers available to handle all customers present at any time. How long does it take for 200 customers to be served?

10. In Exercise 9, what is the maximum number of butchers needed during the course of simulation? Would this number always be sufficient to guarantee that no customer ever had to wait?

11. In Exercise 9, simulate with x butchers, where $x = 1, 2, 3, 4$. When all butchers are busy, a line forms. For each value of x, estimate the mean number of busy butchers.

12. A one-chair unisex hair shop has arrivals at the rate of one every 20 ± 15 minutes. One-half of the arriving customers want a dry cut, 30% want a style, and 20% want a trim only. A dry cut takes 15 ± 5 minutes, a style cut takes 25 ± 10 minutes, and a trim takes 10 ± 3 minutes. Simulate 400 customers coming through the hair shop. Compare the given proportion of service requests of each type with the simulated outcome. Are the results reasonable?

13. An airport has two concourses. Concourse 1 passengers arrive at a rate of one every 15 ± 2 seconds. Concourse 2 passengers arrive at a rate of one every 10 ± 5 seconds. It takes 30 ± 5

seconds to walk down concourse 1 and 35 ± 10 seconds to walk down concourse 2. Both concourses empty into the main lobby, adjacent to the baggage claim. It takes 10 ± 3 seconds to reach the baggage claim area from the main lobby. Only 60% of the passengers go to the baggage claim area. Simulate the passage of 500 passengers through the airport system. How many of these passengers went through the baggage claim area? In this problem, the expected number through the baggage claim area can be computed by $0.60(500) = 300$. How close is the simulation estimate to the expected number? Why the difference?

14. In a multiphasic screening clinic, patients arrive at a rate of one every 5 ± 2 minutes to enter the audiology section. The examination takes 3 ± 1 minutes. Eighty percent of the patients were passed on to the next test with no problems. Of the remaining 20%, one-half require simple procedures that take 2 ± 1 minutes and are then sent for reexamination with the same probability of failure. The other half are sent home with medication. Simulate the system to estimate how long it takes to screen and pass 200 patients. (*Note:* Persons sent home with medication are not considered "passed.")

15. Consider a bank with four tellers. Tellers 3 and 4 deal only with business accounts; Tellers 1 and 2 deal only with general accounts. Clients arrive at the bank at a rate of one every 3 ± 1 minutes. Of the clients, 33% are business accounts. Clients randomly choose between the two tellers available for each type of account. (Assume that a customer chooses a line without regard to its length and does not change lines.) Business accounts take 15 ± 10 minutes to complete, and general accounts take 6 ± 5 minutes to complete. Simulate the system for 500 transactions to be completed. What percentage of time is each type of teller busy? What is the average time that each type of customer spends in the bank?

16. Repeat Exercise 15, but assuming that customers join the shortest line for the teller handling their type of account.

17. In Exercises 15 and 16, estimate the mean delay of business customers and of general customers. (Delay is time spent in the waiting line, and is exclusive of service time.) Also estimate the mean length of the waiting line, and the mean proportion of customers who are delayed longer than 1 minute.

18. Three different machines are available for machining a special type of part for 1 hour of each day. The processing-time data is as follows:

Machine	Time to Machine One Part
1	20 ± 4 seconds
2	10 ± 3 seconds
3	15 ± 5 seconds

Assume that parts arrive by conveyor at a rate of one every 15 ± 5 seconds for the first 3 hours of the day. Machine 1 is available for the first hour, machine 2 for the second hour, and machine 3 for the third hour of each day. How many parts are produced in a day? How large a storage area is needed for parts waiting for a machine? Do parts "pile up" at any particular time? Why?

19. People arrive at a self-service cafeteria at the rate of one every 30 ± 20 seconds. Forty percent go to the sandwich counter, where one worker makes a sandwich in 60 ± 30 seconds. The rest go to the main counter, where one server spoons the prepared meal onto a plate in 45 ± 30 seconds. All customers must pay a single cashier, which takes 25 ± 10 seconds. For all customers, eating takes 20 ± 10 minutes. After eating, 10% of the people go back for dessert, spending an additional 10 ± 2 minutes altogether in the cafeteria. Simulate until 100 people have left the cafeteria. How many people are left in the cafeteria, and what are they doing, at the time the simulation stops?

20. Thirty trucks carrying bits and pieces of a C-5N cargo plane leave Atlanta at the same time for the Port of Savannah. From past experience, it is known that it takes 6 ± 2 hours for a truck to make the trip. Forty percent of the drivers stop for coffee, which takes an additional 15 ± 5 minutes. (a) Model the situation as follows: For each driver, there is a 40% chance of stopping for coffee. When will the last truck reach Savannah? (b) Model it so that exactly 40% of the drivers stop for coffee. When will the last truck reach Savannah?

21. Customers arrive at the Last National Bank every 40 ± 35 seconds. Currently, the customers pick one of two tellers at random. A teller services a customer in 75 ± 25 seconds. Once a customer joins a line, the customer stays in that line until the transaction is complete. Some customers want the bank to change to the single-line method used by the Lotta Trust Bank. Which is the faster method for the customer? Simulate for one hour before collecting any summary statistics, then simulate for an 8-hour period. Compare the two queueing disciplines on the basis of teller utilization (percentage of busy time), mean delay of customers, and proportion of customers who must wait (before service begins) more than 1 minute, and more than 3 minutes.

22. Loana Tool Company rents chain saws. Customers arrive to rent chain saws at the rate of one every 30 ± 30 minutes. Dave and Betty handle these customers. Dave can rent a chain saw in 14 ± 4 minutes. Betty takes 10 ± 5 minutes. Customers returning chain saws arrive at the same rate as those renting chain saws. Dave and Betty spend 2 minutes with a customer to check in the returned chain saw. Service is first-come-first-served. When no customers are present, or Betty alone is busy, Dave gets these returned saws ready for re-renting. For each saw, this maintenance and cleanup takes him 6 ± 4 minutes and 10 ± 6 minutes, respectively. Whenever Dave is idle, he begins the next maintenance or cleanup. Upon finishing a maintenance or cleanup, Dave begins serving customers if one or more is waiting. Betty is always available for serving customers. Simulate the operation of the system starting with an empty shop at 8:00 A.M., closing the doors at 6:00 P.M., and getting chain saws ready for re-renting until 7:00 P.M. From 6:00 until 7:00 P.M., both Dave and Betty do maintenance and cleanup. Estimate the mean delay of customers who are renting chain saws.

23. In Exercise 22, change the shop rule regarding maintenance and cleanup to get a chain saw ready for re-rental. Now Betty does all this work. Upon finishing a cleanup on a saw, she helps Dave if a line is present. (That is, Dave and Betty both serve new customers and check in returned saws until Dave alone is busy, or the shop is empty.) Then she returns to her maintenance and cleanup duties. (a) Estimate the mean delay of customers who are renting chain saws. Compare the two shop rules on this basis. (b) Estimate the proportion of customers who must wait more than 5 minutes. Compare the two shop rules on this basis. (c) Discuss the pros and cons of the two criteria in parts (a) and (b) for comparing the two shop rules. Suggest other criteria.

24. U of LA (University of Lower Altoona) has only one computer for student use. Students arrive at the computer every 15 ± 10 minutes to use the computer for 12 ± 6 minutes. If the computer is busy, 60% will come back in 10 minutes to use the computer. If the computer is still busy, 50% (of the 60%) will return in 15 minutes. How many students fail to use the computer compared to 500 that actually finish? Demand and service occur 24 hours a day.

25. A warehouse holds 1000 cubic meters of cartons. These cartons come in three sizes: little (1 cubic meter), medium (2 cubic meters), and large (3 cubic meters). The cartons arrive at the following rates: little, every 10 ± 10 minutes; medium, every 15 minutes; and large, every 8 ± 8 minutes. If no cartons are removed, how long will it take to fill an empty warehouse?

26. Go Ape! buys a Banana II computer to handle all of its web-browsing needs. Web-browsing employees arrive every 10 ± 10 minutes to use the computer. Web-browsing takes 7 ± 7 minutes. The monkeys that run the computer cause a system failure every 60 ± 60 minutes. The failure lasts for 8 ± 4 minutes. When a failure occurs, the web-browsing that was being done resumes processing from where it was left off. Simulate the operation of this system for 24 hours. Estimate the mean system response time. (A system response time is the length of time from arrival until web-browsing is completed.) Also estimate the mean delay for those web-browsing employees that are in service when a computer system failure occurs.

27. Able, Baker, and Charlie are three carhops at the Sonic Drive-In (service at the speed of sound!). Cars arrive every 5 ± 5 minutes. The carhops service customers at the rate of one every 10 ± 6 minutes. However, the customers prefer Able over Baker, and Baker over Charlie. If the carhop of choice is busy, the customers choose the first available carhop. Simulate the system for 1000 service completions. Estimate Able's, Baker's, and Charlie's utilization (percentage of time busy).

28. Jiffy Car Wash is a five-stage operation that takes 2 ± 1 minutes for each stage. There is room for 6 cars to wait to begin the car wash. The car wash facility holds 5 cars, which move through the system in order, one car not being able to move until the car ahead of it moves. Cars arrive every 2.5 ± 2 minutes for a wash. If the car cannot get into the system, it drives across the street to Speedy Car Wash. Estimate the balking rate per hour. That is, how many cars drive off per hour? Simulate for one 12-hour day.

29. Consider the three machines A, B, and C pictured below. Arrivals of parts and processing times are as indicated (times in minutes).

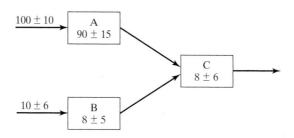

Machine A processes type I parts, machine B processes type II parts, and machine C processes both types of parts. All machines are subject to random breakdown: machine A every 400 ± 350 minutes with a down time of 15 ± 14 minutes, machine B every 200 ± 150 minutes with a downtime of 10 ± 8 minutes, and machine C almost never, so its downtime is ignored. Parts from machine A are processed at machine C as soon as possible, ahead of any type II parts from machine B. When machine A breaks down, any part in it is sent to machine B and processed as soon as B becomes free, but processing begins over again, taking 100 ± 20 minutes. Again, type I parts from machine A are processed ahead of any parts waiting at B, but after any part currently being processed. When machine B breaks down, any part being processed resumes processing as soon as B becomes available. All machines handle one part at a time. Make two independent replications of the simulation. Each replication will consist of an 8-hour initialization phase to load the system with parts, followed by a 40-hour steady-state run. (Independent replications means that each run uses a different stream of random numbers.) Management is interested in the long-run throughput [i.e., the number of parts of each type (I and II) produced per 8-hour day], long-run utilization of each machine, and the existence of bottlenecks (long "lines" of waiting parts, as measured by the queue length at each machine). Report the output data in a table similar to the following:

	Run 1	Run 2	Average of 2 runs
Utilization A			
Utilization B			
Etc.			

Include a brief statement summarizing the important results.

30. Workers come to an internal supply store at the rate of one every 10 ± 4 minutes. Their requisitions are handled by one of three clerks; a clerk takes 22 ± 10 minutes to handle a

requisition. All requisitions are then passed to a single cashier, who spends 7 ± 6 minutes per requisition. Simulate the system for 120 hours. (a) Estimate the utilization of each clerk, based on the 120-hour simulation. (b) How many workers are completely served? How many do the three clerks serve? How many workers arrive? Are all three clerks ever busy at the same time? What is the average number of busy clerks?

31. People arrive at a barbershop at the rate of one every 4.5 minutes. If the shop is full (it can hold five people altogether), 30% of the potential customers leave and come back in 60 ± 20 minutes. The others leave and do not return. One barber gives a haircut in 8 ± 2 minutes; the second talks a lot and takes 12 ± 4 minutes. If both barbers are idle, a customer prefers the first barber. (Treat customers trying to reenter the shop as if they are new customers.) Simulate this system until 300 customers have received a haircut. (a) Estimate the balking rate, that is, the number turned away per minute. (b) Estimate the number turned away per minute who do not try again. (c) What is the average time spent in the shop? (d) What is the average time spent getting a haircut (not including delay)? (e) What is the average number of customers in the shop?

32. People arrive at a microscope exhibit at a rate of one every 8 ± 2 minutes. Only one person can see the exhibit at a time. It takes 5 ± 2 minutes to see the exhibit. A person can buy a "privilege" ticket for $1 which gives him or her priority in line over those who are too cheap to spend the buck. Some 50% of the viewers are willing to do this, but they make their decision to do so only if one or more people are in line when they arrive. The exhibit is open continuously from 10:00 A.M. to 4:00 P.M. Simulate the operation of the system for one complete day. How much money is generated from the sale of privilege tickets?

33. Two machines are available for drilling parts (A-type and B-type). A-type parts arrive at a rate of one every 10 ± 3 minutes, B-type parts at a rate of one every 3 ± 2 minutes. For B-type parts, workers choose an idle machine, or if both drills, the Dewey and the Truman, are busy, they choose a machine at random and stay with their choice. A-type parts must be drilled as soon as possible; therefore, if a machine is available, preferably the Dewey, it is used; otherwise the part goes to the head of the line for the Dewey drill. All jobs take 4 ± 3 minutes to complete. Simulate the completion of 100 A-type parts. Estimate the mean number of A-type parts waiting to be drilled.

34. A computer center has two color printers. Students arrive at a rate of one every 8 ± 2 minutes to use the color printer. They can be interrupted by professors, who arrive at a rate of one every 12 ± 2 minutes. There is one systems analyst who can interrupt anyone, but students are interrupted before professors. The systems analyst spends 6 ± 4 minutes on the color printer and then returns in 20 ± 5 minutes. Professors and students spend 4 ± 2 minutes on the color printer. If a person is interrupted, that person joins the head of the queue and resumes service as soon as possible. Simulate for 50 professor-or-analyst jobs. Estimate the interruption rate per hour, and the mean length of the waiting line of students.

35. Parts are machined on a drill press. They arrive at a rate of one every 5 ± 3 minutes, and it takes 3 ± 2 minutes to machine them. Every 60 ± 60 minutes, a rush job arrives, which takes

12 ± 3 minutes to complete. The rush job interrupts any nonrush job. When the regular job returns to the machine, it stays only for its remaining process time. Simulate the machining of 10 rush jobs. Estimate the mean system response time for each type of part. (A response time is the total time that a part spends in the system.)

36. E-mail messages are generated at a rate of one every 35 ± 10 seconds for transmission one at a time. Transmission takes 20 ± 5 seconds. At intervals of 6 ± 3 minutes, urgent e-mail messages with transmission times of 10 ± 3 seconds take over the transmission line. Any e-mail message in progress must be reprocessed for 2 minutes before it can be resubmitted for transmission. When resubmitted, it goes to the head of the line. Simulate for 90 minutes. Estimate the percentage of time the line is busy with ordinary e-mail messages.

37. A worker packs boxes that arrive at a rate of one every 15 ± 3 minutes. It takes 10 ± 3 minutes to pack a box. Once every hour, the worker is interrupted to wrap specialty orders that take 16 ± 3 minutes to pack. Upon completing the specialty order, the worker then completes the interrupted order. Simulate for 40 hours. Estimate the mean proportion of time that the number of boxes waiting to be packed is more than five and the average number of boxes waiting to be packed.

38. A patient arrives at the Emergency Room at Hello-Hospital about every 40 ± 19 minutes. Each patient will be treated by either Doctor Slipup or Doctor Gutcut. Twenty percent of the patients are classified as NIA (need immediate attention) and the rest as CW (can wait). NIA patients are given the highest priority (3), see a doctor as soon as possible for 40 ± 37 minutes, but then their priority is reduced to 2 and they wait until a doctor is free again, when they receive further treatment for 30 ± 25 minutes and are then discharged. CW patients initially receive the priority 1 and are treated (when their turn comes) for 15 ± 14 minutes; their priority is then increased to 2, they wait again until a doctor is free and receive 10 ± 8 minutes of final treatment, and are then discharged. Simulate for 20 days of continuous operation, 24 hours per day. Precede this by a 2-day initialization period to load the system with patients. Report conditions at times 0 days, 2 days, and 22 days. Does a 2-day initialization appear long enough to load the system to a level reasonably close to steady-state conditions? (a) Measure the average and maximum queue length of NIA patients from arrival to first seeing a doctor. What percent do not have to wait at all? Also tabulate and plot the distribution of this initial waiting time for NIA patients. What percent wait less than 5 minutes before seeing a doctor? (b) Tabulate and plot the distribution of total time in system for all patients. Estimate the 90% quantile—that is, 90% of the patients spend less than x amount of time in the system. Estimate x. (c) Tabulate and plot the distribution of remaining time in system from after the first treatment to discharge, for all patients. Estimate the 90% quantile. (*Note:* Most simulation packages provide the facility to automatically tabulate the distribution of any specified variable.)

39. People arrive at a newspaper stand with an interarrival time that is exponentially distributed with a mean of 0.5 minute. Fifty-five percent of the people buy just the morning paper, 25% buy the morning paper and a *Wall Street Journal*. The remainder buy only the *Wall Street*

Journal. One clerk handles the *Wall Street Journal* sales, another clerk morning-paper sales. A person buying both goes to the *Wall Street Journal* clerk. The time it takes to serve a customer is normally distributed with a mean of 40 seconds and a standard deviation of 4 seconds for all transactions. Collect statistics on queues for each type of transaction. Suggest ways for making the system more efficient. Simulate for 4 hours.

40. Bernie remodels houses and makes room additions. The time it takes to finish a job is normally distributed with a mean of 17 elapsed days and a standard deviation of 3 days. Homeowners sign contracts for jobs at exponentially distributed intervals having a mean of 20 days. Bernie has only one crew. Estimate the mean waiting time (from signing the contract until work begins) for those jobs where a wait occurs. Also estimate the percentage of time the crew is idle. Simulate until 100 jobs have been completed.

41. Parts arrive at a machine in random fashion with exponential interarrival times having a mean of 60 seconds. All parts require 5 seconds to prepare and align for machining. There are three different types of parts, in the proportions shown below. The times to machine each type of part are normally distributed with mean and standard deviation as follows:

Part Type	Percent	Mean (Seconds)	σ (Seconds)
1	50	48	8
2	30	55	9
3	20	85	12

Find the distribution of total time to complete processing for all types of parts. What proportion of parts take more than 60 seconds for complete processing? How long do parts have to wait, on the average? Simulate for one 8-hour day.

42. Shopping times at a department store have been found to have the following distribution:

Shopping Time (Minutes)	Number of Shoppers
0–10	90
10–20	120
20–30	270
30–40	145
40–50	88
50–60	28

After shopping, the customers choose one of six checkout counters. Checkout times are normally distributed with a mean of 5.1 minutes and a standard deviation of 0.7 minutes. Interarrival times are exponentially distributed with a mean of 1 minute. Gather statistics for each checkout counter (including the time waiting for checkout). Tabulate the distribution of time to complete shopping and the distribution of time to complete shopping and checkout procedures. What proportion of customers spend more than 45 minutes in the store? Simulate for one 16-hour day.

43. The interarrival time for parts needing processing is given as follows:

Interarrival Time (Seconds)	Proportion
10–20	0.20
20–30	0.30
30–40	0.50

There are three types of parts: A, B, and C. The proportion of each part, and the mean and standard deviation of the normally distributed processing times are as follows:

Part Type	Proportion	Mean	Standard Deviation
A	0.5	30 seconds	3 seconds
B	0.3	40 seconds	4 seconds
C	0.2	50 seconds	7 seconds

Each machine processes any type of part, one part at a time. Use simulation to compare one with two with three machines working in parallel. What criteria would be appropriate for such a comparison?

44. Orders are received for one of four types of parts. The interarrival time between orders is exponentially distributed with a mean of 10 minutes. The table that follows shows the proportion of the parts by type and the time to fill each type of order by the single clerk.

Part Type	Percentage	Service Time (Minutes)
A	40	$N(6.1, 1.3)$
B	30	$N(9.1, 2.9)$
C	20	$N(11.8, 4.1)$
D	10	$N(15.1, 4.5)$

Orders of types A and B are picked up immediately after they are filled, but orders of types C and D must wait 10 ± 5 minutes to be picked up. Tabulate the distribution of time to complete delivery for all orders combined. What proportion take less than 15 minutes? What proportion take less than 25 minutes? Simulate for an 8-hour initialization period, followed by a 40-hour run. Do not use any data collected in the 8-hour initialization period.

45. Three independent widget-producing machines all require the same type of vital part, which needs frequent maintenance. To increase production it is decided to keep two spare parts on hand (for a total of $2+3 = 5$ parts). After 2 hours of use, the part is removed from the machine and taken to a single technician, who can do the required maintenance in 30 ± 20 minutes. After maintenance, the part is placed in the pool of spare parts, to be put into the first machine that requires it. The technician has other duties, namely, repairing other items which have a higher priority and which arrive every 60 ± 20 minutes requiring 15 ± 15 minutes to repair. Also, the technician takes a 15-minute break in each 2-hour time period. That is, the technician works 1 hour 45 minutes, takes off 15 minutes, works 1 hour 45 minutes, takes off 15 minutes, and so on. (a) What are the model's initial conditions—that is, where are the parts at time 0 and what is their condition? Are these conditions typical of "steady state"? (b) Make each replication of this experiment consist of an 8-hour initialization phase followed by a 40-hour data-collection phase. Make four statistically independent replications of the experiment all in one computer run (i.e., make four runs with each using a different set of random numbers). (c) Estimate the mean number of busy machines and the proportion of time the technician is busy. (d) Parts are estimated to cost the company $50 per part per 8-hour day (regardless of how much they are in use). The cost of the technician is $20 per hour. A working machine produces widgets worth $100 for each hour of production. Develop an expression to represent total cost per hour which can be attributed to widget production (i.e., not all of the technician's time is due to widget production). Evaluate this expression, given the results of the simulation.

46. The Wee Willy Widget Shop overhauls and repairs all types of widgets. The shop consists of five work stations, and the flow of jobs through the shop is as depicted here:

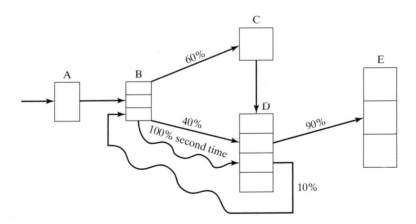

Regular jobs arrive at station A at the rate of one every 15 ± 13 minutes. Rush jobs arrive every 4 ± 3 hours and are given a higher priority except at station C, where they are put on a conveyor and sent through a cleaning and degreasing operation along with all other jobs. For jobs the first time through a station, processing and repair times are as follows:

Station	Number Machines or Workers	Processing and/or Repair Times (Minutes)	Description
A	1	12 ± 2	Receiving clerk
B	3	40 ± 20	Disassembly and parts replacement
C	1	20	Degreaser
D	4	50 ± 40	Reassembly and adjustments
E	3	40 ± 5	Packing and shipping

The times listed above hold for all jobs that follow one of the two sequences $A \longrightarrow B \longrightarrow C \longrightarrow D \longrightarrow E$ or $A \longrightarrow B \longrightarrow D \longrightarrow E$. However, about 10% of the jobs coming out of station D are sent back to B for further work (which takes 30 ± 10 minutes) and then are sent to D and finally to E. The path of these jobs is as follows:

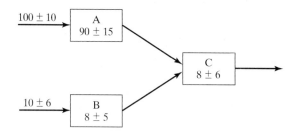

Every 2 hours, beginning 1 hour after opening, the degreasing station C shuts down for routine maintenance, which takes 10 ± 1 minute. However, this routine maintenance does not begin until the current widget, if any, has completed its processing. (a) Make three independent replications of the simulation model, where one replication equals an 8-hour simulation run, preceded by a 2-hour initialization run. The three sets of output represent three typical days. The main performance measure of interest is mean response time per job, where a response time is the total time a job spends in the shop. The shop is never empty in the morning, but the model will be empty without the initialization phase. So run the model for a 2-hour initialization period and collect statistics from time 2 hours to time 10 hours. This "warm-up" period will reduce the downward bias in the estimate of mean response time. Note that the 2-hour warm-up is a device to load a simulation model to some more realistic level than empty. From each of the three independent replications, obtain an estimate of mean response time. Also obtain an overall estimate, the sample average of the three estimates. (b) Management is considering putting one additional worker at the busiest station (A, B, D, or E). Would this significantly improve mean response time? (c) As an alternative to part (b), management is considering replacing machine C with a faster one that processes a widget in only 14 minutes. Would this significantly improve mean response time?

47. A building-materials firm loads trucks with two payloader tractors. The distribution of truck-loading times has been found to be exponential with a mean loading time of 6 minutes. The truck interarrival time is exponentially distributed with an arrival rate of 16 per hour. The waiting time of a truck and driver is estimated to cost $50 per hour. How much (if any) could the firm save (per 10 hour day) if an overhead hopper system that would fill any truck in a constant time of 2 minutes is installed? (Assume that the present tractors could and would adequately service the conveyors loading the hoppers.)

48. A milling-machine department has 10 machines. The runtime until failure occurs on a machine is exponentially distributed with a mean of 20 hours. Repair times are uniformly distributed between 3 and 7 hours. Select an appropriate run length and appropriate initial conditions. (a) How many repair persons are needed to ensure that the mean number of machines running is greater than eight? (b) If there are two repair persons, estimate the number of machines that are either running or being served.

49. Forty people are waiting to pass through a turnstile that takes 2.5 ± 1.0 seconds to traverse. Simulate this system 10 times, each one independent of the others, and determine the range and the average time for all 40 people to traverse.

50. People borrow *Gone with the Wind* from the local library and keep it for 21 ± 10 days. There is only one copy of the book in the library. You are the sixth person on the reservation list (five are ahead of you). Simulate 50 independent cycles of book borrowing to determine the probability that you will receive the book within 100 days.

51. Jobs arrive every 300 ± 30 seconds to be processed through a process that consists of four operations: OP10 requires 50 ± 20 seconds, OP20 requires 70 ± 25 seconds, OP30 requires 60 ± 15 seconds, OP40 requires 90 ± 30 seconds. Simulate this process until 250 jobs are completed; then combine the four operations of the job into one with the distribution 240 ± 100 seconds and simulate the process with this distribution. Does the average time in the system change for the two alternatives?

52. Two types of jobs arrive to be processed on the same machine. Type 1 jobs arrive every 50 ± 30 seconds and require 35 ± 20 seconds for processing. Type 2 jobs arrive every 100 ± 40 seconds and require 20 ± 15 seconds for processing. For an 8-hour simulation, what is the average number of jobs waiting to be processed?

53. Two types of jobs arrive to be processed on the same machine. Type 1 jobs arrive every 80 ± 30 seconds and require 35 ± 20 seconds for processing. Type 2 jobs arrive every 100 ± 40 seconds and require 20 ± 15 seconds for processing. Engineering has judged that there is excess capacity on the machine. For a simulation of 8 hours of operation of the system, find X for Type 3 jobs that arrive every $X \pm 0.4X$ seconds and require a time of 30 seconds on the machine so that the average number of jobs waiting to be processed is two or less.

54. Using spreadsheet software, generate 1000 uniformly distributed random values with mean 10 and spread 2. Plot these values with intervals of width 0.5 between 8 and 12. How close did the simulated set of values come to the expected number in each interval?

55. Using a spreadsheet, generate 1000 exponentially distributed random values with a mean of 10. What is the maximum of the simulated values? What fraction of the generated values is less than the mean of 10? Plot a histogram of the generated values. (Hint: If you cannot find an exponential generator in the spreadsheet you use, use the formula $-10*LOG(1-R)$, where R is a uniformly distributed random number from 0 to 1 and LOG is the natural logarithm. The rationale for this formula is explained in Chapter 8 on random-variate generators.)

Part II

Mathematical and Statistical Models

5

Statistical Models in Simulation

In modeling real-world phenomena, there are few situations where the actions of the entities within the system under study can be predicted completely. The world the model-builder sees is probabilistic rather than deterministic. There are many causes of variation. The time it takes a repairperson to fix a broken machine is a function of the complexity of the breakdown, whether the repairperson brought the proper replacement parts and tools to the site, whether another repairperson asks for assistance during the course of the repair, whether the machine operator receives a lesson in preventive maintenance, and so on. To the model-builder, these variations appear to occur by chance and cannot be predicted. However, some statistical model might well describe the time to make a repair.

An appropriate model can be developed by sampling the phenomenon of interest. Then, through educated guesses (or using software for the purpose), the model-builder would select a known distribution form, make an estimate of the parameter(s) of this distribution, and then test to see how good a fit has been obtained. Through continued efforts in the selection of an appropriate distribution form, a postulated model could be accepted. This multistep process is described in Chapter 9.

Section 5.1 contains a review of probability terminology and concepts. Some typical applications of statistical models, or distribution forms, are given in Section 5.2. Then, a number of selected discrete and continuous distributions are discussed in Sections 5.3 and 5.4. The selected distributions are those that describe a wide variety of probabilistic events and, further, appear in different contexts in other chapters of this text. Additional discussion about the distribution forms appearing in this chapter, and about distribution forms mentioned but not described, is available from a number of sources [Hines and Montgomery, 1990; Ross, 2002; Papoulis, 1990;

Devore, 1999; Walpole and Myers, 2002; Law and Kelton, 2000]. Section 5.5 describes the Poisson process and its relationship to the exponential distribution. Section 5.6 discusses empirical distributions.

5.1 REVIEW OF TERMINOLOGY AND CONCEPTS

1. *Discrete random variables.* Let X be a random variable. If the number of possible values of X is finite, or countably infinite, X is called a discrete random variable. The possible values of X may be listed as $x_1, x_2, \ldots$. In the finite case, the list terminates; in the countably infinite case, the list continues indefinitely.

Example 5.1

The number of jobs arriving each week at a job shop is observed. The random variable of interest is X, where

$$X = \text{number of jobs arriving each week}$$

The possible values of X are given by the range space of X, which is denoted by R_X. Here $R_X = \{0, 1, 2, \ldots\}$.

Let X be a discrete random variable. With each possible outcome x_i in R_X, a number $p(x_i) = P(X = x_i)$ gives the probability that the random variable equals the value of x_i. The numbers $p(x_i), i = 1, 2, \ldots$, must satisfy the following two conditions:

1. $p(x_i) \geq 0$, for all i
2. $\sum_{i=1}^{\infty} p(x_i) = 1$

The collection of pairs $(x_i, p(x_i)), i = 1, 2, \ldots$ is called the probability distribution of X, and $p(x_i)$ is called the probability mass function (pmf) of X.

Example 5.2

Consider the experiment of tossing a single die. Define X as the number of spots on the up face of the die after a toss. Then $R_X = \{1, 2, 3, 4, 5, 6\}$. Assume the die is loaded so that the probability that a given face lands up is proportional to the number of spots showing. The discrete probability distribution for this random experiment is given by

x_i	1	2	3	4	5	6
$p(x_i)$	1/21	2/21	3/21	4/21	5/21	6/21

The conditions stated earlier are satisfied—that is, $p(x_i) \geq 0$ for $i = 1, 2, \ldots, 6$ and $\sum_{i=1}^{\infty} p(x_i) = 1/21 + \cdots + 6/21 = 1$. The distribution is shown graphically in Figure 5.1.

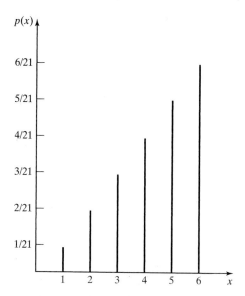

Figure 5.1   Probability mass function for loaded-die example.

2. *Continuous random variables.* If the range space R_X of the random variable X is an interval or a collection of intervals, X is called a continuous random variable. For a continuous random variable X, the probability that X lies in the interval $[a, b]$ is given by

$$P(a \leq X \leq b) = \int_a^b f(x)\, dx \tag{5.1}$$

The function $f(x)$ is called the probability density function (pdf) of the random variable X. The pdf satisfies the following conditions:

a. $f(x) \geq 0$ for all x in R_X
b. $\int_{R_X} f(x)dx = 1$
c. $f(x) = 0$ if x is not in R_X

As a result of Equation (5.1), for any specified value x_0, $P(X = x_0) = 0$, because

$$\int_{x_0}^{x_0} f(x)\, dx = 0$$

$P(X = x_0) = 0$ also means that the following equations hold:

$$P(a \leq X \leq b) = P(a < X \leq b) = P(a \leq X < b) = P(a < X < b) \tag{5.2}$$

The graphical interpretation of Equation (5.1) is shown in Figure 5.2. The shaded area represents the probability that X lies in the interval $[a, b]$.

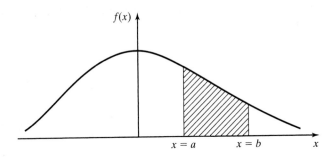

Figure 5.2   Graphical interpretation of $P(a < X < b)$.

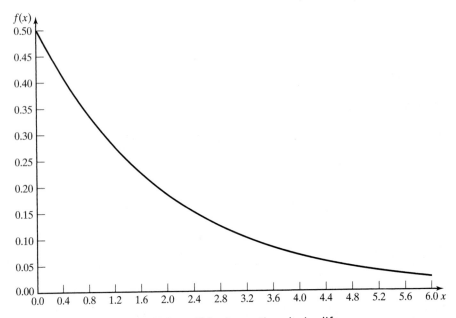

Figure 5.3   pdf for inspection-device life.

Example 5.3

The life of a device used to inspect cracks in aircraft wings is given by X, a continuous random variable assuming all values in the range $x \geq 0$. The pdf of the lifetime, in years, is as follows:

$$f(x) = \begin{cases} \frac{1}{2}e^{-x/2}, & x \geq 0 \\ 0, & \text{otherwise} \end{cases}$$

This pdf is shown graphically in Figure 5.3. The random variable X is said to have an exponential distribution with mean 2 years.

The probability that the life of the device is between 2 and 3 years is calculated as

$$P(2 \leq X \leq 3) = \frac{1}{2} \int_2^3 e^{-x/2}\, dx$$

$$= -e^{-3/2} + e^{-1} = -0.223 + 0.368 = 0.145$$

3. *Cumulative distribution function.* The cumulative distribution function (cdf), denoted by $F(x)$, measures the probability that the random variable X assumes a value less than or equal to x, that is, $F(x) = P(X \leq x)$.

If X is discrete, then

$$F(x) = \sum_{\substack{\text{all} \\ x_i \leq x}} p(x_i) \tag{5.3}$$

If X is continuous, then

$$F(x) = \int_{-\infty}^{x} f(t)\, dt \tag{5.4}$$

Some properties of the cdf are listed here:

a. F is a nondecreasing function. If $a < b$, then $F(a) \leq F(b)$.
b. $\lim_{x \to \infty} F(x) = 1$
c. $\lim_{x \to -\infty} F(x) = 0$

All probability questions about X can be answered in terms of the cdf. For example,

$$P(a < X \leq b) = F(b) - F(a) \quad \text{for all } a < b \tag{5.5}$$

For continuous distributions, not only does Equation (5.5) hold, but also the probabilities in Equation (5.2) are equal to $F(b) - F(a)$.

Example 5.4 _____
The die-tossing experiment described in Example 5.2 has a cdf given as follows:

x	$(-\infty, 1)$	$[1, 2)$	$[2, 3)$	$[3, 4)$	$[4, 5)$	$[5, 6)$	$[6, \infty)$
$F(x)$	0	1/21	3/21	6/21	10/21	15/21	21/21

where $[a, b) = \{a \leq x < b\}$. The cdf for this example is shown graphically in Figure 5.4.

If X is a discrete random variable with possible values $x_1, x_2, \ldots$, where $x_1 < x_2 < \ldots$, the cdf is a step function. The value of the cdf is constant in the interval $[x_{i-1}, x_i)$ and then takes a step, or jump, of size $p(x_i)$ at x_i. Thus, in Example 5.4, $p(3) = 3/21$, which is the size of the step when $x = 3$.

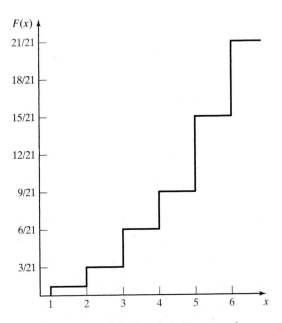

Figure 5.4   cdf for loaded-die example.

Example 5.5

The cdf for the device described in Example 5.3 is given by

$$F(x) = \frac{1}{2} \int_0^x e^{-t/2} dt = 1 - e^{-x/2}$$

The probability that the device will last for less than 2 years is given by

$$P(0 \leq X \leq 2) = F(2) - F(0) = F(2) = 1 - e^{-1} = 0.632$$

The probability that the life of the device is between 2 and 3 years is calculated as

$$P(2 \leq X \leq 3) = F(3) - F(2) = (1 - e^{-3/2}) - (1 - e^{-1})$$
$$= -e^{-3/2} + e^{-1} = -0.223 + 0.368 = 0.145$$

as found in Example 5.3.

4. *Expectation.* An important concept in probability theory is that of the expectation of a random variable. If X is a random variable, the expected value of X, denoted by $E(X)$, for discrete and continuous variables is defined as follows:

$$E(X) = \sum_{\text{all } i} x_i p(x_i) \qquad \text{if } X \text{ is discrete} \tag{5.6}$$

and

$$E(X) = \int_{-\infty}^{\infty} x f(x)\, dx \qquad \text{if } X \text{ is continuous} \tag{5.7}$$

The expected value $E(X)$ of a random variable X is also referred to as the mean, μ, or the first moment of X. The quantity $E(X^n), n \geq 1$, is called the nth moment of X, and is computed as follows:

$$E(X^n) = \sum_{\text{all } i} x_i^n p(x_i) \qquad \text{if } X \text{ is discrete} \tag{5.8}$$

and

$$E(X^n) = \int_{-\infty}^{\infty} x^n f(x)\, dx \qquad \text{if } X \text{ is continuous} \tag{5.9}$$

The variance of a random variable, X, denoted by $V(X)$ or var(X) or σ^2, is defined by

$$V(X) = E[(X - E[X])^2]$$

A useful identity in computing $V(X)$ is given by

$$V(X) = E(X^2) - [E(X)]^2 \tag{5.10}$$

The mean $E(X)$ is a measure of the central tendency of a random variable. The variance of X measures the expected value of the squared difference between the random variable and its mean. Thus, the variance, $V(X)$, is a measure of the spread or variation of the possible values of X around the mean $E(X)$. The standard deviation, σ, is defined to be the square root of the variance, σ^2. The mean, $E(X)$, and the standard deviation, $\sigma = \sqrt{V(X)}$, are expressed in the same units.

Example 5.6
The mean and variance of the die-tossing experiment described in Example 5.2 are computed as follows:

$$E(X) = 1\left(\frac{1}{21}\right) + 2\left(\frac{2}{21}\right) + \cdots + 6\left(\frac{6}{21}\right) = \frac{91}{21} = 4.33$$

To compute $V(X)$ from Equation (5.10), first compute $E(X^2)$ from Equation (5.8) as follows:

$$E(X^2) = 1^2\left(\frac{1}{21}\right) + 2^2\left(\frac{2}{21}\right) + \cdots + 6^2\left(\frac{6}{21}\right) = 21$$

Thus,

$$V(X) = 21 - \left(\frac{91}{21}\right)^2 = 21 - 18.78 = 2.22$$

and

$$\sigma = \sqrt{V(X)} = 1.49$$

Example 5.7

The mean and variance of the life of the device described in Example 5.3 are computed as follows:

$$E(X) = \frac{1}{2} \int_0^\infty x e^{-x/2} \, dx = -x e^{-x/2} \Big|_0^\infty + \int_0^\infty e^{-x/2} \, dx$$

$$= 0 + \frac{1}{1/2} e^{-x/2} \Big|_0^\infty = 2 \text{ years}$$

To compute $V(X)$ from Equation (5.10), first compute $E(X^2)$ from Equation (5.9) as follows:

$$E(X^2) = \frac{1}{2} \int_0^\infty x^2 e^{-x/2} \, dx$$

Thus,

$$E(X^2) = -x^2 e^{-x/2} \Big|_0^\infty + 2 \int_0^\infty x e^{-x/2} \, dx = 8$$

giving

$$V(X) = 8 - 2^2 = 4 \text{ years}^2$$

and

$$\sigma = \sqrt{V(X)} = 2 \text{ years}$$

With a mean life of 2 years and a standard deviation of 2 years, most analysts would conclude that actual lifetimes, X, have a fairly large variability.

5. *The mode.* The mode is used in describing several statistical models that appear in this chapter. In the discrete case, the mode is the value of the random variable that occurs most frequently. In the continuous case, the mode is the value at which the pdf is maximized. The mode might not be unique; if the modal value occurs at two values of the random variable, the distribution is said to be bimodal.

5.2 USEFUL STATISTICAL MODELS

Numerous situations arise in the conduct of a simulation where an investigator may choose to introduce probabilistic events. In Chapter 2, queueing, inventory, and reliability examples were given. In a queueing system, interarrival and service times are often probabilistic. In an inventory model, the time between demands and the lead times (time between placing and receiving an order) can be probabilistic. In a reliability model, the time to failure could be probabilistic. In each of these instances, the simulation analyst desires to generate random events and to use a known statistical model if the underlying distribution can be found. In the following paragraphs, statistical models appropriate to these application areas will be discussed. Additionally, statistical models useful in the case of limited data are mentioned.

1. *Queueing systems.* In Chapter 2, examples of waiting-line problems were given. In Chapters 2, 3, and 4, these problems were solved via simulation. In the queueing examples, interarrival- and

Table 5.1   Arrival Data

Arrival Number	Arrival (Hour:Minutes::Seconds)	Interarrival Time (Minutes::Seconds)
1	10:05::03	—
2	10:12::16	7::13
3	10:15::48	3::32
4	10:24::27	8::39
5	10:32::19	7::52
6	10:35::43	3::24
7	10:39::51	4::08
8	10:40::30	0::39
9	10:41::17	0::47
10	10:44::12	2::55
11	10:45::47	1::35
12	10:50::47	5::00
13	11:00::05	9::18
14	11:04::58	4::53
15	11:06::12	1::14
16	11:11::23	5::11
17	11:16::31	5::08
18	11:17::18	0::47
19	11:21::26	4::08
20	11:24::43	3::17
21	11:31::19	6::36

service-time patterns were given. In these examples, the times between arrivals and the service times were always probabilistic, as is usually the case. However, it is possible to have a constant interarrival time (as in the case of a line moving at a constant speed in the assembly of an automobile), or a constant service time (as in the case of robotized spot welding on the same assembly line). The following example illustrates how probabilistic interarrival times might occur.

Example 5.8

Mechanics arrive at a centralized tool crib as shown in Table 5.1. Attendants check in and check out the requested tools to the mechanics. The collection of data begins at 10:00 A.M. and continues until 20 different interarrival times are recorded. Rather than record the actual time of day, the absolute time from a given origin could have been computed. Thus, the first mechanic could have arrived at time zero, the second mechanic at time 7:13 (7 minutes, 13 seconds), and so on.

Example 5.9

Another way of presenting interarrival data is to find the number of arrivals per time period. Here, such arrivals occur over approximately 1 1/2 hours; it is convenient to look at 10-minute time intervals for the first 20 mechanics. That is, in the first 10-minute time period, one arrival occurred at 10:05::03. In the second time period, two mechanics arrived, and so on. The results are summarized in Table 5.2. This data could then be plotted in a histogram, as shown in Figure 5.5.

Table 5.2   Arrivals in Successive Time Periods

Time Period	Number of Arrivals	Time Period	Number of Arrivals
1	1	6	1
2	2	7	3
3	1	8	3
4	3	9	2
5	4	—	—

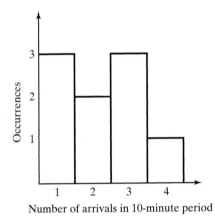

Number of arrivals in 10-minute period

Figure 5.5   Histogram of arrivals per time period.

The distribution of time between arrivals and the distribution of the number of arrivals per time period are important in the simulation of waiting-line systems. "Arrivals" occur in numerous ways; as machine breakdowns, as jobs coming into a jobshop, as units being assembled on a line, as orders to a warehouse, as data packets to a computer system, as calls to a call center, and so on.

Service times could be constant or probabilistic. If service times are completely random, the exponential distribution is often used for simulation purposes; however, there are several other possibilities. It could happen that the service times are constant, but some random variability causes fluctuations in either a positive or a negative way. For example, the time it takes for a lathe to traverse a 10-centimeter shaft should always be the same. However, the material could have slight differences in hardness or the tool might wear; either event could cause different processing times. In these cases, the normal distribution might describe the service time.

A special case occurs when the phenomenon of interest seems to follow the normal probability distribution, but the random variable is restricted to be greater than or less than a certain value. In this case, the truncated normal distribution can be utilized.

The gamma and Weibull distributions are also used to model interarrival and service times. (Actually, the exponential distribution is a special case of both the gamma and the Weibull distributions.) The differences between the exponential, gamma, and Weibull distributions involve the location of the modes of the pdf's and the shapes of their tails for large and small times. The exponential distribution has its mode at the origin, but the gamma and Weibull distributions have their modes at some point (≥ 0) that is a function of the parameter values selected. The tail of the gamma distribution is long, like an exponential distribution; the tail of the Weibull distribution can decline more rapidly or less rapidly than that of an exponential distribution. In practice, this means that, if there are more large service times than an exponential distribution can account for, a Weibull distribution might provide a better model of these service times.

2. *Inventory and supply-chain systems.* In realistic inventory and supply-chain systems, there are at least three random variables: (1) the number of units demanded per order or per time period, (2) the time between demands, and (3) the lead time. (The lead time is defined as the time between the placing of an order for stocking the inventory system and the receipt of that order.) In very simple mathematical models of inventory systems, demand is a constant over time, and lead time is zero, or a constant. However, in most real-world cases, and, hence, in simulation models, demand occurs randomly in time, and the number of units demanded each time a demand occurs is also random, as illustrated by Figure 5.6.

Distributional assumptions for demand and lead time in inventory theory texts are usually based on mathematical tractability, but those assumptions could be invalid in a realistic context. In practice, the lead-time distribution can often be fitted fairly well by a gamma distribution [Hadley and Whitin, 1963]. Unlike analytic models, simulation models can accommodate whatever assumptions appear most reasonable.

The geometric, Poisson, and negative binomial distributions provide a range of distribution shapes that satisfy a variety of demand patterns. The geometric distribution, which is a special case of the negative binomial, has its mode at unity, given that at least one demand has occurred.

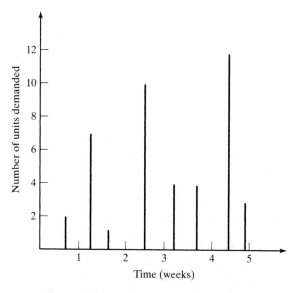

Figure 5.6   Random demands in time.

If demand data are characterized by a long tail, the negative binomial distribution might be appropriate. The Poisson distribution is often used to model demand because it is simple, it is extensively tabulated, and it is well known. The tail of the Poisson distribution is generally shorter than that of the negative binomial, which means that fewer large demands will occur if a Poisson model is used than if a negative binomial distribution is used (assuming that both models have the same mean demand).

3. *Reliability and maintainability.* Time to failure has been modeled with numerous distributions, including the exponential, gamma, and Weilbull. If only random failures occur, the time-to-failure distribution may be modeled as exponential. The gamma distribution arises from modeling standby redundancy, where each component has an exponential time to failure. The Weibull distribution has been extensively used to represent time to failure, and its nature is such that it can be made to approximate many observed phenomena [Hines and Montgomery, 1990]. When there are a number of components in a system and failure is due to the most serious of a large number of defects, or possible defects, the Weibull distribution seems to do particularly well as a model. In situations where most failures are due to wear, the normal distribution might very well be appropriate [Hines and Montgomery, 1990]. The lognormal distribution has been found to be applicable in describing time to failure for some types of components.

4. *Limited data.* In many instances, simulations begin before data collection has been completed. There are three distributions that have application to incomplete or limited data. These are the uniform, triangular, and beta distributions. The uniform distribution can be used when an interarrival or service time is known to be random, but no information is immediately available about the distribution [Gordon, 1975]. However, there are those who do not favor using the uniform distribution, calling it the "distribution of maximum ignorance" because it is not necessary to specify more than the continuous interval in which the random variable may occur. The triangular distribution can be used when assumptions are made about the minimum, maximum, and modal values of the random variable. Finally, the beta distribution provides a variety of distributional forms on the unit interval, ones that, with appropriate modification, can be shifted to any desired interval. The uniform distribution is a special case of the beta distribution. Pegden, Shannon, and Sadowski [1995] discuss the subject of limited data in some detail, and we include further discussion in Chapter 9.

5. *Other distributions.* Several other distributions may be useful in discrete-system simulation. The Bernoulli and binomial distributions are two discrete distributions which might describe phenomena of interest. The hyperexponential distribution is similar to the exponential distribution, but its greater variability might make it useful in certain instances.

5.3 DISCRETE DISTRIBUTIONS

Discrete random variables are used to describe random phenomena in which only integer values can occur. Numerous examples were given in Section 5.2—for example, demands for inventory items. Four distributions are described in the following subsections.

1. *Bernoulli trials and the Bernoulli distribution.* Consider an experiment consisting of n trials, each of which can be a success or a failure. Let $X_j = 1$ if the jth experiment resulted in a success, and let $X_j = 0$ if the jth experiment resulted in a failure. The n Bernoulli trials are

called a Bernoulli process if the trials are independent, each trial has only two possible outcomes (success or failure), and the probability of a success remains constant from trial to trial. Thus,

$$p(x_1, x_2, \ldots, x_n) = p_1(x_1) \cdot p_2(x_2) \cdots p_n(x_n)$$

and

$$p_j(x_j) = p(x_j) = \begin{cases} p, & x_j = 1, j = 1, 2, \ldots, n \\ 1 - p = q, & x_j = 0, j = 1, 2, \ldots, n \\ 0, & \text{otherwise} \end{cases} \tag{5.11}$$

For one trial, the distribution given in Equation (5.11) is called the Bernoulli distribution. The mean and variance of X_j are calculated as follows:

$$E(X_j) = 0 \cdot q + 1 \cdot p = p$$

and

$$V(X_j) = [(0^2 \cdot q) + (1^2 \cdot p)] - p^2 = p(1 - p)$$

2. *Binomial distribution.* The random variable X that denotes the number of successes in n Bernoulli trials has a binomial distribution given by $p(x)$, where

$$p(x) = \begin{cases} \binom{n}{x} p^x q^{n-x}, & x = 0, 1, 2, \ldots, n \\ 0, & \text{otherwise} \end{cases} \tag{5.12}$$

Equation (5.12) is motivated by computing the probability of a particular outcome with all the successes, each denoted by S, occurring in the first x trials, followed by the $n - x$ failures, each denoted by an F—that is,

$$P(\overbrace{SSS\ldots\ldots.SS}^{x \text{ of these}} \overbrace{FF\ldots\ldots.FF}^{n-x \text{ of these}}) = p^x q^{n-x}$$

where $q = 1 - p$. There are

$$\binom{n}{x} = \frac{n!}{x!(n-x)!}$$

outcomes having the required number of S's and F's. Therefore, Equation (5.12) results. An easy approach to calculating the mean and variance of the binomial distribution is to consider X as a sum of n independent Bernoulli random variables, each with mean p and variance $p(1-p) = pq$. Then,

$$X = X_1 + X_2 + \cdots + X_n$$

and the mean, $E(X)$, is given by

$$E(X) = p + p + \cdots + p = np \tag{5.13}$$

and the variance $V(X)$ is given by

$$V(X) = pq + pq + \cdots + pq = npq \qquad (5.14)$$

Example 5.10

A production process manufactures computer chips on the average at 2% nonconforming. Every day, a random sample of size 50 is taken from the process. If the sample contains more than two nonconforming chips, the process will be stopped. Compute the probability that the process is stopped by the sampling scheme.

Consider the sampling process as $n = 50$ Bernoulli trials, each with $p = 0.02$; then the total number of nonconforming chips in the sample, X, would have a binomial distribution given by

$$p(x) = \begin{cases} \binom{50}{x} (0.02)^x (0.98)^{50-x}, & x = 0, 1, 2, \ldots, 50 \\ 0, & \text{otherwise} \end{cases}$$

It is much easier to compute the right-hand side of the following identity to compute the probability that more than two nonconforming chips are found in a sample:

$$P(X > 2) = 1 - P(X \leq 2)$$

The probability $P(X \leq 2)$ is calculated from

$$P(X \leq 2) = \sum_{x=0}^{2} \binom{50}{x} (0.02)^x (0.98)^{50-x}$$

$$= (0.98)^{50} + 50(0.02)(0.98)^{49} + 1225(0.02)^2(0.98)^{48}$$

$$\doteq 0.92$$

Thus, the probability that the production process is stopped on any day, based on the sampling process, is approximately 0.08. The mean number of nonconforming chips in a random sample of size 50 is given by

$$E(X) = np = 50(0.02) = 1$$

and the variance is given by

$$V(X) = npq = 50(0.02)(0.98) = 0.98$$

The cdf for the binomial distribution has been tabulated by Banks and Heikes [1984] and others. The tables decrease the effort considerably for computing probabilities such as $P(a < X \leq b)$. Under certain conditions on n and p, both the Poisson distribution and the normal distribution may be used to approximate the binomial distribution [Hines and Montgomery, 1990].

3. *Geometric and Negative Binomial distributions.* The geometric distribution is related to a sequence of Bernoulli trials; the random variable of interest, X, is defined to be the number of

trials to achieve the first success. The distribution of X is given by

$$p(x) = \begin{cases} q^{x-1}p, & x = 1, 2, \ldots \\ 0, & \text{otherwise} \end{cases} \tag{5.15}$$

The event $\{X = x\}$ occurs when there are $x - 1$ failures followed by a success. Each of the failures has an associated probability of $q = 1 - p$, and each success has probability p. Thus,

$$P(FFF \cdots FS) = q^{x-1}p$$

The mean and variance are given by

$$E(X) = \frac{1}{p} \tag{5.16}$$

and

$$V(X) = \frac{q}{p^2} \tag{5.17}$$

More generally, the negative binomial distribution is the distribution of the number of trials until the kth success, for $k = 1, 2, \ldots$. If Y has a negative binomial distribution with parameters p and k, then the distribution of Y is given by

$$p(y) = \begin{cases} \binom{y-1}{k-1} q^{y-k}p^k, & y = k, k+1, k+2, \ldots \\ 0, & \text{otherwise} \end{cases} \tag{5.18}$$

Because we can think of the negative binomial random variable Y as the sum of k independent geometric random variables, it is easy to see that $E(Y) = k/p$ and $V(X) = kq/p^2$.

Example 5.11

Forty percent of the assembled ink-jet printers are rejected at the inspection station. Find the probability that the first acceptable ink-jet printer is the third one inspected. Considering each inspection as a Bernoulli trial with $q = 0.4$ and $p = 0.6$ yields

$$p(3) = 0.4^2(0.6) = 0.096$$

Thus, in only about 10% of the cases is the first acceptable printer the third one from any arbitrary starting point. To determine the probability that the third printer inspected is the second acceptable printer, we use the negative binomial distribution (5.18),

$$p(3) = \binom{3-1}{2-1} 0.4^{3-2}(0.6)^2 = \binom{2}{1} 0.4(0.6)^2 = 0.288$$

4. *Poisson distribution.* The Poisson distribution describes many random processes quite well and is mathematically quite simple. The Poisson distribution was introduced in 1837 by

S. D. Poisson in a book concerning criminal and civil justice matters. (The title of this rather old text is "Recherches sur la probabilite des jugements en matiere criminelle et en matiere civile." Evidently, the rumor handed down through generations of probability theory professors concerning the origin of the Poisson distribution is just not true. Rumor has it that the Poisson distribution was first used to model deaths from the kicks of horses in the Prussian Army.)

The Poisson probability mass function is given by

$$\text{discrete} \quad p(x) = \begin{cases} \dfrac{e^{-\alpha}\alpha^x}{x!}, & x = 0, 1, \ldots \\ 0, & \text{otherwise} \end{cases} \tag{5.19}$$

where $\alpha > 0$. One of the important properties of the Poisson distribution is that the mean and variance are both equal to α, that is,

$$E(X) = \alpha = V(X)$$

The cumulative distribution function is given by

$$F(x) = \sum_{i=0}^{x} \frac{e^{-\alpha}\alpha^i}{i!} \tag{5.20}$$

The pmf and cdf for a Poisson distribution with $\alpha = 2$ are shown in Figure 5.7. A tabulation of the cdf is given in Table A.4.

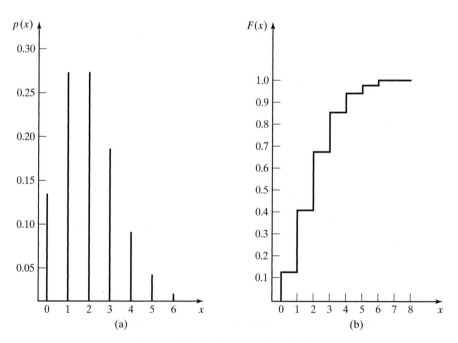

Figure 5.7 Poisson pmf and cdf.

Example 5.12

A computer repair person is "beeped" each time there is a call for service. The number of beeps per hour is known to occur in accordance with a Poisson distribution with a mean of $\alpha = 2$ per hour. The probability of three beeps in the next hour is given by Equation (5.19) with $x = 3$, as follows:

$$p(3) = \frac{e^{-2}2^3}{3!} = \frac{(0.135)(8)}{6} = 0.18$$

This same result can be read from the left side of Figure 5.7 or from Table A.4 by computing

$$F(3) - F(2) = 0.857 - 0.677 = 0.18$$

Example 5.13

In Example 5.12, find the probability of two or more beeps in a 1-hour period.

$$P(2 \text{ or more}) = 1 - p(0) - p(1) = 1 - F(1)$$
$$= 1 - 0.406 = 0.594$$

The cumulative probability, $F(1)$, can be read from the right side of Figure 5.7 or from Table A.4.

Example 5.14

The lead-time demand in an inventory system is the accumulation of demand for an item from the point at which an order is placed until the order is received—that is,

$$L = \sum_{i=1}^{T} D_i \tag{5.21}$$

where L is the lead-time demand, D_i, is the demand during the ith time period, and T is the number of time periods during the lead time. Both D_i and T may be random variables.

An inventory manager desires that the probability of a stockout not exceed a certain fraction during the lead time. For example, it may be stated that the probability of a shortage during the lead time not exceed 5%.

If the lead-time demand is Poisson distributed, the determination of the reorder point is greatly facilitated. The reorder point is the level of inventory at which a new order is placed.

Assume that the lead-time demand is Poisson distributed with a mean of $\alpha = 10$ units and that 95% protection from a stockout is desired. Thus, it is desired to find the smallest value of x such that the probability that the lead-time demand does not exceed x is greater than or equal to 0.95. Using Equation (5.20) requires finding the smallest x such that

$$F(x) = \sum_{i=0}^{x} \frac{e^{-10}10^i}{i!} \geq 0.95$$

The desired result occurs at $x = 15$, which can be found by using Table A.4 or by computation of $p(0), p(1), \ldots$.

5.4 CONTINUOUS DISTRIBUTIONS

Continuous random variables can be used to describe random phenomena in which the variable of interest can take on any value in some interval—for example, the time to failure or the length of a rod. Eight distributions are described in the following subsections.

1. *Uniform distribution.* A random variable X is uniformly distributed on the interval (a, b) if its pdf is given by

$$f(x) = \begin{cases} \dfrac{1}{b-a}, & a \leq x \leq b \\ 0, & \text{otherwise} \end{cases} \qquad (5.22)$$

The cdf is given by

$$F(x) = \begin{cases} 0, & x < a \\ \dfrac{x-a}{b-a}, & a \leq x < b \\ 1, & x \geq b \end{cases} \qquad (5.23)$$

Note that

$$P(x_1 < X < x_2) = F(x_2) - F(x_1) = \frac{x_2 - x_1}{b-a}$$

is proportional to the length of the interval, for all x_1 and x_2 satisfying $a \leq x_1 < x_2 \leq b$. The mean and variance of the distribution are given by

$$E(X) = \frac{a+b}{2} \qquad (5.24)$$

and

$$V(X) = \frac{(b-a)^2}{12} \qquad (5.25)$$

The pdf and cdf when $a = 1$ and $b = 6$ are shown in Figure 5.8.

The uniform distribution plays a vital role in simulation. Random numbers, uniformly distributed between zero and 1, provide the means to generate random events. Numerous methods for generating uniformly distributed random numbers have been devised; some will be discussed in Chapter 7. Uniformly distributed random numbers are then used to generate samples of random variates from all other distributions, as will be discussed in Chapter 8.

Example 5.15

A simulation of a warehouse operation is being developed. About every 3 minutes, a call comes for a forklift truck operator to proceed to a certain location. An initial assumption is made that the time between calls (arrivals) is uniformly distributed with a mean of 3 minutes. By Equation (5.25), the uniform distribution with a mean of 3 and the greatest possible variability would have parameter values of $a = 0$ and $b = 6$ minutes. With very limited data (such as a mean of approximately

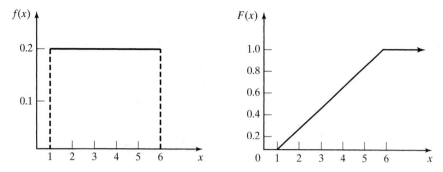

Figure 5.8   pdf and cdf for uniform distribution.

3 minutes) plus the knowledge that the quantity of interest is variable in a random fashion, the uniform distribution with greatest variance can be assumed, at least until more data are available.

Example 5.16

A bus arrives every 20 minutes at a specified stop beginning at 6:40 A.M. and continuing until 8:40 A.M. A certain passenger does not know the schedule, but arrives randomly (uniformly distributed) between 7:00 A.M. and 7:30 A.M. every morning. What is the probability that the passenger waits more than 5 minutes for a bus?

The passenger has to wait more than 5 minutes only if the arrival time is between 7:00 A.M. and 7:15 A.M. or between 7:20 A.M. and 7:30 A.M. If X is a random variable that denotes the number of minutes past 7:00 A.M. that the passenger arrives, the desired probability is

$$P(0 < X < 15) + P(20 < X < 30)$$

Now, X is a uniform random variable on $(0, 30)$. Therefore, the desired probability is given by

$$F(15) + F(30) - F(20) = \frac{15}{30} + 1 - \frac{20}{30} = \frac{5}{6}$$

2. *Exponential distribution.* A random variable X is said to be exponentially distributed with parameter $\lambda > 0$ if its pdf is given by

$$f(x) = \begin{cases} \lambda e^{-\lambda x}, & x \geq 0 \\ 0, & \text{elsewhere} \end{cases} \tag{5.26}$$

The density function is shown in Figures 5.9 and 5.3. Figure 5.9 also shows the cdf.

The exponential distribution has been used to model interarrival times when arrivals are completely random and to model service times that are highly variable. In these instances, λ is a rate: arrivals per hour or services per minute. The exponential distribution has also been used to model the lifetime of a component that fails catastrophically (instantaneously), such as a light bulb; then λ is the failure rate.

Several different exponential pdf's are shown in Figure 5.10. The value of the intercept on the vertical axis is always equal to the value of λ. Note also that all pdf's eventually intersect. (Why?)

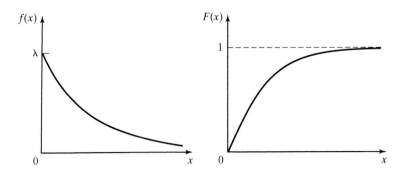

Figure 5.9 Exponential density function and cumulative distribution function.

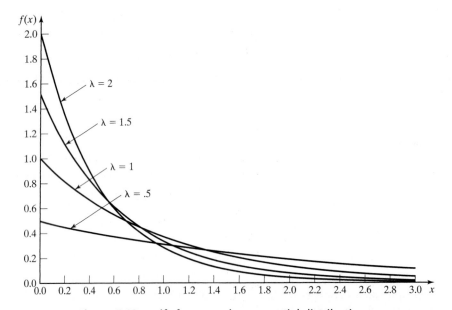

Figure 5.10 pdfs for several exponential distributions.

The exponential distribution has mean and variance given by

$$E(X) = \frac{1}{\lambda} \quad \text{and} \quad V(X) = \frac{1}{\lambda^2} \tag{5.27}$$

Thus, the mean and standard deviation are equal. The cdf can be exhibited by integrating Equation (5.26) to obtain

$$F(x) = \begin{cases} 0, & x < 0 \\ \int_0^x \lambda e^{-\lambda t}\, dt = 1 - e^{-\lambda x}, & x \ge 0 \end{cases} \tag{5.28}$$

Example 5.17 _____

Suppose that the life of an industrial lamp, in thousands of hours, is exponentially distributed with failure rate $\lambda = 1/3$ (one failure every 3000 hours, on the average). The probability that the lamp will last longer than its mean life, 3000 hours, is given by $P(X > 3) = 1 - P(X \leq 3) = 1 - F(3)$. Equation (5.28) is used to compute $F(3)$, obtaining

$$P(X > 3) = 1 - (1 - e^{-3/3}) = e^{-1} = 0.368$$

Regardless of the value of λ, this result will always be the same! That is, the probability that an exponential random variable is greater than its mean is 0.368, for any value of λ.

The probability that the industrial lamp will last between 2000 and 3000 hours is computed as

$$P(2 \leq X \leq 3) = F(3) - F(2)$$

Again, from the cdf given by Equation (5.28),

$$F(3) - F(2) = (1 - e^{-3/3}) - (1 - e^{-2/3})$$
$$= -0.368 + 0.513 = 0.145$$

One of the most important properties of the exponential distribution is that it is "memoryless," which means that, for all $s \geq 0$ and $t \geq 0$,

$$P(X > s + t | X > s) = P(X > t) \tag{5.29}$$

Let X represent the life of a component (a battery, light bulb, computer chip, laser, etc.) and assume that X is exponentially distributed. Equation (5.29) states that the probability that the component lives for at least $s + t$ hours, given that it has survived s hours, is the same as the initial probability that it lives for at least t hours. If the component is alive at time s (if $X > s$), then the distribution of the remaining amount of time that it survives, namely $X - s$, is the same as the original distribution of a new component. That is, the component does not "remember" that it has already been in use for a time s. A used component is as good as new.

That Equation (5.29) holds is shown by examining the conditional probability

$$P(X > s + t | X > s) = \frac{P(X > s + t)}{P(X > s)} \tag{5.30}$$

Equation (5.28) can be used to determine the numerator and denominator of Equation (5.30), yielding

$$P(X > s + t | X > s) = \frac{e^{-\lambda(s+t)}}{e^{-\lambda s}} = e^{-\lambda t}$$
$$= P(X > t)$$

Example 5.18 _____

Find the probability that the industrial lamp in Example 5.17 will last for another 1000 hours, given that it is operating after 2500 hours. This determination can be found using Equations (5.29) and (5.28), as follows:

$$P(X > 3.5 | X > 2.5) = P(X > 1) = e^{-1/3} = 0.717$$

Example 5.18 illustrates the *memoryless* property—namely, that a used component that follows an exponential distribution is as good as a new component. The probability that a new component will have a life greater than 1000 hours is also equal to 0.717. Stated in general, suppose that a component which has a lifetime that follows the exponential distribution with parameter λ is observed and found to be operating at an arbitrary time. Then, the distribution of the remaining lifetime is also exponential with parameter λ. The exponential distribution is the only continuous distribution that has the memoryless property. (The geometric distribution is the only discrete distribution that possesses the memoryless property.)

3. *Gamma distribution.* A function used in defining the gamma distribution is the gamma function, which is defined for all $\beta > 0$ as

$$\Gamma(\beta) = \int_0^\infty x^{\beta-1} e^{-x}\, dx \tag{5.31}$$

By integrating Equation (5.31) by parts, it can be shown that

$$\Gamma(\beta) = (\beta - 1)\Gamma(\beta - 1) \tag{5.32}$$

If β is an integer, then, by using $\Gamma(1) = 1$ and applying Equation (5.32), it can be seen that

$$\Gamma(\beta) = (\beta - 1)! \tag{5.33}$$

The gamma function can be thought of as a generalization of the factorial notion to all positive numbers, not just integers.

A random variable X is gamma distributed with parameters β and θ if its pdf is given by

$$f(x) = \begin{cases} \dfrac{\beta\theta}{\Gamma(\beta)}(\beta\theta x)^{\beta-1} e^{-\beta\theta x}, & x > 0 \\ 0, & \text{otherwise} \end{cases} \tag{5.34}$$

β is called the shape parameter, and θ is called the scale parameter. Several gamma distributions for $\theta = 1$ and various values of β are shown in Figure 5.10a.

The mean and variance of the gamma distribution are given by

$$E(X) = \frac{1}{\theta} \tag{5.35}$$

and

$$V(X) = \frac{1}{\beta\theta^2} \tag{5.36}$$

The cdf of X is given by

$$F(x) = \begin{cases} 1 - \displaystyle\int_x^\infty \dfrac{\beta\theta}{\Gamma(\beta)}(\beta\theta t)^{\beta-1} e^{-\beta\theta t}\, dt, & x > 0 \\ 0, & x \le 0 \end{cases} \tag{5.37}$$

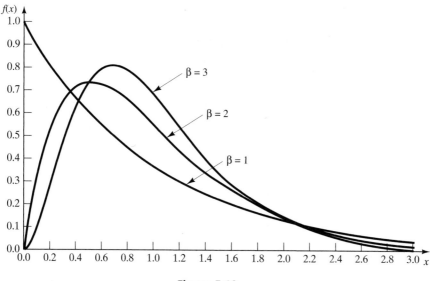

Figure 5.10a

When β is an integer, the gamma distribution is related to the exponential distribution in the following manner: If the random variable, X, is the sum of β independent, exponentially distributed random variables, each with parameter $\beta\theta$, then X has a gamma distribution with parameters β and θ. Thus, if

$$X = X_1 + X_2 + \cdots + X_\beta \qquad (5.38)$$

where the pdf of X_j is given by

$$g(x_j) = \begin{cases} (\beta\theta)e^{-\beta\theta x_j}, & x \geq 0 \\ 0, & \text{otherwise} \end{cases}$$

and the X_j are mutually independent, then X has the pdf given in Equation (5.34). Note that, when $\beta = 1$, an exponential distribution results. This result follows from Equation (5.38) or from letting $\beta = 1$ in Equation (5.34).

4. *Erlang distribution.* The pdf given by Equation (5.34) is often referred to as the Erlang distribution of order (or number of phases) k when $\beta = k$, an integer. Erlang was a Danish telephone engineer who was an early developer of queueing theory. The Erlang distribution could arise in the following context: Consider a series of k stations that must be passed through in order to complete the servicing of a customer. An additional customer cannot enter the first station until the customer in process has negotiated all the stations. Each station has an exponential distribution of service time with parameter $k\theta$. Equations (5.35) and (5.36), which state the mean and variance of a gamma distribution, are valid regardless of the value of β. However, when $\beta = k$, an integer, Equation (5.38) may be used to derive the mean of the distribution in a fairly straightforward manner. The expected value of the sum of random variables is the sum of the expected value of

each random variable. Thus,

$$E(X) = E(X_1) + E(X_2) + \cdots + E(X_k)$$

The expected value of each of the exponentially distributed X_j is given by $1/k\theta$. Thus,

$$E(X) = \frac{1}{k\theta} + \frac{1}{k\theta} + \cdots + \frac{1}{k\theta} = \frac{1}{\theta}$$

If the random variables X_j are independent, the variance of their sum is the sum of the variances, or

$$V(X) = \frac{1}{(k\theta)^2} + \frac{1}{(k\theta)^2} + \cdots + \frac{1}{(k\theta)^2} = \frac{1}{k\theta^2}$$

When $\beta = k$, a positive integer, the cdf given by Equation (5.37) may be integrated by parts, giving

$$F(x) = \begin{cases} 1 - \displaystyle\sum_{i=0}^{k-1} \frac{e^{-k\theta x}(k\theta x)^i}{i!}, & x > 0 \\ 0, & x \le 0 \end{cases} \qquad (5.39)$$

which is the sum of Poisson terms with mean $\alpha = k\theta x$. Tables of the cumulative Poisson distribution may be used to evaluate the cdf when the shape parameter is an integer.

Example 5.19

A college professor of electrical engineering is leaving home for the summer, but would like to have a light burning at all times to discourage burglars. The professor rigs up a device that will hold two light bulbs. The device will switch the current to the second bulb if the first bulb fails. The box in which the light bulbs are packaged says, "Average life 1000 hours, exponentially distributed." The professor will be gone 90 days (2160 hours). What is the probability that a light will be burning when the summer is over and the professor returns?

The probability that the system will operate at least x hours is called the reliability function $R(x)$:

$$R(x) = 1 - F(x)$$

In this case, the total system lifetime is given by Equation (5.38) with $\beta = k = 2$ bulbs and $k\theta = 1/1000$ per hour, so $\theta = 1/2000$ per hour. Thus, $F(2160)$ can be determined from Equation (5.39) as follows:

$$F(2160) = 1 - \sum_{i=0}^{1} \frac{e^{-(2)(1/2000)(2160)}[(2)(1/2000)(2160)]^i}{i!}$$

$$= 1 - e^{-2.16} \sum_{i=0}^{1} \frac{(2.16)^i}{i!} = 0.636$$

Therefore, the chances are about 36% that a light will be burning when the professor returns.

Example 5.20 _____

A medical examination is given in three stages by a physician. Each stage is exponentially distributed with a mean service time of 20 minutes. Find the probability that the exam will take 50 minutes or less. Also, compute the expected length of the exam. In this case, $k = 3$ stages and $k\theta = 1/20$, so that $\theta = 1/60$ per minute. Thus, $F(50)$ can be calculated from Equation (5.39) as follows:

$$F(50) = 1 - \sum_{i=0}^{2} \frac{e^{-(3)(1/60)(50)}[(3)(1/60)(50)]^i}{i!}$$

$$= 1 - \sum_{i=0}^{2} \frac{e^{-5/2}(5/2)^i}{i!}$$

The cumulative Poisson distribution, shown in Table A.4, can be used to calculate that

$$F(50) = 1 - 0.543 = 0.457$$

The probability is 0.457 that the exam will take 50 minutes or less. The expected length of the exam is found from Equation (5.35):

$$E(X) = \frac{1}{\theta} = \frac{1}{1/60} = 60 \text{ minutes}$$

In addition, the variance of X is $V(X) = 1/\beta\theta^2 = 1200$ minutes^2—incidentally, the mode of the Erlang distribution is given by

$$\text{Mode} = \frac{k-1}{k\theta} \tag{5.40}$$

Thus, the modal value in this example is

$$\text{Mode} = \frac{3-1}{3(1/60)} = 40 \text{ minutes}$$

5. *Normal distribution.* A random variable X with mean $-\infty < \mu < \infty$ and variance $\sigma^2 > 0$ has a normal distribution if it has the pdf

$$f(x) = \frac{1}{\sigma\sqrt{2\pi}} \exp\left[-\frac{1}{2}\left(\frac{x-\mu}{\sigma}\right)^2\right], \quad -\infty < x < \infty \tag{5.41}$$

The normal distribution is used so often that the notation $X \sim N(\mu, \sigma^2)$ has been adopted by many authors to indicate that the random variable X is normally distributed with mean μ and variance σ^2. The normal pdf is shown in Figure 5.11.

Some of the special properties of the normal distribution are listed here:

1. $\lim_{x \to -\infty} f(x) = 0$ and $\lim_{x \to \infty} f(x) = 0$; the value of $f(x)$ approaches zero as x approaches negative infinity and, similarly, as x approaches positive infinity.

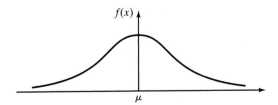

Figure 5.11   pdf of the normal distribution.

2. $f(\mu - x) = f(\mu + x)$; the pdf is symmetric about μ.
3. The maximum value of the pdf occurs at $x = \mu$; the mean and mode are equal.

The cdf for the normal distribution is given by

$$F(x) = P(X \leq x) = \int_{-\infty}^{x} \frac{1}{\sigma\sqrt{2\pi}} \exp\left[-\frac{1}{2}\left(\frac{t-\mu}{\sigma}\right)^2\right] dt \tag{5.42}$$

It is not possible to evaluate Equation (5.42) in closed form. Numerical methods could be used, but it appears that it would be necessary to evaluate the integral for each pair (μ, σ^2). However, a transformation of variables, $z = (t - \mu)/\sigma$, allows the evaluation to be independent of μ and σ. If $X \sim N(\mu, \sigma^2)$, let $Z = (X - \mu)/\sigma$ to obtain

$$F(x) = P(X \leq x) = P\left(Z \leq \frac{x-\mu}{\sigma}\right)$$

$$= \int_{-\infty}^{(x-\mu)/\sigma} \frac{1}{\sqrt{2\pi}} e^{-z^2/2} \, dz \tag{5.43}$$

$$= \int_{-\infty}^{(x-\mu)/\sigma} \phi(z) \, dz = \Phi\left(\frac{x-\mu}{\sigma}\right)$$

The pdf

$$\phi(z) = \frac{1}{\sqrt{2\pi}} e^{-z^2/2}, \quad -\infty < z < \infty \tag{5.44}$$

is the pdf of a normal distribution with mean zero and variance 1. Thus, $Z \sim N(0, 1)$ and it is said that Z has a standard normal distribution. The standard normal distribution is shown in Figure 5.12. The cdf for the standard normal distribution is given by

$$\Phi(z) = \int_{-\infty}^{z} \frac{1}{\sqrt{2\pi}} e^{-t^2/2} \, dt \tag{5.45}$$

Equation (5.45) has been widely tabulated. The probabilities $\Phi(z)$ for $Z \geq 0$ are given in Table A.3. Several examples are now given that indicate how Equation (5.43) and Table A.3 are used.

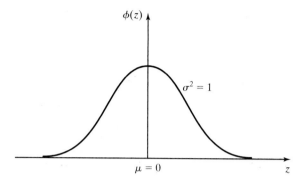

Figure 5.12 pdf of the standard normal distribution.

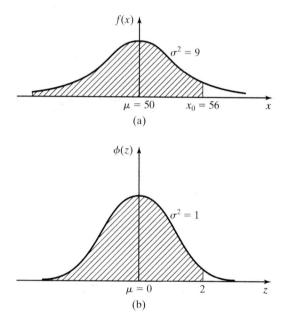

Figure 5.13 Transforming to the standard normal distribution.

Example 5.21

Suppose that it is known that $X \sim N(50, 9)$. Compute $F(56) = P(X \leq 56)$. Using Equation (5.43), we get

$$F(56) = \Phi\left(\frac{56 - 50}{3}\right) = \Phi(2) = 0.9772$$

from Table A.3. The intuitive interpretation is shown in Figure 5.13. Figure 5.13(a) shows the pdf of $X \sim N(50, 9)$ with the specific value, $x_0 = 56$, marked. The shaded portion is the desired probability. Figure 5.13(b) shows the standard normal distribution or $Z \sim N(0, 1)$ with the value

2 marked; $x_0 = 56$ is 2σ (where $\sigma = 3$) greater than the mean. It is helpful to make both sketches such as those in Figure 5.13 to avoid confusion in figuring out required probabilities.

Example 5.22

The time in hours required to load an oceangoing vessel, X, is distributed as $N(12, 4)$. The probability that the vessel will be loaded in less than 10 hours is given by $F(10)$, where

$$F(10) = \Phi\left(\frac{10 - 12}{2}\right) = \Phi(-1) = 0.1587$$

The value of $\Phi(-1) = 0.1587$ is looked up in Table A.3 by using the symmetry property of the normal distribution. Note that $\Phi(1) = 0.8413$. The complement of 0.8413, or 0.1587, is contained in the tail, the shaded portion of the standard normal distribution shown in Figure 5.14(a). In Figure 5.14(b), the symmetry property is used to work out the shaded region to be $\Phi(-1) = 1 - \Phi(1) = 0.1587$. [From this logic, it can be seen that $\Phi(2) = 0.9772$ and $\Phi(-2) = 1 - \Phi(2) = 0.0228$. In general, $\Phi(-x) = 1 - \Phi(x)$.]

The probability that 12 or more hours will be required to load the ship can also be discovered by inspection, by using the symmetry property of the normal pdf and the mean as shown by Figure 5.15. The shaded portion of Figure 5.15(a) shows the problem as originally stated [i.e., evaluate $P(X < 12)$]. Now, $P(X > 12) = 1 - F(12)$. The standardized normal in Figure 5.15(b) is used to evaluate $F(12) = \Phi(0) = 0.50$. Thus, $P(X > 12) = 1 - 0.50 = 0.50$. [The shaded portions in both Figure 5.15(a) and (b) contain 0.50 of the area under the normal pdf.]

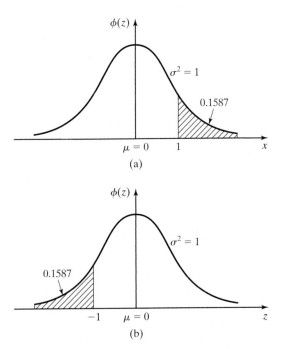

Figure 5.14 Using the symmetry property of the normal distribution.

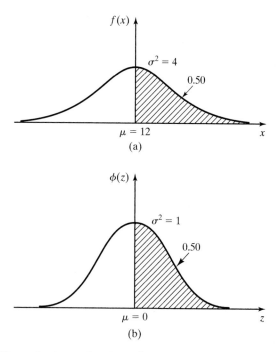

Figure 5.15 Evaluation of probability by inspection.

The probability that between 10 and 12 hours will be required to load a ship is given by

$$P(10 \leq X \leq 12) = F(12) - F(10) = 0.5000 - 0.1587 = 0.3413$$

using earlier results presented in this example. The desired area is shown in the shaded portion of Figure 5.16(a). The equivalent problem shown in terms of the standardized normal distribution is shown in Figure 5.16(b). The probability statement is $F(12) - F(10) = \Phi(0) - \Phi(-1) = 0.5000 - 0.1587 = 0.3413$, from Table A.3.

Example 5.23 _____

The time to pass through a queue to begin self-service at a cafeteria has been found to be $N(15, 9)$. The probability that an arriving customer waits between 14 and 17 minutes is computed as follows:

$$P(14 \leq X \leq 17) = F(17) - F(14) = \Phi\left(\frac{17 - 15}{3}\right) - \Phi\left(\frac{14 - 15}{3}\right)$$

$$= \Phi(0.667) - \Phi(-0.333)$$

The shaded area shown in Figure 5.17(a) represents the probability $F(17) - F(14)$. The shaded area shown in Figure 5.17(b) represents the equivalent probability, $\Phi(0.667) - \Phi(-0.333)$, for the standardized normal distribution. From Table A.3, $\Phi(0.667) = 0.7476$. Now, $\Phi(-0.333) = 1 - \Phi(0.333) = 1 - 0.6304 = 0.3696$. Thus, $\Phi(0.667) - \Phi(-0.333) = 0.3780$. The probability is 0.3780 that the customer will pass through the queue in a time between 14 and 17 minutes.

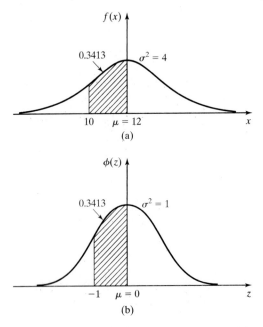

Figure 5.16   Transformation to standard normal for vessel-loading problem.

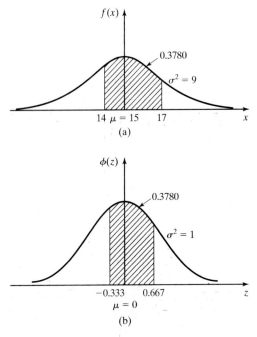

Figure 5.17   Transformation to standard normal for cafeteria problem.

Example 5.24

Lead-time demand, X, for an item is approximated by a normal distribution having mean 25 and variance 9. It is desired to compute the value for lead time that will be exceeded only 5% of the time. Thus, the problem is to find x_0 such that $P(X > x_0) = 0.05$, as shown by the shaded area in Figure 5.18(a). The equivalent problem is shown as the shaded area in Figure 5.18(b). Now,

$$P(X > x_0) = P\left(Z > \frac{x_0 - 25}{3}\right) = 1 - \Phi\left(\frac{x_0 - 25}{3}\right) = 0.05$$

or, equivalently,

$$\Phi\left(\frac{x_0 - 25}{3}\right) = 0.95$$

From Table A.3, it can be seen that $\Phi(1.645) = 0.95$. Thus, x_0 can be found by solving

$$\frac{x_0 - 25}{3} = 1.645$$

or

$$x_0 = 29.935$$

Therefore, in only 5% of the cases will demand during lead time exceed available inventory if an order to purchase is made when the stock level reaches 30.

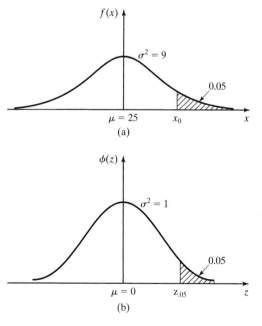

Figure 5.18 Finding x_0 for lead-time–demand problem.

6. *Weibull distribution.* The random variable X has a Weibull distribution if its pdf has the form

$$f(x) = \begin{cases} \dfrac{\beta}{\alpha}\left(\dfrac{x-\nu}{\alpha}\right)^{\beta-1} \exp\left[-\left(\dfrac{x-\nu}{\alpha}\right)^{\beta}\right], & x \geq \nu \\ 0, & \text{otherwise} \end{cases} \qquad (5.46)$$

The three parameters of the Weibull distribution are ν $(-\infty < \nu < \infty)$, which is the location parameter; α $(\alpha > 0)$, which is the scale parameter; and β $(\beta > 0)$, which is the shape parameter. When $\nu = 0$, the Weibull pdf becomes

$$f(x) = \begin{cases} \dfrac{\beta}{\alpha}\left(\dfrac{x}{\alpha}\right)^{\beta-1} \exp\left[-\left(\dfrac{x}{\alpha}\right)^{\beta}\right], & x \geq 0 \\ 0, & \text{otherwise} \end{cases} \qquad (5.47)$$

Figure 5.19 shows several Weibull densities when $\nu = 0$ and $\alpha = 1$. When $\beta = 1$, the Weibull distribution is reduced to

$$f(x) = \begin{cases} \dfrac{1}{\alpha}e^{-x/\alpha}, & x \geq 0 \\ 0, & \text{otherwise} \end{cases}$$

which is an exponential distribution with parameter $\lambda = 1/\alpha$.

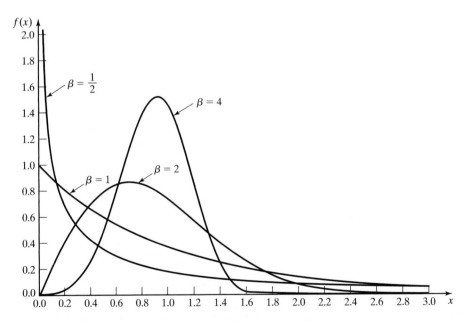

Figure 5.19 Weibull pdfs for $\nu = 0; \alpha = \frac{1}{2}; \beta = \frac{1}{2}, 1, 2, 4$.

The mean and variance of the Weibull distribution are given by the following expressions:

$$E(X) = \nu + \alpha \Gamma \left(\frac{1}{\beta} + 1 \right) \tag{5.48}$$

$$V(X) = \alpha^2 \left[\Gamma \left(\frac{2}{\beta} + 1 \right) - \left[\Gamma \left(\frac{1}{\beta} + 1 \right) \right]^2 \right] \tag{5.49}$$

where $\Gamma(\cdot)$ is defined by Equation (5.31). Thus, the location parameter, ν, has no effect on the variance; however, the mean is increased or decreased by ν. The cdf of the Weibull distribution is given by

$$F(x) = \begin{cases} 0, & x < \nu \\ 1 - \exp \left[-\left(\dfrac{x - \nu}{\alpha} \right)^\beta \right], & x \geq \nu \end{cases} \tag{5.50}$$

Example 5.25

The time to failure for a component screen is known to have a Weibull distribution with $\nu = 0$, $\beta = 1/3$, and $\alpha = 200$ hours. The mean time to failure is given by Equation (5.48) as

$$E(X) = 200\Gamma(3 + 1) = 200(3!) = 1200 \text{ hours}$$

The probability that a unit fails before 2000 hours is computed from Equation (5.50) as

$$F(2000) = 1 - \exp \left[-\left(\frac{2000}{200} \right)^{1/3} \right]$$

$$= 1 - e^{-3\sqrt{10}} = 1 - e^{-2.15} = 0.884$$

Example 5.26

The time it takes for an aircraft to land and clear the runway at a major international airport has a Weibull distribution with $\nu = 1.34$ minutes, $\beta = 0.5$, and $\alpha = 0.04$ minute. Find the probability that an incoming airplane will take more than 1.5 minutes to land and clear the runway. In this case $P(X > 1.5)$ is computed as follows:

$$P(X \leq 1.5) = F(1.5)$$

$$= 1 - \exp \left[-\left(\frac{1.5 - 1.34}{0.04} \right)^{0.5} \right]$$

$$= 1 - e^{-2} = 1 - 0.135 = 0.865$$

Therefore, the probability that an aircraft will require more than 1.5 minutes to land and clear the runway is 0.135.

7. *Triangular distribution.* A random variable X has a triangular distribution if its pdf is given by

$$f(x) = \begin{cases} \dfrac{2(x-a)}{(b-a)(c-a)}, & a \le x \le b \\[2mm] \dfrac{2(c-x)}{(c-b)(c-a)}, & b < x \le c \\[2mm] 0, & \text{elsewhere} \end{cases} \tag{5.51}$$

where $a \le b \le c$. The mode occurs at $x = b$. A triangular pdf is shown in Figure 5.20. The parameters (a, b, c) can be related to other measures, such as the mean and the mode, as follows:

$$E(X) = \frac{a + b + c}{3} \tag{5.52}$$

From Equation (5.52) the mode can be determined as

$$\text{Mode} = b = 3E(X) - (a + c) \tag{5.53}$$

Because $a \le b \le c$,

$$\frac{2a + c}{3} \le E(X) \le \frac{a + 2c}{3}$$

The mode is used more often than the mean to characterize the triangular distribution. As is shown in Figure 5.20, its height is $2/(c - a)$ above the x axis. The variance, $V(X)$, of the triangular distribution is left as an exercise for the student. The cdf for the triangular distribution is given by

$$F(x) = \begin{cases} 0, & x \le a \\[2mm] \dfrac{(x-a)^2}{(b-a)(c-a)}, & a < x \le b \\[2mm] 1 - \dfrac{(c-x)^2}{(c-b)(c-a)}, & b < x \le c \\[2mm] 1, & x > c \end{cases} \tag{5.54}$$

Example 5.27
The central processing unit requirements, for programs that will execute, have a triangular distribution with $a = 0.05$ millisecond, $b = 1.1$ milliseconds, and $c = 6.5$ milliseconds. Find the probability that the CPU requirement for a random program is 2.5 milliseconds or less. The value of $F(2.5)$ is from the portion of the cdf in the interval $(0.05, 1.1)$ plus that portion in the interval $(1.1, 2.5)$. By using Equation (5.54), both portions can be addressed at one time, to yield

$$F(2.5) = 1 - \frac{(6.5 - 2.5)^2}{(6.5 - 0.05)(6.5 - 1.1)} = 0.541$$

Thus, the probability is 0.541 that the CPU requirement is 2.5 milliseconds or less.

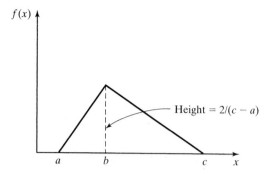

Figure 5.20   pdf of the triangular distribution.

Example 5.28

An electronic sensor evaluates the quality of memory chips, rejecting those that fail. Upon demand, the sensor will give the minimum and maximum number of rejects during each hour of production over the past 24 hours. The mean is also given. Without further information, the quality control department has assumed that the number of rejected chips can be approximated by a triangular distribution. The current dump of data indicates that the minimum number of rejected chips during any hour was zero, the maximum was 10, and the mean was 4. Given that $a = 0$, $c = 10$, and $E(X) = 4$, the value of b can be found from Equation (5.53):

$$b = 3(4) - (0 + 10) = 2$$

The height of the mode is $2/(10 - 0) = 0.2$. Thus, Figure 5.21 can be drawn.

The median is the point at which 0.5 of the area is to the left and 0.5 is to the right. The median in this example is 3.7, also shown on Figure 5.21. Finding the median of the triangular distribution requires an initial location of the value to the left or to the right of the mode. The

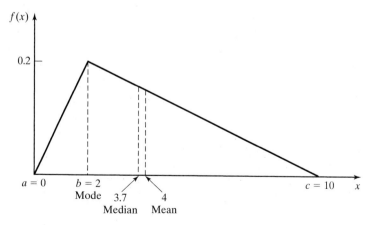

Figure 5.21   Mode, median, and mean for triangular distribution.

area to the left of the mode is computed from Equation (5.54) as

$$F(2) = \frac{2^2}{20} = 0.2$$

Thus, the median is between b and c. Setting $F(x) = 0.5$ in Equation (5.54) and solving for $x = $ median yields

$$0.5 = 1 - \frac{(10 - x)^2}{(10)(8)}$$

with

$$x = 3.7$$

This example clearly shows that the mean, mode, and median are not necessarily equal.

8. *Lognormal distribution.* A random variable X has a lognormal distribution if its pdf is given by

$$f(x) = \begin{cases} \dfrac{1}{\sqrt{2\pi}\sigma x} \exp\left[-\dfrac{(\ln x - \mu)^2}{2\sigma^2}\right], & x > 0 \\ 0, & \text{otherwise} \end{cases} \tag{5.55}$$

where $\sigma^2 > 0$. The mean and variance of a lognormal random variable are

$$E(X) = e^{\mu + \sigma^2/2} \tag{5.56}$$

$$V(X) = e^{2\mu + \sigma^2}(e^{\sigma^2} - 1) \tag{5.57}$$

Three lognormal pdf's, all having mean 1, but variances 1/2, 1, and 2, are shown in Figure 5.22.

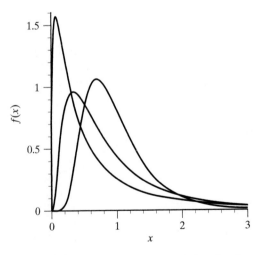

Figure 5.22 pdf of the lognormal distribution.

Notice that the parameters μ and σ^2 are not the mean and variance of the lognormal. These parameters come from the fact that when Y has a $N(\mu, \sigma^2)$ distribution then $X = e^Y$ has a lognormal distribution with parameters μ and σ^2. If the mean and variance of the lognormal are known to be μ_L and σ_L^2, respectively, then the parameters μ and σ^2 are given by

$$\mu = \ln\left(\frac{\mu_L^2}{\sqrt{\mu_L^2 + \sigma_L^2}}\right) \tag{5.58}$$

$$\sigma^2 = \ln\left(\frac{\mu_L^2 + \sigma_L^2}{\mu_L^2}\right) \tag{5.59}$$

Example 5.29

The rate of return on a volatile investment is modeled as having a lognormal distribution with mean 20% and standard deviation 5%. Compute the parameters for the lognormal distribution. From the information given, we have $\mu_L = 20$ and $\sigma_L^2 = 5^2$. Thus, from Equations (5.58) and (5.59),

$$\mu = \ln\left(\frac{20^2}{\sqrt{20^2 + 5^2}}\right) \doteq 2.9654$$

$$\sigma^2 = \ln\left(\frac{20^2 + 5^2}{20^2}\right) \doteq 0.06$$

9. *Beta distribution.* A random variable X is beta-distributed with parameters $\beta_1 > 0$ and $\beta_2 > 0$ if its pdf is given by

$$f(x) = \begin{cases} \dfrac{x^{\beta_1 - 1}(1 - x)^{\beta_2 - 1}}{B(\beta_1, \beta_2)}, & 0 < x < 1 \\ 0, & \text{otherwise} \end{cases} \tag{5.60}$$

where $B(\beta_1, \beta_2) = \Gamma(\beta_1)\Gamma(\beta_2)/\Gamma(\beta_1 + \beta_2)$. The cdf of the beta does not have a closed form in general.

The beta distribution is very flexible and has a finite range from 0 to 1, as shown in Figure 5.23. In practice, we often need a beta distribution defined on a different range, say (a, b), with $a < b$, rather than $(0, 1)$. This is easily accomplished by defining a new random variable

$$Y = a + (b - a)X$$

The mean and variance of Y are given by

$$a + (b - a)\left(\frac{\beta_1}{\beta_1 + \beta_2}\right) \tag{5.61}$$

and

$$(b - a)^2\left(\frac{\beta_1 \beta_2}{(\beta_1 + \beta_2)^2(\beta_1 + \beta_2 + 1)}\right) \tag{5.62}$$

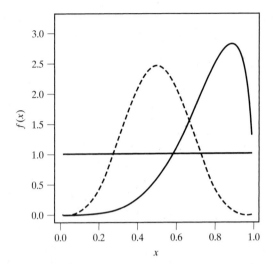

Figure 5.23   The pdf's for several beta distributions.

5.5 POISSON PROCESS

Consider random events such as the arrival of jobs at a job shop, the arrival of e-mail to a mail server, the arrival of boats to a dock, the arrival of calls to a call center, the breakdown of machines in a large factory, and so on. These events may be described by a counting function $N(t)$ defined for all $t \geq 0$. This counting function will represent the number of events that occurred in $[0, t]$. Time zero is the point at which the observation began, regardless of whether an arrival occurred at that instant. For each interval $[0, t]$, the value $N(t)$ is an observation of a random variable where the only possible values that can be assumed by $N(t)$ are the integers $0, 1, 2, \dots$.

The counting process, $\{N(t), t \geq 0\}$, is said to be a Poisson process with mean rate λ if the following assumptions are fulfilled:

1. Arrivals occur one at a time.
2. $\{N(t), t \geq 0\}$ has stationary increments: The distribution of the number of arrivals between t and $t + s$ depends only on the length of the interval s, not on the starting point t. Thus, arrivals are completely at random without rush or slack periods.
3. $\{N(t), t \geq 0\}$ has independent increments: The number of arrivals during nonoverlapping time intervals are independent random variables. Thus, a large or small number of arrivals in one time interval has no effect on the number of arrivals in subsequent time intervals. Future arrivals occur completely at random, independent of the number of arrivals in past time intervals.

If arrivals occur according to a Poisson process, meeting the three preceding assumptions, it can be shown that the probability that $N(t)$ is equal to n is given by

$$P[N(t) = n] = \frac{e^{-\lambda t}(\lambda t)^n}{n!} \quad \text{for } t \geq 0 \text{ and } n = 0, 1, 2, \dots \tag{5.63}$$

Comparing Equation (5.63) to Equation (5.19), it can be seen that $N(t)$ has the Poisson distribution with parameter $\alpha = \lambda t$. Thus, its mean and variance are given by

$$E[N(t)] = \alpha = \lambda t = V[N(t)]$$

For any times s and t such that $s < t$, the assumption of stationary increments implies that the random variable $N(t) - N(s)$, representing the number of arrivals in the interval from s to t, is also Poisson-distributed with mean $\lambda(t - s)$. Thus,

$$P[N(t) - N(s) = n] = \frac{e^{-\lambda(t-s)}[\lambda(t-s)]^n}{n!} \quad \text{for } n = 0, 1, 2, \ldots$$

and

$$E[N(t) - N(s)] = \lambda(t - s) = V[N(t) - N(s)]$$

Now, consider the time at which arrivals occur in a Poisson process. Let the first arrival occur at time A_1, the second occur at time $A_1 + A_2$, and so on, as shown in Figure 5.24. Thus, $A_1, A_2, \ldots$ are successive interarrival times. The first arrival occurs after time t if and only if there are no arrivals in the interval $[0, t]$, so it is seen that

$$\{A_1 > t\} = \{N(t) = 0\}$$

and, therefore,

$$P(A_1 > t) = P[N(t) = 0] = e^{-\lambda t}$$

the last equality following from Equation (5.63). Thus, the probability that the first arrival will occur in $[0, t]$ is given by

$$P(A_1 \le t) = 1 - e^{-\lambda t}$$

which is the cdf for an exponential distribution with parameter λ. Hence, A_1 is distributed exponentially with mean $E(A_1) = 1/\lambda$. It can also be shown that all interarrival times, $A_1, A_2, \ldots$, are exponentially distributed and independent with mean $1/\lambda$. As an alternative definition of a Poisson process, it can be shown that, if interarrival times are distributed exponentially and independently, then the number of arrivals by time t, say $N(t)$, meets the three previously mentioned assumptions and, therefore, is a Poisson process.

Recall that the exponential distribution is memoryless—that is, the probability of a future arrival in a time interval of length s is independent of the time of the last arrival. The probability

Figure 5.24   Arrival process.

of the arrival depends only on the length of the time interval, s. Thus, the memoryless property is related to the properties of independent and stationary increments of the Poisson process.

Additional readings concerning the Poisson process may be obtained from many sources, including Parzen [1999], Feller [1968], and Ross [2002].

Example 5.30
The jobs at a machine shop arrive according to a Poisson process with a mean of $\lambda = 2$ jobs per hour. Therefore, the interarrival times are distributed exponentially, with the expected time between arrivals being $E(A) = 1/\lambda = \frac{1}{2}$ hour.

5.5.1 Properties of a Poisson Process

Several properties of the Poisson process, discussed by Ross [2002] and others, are useful in discrete-system simulation. The first of these properties concerns random splitting. Consider a Poisson process $\{N(t), t \geq 0\}$ having rate λ, as represented by the left side of Figure 5.25.

Suppose that, each time an event occurs, it is classified as either a type I or a type II event. Suppose further that each event is classified as a type I event with probability p and type II event with probability $1 - p$, independently of all other events.

Let $N_1(t)$ and $N_2(t)$ be random variables that denote, respectively, the number of type I and type II events occurring in $[0, t]$. Note that $N(t) = N_1(t) + N_2(t)$. It can be shown that $N_1(t)$ and $N_2(t)$ are both Poisson processes having rates λp and $\lambda(1 - p)$, as shown in Figure 5.25. Furthermore, it can be shown that the two processes are independent.

Example 5.31: (Random Splitting)
Suppose that jobs arrive at a shop in accordance with a Poisson process having rate λ. Suppose further that each arrival is marked "high priority" with probability 1/3 and "low priority" with probability 2/3. Then a type I event would correspond to a high-priority arrival and a type II event would correspond to a low-priority arrival. If $N_1(t)$ and $N_2(t)$ are as just defined, both variables follow the Poisson process, with rates $\lambda/3$ and $2\lambda/3$, respectively.

Example 5.32
The rate in Example 5.31 is $\lambda = 3$ per hour. The probability that no high-priority jobs will arrive in a 2-hour period is given by the Poisson distribution with parameter $\alpha = \lambda pt = 2$. Thus,

$$P(0) = \frac{e^{-2}2^0}{0!} = 0.135$$

Now, consider the opposite situation from random splitting, namely the pooling of two arrival streams. The process of interest is illustrated in Figure 5.26. It can be shown that, if $N_i(t)$ are

Figure 5.25 Random splitting.

Figure 5.26 Pooled process.

random variables representing independent Poisson processes with rates λ_i, for $i = 1$ and 2, then $N(t) = N_1(t) + N_2(t)$ is a Poisson process with rate $\lambda_1 + \lambda_2$.

Example 5.33: (Pooled Process)

A Poisson arrival stream with $\lambda_1 = 10$ arrivals per hour is combined (or pooled) with a Poisson arrival stream with $\lambda_2 = 17$ arrivals per hour. The combined process is a Poisson process with $\lambda = 27$ arrivals per hour.

5.5.2 Nonstationary Poisson Process

If we keep the Poisson Assumptions 1 and 3, but drop Assumption 2 (stationary increments) then we have a *nonstationary Poisson process* (NSPP), which is characterized by $\lambda(t)$, the arrival rate at time t. The NSPP is useful for situations in which the arrival rate varies during the period of interest, including meal times for restaurants, phone calls during business hours, and orders for pizza delivery around 6 P.M.

The key to working with a NSPP is the expected number of arrivals by time t, denoted by

$$\Lambda(t) = \int_0^t \lambda(s)\, ds$$

To be useful as an arrival-rate function, $\lambda(t)$ must be nonnegative and integrable. For a stationary Poisson process with rate λ we have $\Lambda(t) = \lambda t$, as expected.

Let $T_1, T_2, \ldots$ be the arrival times of stationary Poisson process $N(t)$ with $\lambda = 1$, and let $\mathcal{T}_1, \mathcal{T}_2, \ldots$ be the arrival times for a NSPP $\mathcal{N}(t)$ with arrival rate $\lambda(t)$. The fundamental relationship for working with NSPPs is the following:

$$T_i = \Lambda(\mathcal{T}_i)$$
$$\mathcal{T}_i = \Lambda^{-1}(T_i)$$

In words, an NSPP can be transformed into a stationary Poisson process with arrival rate 1, and a stationary Poisson process with arrival rate 1 can be transformed into an NSPP with rate $\lambda(t)$, and the transformation in both cases is related to $\Lambda(t)$.

Example 5.34

Suppose that arrivals to a Post Office occur at a rate of 2 per minute from 8 A.M. until 12 P.M., then drop to 1 every 2 minutes until the day ends at 4 P.M. What is the probability distribution of the number of arrivals between 11 A.M. and 2 P.M?

Let time $t = 0$ correspond to 8 A.M. Then this situation could be modeled as a NSPP $\mathcal{N}(t)$ with rate function

$$\lambda(t) = \begin{cases} 2, & 0 \le t < 4 \\ \dfrac{1}{2}, & 4 \le t \le 8 \end{cases}$$

The expected number of arrivals by time t is therefore

$$\Lambda(t) = \begin{cases} 2t, & 0 \le t < 4 \\ \dfrac{t}{2} + 6, & 4 \le t \le 8 \end{cases}$$

Notice that computing the expected number of arrivals for $4 \le t \le 8$ requires that the integration be done in two parts:

$$\Lambda(t) = \int_0^t \lambda(s)\,ds = \int_0^4 2\,ds + \int_4^t \frac{1}{2}\,ds = \frac{t}{2} + 6$$

Since 2 P.M. and 11 A.M. correspond to times 6 and 3, respectively, we have

$$\begin{aligned} P[\mathcal{N}(6) - \mathcal{N}(3) = k] &= P[N(\Lambda(6)) - N(\Lambda(3)) = k] \\ &= P[N(9) - N(6) = k] \\ &= \frac{e^{9-6}(9-6)^k}{k!} \\ &= \frac{e^3(3)^k}{k!} \end{aligned}$$

where $N(t)$ is a stationary Poisson process with arrival rate 1.

5.6 EMPIRICAL DISTRIBUTIONS

An empirical distribution, which may be either discrete or continuous in form, is a distribution whose parameters are the observed values in a sample of data. This is in contrast to parametric distribution families (such as the exponential, normal, or Poisson), which are characterized by specifying a small number of parameters such as the mean and variance. An empirical distribution may be used when it is impossible or unnecessary to establish that a random variable has any particular parametric distribution. One advantage of an empirical distribution is that nothing is assumed beyond the observed values in the sample; however, this is also a disadvantage because the sample might not cover the entire range of possible values.

Example 5.35:   (Discrete) ——————————————————————————————
Customers at a local restaurant arrive at lunchtime in groups of from one to eight persons. The number of persons per party in the last 300 groups has been observed; the results are summarized

Table 5.3 Arrivals per Party Distribution

Arrivals per Party	Frequency	Relative Frequency	Cumulative Relative Frequency
1	30	0.10	0.10
2	110	0.37	0.47
3	45	0.15	0.62
4	71	0.24	0.86
5	12	0.04	0.90
6	13	0.04	0.94
7	7	0.02	0.96
8	12	0.04	1.00

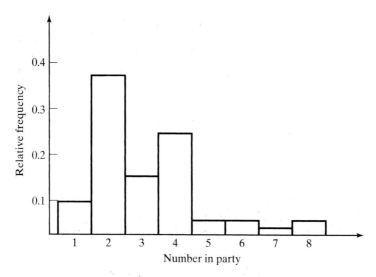

Figure 5.27 Histogram of party size.

in Table 5.3. The relative frequencies appear in Table 5.3 and again in Figure 5.27, which provides a histogram of the data that were gathered. Figure 5.28 provides a cdf of the data. The cdf in Figure 5.28 is called the empirical distribution of the given data.

Example 5.36: (Continuous) _____

The time required to repair a conveyor system that has suffered a failure has been collected for the last 100 instances; the results are shown in Table 5.4. There were 21 instances in which the repair took between 0 and 0.5 hour, and so on. The empirical cdf is shown in Figure 5.29. A piecewise linear curve is formed by the connection of the points of the form $[x, F(x)]$. The points are connected by a straight line. The first connected pair is $(0, 0)$ and $(0.5, 0.21)$; then the points $(0.5, 0.21)$ and $(1.0, 0.33)$ are connected; and so on. More detail on this method is provided in Chapter 8.

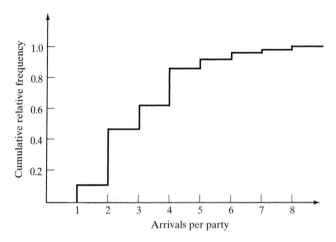

Figure 5.28   Empirical cdf of party size.

Table 5.4   Repair Times for Conveyor

Interval (Hours)	Frequency	Relative Frequency	Cumulative Frequency
$0 < x \le 0.5$	21	0.21	0.21
$0.5 < x \le 1.0$	12	0.12	0.33
$1.0 < x \le 1.5$	29	0.29	0.62
$1.5 < x \le 2.0$	19	0.19	0.81
$2.0 < x \le 2.5$	8	0.08	0.89
$2.5 < x \le 3.0$	11	0.11	1.00

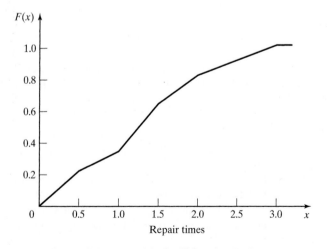

Figure 5.29   Empirical cdf for repair times.

5.7 SUMMARY

In many instances, the world the simulation analyst sees is probabilistic rather than deterministic. The purposes of this chapter were to review several important probability distributions, to familiarize the reader with the notation used in the remainder of the text, and to show applications of the probability distributions in a simulation context.

A major task in simulation is the collection and analysis of input data. One of the first steps in this task is hypothesizing a distributional form for the input data. This is accomplished by comparing the shape of the probability density function or mass function to a histogram of the data and by an understanding that certain physical processes give rise to specific distributions. (Computer software is available to assist in this effort, as will be discussed in Chapter 9.) This chapter was intended to reinforce the properties of various distributions and to give insight into how these distributions arise in practice. In addition, probabilistic models of input data are used in generating random events in a simulation.

Several features that should have made a strong impression on the reader include the differences between discrete, continuous, and empirical distributions; the Poisson process and its properties; and the versatility of the gamma and the Weibull distributions.

REFERENCES

BANKS, J., AND R. G. HEIKES [1984], *Handbook of Tables and Graphs for the Industrial Engineer and Manager*, Reston Publishing, Reston, VA.

DEVORE, J. L. [1999], *Probability and Statistics for Engineers and the Sciences*, 5th ed., Brooks/Cole, Pacific Grove, CA.

FELLER, W. [1968], *An Introduction to Probability Theory and Its Applications*, Vol. I, 3d ed., Wiley, New York.

GORDON, G. [1975], *The Application of GPSS V to Discrete System Simulation*, Prentice-Hall, Englewood Cliffs, NJ.

HADLEY, G., AND T. M. WHITIN [1963], *Analysis of Inventory Systems*, Prentice-Hall, Englewood Cliffs, NJ.

HINES, W. W., AND D. C. MONTGOMERY [1990], *Probability and Statistics in Engineering and Management Science*, 3d ed., Wiley, New York.

LAW, A. M., AND W. D. KELTON [2000], *Simulation Modeling & Analysis*, 3d ed., McGraw–Hill, New York.

PAPOULIS, A. [1990], *Probability and Statistics*, Prentice Hall, Englewood Cliffs, NJ.

PARZEN, E. [1999], *Stochastic Process*, Classics in Applied Mathematics, 24, Society for Industrial & Applied Mathematics, Philadelphia, PA.

PEGDEN, C. D., R. E. SHANNON, AND R. P. SADOWSKI [1995], *Introduction to Simulation Using SIMAN*, 2d ed., McGraw–Hill, New York.

ROSS, S. M. [2002], *Introduction to Probability Models*, 8th ed., Academic Press, New York.

WALPOLE, R. E., AND R. H. MYERS [2002], *Probability and Statistics for Engineers and Scientists*, 7th ed., Prentice Hall, Upper Saddle River, NJ.

EXERCISES

1. A production process manufactures alternators for outboard engines used in recreational boating. On the average, 1% of the alternators will not perform up to the required standards

when tested at the engine assembly plant. When a large shipment of alternators is received at the plant, 100 are tested, and, if more than two are nonconforming, the shipment is returned to the alternator manufacturer. What is the probability of returning a shipment?

2. An industrial chemical that will retard the spread of fire in paint has been developed. The local sales representative has estimated, from past experience, that 48% of the sales calls will result in an order.

(a) What is the probability that the first order will come on the fourth sales call of the day?
(b) If eight sales calls are made in a day, what is the probability of receiving exactly six orders?
(c) If four sales calls are made before lunch, what is the probability that one or fewer results in an order?

3. A recent survey indicated that 82% of single women aged 25 years old will be married in their lifetime. Using the binomial distribution, find the probability that two or three women in a sample of twenty will never be married.

4. The Hawks are currently winning 0.55 of their games. There are 5 games in the next two weeks. What is the probability that they will win more games than they lose?

5. Joe Coledge is the third-string quarterback for the University of Lower Alatoona. The probability that Joe gets into any game is 0.40.

(a) What is the probability that the first game Joe enters is the fourth game of the season?
(b) What is the probability that Joe plays in no more than two of the first five games?

6. For the random variables X_1 and X_2, which are exponentially distributed with parameter $\lambda = 1$, compute $P(X_1 + X_2 > 2)$.

7. Show that the geometric distribution is memoryless.

8. The number of hurricanes hitting the coast of Florida annually has a Poisson distribution with a mean of 0.8.

(a) What is the probability that more than two hurricanes will hit the Florida coast in a year?
(b) What is the probability that exactly one hurricane will hit the coast of Florida in a year?

9. Arrivals at a bank teller's drive-through window are Poisson-distributed at the rate of 1.2 per minute.

(a) What is the probability of zero arrivals in the next minute?
(b) What is the probability of zero arrivals in the next 2 minutes?

10. Records indicate that 1.8% of the entering students at a large state university drop out of school by midterm. What is the probability that three or fewer students will drop out of a random group of 200 entering students?

11. Lane Braintwain is quite a popular student. Lane receives, on the average, four phone calls a night (Poisson distributed). What is the probability that, tomorrow night, the number of calls received will exceed the average by more than one standard deviation?

12. Lead-time demand for condenser units is Poisson-distributed with a mean of 6 units. Prepare a table, for the inventory manager, that will indicate the order level to achieve protection of the following levels: 50%, 80%, 90%, 95%, 97%, 97.5%, 99%, 99.5%, and 99.9%.

13. A random variable X that has pmf given by $p(x) = 1/(n + 1)$ over the range $R_X = \{0, 1, 2, \ldots, n\}$ is said to have a discrete uniform distribution.

   (a) Find the mean and variance of this distribution. (*Hint:*

   $$\sum_{i=1}^{n} i = \frac{n(n + 1)}{2} \text{ and } \sum_{i=1}^{n} i^2 = \frac{n(n + 1)(2n + 1)}{6} \Bigg)$$

   (b) If $R_X = \{a, a + 1, a + 2, \ldots, b\}$, compute the mean and variance of X.

14. The lifetime, in years, of a satellite placed in orbit is given by the following pdf:

   $$f(x) = \begin{cases} 0.4e^{-0.4x}, & x \geq 0 \\ 0, & \text{otherwise} \end{cases}$$

   (a) What is the probability that this satellite is still "alive" after 5 years?
   (b) What is the probability that the satellite dies between 3 and 6 years from the time it is placed in orbit?

15. A mainframe computer crashes in accordance with a Poisson process, with a mean rate of one crash every 36 hours. Determine the probability that the next crash will occur between 24 and 48 hours after the last crash.

16. (The Poisson distribution can be used to approximate the binomial distribution when n is large and p is small—say, p less than 0.1. In utilizing the Poisson approximation, let $\lambda = np$.) In the production of ball bearings, bubbles or depressions occur, rendering the ball bearing unfit for sale. It has been noted that, on the average, one in every 800 of the ball bearings has one or more of these defects. What is the probability that a random sample of 4000 will yield fewer than three ball bearings with bubbles or depressions?

17. For an exponentially distributed random variable X, find the value of λ that satisfies the following relationship:

   $$P(X \leq 3) = 0.9P(X \leq 4)$$

18. Accidents at an industrial site occur one at a time, independently, and completely at random, at a mean rate of one per week. What is the probability that no accidents occur in the next three weeks?

19. A component has an exponential time-to-failure distribution with mean of 10,000 hours.

   (a) The component has already been in operation for its mean life. What is the probability that it will fail by 15,000 hours?
   (b) At 15,000 hours, the component is still in operation. What is the probability that it will operate for another 5000 hours?

20. Suppose that a Die-Hardly-Ever battery has an exponential time-to-failure distribution with a mean of 48 months. At 60 months, the battery is still operating.

    (a) What is the probability that this battery is going to die in the next 12 months?
    (b) What is the probability that the battery dies in an odd year of its life?
    (c) If the battery is operating at 60 months, compute the expected additional months of life.

21. The time to service customers at a bank teller's drive-through window is exponentially distributed with a mean of 50 seconds.

    (a) What is the probability that the two customers in front of an arriving customer will each take less than 60 seconds to complete their transactions?
    (b) What is the probability that the two customers in front will finish their transactions so that an arriving customer can reach the teller's window within 2 minutes?

22. Determine the variance, $V(X)$, of the triangular distribution.

23. The daily use of water, in thousands of liters, at the Hardscrabble Tool and Die Works follows a gamma distribution having shape parameter 2 and scale parameter 1/4. What is the probability that the demand exceeds 4000 liters on any given day?

24. When Admiral Byrd went to the North Pole, he wore battery-powered thermal underwear. The batteries failed instantaneously rather than gradually. The batteries had a life that was exponentially distributed, with a mean of 12 days. The trip took 30 days. Admiral Byrd packed three batteries. What is the probability that three batteries would be a number sufficient to keep the Admiral warm?

25. The time intervals between hits on a web page from remote computers are exponentially distributed, with a mean of 15 seconds. Find the probability that the third hit connection occurs after 30 seconds have elapsed.

26. The rail shuttle cars at Atlanta airport have a dual electrical braking system. A rail car switches to the standby system automatically if the first system fails. If both systems fail, there will be a crash! Assume that the life of a single electrical braking system is exponentially distributed, with a mean of 4,000 operating hours. If the systems are inspected every 5,000 operating hours, what is the probability that a rail car will not crash before that time?

27. Suppose that cars arriving at a toll booth follow a Poisson process with a mean interarrival time of 15 seconds. What is the probability that up to one minute will elapse until three cars have arrived?

28. Suppose that an average of 30 customers per hour arrive at the Sticky Donut Shop in accordance with a Poisson process. What is the probability that more than 5 minutes will elapse before both of the next two customers walk through the door?

29. Professor Dipsy Doodle gives six problems on each exam. Each problem requires an average of 30 minutes grading time for the entire class of 15 students. The grading time for each problem is exponentially distributed, and the problems are independent of each other.

(a) What is the probability that the Professor will finish the grading in $2\frac{1}{2}$ hours or less?

(b) What is the most likely grading time?

(c) What is the expected grading time?

30. An aircraft has dual hydraulic systems. The aircraft switches to the standby system automatically if the first system fails. If both systems have failed, the plane will crash. Assume that the life of a hydraulic system is exponentially distributed, with a mean of 2000 air hours.

(a) If the hydraulic systems are inspected every 2500 hours, what is the probability that an aircraft will crash before that time?

(b) What danger would there be in moving the inspection point to 3000 hours?

31. Show that the beta distribution becomes the uniform distribution over the unit interval when $\beta_1 = \beta_2 = 1$.

32. Lead time is gamma-distributed in 100's of units, with the shape parameter 3 and the scale parameter 1. What is the probability that the lead time exceeds 2 (hundred) units during an upcoming cycle?

33. Lifetime of an inexpensive video card for a PC, in months, denoted by the random variable X, is gamma-distributed with $\beta = 4$ and $\theta = 1/16$. What is the probability that the card will last for at least 2 years?

34. Many states have license plates that conform to the following format:

letter   letter   letter   number   number   number

The numbers are at random, ranging from 100 to 999.

(a) What is the probability that the next two plates seen (at random) will have numbers of 500 or higher?

(b) What is the probability that the sum of the next two plates seen (at random) will have a total of 1000 or higher? (*Hint:* Approximate the discrete uniform distribution with a continuous uniform distribution. The sum of two independent uniform distributions is a triangular distribution.)

35. Let X be a random variable that is normally distributed, with mean 10 and variance 4. Find the values a and b such that $P(a < X < b) = 0.90$ and $|\mu - a| = |\mu - b|$.

36. Given the following distributions,

Normal (10, 4)
Triangular (4, 10, 16)
Uniform (4, 16)
find the probability that $6 < X < 8$ for each of the distributions.

37. Lead time for an item is approximated by a normal distribution with a mean of 20 days and a variance of 4 days^2. Find values of lead time that will be exceeded only 1%, 5%, and 10% of the time.

38. IQ scores are normally distributed throughout society, with mean 100 and standard deviation 15.

    (a) A person with an IQ of 140 or higher is called a "genius." What proportion of society is in the genius category?
    (b) What proportion of society will miss the genius category by 5 or less points?
    (c) Suppose that an IQ of 110 or higher is required to make it through an accredited college or university. What proportion of society could be eliminated from completing a higher education by a low IQ score?

39. Three shafts are made and assembled into a linkage. The length of each shaft, in centimeters, is distributed as follows:

    Shaft 1: $N(60, 0.09)$
    Shaft 2: $N(40, 0.05)$
    Shaft 3: $N(50, 0.11)$

    (a) What is the distribution of the length of the linkage?
    (b) What is the probability that the linkage will be longer than 150.2 centimeters?
    (c) The tolerance limits for the assembly are (149.83, 150.21). What proportion of assemblies are within the tolerance limits? (*Hint:* If $\{X_i\}$ are n independent normal random variables, and if X_i, has mean μ_i and variance σ_i^2, then the sum

    $$Y = X_1 + X_2 + \cdots + X_n$$

    is normal with mean $\sum_{i=1}^{n} \mu_i$ and variance $\sum_{i=1}^{n} \sigma_i^2$.)

40. The circumferences of battery posts in a nickel–cadmium battery are Weibull-distributed with $v = 3.25$ centimeters, $\beta = 1/3$, and $\alpha = 0.005$ centimeters.

    (a) Find the probability that a battery post chosen at random will have a circumference larger than 3.40 centimeters.
    (b) If battery posts are larger than 3.50 centimeters, they will not go through the hole provided; if they are smaller than 3.30 centimeters, the clamp will not tighten sufficiently. What proportion of posts will have to be scrapped for one of these reasons?

41. The time to failure of a nickel–cadmium battery is Weibull distributed with parameters $v = 0$, $\beta = 1/4$, and $\alpha = 1/2$ years.

    (a) Find the fraction of batteries that are expected to fail prior to 1.5 years.
    (b) What fraction of batteries are expected to last longer than the mean life?
    (c) What fraction of batteries are expected to fail between 1.5 and 2.5 years?

42. Demand for electricity at Gipgip Pig Farm for the merry merry month of May has a triangular distribution, with $a = 100$ kwh and $c = 1800$ kwh. The median kwh is 1425. Compute the modal value of kwh for the month.

43. The time to failure on an electronic subassembly can be modeled by a Weibull distribution whose location parameter is zero, $\beta = 1/2$ and $\alpha = 1000$ hours.

    (a) What is the mean time to failure?
    (b) What fraction of these subassemblies will fail by 3000 hours?

44. The gross weight of three-axle trucks that have been checked at the Hahira Inspection Station on Interstate Highway 85 follows a Weibull distribution with parameters $v = 6.8$ tons, $\beta = 1.5$, and $\alpha = 1/2$ ton. Determine the appropriate weight limit such that 0.01 of the trucks will be cited for traveling overweight.

45. The current reading on Sag Revas's gas-mileage indicator is an average of 25.3 miles per gallon. Assume that gas mileage on Sag's car follows a triangular distribution with a minimum value of zero and a maximum value of 50 miles per gallon. What is the value of the median?

46. A postal letter carrier has a route consisting of five segments with the time in minutes to complete each segment being normally distributed, with means and variances as shown:

Tennyson Place	$N(38, 16)$
Windsor Parkway	$N(99, 29)$
Knob Hill Appartments	$N(85, 25)$
Evergreen Drive	$N(73, 20)$
Chastain Shopping Center	$N(52, 12)$

    In addition to the times just mentioned, the letter carrier must organize the mail at the central office, which activity requires a time that is distributed by $N(90, 25)$. The drive to the starting point of the route requires a time that is distributed $N(10, 4)$. The return from the route requires a time that is distributed $N(15, 4)$. The letter carrier then performs administrative tasks with a time that is distributed $N(30, 9)$.

    (a) What is the expected length of the letter carrier's work day?
    (b) Overtime occurs after eight hours of work on a given day. What is the probability that the letter carrier works overtime on any given day?
    (c) What is the probability that the letter carrier works overtime on two or more days in a six-day week?
    (d) What is the probability that the route will be completed within ±24 minutes of eight hours on any given day? (*Hint:* See Exercise 39.)

47. The time to failure of a WD—1 computer chip is known to be Weibull distributed with parameters $v = 0$, $\beta = 1/2$ and $\alpha = 400$ days. Find the fraction expected to survive 600 days.

48. The TV sets on display at Schocker's Department Store are hooked up such that, when one fails, a model exactly like the one that failed will switch on. Three such units are hooked up in

this series arrangement. The lives of these TVs are independent of one another. Each TV has a life which is exponentially distributed, with a mean of 10,000 hours. Find the probability that the combined life of the system is greater than 32,000 hours.

49. High temperature in Biloxi, Mississippi on July 21, denoted by the random variable X, has the following probability density function, where X is in degrees F.

$$f(x) = \begin{cases} \dfrac{2(x - 85)}{119}, & 85 \leq x \leq 92 \\ \dfrac{2(102 - x)}{170}, & 92 < x \leq 102 \\ 0, & \text{otherwise} \end{cases}$$

(a) What is the variance of the temperature, $V(X)$? (If you worked Exercise 22, this is quite easy.)
(b) What is the median temperature?
(c) What is the modal temperature?

50. The time to failure of Eastinghome light bulbs is Weibull distributed with $v = 1.8 \times 10^3$ hours, $\beta = 1/2$, and $\alpha = 1/3 \times 10^3$ hours.

(a) What fraction of bulbs are expected to last longer than the mean lifetime?
(b) What is the median lifetime of a light bulb?

51. Lead-time demand is gamma-distributed in 100's of units, with shape parameter 2 and scale parameter 1/4. What is the probability that the lead time exceeds 4 (hundred) units during an upcoming cycle?

52. Let time $t = 0$ correspond to 6 A.M., and suppose that the arrival rate (in arrivals per hour) of customers to a breakfast restaurant that is open from 6 to 9 A.M. is

$$\lambda(t) = \begin{cases} 30, & 0 \leq t < 1 \\ 45, & 1 \leq t < 2 \\ 20, & 2 \leq t \leq 4 \end{cases}$$

Assuming a NSPP model is appropriate, do the following: (a) Derive $\Lambda(t)$. (b) Compute the expected number of arrivals between 6:30 and 8:30 A.M. (c) Compute the probability that there are fewer than 60 arrivals between 6:30 and 8:30 A.M.

6

Queueing Models

Simulation is often used in the analysis of queueing models. In a simple but typical queueing model, shown in Figure 6.1, customers arrive from time to time and join a *queue* (waiting line), are eventually served, and finally leave the system. The term "customer" refers to any type of entity that can be viewed as requesting "service" from a system. Therefore, many service facilities, production systems, repair and maintenance facilities, communications and computer systems, and transport and material-handling systems can be viewed as queueing systems.

Queueing models, whether solved mathematically or analyzed through simulation, provide the analyst with a powerful tool for designing and evaluating the performance of queueing systems. Typical measures of system performance include server utilization (percentage of time a server is busy), length of waiting lines, and delays of customers. Quite often, when designing or attempting to improve a queueing system, the analyst (or decision maker) is involved in tradeoffs between server utilization and customer satisfaction in terms of line lengths and delays. Queueing theory and simulation analysis are used to predict these measures of system performance as a function of the input parameters. The input parameters include the arrival rate of customers, the service demands of customers, the rate at which a server works, and the number and arrangement of servers. To a certain degree, some of the input parameters are under management's direct control. Consequently, the performance measures could be under their indirect control, provided that the relationship between the performance measures and the input parameters is adequately understood for the given system.

For relatively simple systems, these performance measures can be computed mathematically—at great savings in time and expense as compared with the use of a simulation model—but, for realistic models of complex systems, simulation is usually required. Nevertheless, analytically

Figure 6.1   Simple queueing model.

tractable models, although usually requiring many simplifying assumptions, are valuable for rough-cut estimates of system performance. These rough-cut estimates may then be refined by use of a detailed and more realistic simulation model. Simple models are also useful for developing an understanding of the dynamic behavior of queueing systems and the relationships between various performance measures. This chapter will not develop the mathematical theory of queues but instead will discuss some of the well-known models. For an elementary treatment of queueing theory, the reader is referred to the survey chapters in Hillier and Lieberman [2005], Wagner [1975] or Winston [1997]. More extensive treatments with a view toward applications are given by Cooper [1990], Gross and Harris [1997], Hall [1991] and Nelson [1995]. The latter two texts especially emphasize engineering and management applications.

This chapter discusses the general characteristics of queues, the meanings and relationships of the important performance measures, estimation of the mean measures of performance from a simulation, the effect of varying the input parameters, and the mathematical solution of a small number of important and basic queueing models.

6.1 CHARACTERISTICS OF QUEUEING SYSTEMS

The key elements of a queueing system are the customers and servers. The term "customer" can refer to people, machines, trucks, mechanics, patients, pallets, airplanes, e-mail, cases, orders, or dirty clothes—anything that arrives at a facility and requires service. The term "server" might refer to receptionists, repairpersons, mechanics, tool-crib clerks, medical personnel, automatic storage and retrieval machines (e.g., cranes), runways at an airport, automatic packers, order pickers, CPUs in a computer, or washing machines—any resource (person, machine, etc.) that provides the requested service. Although the terminology employed will be that of a customer arriving to a service facility, sometimes the server moves to the customer; for example, a repairperson moving to a broken machine. This in no way invalidates the models but is merely a matter of terminology. Table 6.1 lists a number of different systems together with a subsystem consisting of "arriving customers" and one or more "servers." The remainder of this section describes the elements of a queueing system in more detail.

6.1.1 The Calling Population

The population of potential customers, referred to as the *calling population*, may be assumed to be finite or infinite. For example, consider a bank of five machines that are curing tires. After an

Table 6.1   Examples of Queueing Systems

System	Customers	Server(s)
Reception desk	People	Receptionist
Repair facility	Machines	Repairperson
Garage	Trucks	Mechanic
Tool crib	Mechanics	Tool-crib clerk
Hospital	Patients	Nurses
Warehouse	Pallets	Fork-lift Truck
Airport	Airplanes	Runway
Production line	Cases	Case-packer
Warehouse	Orders	Order-picker
Road network	Cars	Traffic light
Grocery	Shoppers	Checkout station
Laundry	Dirty linen	Washing machines/dryers
Job shop	Jobs	Machines/workers
Lumberyard	Trucks	Overhead crane
Sawmill	Logs	Saws
Computer	Jobs	CPU, disk, CDs
Telephone	Calls	Exchange
Ticket office	Football fans	Clerk
Mass transit	Riders	Buses, trains

interval of time, a machine automatically opens and must be attended by a worker who removes the tire and puts an uncured tire into the machine. The machines are the "customers," who "arrive" at the instant they automatically open. The worker is the "server," who "serves" an open machine as soon as possible. The calling population is finite and consists of the five machines.

In systems with a large population of potential customers, the calling population is usually assumed to be infinite. For such systems, this assumption is usually innocuous and, furthermore, it might simplify the model. Examples of infinite populations include the potential customers of a restaurant, bank, or other similar service facility and also very large groups of machines serviced by a technician. Even though the actual population could be finite but large, it is generally safe to use infinite population models—provided that the number of customers being served or waiting for service at any given time is a small proportion of the population of potential customers.

The main difference between finite and infinite population models is how the arrival rate is defined. In an infinite population model, the arrival rate (i.e., the average number of arrivals per unit of time) is not affected by the number of customers who have left the calling population and joined the queueing system. When the arrival process is homogeneous over time (e.g., there are no "rush hours"), the arrival rate is usually assumed to be constant. On the other hand, for finite calling-population models, the arrival rate to the queueing system does depend on the number of customers being served and waiting. To take an extreme case, suppose that the calling population has one member, for example, a corporate jet. When the corporate jet is being serviced by the team of mechanics who are on duty 24 hours per day, the arrival rate is zero, because there are no other potential customers (jets) who can arrive at the service facility (team of mechanics). A more typical example is that of the five tire-curing machines serviced by a single worker. When all five are closed and curing a tire, the worker is idle and the arrival rate is at a maximum, but the instant a machine opens and requires service, the arrival rate decreases. At those times when all five are open (so four machines are waiting for service while the worker is attending the other one), the arrival rate is zero; that is, no arrival is possible until the worker finishes with a machine, in which case it returns to the calling population and becomes a potential arrival. It may seem odd that the arrival rate is at its maximum when all five machines are closed. But the arrival rate is defined as the expected number of arrivals in the next unit of time, so it becomes clear that this expectation is largest when all machines could potentially open in the next unit of time.

6.1.2 System Capacity

In many queueing systems, there is a limit to the number of customers that may be in the waiting line or system. For example, an automatic car wash might have room for only 10 cars to wait in line to enter the mechanism. It might be too dangerous (or illegal) for cars to wait in the street. An arriving customer who finds the system full does not enter but returns immediately to the calling population. Some systems, such as concert ticket sales for students, may be considered as having unlimited capacity, since there are no limits on the number of students allowed to wait to purchase tickets. As will be seen later, when a system has limited capacity, a distinction is made between the arrival rate (i.e., the number of arrivals per time unit) and the effective arrival rate (i.e., the number who arrive and enter the system per time unit).

6.1.3 The Arrival Process

The arrival process for infinite-population models is usually characterized in terms of interarrival times of successive customers. Arrivals may occur at scheduled times or at random times. When at random times, the interarrival times are usually characterized by a probability distribution. In addition, customers may arrive one at a time or in batches. The batch may be of constant size or of random size.

The most important model for random arrivals is the Poisson arrival process. If A_n represents the interarrival time between customer $n - 1$ and customer n (A_1 is the actual arrival time of the first customer), then, for a Poisson arrival process, A_n is exponentially distributed with mean $1/\lambda$ time units. The arrival rate is λ customers per time unit. The number of arrivals in a time interval of length t, say $N(t)$, has the Poisson distribution with mean λt customers. For further discussion of the relationship between the Poisson distribution and the exponential distribution, the reader is referred to Section 5.5.

The Poisson arrival process has been employed successfully as a model of the arrival of people to restaurants, drive-in banks, and other service facilities; the arrival of telephone calls to a call center; the arrival of demands, or orders for a service or product; and the arrival of failed components or machines to a repair facility.

A second important class of arrivals is scheduled arrivals, such as patients to a physician's office or scheduled airline flight arrivals to an airport. In this case, the interarrival times $\{A_n, n = 1, 2, \ldots\}$ could be either *constant* or *constant plus or minus a small random amount* to represent early or late arrivals.

A third situation occurs when at least one customer is assumed to always be present in the queue, so that the server is never idle because of a lack of customers. For example, the "customers" may represent raw material for a product, and sufficient raw material is assumed to be always available.

For finite-population models, the arrival process is characterized in a completely different fashion. Define a customer as *pending* when that customer is outside the queueing system and a member of the potential calling population. For example, a tire-curing machine is "pending" when it is closed and curing a tire, and it becomes "not pending" the instant it opens and demands service from the worker. Define a *runtime* of a given customer as the length of time from departure from the queueing system until that customer's next arrival to the queue. Let $A_1^{(i)}, A_2^{(i)}, \ldots$ be the successive runtimes of customer i, and let $S_1^{(i)}, S_2^{(i)}, \ldots$ be the corresponding successive system times; that is, $S_n^{(i)}$ is the total time spent in system by customer i during the nth visit. Figure 6.2 illustrates these concepts for machine 3 in the tire-curing example. The total arrival process is the superposition of the arrival times of all customers. Figure 6.2 shows the first and second arrival of machine 3, but these two times are not necessarily two successive arrivals to the system. For instance, if it is assumed that all machines are pending at time 0, the first arrival to the system occurs at time $A_1 = \min\{A_1^{(1)}, A_1^{(2)}, A_1^{(3)}, A_1^{(4)}, A_1^{(5)}\}$. If $A_1 = A_1^{(2)}$, then machine 2 is the first arrival (i.e., the first to open) after time 0. As discussed earlier, the arrival rate is not constant but is a function of the number of pending customers.

One important application of finite-population models is the machine-repair problem. The machines are the customers, and a runtime is also called time to failure. When a machine fails, it "arrives" at the queueing system (the repair facility) and remains there until it is "served" (repaired). Times to failure for a given class of machine have been characterized by the exponential, the Weibull, and the gamma distributions. Models with an exponential runtime are sometimes analytically tractable; an example is given in Section 6.5. Successive times to failure are usually assumed to be statistically independent, but they could depend on other factors, such as the age of a machine since its last major overhaul.

6.1.4 Queue Behavior and Queue Discipline

Queue behavior refers to the actions of customers while in a queue waiting for service to begin. In some situations, there is a possibility that incoming customers will balk (leave when they see that the line is too long), renege (leave after being in the line when they see that the line is

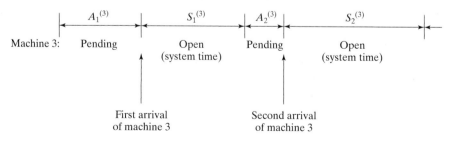

Figure 6.2   Arrival process for a finite-population model.

moving too slowly), or jockey (move from one line to another if they think they have chosen a slow line).

Queue discipline refers to the logical ordering of customers in a queue and determines which customer will be chosen for service when a server becomes free. Common queue disciplines include first-in–first-out (FIFO); last-in–first-out (LIFO); service in random order (SIRO); shortest processing time first (SPT); and service according to priority (PR). In a job shop, queue disciplines are sometimes based on due dates and on expected processing time for a given type of job. Notice that a FIFO queue discipline implies that services begin in the same order as arrivals, but that customers could leave the system in a different order because of different-length service times.

6.1.5 Service Times and the Service Mechanism

The service times of successive arrivals are denoted by $S_1, S_2, S_3, \ldots$ They may be constant or of random duration. In the latter case, $\{S_1, S_2, S_3, \ldots\}$ is usually characterized as a sequence of independent and identically distributed random variables. The exponential, Weibull, gamma, lognormal and truncated normal distributions have all been used successfully as models of service times in different situations. Sometimes services are identically distributed for all customers of a given type or class or priority, whereas customers of different types might have completely different service-time distributions. In addition, in some systems, service times depend upon the time of day or upon the length of the waiting line. For example, servers might work faster than usual when the waiting line is long, thus effectively reducing the service times.

A queueing system consists of a number of service centers and interconnecting queues. Each service center consists of some number of servers, c, working in parallel; that is, upon getting to the head of the line, a customer takes the first available server. Parallel service mechanisms are either single server ($c = 1$), multiple server ($1 < c < \infty$), or unlimited servers ($c = \infty$). A self-service facility is usually characterized as having an unlimited number of servers.

Example 6.1 _____
Consider a discount warehouse where customers may either serve themselves or wait for one of three clerks, then finally leave after paying a single cashier. The system is represented by the flow diagram in Figure 6.3. The subsystem, consisting of queue 2 and service center 2, is shown in more detail in Figure 6.4. Other variations of service mechanisms include batch service (a server serving several customers simultaneously), and a customer's requiring several servers simultaneously. In the discount warehouse, a clerk might pick several small orders at the same time, but it may take two of the clerks to handle one heavy item.

Example 6.2 _____
A candy manufacturer has a production line that consists of three machines separated by inventory-in-process buffers. The first machine makes and wraps the individual pieces of candy, the second packs 50 pieces in a box, and the third machine seals and wraps the box. The two inventory buffers have capacities of 1000 boxes each. As illustrated by Figure 6.5, the system is modeled as having three service centers, each center having $c = 1$ server (a machine), with queue capacity constraints between machines. It is assumed that a sufficient supply of raw material is always available at the first queue. Because of the queue capacity constraints, machine 1 shuts down

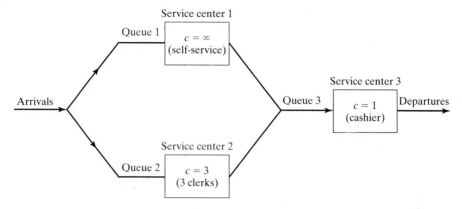

Figure 6.3 Discount warehouse with three service centers.

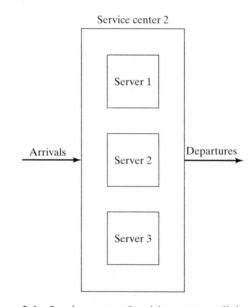

Figure 6.4 Service center 2, with $c = 3$ parallel servers.

Figure 6.5 Candy-production line.

whenever its inventory buffer (queue 2) fills to capacity, and machine 2 shuts down whenever its buffer empties. In brief, the system consists of three single-server queues in series with queue capacity constraints and a continuous arrival stream at the first queue.

6.2 QUEUEING NOTATION

Recognizing the diversity of queueing systems, Kendall [1953] proposed a notational system for parallel server systems which has been widely adopted. An abridged version of this convention is based on the format $A/B/c/N/K$. These letters represent the following system characteristics:

A represents the interarrival-time distribution.

B represents the service-time distribution.

c represents the number of parallel servers.

N represents the system capacity.

K represents the size of the calling population.

Common symbols for A and B include M (exponential or Markov), D (constant or deterministic), E_k (Erlang of order k), PH (phase-type), H (hyperexponential), G (arbitrary or general), and GI (general independent).

For example, $M/M/1/\infty/\infty$ indicates a single-server system that has unlimited queue capacity and an infinite population of potential arrivals. The interarrival times and service times are exponentially distributed. When N and K are infinite, they may be dropped from the notation. For example, $M/M/1/\infty/\infty$ is often shortened to $M/M/1$. The tire-curing system can be initially represented by $G/G/1/5/5$.

Additional notation used throughout the remainder of this chapter for parallel server systems is listed in Table 6.2. The meanings may vary slightly from system to system. All systems will be assumed to have a FIFO queue discipline.

6.3 LONG-RUN MEASURES OF PERFORMANCE OF QUEUEING SYSTEMS

The primary long-run measures of performance of queueing systems are the long-run time-average number of customers in the system (L) and in the queue (L_Q), the long-run average time spent in system (w) and in the queue (w_Q) per customer, and the server utilization, or proportion of time that a server is busy (ρ). The term "system" usually refers to the waiting line plus the service mechanism, but, in general, can refer to any subsystem of the queueing system; whereas the term "queue" refers to the waiting line alone. Other measures of performance of interest include the long-run proportion of customers who are delayed in queue longer than t_0 time units, the long-run proportion of customers turned away because of capacity constraints, and the long-run proportion of time the waiting line contains more than k_0 customers.

Table 6.2 Queueing Notation for Parallel Server Systems

P_n	Steady-state probability of having n customers in system
$P_n(t)$	Probability of n customers in system at time t
λ	Arrival rate
λ_e	Effective arrival rate
μ	Service rate of one server
ρ	Server utilization
A_n	Interarrival time between customers $n-1$ and n
S_n	Service time of the nth arriving customer
W_n	Total time spent in system by the nth arriving customer
W_n^Q	Total time spent in the waiting line by customer n
$L(t)$	The number of customers in system at time t
$L_Q(t)$	The number of customers in queue at time t
L	Long-run time-average number of customers in system
L_Q	Long-run time-average number of customers in queue
w	Long-run average time spent in system per customer
w_Q	Long-run average time spent in queue per customer

This section defines the major measures of performance for a general $G/G/c/N/K$ queueing system, discusses their relationships, and shows how they can be estimated from a simulation run. There are two types of estimators: an ordinary sample average, and a time-integrated (or time-weighted) sample average.

6.3.1 Time-Average Number in System L

Consider a queueing system over a period of time T, and let $L(t)$ denote the number of customers in the system at time t. A simulation of such a system is shown in Figure 6.6.

Let T_i denote the total time during $[0, T]$ in which the system contained exactly i customers. In Figure 6.6, it is seen that $T_0 = 3$, $T_1 = 12$, $T_2 = 4$, and $T_3 = 1$. (The line segments whose lengths total $T_1 = 12$ are labelled "T_1" in Figure 6.6, etc.) In general, $\sum_{i=0}^{\infty} T_i = T$. The time-weighted-average number in a system is defined by

$$\widehat{L} = \frac{1}{T} \sum_{i=0}^{\infty} iT_i = \sum_{i=0}^{\infty} i \left(\frac{T_i}{T} \right) \tag{6.1}$$

For Figure 6.6, $\widehat{L} = [0(3) + 1(12) + 2(4) + 3(1)]/20 = 23/20 = 1.15$ customers. Notice that T_i/T is the proportion of time the system contains exactly i customers. The estimator $\widehat{L}$ is an example of a time-weighted average.

By considering Figure 6.6, it can be seen that the total area under the function $L(t)$ can be decomposed into rectangles of height i and length T_i. For example, the rectangle of area $3 \times T_3$ has base running from $t = 7$ to $t = 8$ (thus $T_3 = 1$); however, most of the rectangles are broken into parts, such as the rectangle of area $2 \times T_2$ which has part of its base between $t = 5$ and $t = 7$ and the remainder from $t = 8$ to $t = 10$ (thus $T_2 = 2 + 2 = 4$). It follows that the total area is

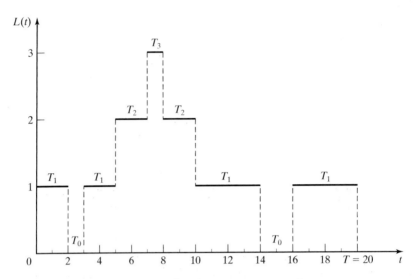

Figure 6.6   Number in system, $L(t)$, at time t.

given by $\sum_{i=0}^{\infty} iT_i = \int_0^T L(t)\,dt$, and, therefore, that

$$\widehat{L} = \frac{1}{T}\sum_{i=0}^{\infty} iT_i = \frac{1}{T}\int_0^T L(t)\,dt \tag{6.2}$$

The expressions in Equations (6.1) and (6.2) are always equal for any queueing system, regardless of the number of servers, the queue discipline, or any other special circumstances. Equation (6.2) justifies the terminology *time-integrated average*.

Many queueing systems exhibit a certain kind of long-run stability in terms of their average performance. For such systems, as time T gets large, the observed time-average number in the system $\widehat{L}$ approaches a limiting value, say L, which is called the long-run time-average number in system—that is, with probability 1,

$$\widehat{L} = \frac{1}{T}\int_0^T L(t)\,dt \longrightarrow L \text{ as } T \longrightarrow \infty \tag{6.3}$$

The estimator $\widehat{L}$ is said to be strongly consistent for L. If simulation run length T is sufficiently long, the estimator $\widehat{L}$ becomes arbitrarily close to L. Unfortunately, for $T < \infty$, $\widehat{L}$ depends on the initial conditions at time 0.

Equations (6.2) and (6.3) can be applied to any subsystem of a queueing system as well as they can to the whole system. If $L_Q(t)$ denotes the number of customers waiting in line, and T_i^Q denotes the total time during $[0, T]$ in which exactly i customers are waiting in line, then

$$\widehat{L_Q} = \frac{1}{T}\sum_{i=0}^{\infty} iT_i^Q = \frac{1}{T}\int_0^T L_Q(t)\,dt \longrightarrow L_Q \text{ as } T \longrightarrow \infty \tag{6.4}$$

where $\widehat{L_Q}$ is the observed time-average number of customers waiting in line from time 0 to time T and L_Q is the long-run time-average number waiting in line.

Example 6.3

Suppose that Figure 6.6 represents a single-server queue—that is, a $G/G/1/N/K$ queueing system ($N \geq 3, K \geq 3$). Then the number of customers waiting in line is given by $L_Q(t)$, defined by

$$L_Q(t) = \begin{cases} 0 & \text{if } L(t) = 0 \\ L(t) - 1 & \text{if } L(t) \geq 1 \end{cases}$$

and shown in Figure 6.7. Thus, $T_0^Q = 5 + 10 = 15$, $T_1^Q = 2 + 2 = 4$, and $T_2^Q = 1$. Therefore,

$$\widehat{L_Q} = \frac{0(15) + 1(4) + 2(1)}{20} = 0.3 \text{ customers}$$

6.3.2 Average Time Spent in System Per Customer *w*

If we simulate a queueing system for some period of time, say T, then we can record the time each customer spends in the system during $[0, T]$, say $W_1, W_2, \ldots, W_N$, where N is the number of arrivals during $[0, T]$. The average time spent in system per customer, called the average system time, is given by the ordinary sample average

$$\widehat{w} = \frac{1}{N} \sum_{i=1}^{N} W_i \tag{6.5}$$

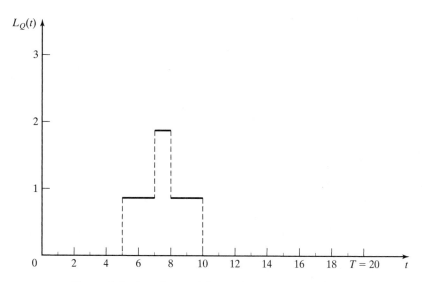

Figure 6.7   Number waiting in line, $L_Q(t)$, at time t.

For stable systems, as $N \longrightarrow \infty$,

$$\widehat{w} \longrightarrow w \tag{6.6}$$

with probability 1, where w is called the long-run average system time.

   If the system under consideration is the queue alone, Equations (6.5) and (6.6) are written as

$$\widehat{w}_Q = \frac{1}{N} \sum_{i=1}^{N} W_i^Q \longrightarrow w_Q \quad \text{as } N \longrightarrow \infty \tag{6.7}$$

where W_i^Q is the total time customer i spends waiting in queue, $\widehat{w}_Q$ is the observed average time spent in queue (called delay), and w_Q is the long-run average delay per customer. The estimators $\widehat{w}$ and $\widehat{w}_Q$ are influenced by initial conditions at time 0 and the run length T, analogously to $\widehat{L}$.

Example 6.4

For the system history shown in Figure 6.6, $N = 5$ customers arrive, $W_1 = 2$, and $W_5 = 20-16 = 4$, but W_2, W_3, and W_4 cannot be computed unless more is known about the system. Assume that the system has a single server and a FIFO queue discipline. This implies that customers will depart from the system in the same order in which they arrived. Each jump upward of $L(t)$ in Figure 6.6 represents an arrival. Arrivals occur at times 0, 3, 5, 7, and 16. Similarly, departures occur at times 2, 8, 10, and 14. (A departure may or may not have occurred at time 20.) Under these assumptions, it is apparent that $W_2 = 8 - 3 = 5$, $W_3 = 10 - 5 = 5$, $W_4 = 14 - 7 = 7$, and therefore

$$\widehat{w} = \frac{2 + 5 + 5 + 7 + 4}{5} = \frac{23}{5} = 4.6 \text{ time units}$$

Thus, on the average, an arbitrary customer spends 4.6 time units in the system. As for time spent in the waiting line, it can be computed that $W_1^Q = 0$, $W_2^Q = 0$, $W_3^Q = 8 - 5 = 3$, $W_4^Q = 10 - 7 = 3$, and $W_5^Q = 0$; thus,

$$\widehat{w}_Q = \frac{0 + 0 + 3 + 3 + 0}{5} = 1.2 \text{ time units}$$

6.3.3 The Conservation Equation: $L = \lambda w$

For the system exhibited in Figure 6.6, there were $N = 5$ arrivals in $T = 20$ time units, and thus the observed arrival rate was $\lambda = N/T = 1/4$ customer per time unit. Recall that $\widehat{L} = 1.15$ and $\widehat{w} = 4.6$; hence, it follows that

$$\widehat{L} = \widehat{\lambda}\widehat{w} \tag{6.8}$$

This relationship between L, λ, and w is not coincidental; it holds for almost all queueing systems or subsystems regardless of the number of servers, the queue discipline, or any other special circumstances. Allowing $T \longrightarrow \infty$ and $N \longrightarrow \infty$, Equation (6.8) becomes

$$L = \lambda w \tag{6.9}$$

where $\widehat{\lambda} \longrightarrow \lambda$, and λ is the long-run average arrival rate. Equation (6.9) is called a conservation equation and is usually attributed to Little [1961]. It says that the average number of customers in the system at an arbitrary point in time is equal to the average number of arrivals per time unit, times the average time spent in the system. For Figure 6.6, there is one arrival every 4 time units (on the average) and each arrival spends 4.6 time units in the system (on the average), so at an arbitrary point in time there will be $(1/4)(4.6) = 1.15$ customers present (on the average).

Equation (6.8) can also be derived by reconsidering Figure 6.6 in the following manner: Figure 6.8 shows system history, $L(t)$, exactly as in Figure 6.6, with each customer's time in the system, W_i, represented by a rectangle. This representation again assumes a single-server system with a FIFO queue discipline. The rectangles for the third and fourth customers are in two and three separate pieces, respectively. The ith rectangle has height 1 and length W_i for each $i = 1, 2, \ldots, N$. It follows that the total system time of all customers is given by the total area under the number-in-system function, $L(t)$; that is,

$$\sum_{i=1}^{N} W_i = \int_0^T L(t)\, dt \tag{6.10}$$

Therefore, by combining Equations (6.2) and (6.5) with $\widehat{\lambda} = N/T$, it follows that

$$\widehat{L} = \frac{1}{T} \int_0^T L(t)\, dt = \frac{N}{T} \frac{1}{N} \sum_{i=1}^{N} W_i = \widehat{\lambda}\widehat{w}$$

which is Little's equation (6.8). The intuitive and informal derivation presented here depended on the single-server FIFO assumptions, but these assumptions are not necessary. In fact, Equation (6.10), which was the key to the derivation, holds (at least approximately) in great generality, and thus so do Equations (6.8) and (6.9). Exercises 14 and 15 ask the reader to derive Equations (6.10) and (6.8) under different assumptions.

Technical note: If, as defined in Section 6.3.2, W_i is the system time for customer i during $[0, T]$, then Equation (6.10) and hence Equation (6.8) hold exactly. Some authors choose to define W_i as total system time for customer i; this change will affect the value of W_i only for those customers i who arrive before time T but do not depart until after time T (possibly customer 5 in Figure 6.8). With this change in definition, Equations (6.10) and (6.8) hold only approximately. Nevertheless, as $T \longrightarrow \infty$ and $N \longrightarrow \infty$, the error in Equation (6.8) decreases to zero, and, therefore, the conservation equation (6.9) for long-run measures of performance—namely, $L = \lambda w$—holds exactly.

6.3.4 Server Utilization

Server utilization is defined as the proportion of time that a server is busy. Observed server utilization, denoted by $\widehat{\rho}$, is defined over a specified time interval $[0, T]$. Long-run server utilization is denoted by ρ. For systems that exhibit long-run stability,

$$\widehat{\rho} \longrightarrow \rho \text{ as } T \longrightarrow \infty$$

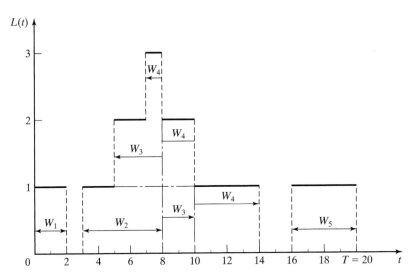

Figure 6.8   System times, W_i, for single-server FIFO system.

Example 6.5

Per Figure 6.6 or 6.8, and assuming that the system has a single server, it can be seen that the server utilization is $\widehat{\rho} = $ (total busy time)$/T = (\sum_{i=1}^{\infty} T_i)/T = (T - T_0)/T = 17/20$.

Server utilization in G/G/1/∞/∞ queues

Consider any single-server queueing system with average arrival rate λ customers per time unit, average service time $E(S) = 1/\mu$ time units, and infinite queue capacity and calling population Notice that $E(S) = 1/\mu$ implies that, when busy, the server is working at the rate μ customers per time unit, on the average; μ is called the service rate. The server alone is a subsystem that can be considered as a queueing system in itself; hence, the conservation Equation (6.9), $L = \lambda w$, can be applied to the server. For stable systems, the average arrival rate to the server, say λ_s, must be identical to the average arrival rate to the system, λ (certainly $\lambda_s \leq \lambda$—customers cannot be served faster than they arrive—but, if $\lambda_s < \lambda$, then the waiting line would tend to grow in length at an average rate of $\lambda - \lambda_s$ customers per time unit, and so we would have an unstable system). For the server subsystem, the average system time is $w = E(S) = \mu^{-1}$. The actual number of customers in the server subsystem is either 0 or 1, as shown in Figure 6.9 for the system represented by Figure 6.6. Hence, the average number, $\widehat{L}_s$, is given by

$$\widehat{L}_s = \frac{1}{T} \int_0^T (L(t) - L_Q(t))\, dt = \frac{T - T_0}{T}$$

In this case, $\widehat{L}_s = 17/20 = \widehat{\rho}$. In general, for a single-server queue, the average number of customers being served at an arbitrary point in time is equal to server utilization. As $T \longrightarrow$

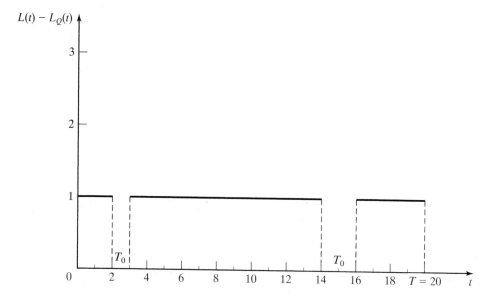

Figure 6.9 Number being served, $L(t) - L_Q(t)$, at time t.

∞, $\widehat{L_s} = \widehat{\rho} \longrightarrow L_s = \rho$. Combining these results into $L = \lambda w$ for the server subsystem yields

$$\rho = \lambda E(S) = \frac{\lambda}{\mu} \tag{6.11}$$

—that is, the long-run server utilization in a single-server queue is equal to the average arrival rate divided by the average service rate. For a single-server queue to be stable, the arrival rate λ must be less than the service rate μ:

$$\lambda < \mu$$

or

$$\rho = \frac{\lambda}{\mu} < 1 \tag{6.12}$$

If the arrival rate is greater than the service rate ($\lambda > \mu$), the server will eventually get further and further behind. After a time, the server will always be busy, and the waiting line will tend to grow in length at an average rate of ($\lambda - \mu$) customers per time unit, because departures will be occurring at rate μ per time unit. For stable single-server systems ($\lambda < \mu$ or $\rho < 1$), long-run measures of performance such as average queue length L_Q (and also L, w, and w_Q) are well defined and have meaning. For unstable systems ($\lambda > \mu$), long-run server utilization is 1, and long-run average queue length is infinite; that is,

$$\frac{1}{T} \int_0^T L_Q(t)\, dt \longrightarrow +\infty \text{ as } T \longrightarrow \infty$$

Similarly, $L = w = w_Q = \infty$. Therefore these long-run measures of performance are meaningless for unstable queues. The quantity λ/μ is also called the offered load and is a measure of the workload imposed on the system.

Server utilization in G/G/c/∞/∞ queues

Consider a queueing system with c identical servers in parallel. If an arriving customer finds more than one server idle, the customer chooses a server without favoring any particular server. (For example, the choice of server might be made at random.) Arrivals occur at rate λ from an infinite calling population, and each server works at rate μ customers per time unit. From Equation (6.9), $L = \lambda w$, applied to the server subsystem alone, an argument similar to the one given for a single server leads to the result that, for systems in statistical equilibrium, the average number of busy servers, say L_s, is given by

$$L_s = \lambda E(S) = \frac{\lambda}{\mu} \tag{6.13}$$

Clearly, $0 \le L_s \le c$. The long-run average server utilization is defined by

$$\rho = \frac{L_s}{c} = \frac{\lambda}{c\mu} \tag{6.14}$$

and so $0 \le \rho \le 1$. The utilization ρ can be interpreted as the proportion of time an arbitrary server is busy in the long run.

The maximum service rate of the $G/G/c/\infty/\infty$ system is $c\mu$, which occurs when all servers are busy. For the system to be stable, the average arrival rate λ must be less than the maximum service rate $c\mu$; that is, the system is stable if and only if

$$\lambda < c\mu \tag{6.15}$$

or, equivalently, if the offered load λ/μ is less than the number of servers c. If $\lambda > c\mu$, then arrivals are occurring, on the average, faster than the system can handle them, all servers will be continuously busy, and the waiting line will grow in length at an average rate of $(\lambda - c\mu)$ customers per time unit. Such a system is unstable, and the long-run performance measures (L, L_Q, w, and w_Q) are again meaningless for such systems.

Notice that Condition (6.15) generalizes Condition (6.12), and the equation for utilization for stable systems, Equation (6.14), generalizes Equation (6.11).

Equations (6.13) and (6.14) can also be applied when some servers work more than others, for example, when customers favor one server over others, or when certain servers serve customers only if all other servers are busy. In this case, the L_s given by Equation (6.13) is still the average number of busy servers, but ρ, as given by Equation (6.14), cannot be applied to an individual server. Instead, ρ must be interpreted as the average utilization of all servers.

Example 6.6

Customers arrive at random to a license bureau at a rate of $\lambda = 50$ customers per hour. Currently, there are 20-clerks, each serving $\mu = 5$ customers per hour on the average. Therefore the long-run, or steady-state, average utilization of a server, given by Equation (6.14), is

$$\rho = \frac{\lambda}{c\mu} = \frac{50}{20(5)} = 0.5$$

and the average number of busy servers is

$$L_s = \frac{\lambda}{\mu} = \frac{50}{5} = 10$$

Thus, in the long run, a typical clerk is busy serving customers only 50% of the time. The office manager asks whether the number of servers can be decreased. By Equation (6.15), it follows that, for the system to be stable, it is necessary for the number of servers to satisfy

$$c > \frac{\lambda}{\mu}$$

or $c > 50/5 = 10$. Thus, possibilities for the manager to consider include $c = 11$, or $c = 12$, or $c = 13, \ldots$. Notice that $c \geq 11$ guarantees long-run stability only in the sense that all servers, when busy, can handle the incoming work load (i.e., $c\mu > \lambda$) on average. The office manager could well desire to have more than the minimum number of servers ($c = 11$) because of other factors, such as customer delays and length of the waiting line. A stable queue can still have very long lines on average.

Server utilization and system performance

As will be illustrated here and in later sections, system performance can vary widely for a given value of utilization, ρ. Consider a $G/G/1/\infty/\infty$ queue: that is, a single-server queue with arrival rate λ, service rate μ, and utilization $\rho = \lambda/\mu < 1$.

At one extreme, consider the $D/D/1$ queue, which has deterministic arrival and service times. Then all interarrival times $\{A_1, A_2, \ldots\}$ are equal to $E(A) = 1/\lambda$, and all service times $\{S_1, S_2, \ldots\}$ are equal to $E(S) = 1/\mu$. Assuming that a customer arrives to an empty system at time 0, the system evolves in a completely deterministic and predictable fashion, as shown in Figure 6.10. Observe that $L = \rho = \lambda/\mu, w = E(S) = \mu^{-1}$, and $L_Q = w_Q = 0$. By varying λ and μ, server utilization can assume any value between 0 and 1, yet there is never any line whatsoever. What, then, causes lines to build, if not a high server utilization? In general, it is the variability of interarrival and service times that causes lines to fluctuate in length.

Example 6.7

Consider a physician who schedules patients every 10 minutes and who spends S_i minutes with the ith patient, where

$$S_i = \begin{cases} 9 \text{ minutes with probability } 0.9 \\ 12 \text{ minutes with probability } 0.1 \end{cases}$$

Thus, arrivals are deterministic ($A_1 = A_2 = \cdots = \lambda^{-1} = 10$) but services are stochastic (or probabilistic), with mean and variance given by

$$E(S_i) = 9(0.9) + 12(0.1) = 9.3 \text{ minutes}$$

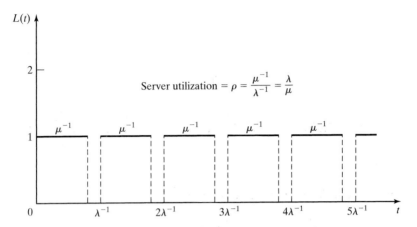

Figure 6.10   Deterministic queue $(D/D/1)$.

and

$$V(S_i) = E(S_i^2) - [E(S_i)]^2$$
$$= 9^2(0.9) + 12^2(0.1) - (9.3)^2$$
$$= 0.81 \text{ minutes}^2$$

Here, $\rho = \lambda/\mu = E(S)/E(A) = 9.3/10 = 0.93 < 1$, the system is stable, and the physician will be busy 93% of the time in the long run. In the short run, lines will not build up as long as patients require only 9 minutes of service, but, because of the variability in the service times, 10% of the patients will require 12 minutes, which in turn will cause a temporary line to form.

Suppose the system is simulated with service times, $S_1 = 9, S_2 = 12, S_3 = 9, S_4 = 9, S_5 = 9, \ldots$. Assuming that at time 0 a patient arrived to find the doctor idle and subsequent patients arrived precisely at times 10, 20, 30, ..., the system evolves as in Figure 6.11. The delays in queue are $W_1^Q = W_2^Q = 0, W_3^Q = 22 - 20 = 2, W_4^Q = 31 - 30 = 1, W_5^Q = 0$. The occurrence of a relatively long service time (here $S_2 = 12$) caused a waiting line to form temporarily. In general, because of the variability of the interarrival and service distributions, relatively small interarrival times and relatively large service times occasionally do occur, and these in turn cause lines to lengthen. Conversely, the occurrence of a large interarrival time or a small service time will tend to shorten an existing waiting line. The relationship between utilization, service and interarrival variability, and system performance will be explored in more detail in Section 6.4.

6.3.5 Costs in Queueing Problems

In many queueing situations, costs can be associated with various aspects of the waiting line or servers. Suppose that the system incurs a cost for each customer in the queue, say at a rate of $10 per hour per customer. If customer j spends W_j^Q hours in the queue, then $\sum_{j=1}^{N}(\$10 \cdot W_j^Q)$

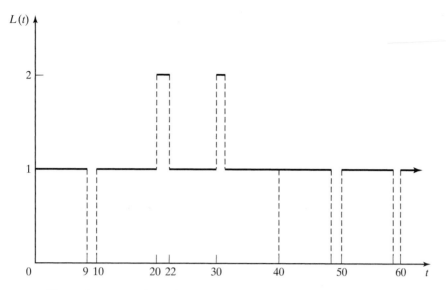

Figure 6.11   Number of patients in the doctor's office at time t.

is the total cost of the N customers who arrive during the simulation. Thus, the average cost per customer is

$$\sum_{j=1}^{N} \frac{\$10 \cdot W_j^Q}{N} = \$10 \cdot \widehat{w}_Q$$

by Equation (6.7). If $\widehat{\lambda}$ customers per hour arrive (on the average), the average cost per hour is

$$\left(\widehat{\lambda} \frac{\text{customers}}{\text{hour}}\right) \left(\frac{\$10 \cdot \widehat{w}_Q}{\text{customer}}\right) = \$10 \cdot \widehat{\lambda} \widehat{w}_Q = \$10 \cdot \widehat{L}_Q/\text{hour}$$

the last equality following by Little's equation (6.8). An alternative way to derive the average cost per hour is to consider Equation (6.2). If T_i^Q is the total time over the interval $[0, T]$ that the system contains exactly i customers, then $\$10 \, iT_i^Q$ is the cost incurred by the system during the time exactly i customers are present. Thus, the total cost is $\sum_{i=1}^{\infty} (\$10 \cdot iT_i^Q)$, and the average cost per hour is

$$\sum_{i=1}^{\infty} \frac{\$10 \cdot iT_i^Q}{T} = \$10 \cdot \widehat{L}_Q/ \text{ hour}$$

by Equation (6.2). In these cost expressions, $\widehat{L}_Q$ may be replaced by L_Q (if the long-run number in queue is known), or by L or $\widehat{L}$ (if costs are incurred while the customer is being served in addition to being delayed).

The server may also impose costs on the system. If a group of c parallel servers ($1 \leq c < \infty$) have utilization ρ, and each server imposes a cost of \$5 per hour while busy, the total server cost per hour is

$$\$5 \cdot c\rho$$

because $c\rho$ is the average number of busy servers. If server cost is imposed only when the servers are idle, then the server cost per hour would be

$$\$5 \cdot c(1 - \rho)$$

because $c(1 - \rho) = c - c\rho$ is the average number of idle servers. In many problems, two or more of these various costs are combined into a total cost. Such problems are illustrated by Exercises 1, 8, 12, and 19. In most cases, the objective is to minimize total costs (given certain constraints) by varying those parameters that are under management's control, such as the number of servers, the arrival rate, the service rate, and the system capacity.

6.4 STEADY-STATE BEHAVIOR OF INFINITE-POPULATION MARKOVIAN MODELS

This section presents the steady-state solution of a number of queueing models that can be solved mathematically. For the infinite-population models, the arrivals are assumed to follow a Poisson process with rate λ arrivals per time unit—that is, the interarrival times are assumed to be exponentially distributed with mean $1/\lambda$. Service times may be exponentially distributed (M) or arbitrarily (G). The queue discipline will be FIFO. Because of the exponential-distributional assumptions on the arrival process, these models are called Markovian models.

A queueing system is said to be in statistical equilibrium, or steady state, if the probability that the system is in a given state is not time dependent—that is,

$$P(L(t) = n) = P_n(t) = P_n$$

is independent of time t. Two properties—approaching statistical equilibrium from any starting state, and remaining in statistical equilibrium once it is reached—are characteristic of many stochastic models, and, in particular, of all the systems studied in the following subsections. On the other hand, if an analyst were interested in the transient behavior of a queue over a relatively short period of time and were given some specific initial conditions (such as idle and empty), the results to be presented here would be inappropriate. A transient mathematical analysis or, more likely, a simulation model would be the chosen tool of analysis.

The mathematical models whose solutions are shown in the following subsections can be used to obtain approximate results even when the assumptions of the model do not strictly hold. These results may be considered as a rough guide to the behavior of the system. A simulation may then be used for a more refined analysis. However, it should be remembered that a mathematical analysis (when it is applicable) provides the true value of the model parameter (e.g., L), whereas a simulation analysis delivers a statistical estimate (e.g., $\widehat{L}$) of the parameter. On the other hand, for complex systems, a simulation model is often a more faithful representation than a mathematical model.

For the simple models studied here, the steady-state parameter L, the time-average number of customers in the system, can be computed as

$$L = \sum_{n=0}^{\infty} nP_n \tag{6.16}$$

where $\{P_n\}$ are the steady-state probabilities of finding n customers in the system (as defined in Table 6.2). As was discussed in Section 6.3 and was expressed in Equation (6.3), L can also be interpreted as a long-run measure of performance of the system. Once L is given, the other steady-state parameters can be computed readily from Little's equation (6.9) applied to the whole system and to the queue alone:

$$w = \frac{L}{\lambda}$$

$$w_Q = w - \frac{1}{\mu} \tag{6.17}$$

$$L_Q = \lambda w_Q$$

where λ is the arrival rate and μ is the service rate per server.

For the $G/G/c/\infty/\infty$ queues considered in this section to have a statistical equilibrium, a necessary and sufficient condition is that $\lambda/(c\mu) < 1$, where λ is the arrival rate, μ is the service rate of one server, and c is the number of parallel servers. For these unlimited capacity, infinite-calling-population models, it shall be assumed that the theoretical server utilization, $\rho = \lambda/(c\mu)$, satisfies $\rho < 1$. For models with finite system capacity or finite calling population, the quantity $\lambda/(c\mu)$ may assume any positive value.

6.4.1 Single-Server Queues with Poisson Arrivals and Unlimited Capacity: M/G/1

Suppose that service times have mean $1/\mu$ and variance σ^2 and that there is one server. If $\rho = \lambda/\mu < 1$, then the $M/G/1$ queue has a steady-state probability distribution with steady-state characteristics, as given in Table 6.3. In general, there is no simple expression for the steady-state probabilities $P_0, P_1, P_2, \ldots$. When $\lambda < \mu$, the quantity $\rho = \lambda/\mu$ is the server utilization, or long-run proportion of time the server is busy. As is seen in Table 6.3, $1 - P_0 = \rho$ can also be interpreted as the steady-state probability that the system contains one or more customers. Notice also that $L - L_Q = \rho$ is the time-average number of customers being served.

Example 6.8

Widget-making machines malfunction apparently at random and then require a mechanic's attention. It is assumed that malfunctions occur according to a Poisson process, at the rate $\lambda = 1.5$ per hour. Observation over several months has found that repair times by the single mechanic take an average time of 30 minutes, with a standard deviation of 20 minutes. Thus the mean service time $1/\mu = 1/2$ hour, the service rate is $\mu = 2$ per hour and $\sigma^2 = (20)^2$ minutes$^2 = 1/9$ hour^2. The "customers" are the widget makers, and the appropriate model is the $M/G/1$ queue, because only the mean and variance of service times are known, not their distribution. The proportion of

Table 6.3   Steady-State Parameters of the $M/G/1$
Queue

ρ	$\dfrac{\lambda}{\mu}$
L	$\rho + \dfrac{\lambda^2(1/\mu^2 + \sigma^2)}{2(1-\rho)} = \rho + \dfrac{\rho^2(1+\sigma^2\mu^2)}{2(1-\rho)}$
w	$\dfrac{1}{\mu} + \dfrac{\lambda(1/\mu^2 + \sigma^2)}{2(1-\rho)}$
w_Q	$\dfrac{\lambda(1/\mu^2 + \sigma^2)}{2(1-\rho)}$
L_Q	$\dfrac{\lambda^2(1/\mu^2 + \sigma^2)}{2(1-\rho)} = \dfrac{\rho^2(1+\sigma^2\mu^2)}{2(1-\rho)}$
P_0	$1-\rho$

time the mechanic is busy is $\rho = \lambda/\mu = 1.5/2 = 0.75$, and, by Table 6.3, the steady-state time average number of broken machines is

$$L = 0.75 + \frac{(1.5)^2[(0.5)^2 + 1/9]}{2(1 - 0.75)}$$

$$= 0.75 + 1.625 = 2.375 \text{ machines}$$

Thus, an observer who notes the state of the repair system at arbitrary times would find an average of 2.375 broken machines (over the long run).

A closer look at the formulas in Table 6.3 reveals the source of the waiting lines and delays in an $M/G/1$ queue. For example, L_Q may be rewritten as

$$L_Q = \frac{\rho^2}{2(1 - \rho)} + \frac{\lambda^2\sigma^2}{2(1 - \rho)}$$

The first term involves only the ratio of the mean arrival rate, λ, to the mean service rate, μ. As shown by the second term, if λ and μ are held constant, the average length of the waiting line (L_Q) depends on the variability, σ^2, of the service times. If two systems have identical mean service times and mean interarrival times, the one with the more variable service times (larger σ^2) will tend to have longer lines on the average. Intuitively, if service times are highly variable, there is a high probability that a large service time will occur (say, much larger than the mean service time), and, when large service times do occur, there is a higher-than-usual tendency for lines to form and delays of customers to increase. (The reader should not confuse "steady state" with low variability or short lines; a system in steady-state or statistical equilibrium can be highly variable and can have long waiting lines.)

Example 6.9

There are two workers competing for a job. Able claims an average service time that is faster than Baker's, but Baker claims to be more consistent, even if not as fast. The arrivals occur according to a Poisson process at the rate $\lambda = 2$ per hour (1/30 per minute). Able's service statistics are an average service time of 24 minutes with a standard deviation of 20 minutes. Baker's service statistics are an average service time of 25 minutes, but a standard deviation of only 2 minutes. If the average length of the queue is the criterion for hiring, which worker should be hired? For Able, $\lambda = 1/30$ per minute, $1/\mu = 24$ minutes, $\sigma^2 = 20^2 = 400$ minutes^2, $\rho = \lambda/\mu = 24/30 = 4/5$, and the average queue length is computed as

$$L_Q = \frac{(1/30)^2[24^2 + 400]}{2(1 - 4/5)} = 2.711 \text{ customers}$$

For Baker, $\lambda = 1/30$ per minute, $1/\mu = 25$ minutes, $\sigma^2 = 2^2 = 4$ minutes^2, $\rho = 25/30 = 5/6$, and the average queue length is

$$L_Q = \frac{(1/30)^2[25^2 + 4]}{2(1 - 5/6)} = 2.097 \text{ customers}$$

Although working faster on the average, Able's greater service variability results in an average queue length about 30% greater than Baker's. On the basis of average queue length, L_Q, Baker wins. On the other hand, the proportion of arrivals who would find Able idle and thus experience no delay is $P_0 = 1 - \rho = 1/5 = 20\%$, but the proportion who would find Baker idle and thus experience no delay is $P_0 = 1 - \rho = 1/6 = 16.7\%$.

One case of the $M/G/1$ queue that is of special note occurs when service times are exponential, which we describe next.

The M/M/1 queue. Suppose that service times in an $M/G/1$ queue are exponentially distributed, with mean $1/\mu$; then the variance as given by Equation (5.27) is $\sigma^2 = 1/\mu^2$. The mean and standard deviation of the exponential distribution are equal, so the $M/M/1$ queue will often be a useful approximate model when service times have standard deviations approximately equal to their means. The steady-state parameters, given in Table 6.4, may be computed by substituting $\sigma^2 = 1/\mu^2$ into the formulas in Table 6.3. Alternatively, L may be computed by Equation (6.16) from the steady-state probabilities P_n given in Table 6.4, and then w, w_Q, and L_Q may be computed from Equations (6.17). The student can show that the two expressions for each parameter are equivalent by substituting $\rho = \lambda/\mu$ into the right-hand side of each equation in Table 6.4.

Example 6.10

Suppose that the interarrival times and service times at a single-chair unisex hair-styling shop have been shown to be exponentially distributed. The values of λ and μ are 2 per hour and 3 per hour, respectively—that is, the time between arrivals averages 1/2 hour, exponentially distributed, and the service time averages 20 minutes, also exponentially distributed. The server utilization and

Table 6.4 Steady-State Parameters of the $M/M/1$ Queue

L	$\dfrac{\lambda}{\mu - \lambda} = \dfrac{\rho}{1 - \rho}$
w	$\dfrac{1}{\mu - \lambda} = \dfrac{1}{\mu(1 - \rho)}$
w_Q	$\dfrac{\lambda}{\mu(\mu - \lambda)} = \dfrac{\rho}{\mu(1 - \rho)}$
L_Q	$\dfrac{\lambda^2}{\mu(\mu - \lambda)} = \dfrac{\rho^2}{1 - \rho}$
P_n	$\left(1 - \dfrac{\lambda}{\mu}\right)\left(\dfrac{\lambda}{\mu}\right)^n = (1 - \rho)\rho^n$

the probabilities for zero, one, two, three, and four or more customers in the shop are computed as follows:

$$\rho = \frac{\lambda}{\mu} = \frac{2}{3}$$

$$P_0 = 1 - \frac{\lambda}{\mu} = \frac{1}{3}$$

$$P_1 = \left(\frac{1}{3}\right)\left(\frac{2}{3}\right) = \frac{2}{9}$$

$$P_2 = \left(\frac{1}{3}\right)\left(\frac{2}{3}\right)^2 = \frac{4}{27}$$

$$P_3 = \left(\frac{1}{3}\right)\left(\frac{2}{3}\right)^3 = \frac{8}{81}$$

$$P_{\geq 4} = 1 - \sum_{n=0}^{3} P_n = 1 - \frac{1}{3} - \frac{2}{9} - \frac{4}{27} - \frac{8}{81} = \frac{16}{81}$$

From the calculations, the probability that the hair stylist is busy is $1 - P_0 = \rho = 0.67$; thus, the probability that the hair stylist is idle is 0.33. The time-average number of customers in the system is given by Table 6.4 as

$$L = \frac{\lambda}{\mu - \lambda} = \frac{2}{3 - 2} = 2 \text{ customers}$$

The average time an arrival spends in the system can be obtained from Table 6.4 or Equation (6.17) as

$$w = \frac{L}{\lambda} = \frac{2}{2} = 1 \text{ hour}$$

The average time the customer spends in the queue can be obtained From Equation (6.17) as

$$w_Q = w - \frac{1}{\mu} = 1 - \frac{1}{3} = \frac{2}{3} \text{ hour}$$

From Table 6.4, the time-average number in the queue is given by

$$L_Q = \frac{\lambda^2}{\mu(\mu - \lambda)} = \frac{4}{3(1)} = \frac{4}{3} \text{ customers}$$

Finally, notice that multiplying $w = w_Q + 1/\mu$ through by λ and using Little's equation (6.9) yields

$$L = L_Q + \frac{\lambda}{\mu} = \frac{4}{3} + \frac{2}{3} = 2 \text{ customers}$$

Example 6.11

For the $M/M/1$ queue with service rate $\mu = 10$ customers per hour, consider how L and w increase as the arrival rate, λ, increases from 5 to 8.64 by increments of 20%, and then to $\lambda = 10$.

λ	5.0	6.0	7.2	8.64	10.0
ρ	0.500	0.600	0.720	0.864	1.0
L	1.00	1.50	2.57	6.35	∞
w	0.20	0.25	0.36	0.73	∞

For any $M/G/1$ queue, if $\lambda/\mu \geq 1$, waiting lines tend to continually grow in length; the long-run measures of performance, $L, w, w_Q,$ and L_Q are all infinite ($L = w = w_Q = L_Q = \infty$); and a steady-state probability distribution does not exist. As is shown here for $\lambda < \mu$, if ρ is close to 1, waiting lines and delays will tend to be long. Notice that the increase in average system time, w, and average number in system, L, is highly nonlinear as a function of ρ. For example, as λ increases by 20%, L increases first by 50% (from 1.00 to 1.50), then by 71% (to 2.57), and then by 147% (to 6.35).

Example 6.12

If arrivals are occurring at rate $\lambda = 10$ per hour, and management has a choice of two servers, one who works at rate $\mu_1 = 11$ customers per hour and the second at rate $\mu_2 = 12$ customers per hour, the respective utilizations are $\rho_1 = \lambda/\mu_1 = 10/11 = 0.909$ and $\rho_2 = \lambda/\mu_2 = 10/12 = 0.833$. If the $M/M/1$ queue is used as an approximate model, then, with the first server, the average number in the system would be, by Table 6.4,

$$L_1 = \frac{\rho_1}{1 - \rho_1} = 10$$

and, with the second server, the average number in the system would be

$$L_2 = \frac{\rho_2}{1 - \rho_2} = 5$$

Thus, a decrease in service rate from 12 to 11 customers per hour, a mere 8.3% decrease, would result in an increase in average number in system from 5 to 10, which is a 100% increase.

The effect of utilization and service variability

For any $M/G/1$ queue, if lines are too long, they can be reduced by decreasing the server utilization ρ or by decreasing the service time variability, σ^2. These remarks hold for almost all queues, not just the $M/G/1$ queue. The utilization factor ρ can be reduced by decreasing the arrival rate λ, by increasing the service rate μ, or by increasing the number of servers, because, in general, $\rho = \lambda/(c\mu)$, where c is the number of parallel servers. The effect of additional servers will be studied in the following subsections. Figure 6.12 illustrates the effect of service variability. The mean steady-state number in the queue, L_Q, is plotted versus utilization ρ for a number of different coefficients of variation. The coefficient of variation (cv) of a positive random variable X is defined as

$$(\text{cv})^2 = \frac{V(X)}{[E(X)]^2}$$

It is a measure of the variability of a distribution. The larger its value, the more variable is the distribution relative to its expected value. For deterministic service times, $V(X) = 0$, so cv $= 0$. For Erlang service times of order k, $V(X) = 1/(k\mu^2)$ and $E(X) = 1/\mu$, so cv $= 1/\sqrt{k}$. For exponential service times at service rate μ, the mean service time is $E(X) = 1/\mu$ and the variance is $V(X) = 1/\mu^2$, so cv $= 1$. If service times have standard deviation greater than their mean (i.e., if cv > 1), then the hyperexponential distribution, which can achieve any desired coefficient of variation greater than 1, provides a good model. One occasion where it arises is given in Exercise 16.

The formula for L_Q for any $M/G/1$ queue can be rewritten in terms of the coefficient of variation by noticing that $(\text{cv})^2 = \sigma^2/(1/\mu)^2 = \sigma^2\mu^2$. Therefore,

$$L_Q = \frac{\rho^2(1 + \sigma^2\mu^2)}{2(1 - \rho)}$$

$$= \frac{\rho^2(1 + (\text{cv})^2)}{2(1 - \rho)}$$

$$= \left(\frac{\rho^2}{1 - \rho}\right)\left(\frac{1 + (\text{cv})^2}{2}\right) \tag{6.18}$$

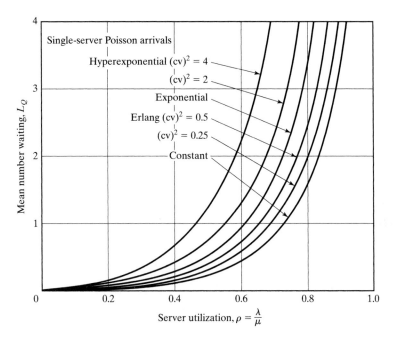

Figure 6.12   Mean number of customers waiting, L_Q, in $M/G/1$ queue having service distributions with given cv. (Adapted from Geoffrey Gordon, *System Simulation*, 2d ed., Prentice-Hall, Englewood Cliffs, NJ, 1978.)

The first term, $\rho^2/(1-\rho)$, is L_Q for an $M/M/1$ queue. The second term, $\left(1 + (\text{cv})^2\right)/2$, corrects the $M/M/1$ formula to account for a nonexponential service-time distribution. The formula for w_Q can be obtained from the corresponding $M/M/1$ formula by applying the same correction factor.

6.4.2 Multiserver Queue: M/M/c/∞/∞

Suppose that there are c channels operating in parallel. Each of these channels has an independent and identical exponential service-time distribution, with mean $1/\mu$. The arrival process is Poisson with rate λ. Arrivals will join a single queue and enter the first available service channel. The queueing system is shown in Figure 6.13. If the number in system is $n < c$, an arrival will enter an available channel. However, when $n \geq c$, a queue will build if arrivals occur.

The offered load is defined by λ/μ. If $\lambda \geq c\mu$, the arrival rate is greater than or equal to the maximum service rate of the system (the service rate when all servers are busy); thus, the system cannot handle the load put upon it, and therefore it has no statistical equilibrium. If $\lambda > c\mu$, the waiting line grows in length at the rate $(\lambda - c\mu)$ customers per time unit, on the average. Customers are entering the system at rate λ per time unit but are leaving the system at a maximum rate of $c\mu$ per time unit.

For the $M/M/c$ queue to have statistical equilibrium, the offered load must satisfy $\lambda/\mu < c$, in which case $\lambda/(c\mu) = \rho$, the server utilization. The steady-state parameters are listed in Table 6.5. Most of the measures of performance can be expressed fairly simply in terms of P_0,

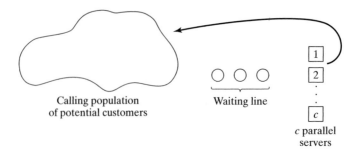

Figure 6.13   Multiserver queueing system.

Table 6.5   Steady-State parameters for the $M/M/c$ Queue

ρ	$\dfrac{\lambda}{c\mu}$
P_0	$\left\{\left[\displaystyle\sum_{n=0}^{c-1}\dfrac{(\lambda/\mu)^n}{n!}\right]+\left[\left(\dfrac{\lambda}{\mu}\right)^c\left(\dfrac{1}{c!}\right)\left(\dfrac{c\mu}{c\mu-\lambda}\right)\right]\right\}^{-1}$
	$=\left\{\left[\displaystyle\sum_{n=0}^{c-1}\dfrac{(c\rho)^n}{n!}\right]+\left[(c\rho)^c\left(\dfrac{1}{c!}\right)\dfrac{1}{1-\rho}\right]\right\}^{-1}$
$P(L(\infty)\geq c)$	$\dfrac{(\lambda/\mu)^c P_0}{c!(1-\lambda/c\mu)}=\dfrac{(c\rho)^c P_0}{c!(1-\rho)}$
L	$c\rho+\dfrac{(c\rho)^{c+1}P_0}{c(c!)(1-\rho)^2}=c\rho+\dfrac{\rho P(L(\infty)\geq c)}{1-\rho}$
w	$\dfrac{L}{\lambda}$
w_Q	$w-\dfrac{1}{\mu}$
L_Q	$\lambda w_Q=\dfrac{(c\rho)^{c+1}P_0}{c(c!)(1-\rho)^2}=\dfrac{\rho P(L(\infty)\geq c)}{1-\rho}$
$L-L_Q$	$\dfrac{\lambda}{\mu}=c\rho$

the probability that the system is empty, or $\sum_{n=c}^{\infty}P_n$, the probability that all servers are busy, denoted by $P(L(\infty)\geq c)$, where $L(\infty)$ is a random variable representing the number in system in statistical equilibrium (after a very long time). Thus, $P(L(\infty)=n)=P_n, n=0,1,2,\ldots$. The value of P_0 is necessary for computing all the measures of performance, and the equation for P_0 is somewhat more complex than in the previous cases. However, P_0 depends only on c and ρ. A good approximation to P_0 can be obtained by using Figure 6.14, where P_0 is plotted versus

Figure 6.14   Values of P_0 for $M/M/c/\infty$ model. (From F. S. Hillier and G. J. Lieberman, *Introduction to Operations Research*, 5th ed., 1990, p. 616. Adapted with permission of McGraw–Hill, Inc., New York.)

ρ on semilog paper for various values c. Figure 6.15 is a plot of L versus ρ for different values of c.

The results in Table 6.5 simplify to those in Table 6.4 when $c = 1$, the case of a single server. Notice that the average number of busy servers, or the average number of customers being served, is given by the simple expression $L - L_Q = \lambda/\mu = c\rho$.

Example 6.13
Many early examples of queueing theory applied to practical problems concerning tool cribs. Attendants manage the tool cribs as mechanics, assumed to be from an infinite calling population, arrive for service. Assume Poisson arrivals at rate 2 mechanics per minute and exponentially distributed service times with mean 40 seconds.

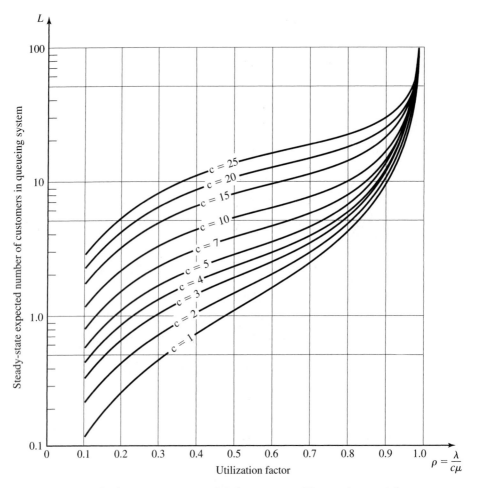

Figure 6.15   Values of L for $M/M/c/\infty$ model. (From F. S. Hillier and G. J. Lieberman, *Introduction to Operations Research*, 5th ed., 1990, p. 617. Adapted with permission of McGraw–Hill, Inc., New York.)

Now, $\lambda = 2$ per minute, and $\mu = 60/40 = 3/2$ per minute. The offered load is greater than 1:

$$\frac{\lambda}{\mu} = \frac{2}{3/2} = \frac{4}{3} > 1$$

so more than one server is needed if the system is to have a statistical equilibrium. The requirement for steady state is that $c > \lambda/\mu = 4/3$. Thus at least $c = 2$ attendants are needed. The quantity 4/3 is the expected number of busy servers, and for $c \ge 2$, $\rho = 4/(3c)$ is the long-run proportion of time each server is busy. (What would happen if there were only $c = 1$ server?)

Let there be $c = 2$ attendants. First, P_0 is calculated as

$$P_0 = \left\{ \sum_{n=0}^{1} \frac{(4/3)^n}{n!} + \left(\frac{4}{3}\right)^2 \left(\frac{1}{2!}\right) \left[\frac{2(3/2)}{2(3/2) - 2}\right] \right\}^{-1}$$

$$= \left\{1 + \frac{4}{3} + \left(\frac{16}{9}\right)\left(\frac{1}{2}\right)(3)\right\}^{-1} = \left(\frac{15}{3}\right)^{-1} = \frac{1}{5} = 0.2$$

Next, the probability that all servers are busy is computed as

$$P(L(\infty) \geq 2) = \frac{(4/3)^2}{2!(1 - 2/3)} \left(\frac{1}{5}\right) = \left(\frac{8}{3}\right)\left(\frac{1}{5}\right) = \frac{8}{15} = 0.533$$

Thus, the time-average length of the waiting line of mechanics is

$$L_Q = \frac{(2/3)(8/15)}{1 - 2/3} = 1.07 \text{ mechanics}$$

and the time-average number in system is given by

$$L = L_Q + \frac{\lambda}{\mu} = \frac{16}{15} + \frac{4}{3} = \frac{12}{5} = 2.4 \text{ mechanics}$$

From Little's relationships, the average time a mechanic spends at the tool crib is

$$w = \frac{L}{\lambda} = \frac{2.4}{2} = 1.2 \text{ minutes}$$

and the average time spent waiting for an attendant is

$$w_Q = w - \frac{1}{\mu} = 1.2 - \frac{2}{3} = 0.533 \text{ minute}$$

Example 6.14

Using the data of Example 6.13, compute P_0 and L from Figures 6.14 and 6.15. First, compute

$$\rho = \frac{\lambda}{c\mu} = \frac{2}{2(3/2)} = \frac{2}{3} = 0.667$$

Entering the utilization factor 0.667 on the horizontal axis of Figure 6.14 gives the value 0.2 for P_0 on the vertical axis. Similarly, the value $L = 2.4$ is read from the vertical axis of Figure 6.15.

An Approximation for the M/G/c/∞ Queue

Recall that formulas for L_Q and w_Q for the $M/G/1$ queue can be obtained from the corresponding $M/M/1$ formulas by multiplying them by the correction factor $\left(1 + (\text{cv})^2\right)/2$, as in Equation (6.18). *Approximate* formulas for the $M/G/c$ queue can be obtained by applying the same correction factor to the $M/M/c$ formulas for L_Q and w_Q (no exact formula exists for $1 < c < \infty$). The nearer the cv is to 1, the better the approximation.

Example 6.15 _____

Recall Example 6.13. Suppose that the service times for the mechanics at the tool crib are not exponentially distributed, but are known to have a standard deviation of 30 seconds. Then we have an $M/G/c$ model, rather than an $M/M/c$. The mean service time is 40 seconds, so the coefficient of variation of the service time is

$$cv = \frac{30}{40} = \frac{3}{4} < 1$$

Therefore, the accuracy of L_Q and w_Q can be improved by the correction factor

$$\frac{1 + (cv)^2}{2} = \frac{1 + (3/4)^2}{2} = \frac{25}{32} = 0.78$$

For example, when there are $c = 2$ attendants,

$$L_Q = (0.78)(1.07) = 0.83 \text{ mechanics}$$

Notice that, because the coefficient of variation of the service time is less than 1, the congestion in the system, as measured by L_Q, is less than in the corresponding $M/M/2$ model.

The correction factor applies only to the formulas for L_Q and w_Q. Little's formula can then be used to calculate L and w. Unfortunately, there is no general method for correcting the steady-state probabilities, P_n.

_____

When the Number of Servers is Infinite (M/G/∞/∞)

There are at least three situations in which it is appropriate to treat the number of servers as infinite:

1. when each customer is its own server—in other words, in a self-service system;
2. when service capacity far exceeds service demand, as in a so-called ample-server system; and
3. when we want to know how many servers are required so that customers will rarely be delayed.

The steady-state parameters for the $M/G/\infty$ queue are listed in Table 6.6. In the table, λ is the arrival rate of the Poisson arrival process, and $1/\mu$ is the expected service time of the general service-time distribution (including exponential, constant, or any other).

Example 6.16 _____

Prior to introducing their new, subscriber-only, on-line computer information service, The Connection must plan their system capacity in terms of the number of users that can be logged in simultaneously. If the service is successful, customers are expected to log on at a rate of $\lambda = 500$ per hour, according to a Poisson process, and stay connected for an average of $1/\mu = 180$ minutes (or 3 hours). In the real system, there will be an upper limit on simultaneous users, but, for

Table 6.6 Steady-State Parameters for the $M/G/\infty$ Queue

P_0	$e^{-\lambda/\mu}$
w	$\dfrac{1}{\mu}$
w_Q	0
L	$\dfrac{\lambda}{\mu}$
L_Q	0
P_n	$\dfrac{e^{-\lambda/\mu}(\lambda/\mu)^n}{n!}, n = 0, 1, \ldots$

planning purposes, The Connection can pretend that the number of simultaneous users is infinite. An $M/G/\infty$ model of the system implies that the expected number of simultaneous users is $L = \lambda/\mu = 500(3) = 1500$, so a capacity greater than 1500 is certainly required. To ensure providing adequate capacity 95% of the time, The Connection could allow the number of simultaneous users to be the smallest value c such that

$$P(L(\infty) \leq c) = \sum_{n=0}^{c} P_n = \sum_{n=0}^{c} \frac{e^{-1500}(1500)^n}{n!} \geq 0.95$$

The capacity $c = 1564$ simultaneous users satisfies this requirement.

6.4.3 Multiserver Queues with Poisson Arrivals and Limited Capacity: M/M/c/N/∞

Suppose that service times are exponentially distributed at rate μ, that there are c servers, and that the total system capacity is $N \geq c$ customers. If an arrival occurs when the system is full, that arrival is turned away and does not enter the system. As in the preceding section, suppose that arrivals occur randomly according to a Poisson process with rate λ arrivals per time unit. For any values of λ and μ such that $\rho \neq 1$, the $M/M/c/N$ queue has a statistical equilibrium with steady-state characteristics as given in Table 6.7 (formulas for the case $\rho = 1$ can be found in Hillier and Lieberman [2005]).

The effective arrival rate, λ_e, is defined as the mean number of arrivals per time unit who enter and remain in the system. For all systems, $\lambda_e \leq \lambda$; for the unlimited-capacity systems, $\lambda_e = \lambda$; but, for systems such as the present one, which turn customers away when full, $\lambda_e < \lambda$. The effective arrival rate is computed by

$$\lambda_e = \lambda(1 - P_N)$$

because $1 - P_N$ is the probability that a customer, upon arrival, will find space and be able to enter the system. When one is using Little's equations (6.17) to compute mean time spent in system w and in queue w_Q, λ must be replaced by λ_e.

Table 6.7 Steady-State Parameters for the $M/M/c/N$
Queue ($N =$ System Capacity, $a = \lambda/\mu$, $\rho = \lambda/(c\mu)$)

P_0	$\left[1 + \displaystyle\sum_{n=1}^{c} \frac{a^n}{n!} + \frac{a^c}{c!} \sum_{n=c+1}^{N} \rho^{n-c}\right]^{-1}$
P_N	$\dfrac{a^N}{c!c^{N-c}} P_0$
L_Q	$\dfrac{P_0 a^c \rho}{c!(1-\rho)^2}\left[1 - \rho^{N-c} - (N-c)\rho^{N-c}(1-\rho)\right]$
λ_e	$\lambda(1 - P_N)$
w_Q	$\dfrac{L_Q}{\lambda_e}$
w	$w_Q + \dfrac{1}{\mu}$
L	$\lambda_e w$

Example 6.17

The unisex hair-styling shop described in Example 6.10 can hold only three customers: one in service, and two waiting. Additional customers are turned away when the system is full. The offered load is as previously determined, namely $\lambda/\mu = 2/3$.

In order to calculate the performance measures, first compute P_0:

$$P_0 = \left[1 + \frac{2}{3} + \frac{2}{3}\sum_{n=2}^{3}\left(\frac{2}{3}\right)^{n-1}\right]^{-1} = 0.415$$

The probability that there are three customers in the system (the system is full) is

$$P_N = P_3 = \frac{(2/3)^3}{1!1^2}P_0 = \frac{8}{65} = 0.123$$

Then, the average length of the queue (customers waiting for a haircut) is given by

$$L_Q = \frac{(27/65)(2/3)(2/3)}{(1-2/3)^2}\left[1 - (2/3)^2 - 2(2/3)^2(1 - 2/3)\right] = 0.431 \text{ customer}$$

Now, the effective arrival rate, λ_e, is given by

$$\lambda_e = 2\left(1 - \frac{8}{65}\right) = \frac{114}{65} = 1.754 \text{ customers per hour}$$

Therefore, from Little's equation, the expected time spent waiting in queue is

$$w_Q = \frac{L_Q}{\lambda_e} = \frac{28}{114} = 0.246 \text{ hour}$$

and the expected total time in the shop is

$$w = w_Q + \frac{1}{\mu} = \frac{66}{114} = 0.579 \text{ hour}$$

One last application of Little's equation gives the expected number of customers in the shop (in queue and getting a haircut) as

$$L = \lambda_e w = \frac{66}{65} = 1.015 \text{ customers}$$

Notice that $1 - P_0 = 0.585$ is the average number of customers being served or, equivalently, the probability that the single server is busy. Thus, the server utilization, or proportion of time the server is busy in the long run, is given by

$$1 - P_0 = \frac{\lambda_e}{\mu} = 0.585$$

The reader should compare these results to those of the unisex hair-styling shop before the capacity constraint was placed on the system. Specifically, in systems with limited capacity, the offered load λ/μ can assume any positive value and no longer equals the server utilization $\rho = \lambda_e/\mu$. Notice that server utilization decreases from 67% to 58.5% when the system imposes a capacity constraint.

6.5 STEADY-STATE BEHAVIOR OF FINITE-POPULATION MODELS (M/M/C/K/K)

In many practical problems, the assumption of an infinite calling population leads to invalid results because the calling population is, in fact, small. When the calling population is small, the presence of one or more customers in the system has a strong effect on the distribution of future arrivals, and the use of an infinite-population model can be misleading. Typical examples include a small group of machines that break down from time to time and require repair, or a small group of mechanics who line up at a counter for parts or tools. In the extreme case, if all the machines are broken, no new "arrivals" (breakdowns) of machines can occur; similarly, if all the mechanics are in line, no arrival is possible to the tool and parts counter. Contrast this to the infinite-population models, in which the arrival rate, λ, of customers to the system is assumed to be independent of the state of the system.

Consider a finite-calling-population model with K customers. The time between the end of one service visit and the next call for service for each member of the population is assumed to be exponentially distributed, with mean $1/\lambda$ time units; service times are also exponentially

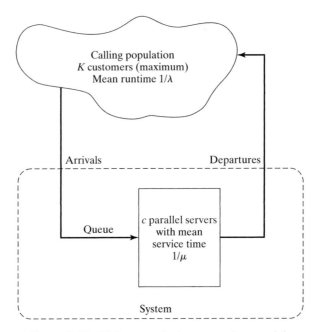

Figure 6.16   Finite-population queueing model.

distributed, with mean $1/\mu$ time units; there are c parallel servers, and system capacity is K, so that all arrivals remain for service. Such a system is depicted in Figure 6.16.

    The steady-state parameters for this model are listed in Table 6.8. An electronic spreadsheet or a symbolic calculation program is useful for evaluating these complex formulas. For example, Figure 6.17 is a procedure written for the symbolic calculation program Maple to calculate the steady-state probabilities for the $M/M/c/K/K$ queue. Another approach is to use precomputed queueing tables, such as those found in Banks and Heikes [1984], Hillier and Yu [1981], Peck and Hazelwood [1958] or Descloux [1962].

    The effective arrival rate λ_e has several valid interpretations:

$$\lambda_e = \text{long-run effective arrival rate of customers to the queue}$$

$$= \text{long-run effective arrival rate of customers entering service}$$

$$= \text{long-run rate at which customers exit from service}$$

$$= \text{long-run rate at which customers enter the calling population}$$

$$\text{(and begin a new runtime)}$$

$$= \text{long-run rate at which customers exit from the calling population}$$

Example 6.18 _____

There are two workers who are responsible for 10 milling machines. The machines run on the average for 20 minutes, then require an average 5-minute service period, both times exponentially

Table 6.8 Steady-State Parameters for the M/M/c/K/K Queue

P_0	$\left[\displaystyle\sum_{n=0}^{c-1}\binom{K}{n}\left(\frac{\lambda}{\mu}\right)^n + \sum_{n=c}^{K}\frac{K!}{(K-n)!c!c^{n-c}}\left(\frac{\lambda}{\mu}\right)^n\right]^{-1}$
P_n	$\begin{cases}\dbinom{K}{n}\left(\dfrac{\lambda}{\mu}\right)^n P_0, & n=0,1,\ldots,c-1 \\[2ex] \dfrac{K!}{(K-n)!c!c^{n-c}}\left(\dfrac{\lambda}{\mu}\right)^n P_0, & n=c,c+1,\ldots,K\end{cases}$
L	$\displaystyle\sum_{n=0}^{K} nP_n$
L_Q	$\displaystyle\sum_{n=c+1}^{K}(n-c)P_n$
λ_e	$\displaystyle\sum_{n=0}^{K}(K-n)\lambda P_n$
w	L/λ_e
w_Q	L_Q/λ_e
ρ	$\dfrac{L-L_Q}{c}=\dfrac{\lambda_e}{c\mu}$

```
mmcKK := proc(lambda, mu, c, K)
      # return steady-state probabilities for M/M/c/K/K queue
      # notice that p[n+1] is P_n, n=0,..,K
      local crho, Kfac, cfac, p, n;
      p := vector(K+1,0);
      crho := lambda/mu;
      Kfac := K!;
      cfac := c!;
      p[1] := sum((Kfac/(n!*(K-n)!))*crho^n, n=0..c-1) +
          sum((Kfac/(c^(n-c)*(K-n)!*cfac))*crho^n, n=c..K);
      p[1] := 1/p[1];
      for n from 1 to c-1
      do
          p[n+1] := p[1]*(Kfac/(n!*(K-n)!))*crho^n;
      od;
      for n from c to K
      do
          p[n+1] := p[1]*(Kfac/(c^(n-c)*(K-n)!*cfac))*crho^n;
      od;
      RETURN(evalm(p));
      end;
```

Figure 6.17 Maple procedure to calculate P_n for the M/M/c/K/K queue.

distributed. Therefore, $\lambda = 1/20$ and $\mu = 1/5$. Compute the various measures of performance for this system.

All of the performance measures depend on P_0, which is

$$\left[\sum_{n=0}^{2-1} \binom{10}{n} \left(\frac{5}{20}\right)^n + \sum_{n=c}^{10} \frac{10!}{(10-n)!2!2^{n-2}} \left(\frac{5}{20}\right)^n \right]^{-1} = 0.065$$

From P_0, we can obtain the other P_n, from which we can compute the average number of machines waiting for service,

$$L_Q = \sum_{n=3}^{10} (n-2)P_n = 1.46 \text{ machines}$$

the effective arrival rate,

$$\lambda_e = \sum_{n=0}^{10} (10-n) \left(\frac{1}{20}\right) P_n = 0.342 \text{ machines/minute}$$

and the average waiting time in the queue,

$$w_Q = L_Q/\lambda_e = 4.27 \text{ minutes}$$

Similarly, we can compute the expected number of machines being serviced or waiting to be serviced,

$$L = \sum_{n=0}^{10} nP_n = 3.17 \text{ machines}$$

The average number of machines being serviced is given by

$$L - L_Q = 3.17 - 1.46 = 1.71 \text{ machines}$$

Each machine must be either running, waiting to be serviced, or in service, so the average number of running machines is given by

$$K - L = 10 - 3.17 = 6.83 \text{ machines}$$

A question frequently asked is this: What will happen if the number of servers is increased or decreased? If the number of workers in this example increases to three ($c = 3$), then the time-average number of running machines increases to

$$K - L = 7.74 \text{ machines}$$

an increase of 0.91 machine, on the average.

Conversely, what happens if the number of servers decreases to one? Then the time-average number of running machines decreases to

$$K - L = 3.98 \text{ machines}$$

The decrease from two servers to one has resulted in a drop of nearly three machines running, on the average. Exercise 17 asks the reader to examine the effect on server utilization of adding or deleting one server.

Example 6.18 illustrates several general relationships that have been found to hold for almost all queues. If the number of servers is decreased, delays, server utilization, and the probability of an arrival having to wait to begin service all increase.

6.6 NETWORKS OF QUEUES

In this chapter, we have emphasized the study of single queues of the $G/G/c/N/K$ type. However, many systems are naturally modeled as networks of single queues in which customers departing from one queue may be routed to another. Example 6.1 (see, in particular, Figure 6.3) and Example 6.2 (see Figure 6.5) are illustrations.

The study of mathematical models of networks of queues is beyond the scope of this chapter; see, for instance, Gross and Harris [1997], Nelson [1995], and Kleinrock [1976]. However, a few fundamental principles are very useful for rough-cut modeling, perhaps prior to a simulation study. The following results assume a stable system with infinite calling population and no limit on system capacity:

1. Provided that no customers are created or destroyed in the queue, then the departure rate out of a queue is the same as the arrival rate into the queue, over the long run.
2. If customers arrive to queue i at rate λ_i, and a fraction $0 \leq p_{ij} \leq 1$ of them are routed to queue j upon departure, then the arrival rate from queue i to queue j is $\lambda_i p_{ij}$ over the long run.
3. The overall arrival rate into queue j, λ_j, is the sum of the arrival rate from all sources. If customers arrive from outside the network at rate a_j, then

$$\lambda_j = a_j + \sum_{\text{all } i} \lambda_i p_{ij}$$

4. If queue j has $c_j < \infty$ parallel servers, each working at rate μ_j, then the long-run utilization of each server is

$$\rho_j = \frac{\lambda_j}{c_j \mu_j}$$

and $\rho_j < 1$ is required for the queue to be stable.

5. If, for each queue j, arrivals from outside the network form a Poisson process with rate a_j, and if there are c_j identical servicers delivering exponentially distributed service times with mean $1/\mu_j$ (where c_j may be ∞), then, in steady state, queue j behaves like an $M/M/c_j$ queue with arrival rate $\lambda_j = a_j + \sum_{\text{all } i} \lambda_i p_{ij}$.

Example 6.19

Consider again the discount store described in Example 6.1 and shown in Figure 6.3. Suppose that customers arrive at the rate 80 per hour and that, of those arrivals, 40% choose self-service; then, the arrival rate to service center 1 is $\lambda_1 = (80)(0.40) = 32$ per hour, and the arrival rate to service center 2 is $\lambda_2 = (80)(0.6) = 48$ per hour. Suppose that each of the $c_2 = 3$ clerks at service center 2 works at the rate $\mu_2 = 20$ customers per hour. Then the long-run utilization of the clerks is

$$\rho_2 = \frac{48}{(3)(20)} = 0.8$$

All customers must see the cashier at service center 3. The overall arrival rate to service center 3 is $\lambda_3 = \lambda_1 + \lambda_2 = 80$ per hour, regardless of the service rate at service center 1, because, over the long run, the departure rate out of each service center must be equal to the arrival rate into it. If the cashier works at rate $\mu_3 = 90$ per hour, then the utilization of the cashier is

$$\rho_3 = \frac{80}{90} = 0.89$$

Example 6.20

At a Driver's License branch office, drivers arrive at the rate 50 per hour. All arrivals must first check in with one of two clerks, with the average check-in time being 2 minutes. After check in, 15% of the drivers need to take a written test that lasts approximately 20 minutes. All arrivals must wait to have their picture taken and their license produced; this station can process about 60 drivers per hour. The branch manager wants to know whether it is adding a check-in clerk or adding a new photo station that will lead to a greater reduction in customer delay.

To solve the problem, let the check-in clerks be queue 1 (with $c_1 = 2$ servers, each working at rate $\mu_1 = 30$ drivers per hour), let the testing station be queue 2 (with $c_2 = \infty$ servers, because any number of people can be taking the written test simultaneously, and service rate $\mu_2 = 3$ drivers per hour), and let the photo station be queue 3 (with $c_3 = 1$ server working at rate $\mu_3 = 60$ drivers per hour). The arrival rates to each queue are as follows:

$$\lambda_1 = a_1 + \sum_{i=1}^{3} p_{i1}\lambda_i = 50 \text{ drivers per hour}$$

$$\lambda_2 = a_2 + \sum_{i=1}^{3} p_{i2}\lambda_i = (0.15)\lambda_1 \text{ drivers per hour}$$

$$\lambda_3 = a_3 + \sum_{i=1}^{3} p_{i3}\lambda_i = (1)\lambda_2 + (0.85)\lambda_1 \text{ drivers per hour}$$

Notice that arrivals from outside the network occur only at queue 1, so $a_1 = 50$ and $a_2 = a_3 = 0$. Solving this system of equations gives $\lambda_1 = \lambda_3 = 50$ and $\lambda_2 = 7.5$.

If we approximate the arrival process as Poisson, and the service times at each queue as exponentially distributed, then the check-in clerks can be approximated as an $M/M/c_1$ queue, the testing station as an $M/M/\infty$ queue, and the photo station as an $M/M/c_3$ queue. Thus, under the current set-up, the check-in station is an $M/M/2$; using the formulas in Table 6.5 gives $w_Q = 0.0758$ hours. If we add a clerk, so that the model is $M/M/3$, the waiting time in queue drops to 0.0075 hours, a savings of 0.0683 hours or about 4.1 minutes.

The current photo station can be modeled as an $M/M/1$ queue, giving $w_Q = 0.0833$ hours; adding a second photo station ($M/M/2$) causes the time in queue to drop to 0.0035 hours, a savings of 0.0798 hours, or about 4.8 minutes. Therefore, a second photo station offers a slightly greater reduction in waiting time than does adding a third clerk.

If desired, the testing station can be analyzed by using the results for an $M/M/\infty$ queue in Table 6.6. For instance, the expected number of people taking the test at any time is $L = \lambda_2/\mu_2 = 7.5/3 = 2.5$.

6.7 SUMMARY

Queueing models have found widespread use in the analysis of service facilities, production and material-handling systems, telephone and communications systems, and many other situations where congestion or competition for scarce resources can occur. This chapter has introduced the basic concepts of queueing models and shown how simulation, and in some cases a mathematical analysis, can be used to estimate the performance measures of a system.

A simulation may be used to generate one or more artificial histories of a complex system. This simulation-generated data may, in turn, be used to estimate desired performance measures of the system. Commonly used performance measures, including L, L_Q, w, w_Q, ρ, and λ_e were introduced, and formulas were given for their estimation from data.

When simulating any system that evolves over time, the analyst must decide whether transient behavior or steady-state performance is to be studied. Simple formulas exist for the steady-state behavior of some queues, but estimating steady-state performance measures from simulation-generated data requires recognizing and dealing with the possibly deleterious effect of the initial conditions on the estimators of steady-state performance. These estimators could be severely biased (either high or low), if the initial conditions are unrepresentative of steady state or if simulation run length is too short. These estimation problems are discussed at greater length in Chapter 11.

Whether the analyst is interested in transient or in steady-state performance of a system, it should be recognized that the estimates obtained from a simulation of a stochastic queue are exactly that—estimates. Every such estimate contains random error, and a proper statistical analysis is required to assess the accuracy of the estimate. Methods for conducting such a statistical analysis are discussed in Chapters 11 and 12.

In the last three sections of this chapter, it was shown that a number of simple models can be solved mathematically. Although the assumptions behind such models might not be met exactly in a practical application, these models can still be useful in providing a rough estimate of a performance measure. In many cases, models with exponentially distributed interarrival and service times will provide a conservative estimate of system behavior. For example, if the model predicts that average waiting time, w, will be 12.7 minutes, then average waiting time in the real system is likely to be less than 12.7 minutes. The conservative nature of exponential models arises because (a) performance measures, such as w and L, are generally increasing functions of the variance of interarrival times and service times (recall the $M/G/1$ queue), and (b) the exponential distribution is fairly highly variable, having its standard deviation always equal to its mean. Thus, if the arrival process or service mechanism of the real system is less variable than exponentially distributed interarrival or service times, it is likely that the average number in the system, L, and the average time spent in system, w, will be less than what is predicted by the exponential model. Of course, if the interarrival and service times are *more* variable than exponential random variables, then the M/M queueing models could underestimate congestion.

An important application of mathematical queueing models is determining the minimum number of servers needed at a work station or service center. Quite often, if the arrival rate λ and the service rate μ are known or can be estimated, then the simple inequality $\lambda/(c\mu) < 1$ can be used to provide an initial estimate for the number of servers, c, at a work station. For a large system with many work stations, it could be quite time consuming to have to simulate every possibility $(c_1, c_2, \dots)$ for the number of servers, c_i, at work station i. Thus, a bit of mathematical analysis rough estimates could save a great deal of computer time and analyst's time.

Finally, the qualitative behavior of the simple exponential models of queueing carries over to more complex systems. In general, it is the variability of service times and the variability of the arrival process that causes waiting lines to build up and congestion to occur. For most systems, if the arrival rate increases, or if the service rate decreases, or if the variance of service times or interarrival times increases, then the system will become more congested. Congestion can be decreased by adding more servers or by reducing the mean value and variability of service times. Simple queueing models can be a great aid in quantifying these relationships and in evaluating alternative system designs.

REFERENCES

BANKS, J., AND R. G. HEIKES [1984], *Handbook of Tables and Graphs for the Industrial Engineer and Manager*, Reston, Reston, VA.

COOPER, R. B. [1990], *Introduction to Queueing Theory*, 3d ed., George Washington University, Washington, DC.

DESCLOUX, A. [1962], *Delay Tables for Finite- and Infinite-Source Systems*, McGraw–Hill, New York.

GROSS, D., AND C. HARRIS [1997], *Fundamentals of Queueing Theory*, 3d ed., Wiley, New York.

HALL, R. W. [1991], *Queueing Methods: For Services and Manufacturing*, Prentice Hall, Englewood Cliffs, NJ.

HILLIER, F. S., AND G. J. LIEBERMAN [2005], *Introduction to Operations Research*, 8th ed., McGraw–Hill, New York.

HILLIER, F. S., AND O. S. YU [1981], *Queueing Tables and Graphs*, Elsevier North-Holland, New York.

KENDALL, D. G. [1953], "Stochastic Processes Occurring in the Theory of Queues and Their Analysis by the Method of Imbedded Markov Chains," *Annals of Mathematical Statistics*, Vol. 24, pp. 338–354.

KLEINROCK, L. [1976], *Queueing Systems, Vol 2: Computer Applications*, Wiley, New York.

LITTLE, J. D. C. [1961], "A Proof for the Queueing Formula $L = \lambda w$," *Operations Research*, Vol. 16, pp. 651–665.

NELSON, B. L. [1995], *Stochastic Modeling: Analysis & Simulation*, Dover Publications, Mineola, NY.

PECK, L. G., AND R. N. HAZELWOOD [1958], *Finite Queueing Tables*, Wiley, New York.

WAGNER, H. M. [1975], *Principles of Operations Research*, 2d ed., Prentice-Hall, Englewood Cliffs, NJ.

WINSTON, W. L. [1997], *Operations Research: Applications and Algorithms*, Duxbury Press, Pacific Grove, CA.

EXERCISES

1. A tool crib has exponential interarrival and service times and serves a very large group of mechanics. The mean time between arrivals is 4 minutes. It takes 3 minutes on the average for a tool-crib attendant to service a mechanic. The attendant is paid $10 per hour and the mechanic is paid $15 per hour. Would it be advisable to have a second tool-crib attendant?

2. A two-runway (one runway for landing, one runway for taking off) airport is being designed for propeller-driven aircraft. The time to land an airplane is known to be exponentially distributed, with a mean of 1-1/2 minutes. If airplane arrivals are assumed to occur at random, what arrival rate can be tolerated if the average wait in the sky is not to exceed 3 minutes?

3. The Port of Trop can service only one ship at a time. However, there is mooring space for three more ships. Trop is a favorite port of call, but, if no mooring space is available, the ships have to go to the Port of Poop. An average of seven ships arrive each week, according to a Poisson process. The Port of Trop has the capacity to handle an average of eight ships a week, with service times exponentially distributed. What is the expected number of ships waiting or in service at the Port of Trop?

4. At Metropolis City Hall, two workers "pull strings" (make deals) every day. Strings arrive to be pulled on an average of one every 10 minutes throughout the day. It takes an average of 15 minutes to pull a string. Both times between arrivals and service times are exponentially distributed. What is the probability that there are no strings to be pulled in the system at a random point in time? What is the expected number of strings waiting to be pulled? What is the probability that both string-pullers are busy? What is the effect on performance if a third string-puller, working at the same speed as the first two, is added to the system?

5. At Tony and Cleo's bakery, one kind of birthday cake is offered. It takes 15 minutes to decorate this particular cake, and the job is performed by one particular baker. In fact, it is all this baker does. What mean time between arrivals (exponentially distributed) can be accepted if the mean length of the queue for decorating is not to exceed five cakes?

6. Patients arrive for a physical examination according to a Poisson process at the rate 1 per hour. The physical examination requires three stages, each one independently exponentially distributed, with a service time of 15 minutes. A patient must go through all three stages before the next patient is admitted to the treatment facility. Compute the average number of delayed patients, L_Q, for this system. (Hint: The variance of the sum of independent random variables is the sum of the variance.)

7. Suppose that mechanics arrive randomly at a tool crib according to a Poisson process with rate $\lambda = 10$ per hour. It is known that the single tool clerk serves a mechanic in 4 minutes on the

average, with a standard deviation of approximately 2 minutes. Suppose that mechanics make $15.00 per hour. Estimate the steady-state average cost per hour of mechanics waiting for tools.

8. Arrivals to an airport are all directed to the same runway. At a certain time of the day, these arrivals form a Poisson process with rate 30 per hour. The time to land an aircraft is a constant 90 seconds. Determine L_Q, w_Q, L, and w for this airport. If a delayed aircraft burns $5000 worth of fuel per hour on the average, determine the average cost per aircraft of delay in waiting to land.

9. A machine shop repairs small electric motors, which arrive according to a Poisson process at the rate 12 per week (5-day, 40-hour workweek). An analysis of past data indicates that engines can be repaired, on the average, in 2.5 hours, with a variance of 1 hour^2. How many working hours should a customer expect to leave a motor at the repair shop (not knowing the status of the system)? If the variance of the repair time could be controlled, what variance would reduce the expected waiting time to 6.5 hours?

10. Arrivals to a self-service gasoline pump occur in a Poisson fashion at the rate 12 per hour. Service time has a distribution that averages 4 minutes, with a standard deviation of 1-1/3 minutes. What is the expected number of vehicles in the system?

11. Classic Car Care has one worker who washes cars in a four-step method—soap, rinse, dry, vacuum. The time to complete each step is exponentially distributed, with mean 9 minutes. Every car goes through every step before another car begins the process. On the average, one car every 45 minutes arrives for a wash job, according to a Poisson process. What is the average time a car waits to begin the wash job? What is the average number of cars in the car wash system? What is the average time required to wash a car?

12. A room has 10 cotton spinning looms. Once the looms are set up, they run automatically. The setup time is exponentially distributed, with mean 10 minutes. The machines run for an average of 40 minutes, also exponentially distributed. Loom operators are paid $10 an hour, and looms not running incur a cost of $40 an hour. How many loom operators should be employed to minimize the total cost of the loom room? If the objective becomes "on the average, no loom should wait more than 1 minute for an operator," how many persons should be employed? How many operators should be employed to ensure that, on average, at least 7.5 looms are running at all times?

13. Given the following information for a finite calling population problem with exponentially distributed runtimes and service times:

$$K = 10$$

$$\frac{1}{\mu} = 15$$

$$\frac{1}{\lambda} = 82$$

$$c = 2$$

Compute L_Q and w_Q. Find the value of λ such that $L_Q = L/2$.

14. Suppose that Figure 6.6 represents the number in system for a last-in–first-out (LIFO) single-server system. Customers are not preempted (i.e., kicked out of service), but, upon service completion, the most recent arrival next begins service. For this LIFO system, apportion the total area under $L(t)$ to each individual customer, as was done in Figure 6.8 for the FIFO system. Using the figure, show that Equations (6.10) and (6.8) hold for the single-server LIFO system.

15. Repeat Exercise 14, but assuming that

    (a) Figure 6.6 represents a FIFO system with $c = 2$ servers;
    (b) Figure 6.6 represents a LIFO system with $c = 2$ servers:

16. Consider a $M/G/1$ queue with the following type of service distribution: Customers request one of two types of service, in the proportions p and $1 - p$. Type i service is exponentially distributed at rate μ_i, $i = 1, 2$. Let X_i denote a type-i service time and X an arbitrary service time. Then $E(X_i) = 1/\mu_i$, $V(X_i) = 1/\mu_i^2$ and

$$
X = \begin{cases} X_1 & \text{with probability } p \\ X_2 & \text{with probability } (1 - p) \end{cases}
$$

The random variable X is said to have a hyperexponential distribution with parameters (μ_1, μ_2, p).

    (a) Show that $E(X) = p/\mu_1 + (1 - p)/\mu_2$ and $E(X^2) = 2p/\mu_1^2 + 2(1 - p)/\mu_2^2$.
    (b) Use $V(X) = E(X^2) - [E(X)]^2$ to show $V(X) = 2p/\mu_1^2 + 2(1 - p)/\mu_2^2 - [p/\mu_1 + (1 - p)/\mu_2]^2$.
    (c) For any hyperexponential random variable, if $\mu_1 \neq \mu_2$ and $0 < p < 1$, show that its coefficient of variation is greater than 1—that is, $(\text{cv})^2 = V(X)/[E(X)]^2 > 1$. Thus, the hyperexponential distribution provides a family of statistical models for service times that are more variable than exponentially distributed service times. *Hint*: The algebraic expression for $(\text{cv})^2$, by using parts (a) and (b), can be manipulated into the form $(\text{cv})^2 = 2p(1 - p)(1/\mu_1 - 1/\mu_2)^2/[E(X)]^2 + 1$.
    (d) Many choices of μ_1, μ_2, and p lead to the same overall mean $E(X)$ and $(\text{cv})^2$. If a distribution with mean $E(X) = 1$ and coefficient of variation $\text{cv} = 2$ is desired, find values of μ_1, μ_2, and p to achieve this. *Hint*: Choose $p = 1/4$ arbitrarily; then solve the following equations for μ_1 and μ_2.

$$
\frac{1}{4\mu_1} + \frac{3}{4\mu_2} = 1
$$

$$
\frac{3}{8} \left(\frac{1}{\mu_1} - \frac{1}{\mu_2} \right)^2 + 1 = 4
$$

17. In Example 6.18, compare the systems with $c = 1$, $c = 2$, and $c = 3$ servers on the basis of server utilization ρ (the proportion of time a typical server is busy).

18. In Example 6.18, increase the number of machines by 2, then compare the systems with $c = 1$, $c = 2$, and $c = 3$ servers on the basis of server utilization ρ (the proportion of time a typical server is busy).

19. A small lumberyard is supplied by a fleet of 10 trucks. One overhead crane is available to unload the long logs from the trucks. It takes an average of 1 hour to unload a truck. After unloading, a truck takes an average of 3 hours to get the next load of logs and return to the lumberyard.

    (a) Certain distributional assumptions are needed to analyze this problem with the models of this chapter. State them and discuss their reasonableness.
    (b) With one crane, what is the average number of trucks waiting to be unloaded? On average, how many trucks arrive at the yard each hour? What percentage of trucks upon arrival find the crane busy? Is this the same as the long-run proportion of time the crane is busy?
    (c) Suppose that a second crane is installed at the lumberyard. Answer the same questions as in part (b). Make a chart comparing one crane to two cranes.
    (d) If the value of the logs brought to the yard is approximately $200 per truck load and if long-run crane costs are $50 per hour per crane (whether busy or idle), compute an optimal number of cranes on the basis of cost per hour.
    (e) In addition to the costs in part (d), if management decides to consider the cost of idle trucks and drivers, what is the optimal number of cranes? A truck and its driver together are estimated to cost approximately $40 per hour and are considered to be idle when they are waiting for a crane.

20. A tool crib with one attendant serves a group of 10 mechanics. Mechanics work for an exponentially distributed amount of time with mean 20 minutes, then go to the crib to request a special tool. Service times by the attendant are exponentially distributed, with mean 3 minutes. If the attendant is paid $6 per hour and the mechanic is paid $10 per hour, would it be advisable to have a second attendant? (Compare to Exercise 1.)

21. This problem is based on Case 8.1 in Nelson [1995]. A large consumer shopping mall is to be constructed. During busy times, the arrival rate of cars is expected to be 1000 per hour, and studies at other malls suggest that customers will spend 3 hours, on average, shopping. The mall designers would like to have sufficient parking so that there are enough spaces 99.9% of the time. How many spaces should they have? Hint: Model the system as an $M/G/\infty$ queue where the spaces are servers, and find out how many spaces are adequate with probability 0.999.

22. In Example 6.19, suppose that the overall arrival rate is expected to increase to 160 per hour. If the service rates do not change, how many clerks will be needed at service centers 2 and 3, just to keep up with the customer load?

23. A small copy shop has a self-service copier. Currently there is room for only 4 people to line up for the machine (including the person using the machine); when there are more than 4 people, then the additional people must line up outside the shop. The owners would like to avoid having people line up outside the shop, as much as possible. For that reason, they are thinking about adding a second self-service copier. Self-service customers have been observed

to arrive at the rate 24 per hour, and they use the machine 2 minutes, on average. Assess the impact of adding another copier. Carefully state any assumptions or approximations you make.

24. A self-service car wash has 4 washing stalls. When in a stall, a customer may choose from among three options: rinse only; wash and rinse; and wash, rinse, and wax. Each option has a fixed time to complete: rinse only, 3 minutes; wash and rinse, 7 minutes; wash, rinse, and wax, 12 minutes. The owners have observed that 20% of customers rinse only; 70% wash and rinse; and 10% wash, rinse, and wax. There are no scheduled appointments, and customers arrive at a rate of about 34 cars per hour. There is room for only 3 cars to wait in the parking lot, so, currently, many customers are lost. The owners wants to know how much more business they will do if they add another stall. Adding a stall will take away one space in the parking lot.

Develop a queueing model of the system. Estimate the rate at which customers will be lost in the current and proposed system. Carefully state any assumptions or approximations you make.

25. Find examples of queueing models used in real applications. A good source is the journal *Interfaces*.

26. Study the effect of *pooling servers* (having multiple servers draw from a single queue, rather than each having its own queue) by comparing the performance measures for two M/M/1 queues, each with arrival rate λ and service rate μ, to an M/M/2 queue with arrival rate 2λ and service rate μ for each server.

27. A repair and inspection facility consists of two stations: a repair station with two technicians, and an inspection station with 1 inspector. Each repair technician works at the rate 3 items per hour; the inspector can inspect 8 items per hour. Approximately 10% of all items fail inspection and are sent back to the repair station. (This percentage holds even for items that have been repaired two or more times.) If items arrive at the rate 5 per hour, what is the long-run expected delay that items experience at each of the two stations, assuming a Poisson arrival process and exponentially distributed service times? What is the maximum arrival rate that the system can handle without adding personnel?

Part III

Random Numbers

Part III

Random Numbers

7

Random-Number Generation

Random numbers are a necessary basic ingredient in the simulation of almost all discrete systems. Most computer languages have a subroutine, object, or function that will generate a random number. Similarly, simulation languages generate random numbers that are used to generate event times and other random variables. In this chapter, the generation of random numbers and their subsequent testing for randomness is described. Chapter 8 shows how random numbers are used to generate a random variable with any desired probability distribution.

7.1 PROPERTIES OF RANDOM NUMBERS

A sequence of random numbers, $R_1, R_2, \ldots$, must have two important statistical properties: uniformity and independence. Each random number R_i must be an independent sample drawn from a continuous uniform distribution between zero and 1—that is, the pdf is given by

$$f(x) = \begin{cases} 1, & 0 \leq x \leq 1 \\ 0, & \text{otherwise} \end{cases}$$

This density function is shown in Figure 7.1. The expected value of each R_i is given by

$$E(R) = \int_0^1 x \, dx = \left. \frac{x^2}{2} \right|_0^1 = \frac{1}{2}$$

and the variance is given by

$$V(R) = \int_0^1 x^2 \, dx - [E(R)]^2 = \left. \frac{x^3}{3} \right|_0^1 - \left(\frac{1}{2}\right)^2 = \frac{1}{3} - \frac{1}{4} = \frac{1}{12}$$

Figure 7.1   pdf for random numbers.

Some consequences of the uniformity and independence properties are the following:

1. If the interval [0, 1] is divided into n classes, or subintervals of equal length, the expected number of observations in each interval is N/n, where N is the total number of observations.
2. The probability of observing a value in a particular interval is independent of the previous values drawn. *equally likely*

7.2 GENERATION OF PSEUDO-RANDOM NUMBERS

Notice that the title of this section has the word "pseudo" in it. "Pseudo" means false, so false random numbers are being generated! In this instance, "pseudo" is used to imply that the very act of generating random numbers by a known method removes the potential for true randomness. If the method is known, the set of random numbers can be replicated. Then an argument can be made that the numbers are not truly random. The goal of any generation scheme, however, is to produce a sequence of numbers between 0 and 1 that simulates, or imitates, the ideal properties of uniform distribution and independence as closely as possible.

To be sure, in the generation of pseudo-random numbers, certain problems or errors can occur. These errors, or departures from ideal randomness, are all related to the properties stated previously. Some examples of such departures include the following:

1. The generated numbers might not be uniformly distributed.
2. The generated numbers might be discrete-valued instead of continuous-valued.
3. The mean of the generated numbers might be too high or too low.
4. The variance of the generated numbers might be too high or too low.
5. There might be dependence. The following are examples:
   (a) autocorrelation between numbers;
   (b) numbers successively higher or lower than adjacent numbers;
   (c) several numbers above the mean followed by several numbers below the mean.

Departures from uniformity and independence for a particular generation scheme often can be detected by such tests as those described in Section 7.4. If such departures are detected, the generation scheme should be dropped in favor of an acceptable generator. Generators that pass the tests in Section 7.4 and tests even more stringent have been developed; thus, there is no excuse for using a generator that has been found to be defective.

Usually, random numbers are generated by a digital computer, as part of the simulation. There are numerous methods that can be used to generate the values. Before we describe some of these methods, or *routines*, there are a number of important considerations that we should mention:

1. The routine should be fast. Individual computations are inexpensive, but simulation could require many millions of random numbers. The total cost can be managed by selecting a computationally efficient method of random-number generation.
2. The routine should be portable to different computers—and, ideally, to different programming languages. This is desirable so that the simulation program will produce the same results wherever it is executed.
3. The routine should have a sufficiently long cycle. The cycle length, or period, represents the length of the random number sequence before previous numbers begin to repeat themselves in an earlier order. Thus, if 10,000 events are to be generated, the period should be many times that long.

   A special case of cycling is degenerating. A routine degenerates when the same random numbers appear repeatedly. Such an occurrence is certainly unacceptable. This can happen rapidly with some methods.
4. The random numbers should be replicable. Given the starting point (or conditions) it should be possible to generate the same set of random numbers, completely independent of the system that is being simulated. This is helpful for debugging purposes and is a means of facilitating comparisons between systems (see Chapter 12). For the same reasons, it should be possible to easily specify different starting points, widely separated, within the sequence.
5. Most important, the generated random numbers should closely approximate the ideal statistical properties of uniformity and independence.

Inventing techniques that seem to generate random numbers is easy; inventing techniques that really do produce sequences that appear to be independent, uniformly distributed random numbers is incredibly difficult. There is now a vast literature and rich theory on the topic, and many hours of testing have been devoted to establishing the properties of various generators. Even when a technique is known to be theoretically sound, it is seldom easy to implement it in a way that will be fast and portable. The goal of this chapter is to make the reader aware of the central issues in random-number generation, to enhance understanding and to show some of the techniques that are used by those working in this area.

7.3 TECHNIQUES FOR GENERATING RANDOM NUMBERS

The linear congruential method of Section 7.3.1 is the most widely used technique for generating random numbers, so we describe it in detail. We also report an extension of this method that yields sequences with a longer period. Many other methods have been proposed, and they are reviewed in Bratley, Fox, and Schrage [1996], Law and Kelton [2000], and Ripley [1987].

7.3.1 Linear Congruential Method

The linear congruential method, initially proposed by Lehmer [1951], produces a sequence of integers, $X_1, X_2, \ldots$ between zero and $m - 1$ by following a recursive relationship:

$$X_{i+1} = (aX_i + c) \bmod m, \quad i = 0, 1, 2, \ldots \tag{7.1}$$

The initial value X_0 is called the seed, a is called the multiplier, c is the increment, and m is the modulus. If $c \neq 0$ in Equation (7.1), then the form is called the *mixed congruential method*. When $c = 0$, the form is known as the *multiplicative congruential method*. The selection of the values for $a, c, m,$ and X_0 drastically affects the statistical properties and the cycle length. Variations of Equation (7.1) are quite common in the computer generation of random numbers. An example will illustrate how this technique operates.

Example 7.1

Use the linear congruential method to generate a sequence of random numbers with $X_0 = 27$, $a = 17$, $c = 43$, and $m = 100$. Here, the integer values generated will all be between zero and 99 because of the value of the modulus. Also, notice that random integers are being generated rather than random numbers. These random integers should appear to be uniformly distributed on the integers zero to 99. Random numbers between zero and 1 can be generated by

$$R_i = \frac{X_i}{m}, \quad i = 1, 2, \ldots \tag{7.2}$$

The sequence of X_i and subsequent R_i values is computed as follows:

$X_0 = 27$

$X_1 = (17 \cdot 27 + 43) \bmod 100 = 502 \bmod 100 = 2$

$R_1 = \dfrac{2}{100} = 0.02$

$X_2 = (17 \cdot 2 + 43) \bmod 100 = 77 \bmod 100 = 77$

$R_2 = \dfrac{77}{100} = 0.77$

$X_3 = (17 \cdot 77 + 43) \bmod 100 = 1352 \bmod 100 = 52$

$R_3 = \dfrac{52}{100} = 0.52$

$\vdots$

Recall that $a = b \bmod m$ provided that $(b - a)$ is divisible by m with no remainder. Thus, $X_1 = 502 \bmod 100$, but $502/100$ equals 5 with a remainder of 2, so that $X_1 = 2$. In other words, $(502 - 2)$ is evenly divisible by $m = 100$, so $X_1 = 502$ "reduces" to $X_1 = 2 \bmod 100$. (A shortcut for the modulo, or reduction operation for the case $m = 10^b$, a power of 10, is illustrated in Example 7.3.)

    The ultimate test of the linear congruential method, as of any generation scheme, is how closely the generated numbers $R_1, R_2, \ldots$ approximate uniformity and independence. There are, however, several secondary properties that must be considered. These include *maximum density* and *maximum period*.

First, notice that the numbers generated from Equation (7.2) assume values only from the set $I = \{0, 1/m, 2/m, \ldots, (m-1)/m\}$, because each X_i is an integer in the set $\{0, 1, 2, \ldots, m-1\}$. Thus, each R_i is discrete on I, instead of continuous on the interval [0, 1]. This approximation appears to be of little consequence if the modulus m is a very large integer. (Values such as $m = 2^{31} - 1$ and $m = 2^{48}$ are in common use in generators appearing in many simulation languages.) By maximum density is meant that the values assumed by $R_i, i = 1, 2, \ldots$, leave no large gaps on [0, 1].

Second, to help achieve maximum density, and to avoid cycling (i.e., recurrence of the same sequence of generated numbers) in practical applications, the generator should have the largest possible period. Maximal period can be achieved by the proper choice of a, c, m, and X_0 [Fishman, 1978; Law and Kelton, 2000].

- For m a power of 2, say $m = 2^b$, and $c \neq 0$, the longest possible period is $P = m = 2^b$, which is achieved whenever c is relatively prime to m (that is, the greatest common factor of c and m is 1) and $a = 1 + 4k$, where k is an integer.
- For m a power of 2, say $m = 2^b$, and $c = 0$, the longest possible period is $P = m/4 = 2^{b-2}$, which is achieved if the seed X_0 is odd and if the multiplier, a, is given by $a = 3 + 8k$ or $a = 5 + 8k$, for some $k = 0, 1, \ldots$.
- For m a prime number and $c = 0$, the longest possible period is $P = m - 1$, which is achieved whenever the multiplier, a, has the property that the smallest integer k such that $a^k - 1$ is divisible by m is $k = m - 1$.

Example 7.2

Using the multiplicative congruential method, find the period of the generator for $a = 13, m = 2^6 = 64$, and $X_0 = 1, 2, 3$, and 4. The solution is given in Table 7.1. When the seed is 1 or 3, the sequence has period 16. However, a period of length eight is achieved when the seed is 2 and a period of length four occurs when the seed is 4.

In Example 7.2, $m = 2^6 = 64$ and $c = 0$. The maximal period is therefore $P = m/4 = 16$. Notice that this period is achieved by using odd seeds, $X_0 = 1$ and $X_0 = 3$; even seeds, $X_0 = 2$ and $X_0 = 4$, yield the periods eight and four, respectively, both less than the maximum. Notice that $a = 13$ is of the form $5 + 8k$ with $k = 1$, as is required to achieve maximal period.

When $X_0 = 1$, the generated sequence assumes values from the set $\{1, 5, 9, 13, \ldots, 53, 57, 61\}$. The "gaps" in the sequence of generated random numbers, R_i, are quite large (i.e., the gap is $5/64 - 1/64$ or 0.0625). Such a gap gives rise to concern about the density of the generated sequence.

The generator in Example 7.2 is not viable for any application—its period is too short, and its density is insufficient. However, the example shows the importance of properly choosing a, c, m, and X_0.

Speed and efficiency in using the generator on a digital computer is also a selection consideration. Speed and efficiency are aided by use of a modulus, m, which is either a power of 2 or close to a power of 2. Since most digital computers use a binary representation of numbers, the modulo, or remaindering, operation of Equation (7.1) can be conducted efficiently when the modulo is a power of 2 (i.e., $m = 2^b$). After ordinary arithmetic yields a value for $aX_i + c$, X_{i+1} is obtained

Table 7.1 Period Determination Using Various Seeds

i	X_i	X_i	X_i	X_i
0	1	2	3	4
1	13	26	39	52
2	41	18	59	36
3	21	42	63	20
4	17	34	51	4
5	29	58	23	
6	57	50	43	
7	37	10	47	
8	33	2	35	
9	45		7	
10	9		27	
11	53		31	
12	49		19	
13	61		55	
14	25		11	
15	5		15	
16	1		3	

by dropping the leftmost binary digits in $aX_i + c$ and then using only the b rightmost binary digits. The following example illustrates, by analogy, this operation using $m = 10^b$, because most human beings think in decimal representation.

Example 7.3

Let $m = 10^2 = 100$, $a = 19$, $c = 0$, and $X_0 = 63$, and generate a sequence of random integers using Equation (7.1).

$$X_0 = 63$$

$$X_1 = (19)(63) \bmod 100 = 1197 \bmod 100 = 97$$

$$X_2 = (19)(97) \bmod 100 = 1843 \bmod 100 = 43$$

$$X_3 = (19)(43) \bmod 100 = 817 \bmod 100 = 17$$

$$\vdots$$

When m is a power of 10, say $m = 10^b$, the modulo operation is accomplished by saving the b rightmost (decimal) digits. By analogy, the modulo operation is most efficient for binary computers when $m = 2^b$ for some $b > 0$.

Example 7.4

The last example in this section is in actual use. It has been extensively tested [Learmonth and Lewis, 1973; Lewis et al., 1969]. The values for a, c, and m have been selected to ensure that the characteristics desired in a generator are most likely to be achieved. By changing X_0, the user can control the repeatability of the stream.

Let $a = 7^5 = 16,807$; $m = 2^{31} - 1 = 2,147,483,647$ (a prime number); and $c = 0$. These choices satisfy the conditions that insure a period of $P = m - 1$ (well over 2 billion). Further, specify the seed $X_0 = 123,457$. The first few numbers generated are as follows:

$$X_1 = 7^5(123,457) \bmod (2^{31} - 1) = 2,074,941,799 \bmod (2^{31} - 1)$$

$$X_1 = 2,074,941,799$$

$$R_1 = \frac{X_1}{2^{31}} = 0.9662$$

$$X_2 = 7^5(2,074,941,799) \bmod (2^{31} - 1) = 559,872,160$$

$$R_2 = \frac{X_2}{2^{31}} = 0.2607$$

$$X_3 = 7^5(559,872,160) \bmod (2^{31} - 1) = 1,645,535,613$$

$$R_3 = \frac{X_3}{2^{31}} = 0.7662$$

$$\vdots$$

Notice that this routine divides by $m + 1$ instead of m; however, for such a large value of m, the effect is negligible.

7.3.2 Combined Linear Congruential Generators

As computing power has increased, the complexity of the systems that we are able to simulate has also increased. A random-number generator with period $2^{31} - 1 \approx 2 \times 10^9$, such as the popular generator described in Example 7.4, is no longer adequate for all applications. Examples include the simulation of highly reliable systems, in which hundreds of thousands of elementary events must be simulated to observe even a single failure event, and the simulation of complex computer networks, in which thousands of users are executing hundreds of programs. An area of current research is the deriving of generators with substantially longer periods.

One fruitful approach is to combine two or more multiplicative congruential generators in such a way that the combined generator has good statistical properties and a longer period. The following result from L'Ecuyer [1988] suggests how this can be done:

If $W_{i,1}, W_{i,2}, \ldots, W_{i,k}$ are any independent, discrete-valued random variables (not necessarily identically distributed), but one of them, say $W_{i,1}$, is uniformly distributed on the integers from 0 to $m_1 - 2$, then

$$W_i = \left(\sum_{j=1}^{k} W_{i,j} \right) \bmod m_1 - 1$$

is uniformly distributed on the integers from 0 to $m_1 - 2$.

To see how this result can be used to form combined generators, let $X_{i,1}, X_{i,2}, \ldots, X_{i,k}$ be the ith output from k different multiplicative congruential generators, where the jth generator has prime modulus m_j and the multiplier a_j is chosen so that the period is $m_j - 1$. Then the jth generator is producing integers $X_{i,j}$ that are approximately uniformly distributed on the integers from 1 to $m_j - 1$, and $W_{i,j} = X_{i,j} - 1$ is approximately uniformly distributed on the integers from 0 to $m_j - 2$. L'Ecuyer [1988] therefore suggests combined generators of the form

$$X_i = \left(\sum_{j=1}^{k} (-1)^{j-1} X_{i,j} \right) \bmod m_1 - 1$$

with

$$R_i = \begin{cases} \dfrac{X_i}{m_1}, & X_i > 0 \\[2mm] \dfrac{m_1 - 1}{m_1}, & X_i = 0 \end{cases}$$

Notice that the "$(-1)^{j-1}$" coefficient implicitly performs the subtraction $X_{i,1} - 1$; for example, if $k = 2$ then $(-1)^0 (X_{i,1} - 1) - (-1)^1 (X_{i,2} - 1) = \sum_{j=1}^{2} (-1)^{j-1} X_{i,j}$.

The maximum possible period for such a generator is

$$P = \frac{(m_1 - 1)(m_2 - 1) \cdots (m_k - 1)}{2^{k-1}}$$

which is achieved by the generator described in the next example.

Example 7.5

For 32-bit computers, L'Ecuyer [1988] suggests combining $k = 2$ generators with $m_1 = 2,147,483,563$, $a_1 = 40,014$, $m_2 = 2,147,483,399$ and $a_2 = 40,692$. This leads to the following algorithm:

1. Select seed $X_{1,0}$ in the range [1, 2,14,74,83,562] for the first generator, and seed $X_{2,0}$ in the range [1, 2,14,74,83,398] for the second.
    Set $j = 0$.
2. Evaluate each individual generator.

$$X_{1,j+1} = 40,014 \, X_{1,j} \bmod 2,147,483,563$$

$$X_{2,j+1} = 40,692 \, X_{2,j} \bmod 2,147,483,399$$

3. Set

$$X_{j+1} = (X_{1,j+1} - X_{2,j+1}) \bmod 2,147,483,562$$

4. Return

$$
R_{j+1} = \begin{cases} \dfrac{X_{j+1}}{2,147,483,563}, & X_{j+1} > 0 \\[2ex] \dfrac{2,147,483,562}{2,147,483,563}, & X_{j+1} = 0 \end{cases}
$$

5. Set $j = j + 1$ and go to step 2.

This combined generator has period $(m_1 - 1)(m_2 - 1)/2 \approx 2 \times 10^{18}$. Perhaps surprisingly, even such a long period might not be adequate for all applications. See L'Ecuyer [1996, 1999] and L'Ecuyer *et al.* [2002] for combined generators with periods as long as $2^{191} \approx 3 \times 10^{57}$.

7.3.3 Random-Number Streams

The seed for a linear congruential random-number generator (seeds, in the case of a combined linear congruential generator) is the integer value X_0 that initializes the random-number sequence. Since the sequence of integers $X_0, X_1, \ldots, X_P, X_0, X_1, \ldots$ produced by a generator repeats, any value in the sequence could be used to "seed" the generator.

For a linear congruential generator, a random-number *stream* is nothing more than a convenient way to refer to a starting seed taken from the sequence $X_0, X_1, \ldots, X_P$ (for a combined generator, starting seeds for all of the basic generators are required); typically these starting seeds are far apart in the sequence. For instance, if the streams are b values apart, then stream i could be defined by starting seed

$$
S_i = X_{b(i-1)}
$$

for $i = 1, 2, \ldots, \lfloor P/b \rfloor$. Values of $b = 100,000$ were common in older generators, but values as large as $b = 10^{37}$ are in use in modern combined linear congruential generators. (See, for instance, L'Ecuyer *et al.* [2002] for the implementation of such a generator.) Thus, a single random-number generator with k streams acts like k distinct virtual random-number generators, provided that the current value of seed for each stream is maintained. Exercise 21 illustrates one way to create streams that are widely separated in the random-number sequence.

In Chapter 12, we will consider the problem of comparing two or more alternative systems via simulation, and we will show that there are advantages to dedicating portions of the pseudorandom number sequence to the same purpose in each of the simulated systems. For instance, in comparing the efficiency of several queueing systems, a fairer comparison will be achieved if all of the simulated systems experience exactly the same sequence of customer arrivals. Such synchronization can be achieved by assigning a specific stream to generate arrivals in each of the queueing simulations. If the starting seeds for the streams are spaced far enough apart, then this has the same effect as having a distinct random-number generator whose only purpose is to generate customer arrivals.

7.4 TESTS FOR RANDOM NUMBERS

The desirable properties of random numbers—uniformity and independence—were discussed in Section 7.1. To check on whether these desirable properties have been achieved, a number of tests can be performed. (Fortunately, the appropriate tests have already been conducted for most commercial simulation software.) The tests can be placed in two categories, according to the properties of interest: uniformity, and independence. A brief description of two types of tests is given in this chapter:

1. *Frequency test.* Uses the Kolmogorov–Smirnov or the chi-square test to compare the distribution of the set of numbers generated to a uniform distribution.
2. *Autocorrelation test.* Tests the correlation between numbers and compares the sample correlation to the expected correlation, zero.

In testing for uniformity, the hypotheses are as follows:

$$H_0 : \quad R_i \sim U[0, 1]$$
$$H_1 : \quad R_i \not\sim U[0, 1]$$

The null hypothesis, H_0, reads that the numbers are distributed uniformly on the interval $[0, 1]$. Failure to reject the null hypothesis means that evidence of nonuniformity has not been detected by this test. This does not imply that further testing of the generator for uniformity is unnecessary.

In testing for independence, the hypotheses are as follows:

$$H_0 : \quad R_i \sim \text{independently}$$
$$H_1 : \quad R_i \not\sim \text{independently}$$

This null hypothesis, H_0, reads that the numbers are independent. Failure to reject the null hypothesis means that evidence of dependence has not been detected by this test. This does not imply that further testing of the generator for independence is unnecessary.

For each test, a level of significance α must be stated. The level α is the probability of rejecting the null hypothesis when the null hypothesis is true:

$$\alpha = P(\text{reject } H_0 | H_0 \text{ true})$$

The decision maker sets the value of α for any test. Frequently, α is set to 0.01 or 0.05.

If several tests are conducted on the same set of numbers, the probability of rejecting the null hypothesis on at least one test, by chance alone [i.e., making a Type I (α) error], increases. Say that $\alpha = 0.05$ and that five different tests are conducted on a sequence of numbers. The probability of rejecting the null hypothesis on at least one test, by chance alone, could be as large as 0.25.

Similarly, if one test is conducted on many sets of numbers from a generator, the probability of rejecting the null hypothesis on at least one test by chance alone [i.e., making a Type I (α) error], increases as more sets of numbers are tested. For instance, if 100 sets of numbers were subjected to the test, with $\alpha = 0.05$, it would be expected that five of those tests would be rejected by chance alone. If the number of rejections in 100 tests is close to 100α, then there is no compelling reason to discard the generator. The concept discussed in this and the preceding paragraph is discussed further at the conclusion of Example 7.8.

If one of the well-known simulation languages or random-number generators is used, it is probably unnecessary to apply the tests just mentioned and described in Sections 7.4.1 and 7.4.2. However, random-number generators frequently are added to software that is not specifically developed for simulation, such as spreadsheet programs, symbolic/numerical calculators, and programming languages. If the generator that is at hand is not explicitly known or documented, then the tests in this chapter should be applied to many samples of numbers from the generator. Some additional tests that are commonly used, but are not covered here, are Good's serial test for sampling numbers [1953, 1967], the median-spectrum test [Cox and Lewis, 1966; Durbin, 1967], the runs test [Law and Kelton 2000] and a variance heterogeneity test [Cox and Lewis, 1966]. Even if a set of numbers passes all the tests, there is no guarantee of randomness; it is always possible that some underlying pattern has gone undetected.

In this book, we emphasize empirical tests that are applied to actual sequences of numbers produced by a generator. Because of the extremely long period of modern pseudo-random-number generators, as described in Section 7.3.2, it is no longer possible to apply these tests to a significant portion of the period of such generators. The tests can be used as a check if one encounters a generator with completely unknown properties (perhaps one that is undocumented and buried deep in a software package), but they cannot be used to establish the quality of a generator throughout its period. Fortunately, there are also families of theoretical tests that evaluate the choices for m, a, and c without actually generating any numbers, the most common being the spectral test. Many of these tests assess how k-tuples of random numbers fill up a k-dimensional unit cube. These tests are beyond the scope of this book; see, for instance, Ripley [1987].

In the examples of tests that follow, the hypotheses are not restated. The hypotheses are as indicated in the foregoing paragraphs. Although few simulation analysts will need to perform these tests, every simulation user should be aware of the qualities of a good random-number generator.

7.4.1 Frequency Tests

A basic test that should always be performed to validate a new generator is the test of uniformity. Two different methods of testing are available. They are the Kolmogorov-Smirnov and the chi-square test. Both of these tests measure the degree of agreement between the distribution of a sample of generated random numbers and the theoretical uniform distribution. Both tests are based on the null hypothesis of no significant difference between the sample distribution and the theoretical distribution.

1. *The Kolmogorov–Smirnov test.* This test compares the continuous cdf, $F(x)$, of the uniform distribution with the empirical cdf, $S_N(x)$, of the sample of N observations. By definition,

$$F(x) = x, \quad 0 \le x \le 1$$

If the sample from the random-number generator is $R_1, R_2, \ldots, R_N$, then the empirical cdf, $S_N(x)$, is defined by

$$S_N(x) = \frac{\text{number of } R_1, R_2, \ldots, R_N \text{ which are } \le x}{N}$$

As N becomes larger, $S_N(x)$ should become a better approximation to $F(x)$, provided that the null hypothesis is true.

In Section 5.6, empirical distributions were described. The cdf of an empirical distribution is a step function with jumps at each observed value. This behavior was illustrated by Example 5.35.

The Kolmogorov-Smirnov test is based on the largest absolute deviation between $F(x)$ and $S_N(x)$ over the range of the random variable—that is, it is based on the statistic

$$D = \max |F(x) - S_N(x)| \tag{7.3}$$

The sampling distribution of D is known; it is tabulated as a function of N in Table A.8. For testing against a uniform cdf, the test procedure follows these steps:

Step 1. Rank the data from smallest to largest. Let $R_{(i)}$, denote the ith smallest observation, so that

$$R_{(1)} \le R_{(2)} \le \cdots \le R_{(N)}$$

Step 2. Compute

$$D^+ = \max_{1 \le i \le N} \left\{ \frac{i}{N} - R_{(i)} \right\}$$

$$D^- = \max_{1 \le i \le N} \left\{ R_{(i)} - \frac{i-1}{N} \right\}$$

Step 3. Compute $D = \max(D^+, D^-)$.

Step 4. Locate in Table A.8 the critical value, D_α, for the specified significance level α and the given sample size N.

Step 5. If the sample statistic D is greater than the critical value, D_α, the null hypothesis that the data are a sample from a uniform distribution is rejected. If $D \le D_\alpha$, conclude that no difference has been detected between the true distribution of $\{R_1, R_2, \ldots, R_N\}$ and the uniform distribution.

Example 7.6

Suppose that the five numbers 0.44, 0.81, 0.14, 0.05, 0.93 were generated, and it is desired to perform a test for uniformity by using the Kolmogorov–Smirnov test with the level of significance $\alpha = 0.05$. First, the numbers must be ranked from smallest to largest. The calculations can be facilitated by use of Table 7.2. The top row lists the numbers from smallest ($R_{(1)}$) to largest ($R_{(5)}$). The computations for D^+, namely $i/N - R_{(i)}$, and for D^-, namely $R_{(i)} - (i-1)/N$, are easily accomplished by using Table 7.2. The statistics are computed as $D^+ = 0.26$ and $D^- = 0.21$. Therefore, $D = \max\{0.26, 0.21\} = 0.26$. The critical value of D, obtained from Table A.8 for $\alpha = 0.05$ and $N = 5$, is 0.565. Since the computed value, 0.26, is less than the tabulated critical value, 0.565, the hypothesis that the distribution of the generated numbers is the uniform distribution is not rejected.

The calculations in Table 7.2 are illustrated in Figure 7.2, where the empirical cdf, $S_N(x)$, is compared to the uniform cdf, $F(x)$. It can be seen that D^+ is the largest deviation of $S_N(x)$

Table 7.2 Calculations for Kolmogorov–Smirnov Test

$R_{(i)}$	0.05	0.14	0.44	0.81	0.93
i/N	0.20	0.40	0.60	0.80	1.00
$i/N - R_{(i)}$	0.15	0.26	0.16	—	0.07
$R_{(i)} - (i-1)/N$	0.05	—	0.04	0.21	0.13

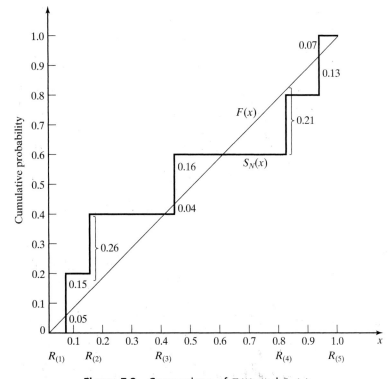

Figure 7.2 Comparison of $F(x)$ and $S_N(x)$.

above $F(x)$, and that D^- is the largest deviation of $S_N(x)$ below $F(x)$. For example, at $R_{(3)}$, the value of D^+ is given by $3/5 - R_{(3)} = 0.60 - 0.44 = 0.16$, and that of D^- is given by $R_{(3)} - 2/5 = 0.44 - 0.40 = 0.04$. Although the test statistic D is defined by Equation (7.3) as the maximum deviation over all x, it can be seen from Figure 7.2 that the maximum deviation will always occur at one of the jump points $R_{(1)}, R_{(2)}, \ldots$; thus, the deviation at other values of x need not be considered.

2. *The chi-square test.* The chi-square test uses the sample statistic

$$\chi_0^2 = \sum_{i=1}^{n} \frac{(O_i - E_i)^2}{E_i}$$

where O_i is the observed number in the ith class, E_i is the expected number in the ith class, and n is the number of classes. For the uniform distribution, E_i, the expected number in each class is given by

$$E_i = \frac{N}{n}$$

for equally spaced classes, where N is the total number of observations. It can be shown that the sampling distribution of χ_0^2 is approximately the chi-square distribution with $n - 1$ degrees of freedom.

Example 7.7

Use the chi-square test with $\alpha = 0.05$ to test for whether the data shown next are uniformly distributed. Table 7.3 contains the essential computations. The test uses $n = 10$ intervals of equal length, namely $[0, 0.1), [0.1, 0.2), \ldots, [0.9, 1.0)$. The value of χ_0^2 is 3.4. This is compared with the critical value $\chi_{0.05,9}^2 = 16.9$ from Table A.6. Since χ_0^2 is much smaller than the tabulated value of $\chi_{0.05,9}^2$, the null hypothesis of a uniform distribution is not rejected.

0.34	0.90	0.25	0.89	0.87	0.44	0.12	0.21	0.46	0.67
0.83	0.76	0.79	0.64	0.70	0.81	0.94	0.74	0.22	0.74
0.96	0.99	0.77	0.67	0.56	0.41	0.52	0.73	0.99	0.02
0.47	0.30	0.17	0.82	0.56	0.05	0.45	0.31	0.78	0.05
0.79	0.71	0.23	0.19	0.82	0.93	0.65	0.37	0.39	0.42
0.99	0.17	0.99	0.46	0.05	0.66	0.10	0.42	0.18	0.49
0.37	0.51	0.54	0.01	0.81	0.28	0.69	0.34	0.75	0.49
0.72	0.43	0.56	0.97	0.30	0.94	0.96	0.58	0.73	0.05
0.06	0.39	0.84	0.24	0.40	0.64	0.40	0.19	0.79	0.62
0.18	0.26	0.97	0.88	0.64	0.47	0.60	0.11	0.29	0.78

Table 7.3 Computations for Chi-Square Test

Interval	O_i	E_i	$O_i - E_i$	$(O_i - E_i)^2$	$\dfrac{(O_i - E_i)^2}{E_i}$
1	8	10	−2	4	0.4
2	8	10	−2	4	0.4
3	10	10	0	0	0.0
4	9	10	−1	1	0.1
5	12	10	2	4	0.4
6	8	10	−2	4	0.4
7	10	10	0	0	0.0
8	14	10	4	16	1.6
9	10	10	0	0	0.0
10	11	10	1	1	0.1
	$\overline{100}$	$\overline{100}$	$\overline{0}$		$\overline{3.4}$

Different authors have offered considerations concerning the application of the χ^2 test. In the application to a data set the size of that in Example 7.7, the considerations do not apply—that is, if 100 values are in the sample and from 5 to 10 intervals of equal length are used, the test will be acceptable. In general, it is recommended that n and N be chosen so that each $E_i \geq 5$.

Both the Kolmogorov–Smirnov test and the chi-square test are acceptable for testing the uniformity of a sample of data, provided that the sample size is large. However, the Kolmogorov–Smirnov test is the more powerful of the two and is recommended. Furthermore, the Kolmogorov–Smirnov test can be applied to small sample sizes, whereas the chi-square is valid only for large samples, say $N \geq 50$.

Imagine a set of 100 numbers which are being tested for independence, one where the first 10 values are in the range 0.01–0.10, the second 10 values are in the range 0.11–0.20, and so on. This set of numbers would pass the frequency tests with ease, but the ordering of the numbers produced by the generator would not be random. The test in the next section of this chapter is concerned with the independence of random numbers that are generated.

7.4.2 Tests for Autocorrelation

The tests for autocorrelation are concerned with the dependence between numbers in a sequence. As an example, consider the following sequence of numbers:

0.12	0.01	0.23	0.28	0.89	0.31	0.64	0.28	0.83	0.93
0.99	0.15	0.33	0.35	0.91	0.41	0.60	0.27	0.75	0.88
0.68	0.49	0.05	0.43	0.95	0.58	0.19	0.36	0.69	0.87

From a visual inspection, these numbers appear random, and they would probably pass all the tests presented to this point. However, an examination of the 5th, 10th, 15th (every five numbers beginning with the fifth), and so on, indicates a very large number in that position. Now, 30 numbers is a rather small sample size on which to reject a random number generator, but the notion is that numbers in the sequence might be related. In this particular section, a method for discovering whether such a relationship exists is described. The relationship would not have to be all high numbers. It is possible to have all low numbers in the locations being examined, or the numbers could alternate from very high to very low.

The test to be described shortly requires the computation of the autocorrelation between every m numbers (m is also known as the lag), starting with the ith number. Thus, the autocorrelation ρ_{im} between the following numbers would be of interest: $R_i, R_{i+m}, R_{i+2m}, \ldots, R_{i+(M+1)m}$. The value M is the largest integer such that $i + (M+1)m \leq N$, where N is the total number of values in the sequence. (Thus, a subsequence of length $M + 2$ is being tested.)

A nonzero autocorrelation implies a lack of independence, so the following two-tailed test is appropriate:

$$H_0 : \quad \rho_{im} = 0$$
$$H_1 : \quad \rho_{im} \neq 0$$

For large values of M, the distribution of the estimator of ρ_{im}, denoted $\widehat{\rho}_{im}$, is approximately normal if the values $R_i, R_{i+m}, R_{i+2m}, \ldots, R_{i+(M+1)m}$ are uncorrelated. Then the test statistic can

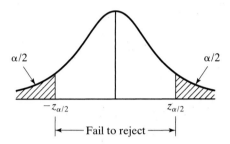

Figure 7.3   Failure to reject hypothesis.

be formed as follows:

$$Z_0 = \frac{\widehat{\rho}_{im}}{\sigma_{\widehat{\rho}_{im}}}$$

which is distributed normally with a mean of zero and a variance of 1, under the assumption of independence, for large M.

The formula for $\widehat{\rho}_{im}$, in a slightly different form, and the standard deviation of the estimator, $\sigma_{\widehat{\rho}_{im}}$, are given by Schmidt and Taylor [1970] as follows:

$$\widehat{\rho}_{im} = \frac{1}{M+1}\left[\sum_{k=0}^{M} R_{i+km}R_{i+(k+1)m}\right] - 0.25$$

and

$$\sigma_{\widehat{\rho}_{im}} = \frac{\sqrt{13M+7}}{12(M+1)}$$

After computing Z_0, do not reject the null hypothesis of independence if $-z_{\alpha/2} \leq Z_0 \leq z_{\alpha/2}$ where α is the level of significance and $z_{\alpha/2}$ is obtained from Table A.3. Figure 7.3 illustrates this test.

If $\rho_{im} > 0$, the subsequence is said to exhibit positive autocorrelation. In this case, successive values at lag m have a higher probability than expected of being close in value (i.e., high random numbers in the subsequence followed by high, and low followed by low). On the other hand, if $\rho_{im} < 0$, the subsequence is exhibiting negative autocorrelation, which means that low random numbers tend to be followed by high ones, and vice versa. The desired property, independence (which implies zero autocorrelation), means that there is no discernible relationship of the nature discussed here between successive random numbers at lag m.

Example 7.8 _____

Test for whether the 3rd, 8th, 13th, and so on, numbers in the sequence at the beginning of this section are autocorrelated using $\alpha = 0.05$. Here, $i = 3$ (beginning with the third number), $m = 5$ (every five numbers), $N = 30$ (30 numbers in the sequence), and $M = 4$ (largest integer such that $3 + (M + 1)5 \leq 30$). Then,

$$\widehat{\rho}_{35} = \frac{1}{4+1}[(0.23)(0.28) + (0.28)(0.33) + (0.33)(0.27) + (0.27)(0.05)$$

$$+(0.05)(0.36)] - 0.25$$

$$= -0.1945$$

and

$$\sigma_{\widehat{\rho}_{35}} = \frac{\sqrt{13(4)+7}}{12(4+1)} = 0.1280$$

Then, the test statistic assumes the value

$$Z_0 = -\frac{0.1945}{0.1280} = -1.516$$

Now, the critical value from Table A.3 is

$$z_{0.025} = 1.96$$

Therefore, the hypothesis of independence cannot be rejected on the basis of this test.

It can be observed that this test is not very sensitive for small values of M, particularly when the numbers being tested are on the low side. Imagine what would happen if each of the entries in the foregoing computation of $\widehat{\rho}_{im}$ were equal to zero. Then $\widehat{\rho}_{im}$ would be equal to -0.25 and the calculated Z would have the value of -1.95, not quite enough to reject the hypothesis of independence.

There are many sequences that can be formed in a set of data, given a large value of N. For example, beginning with the first number in the sequence, possibilities include (1) the sequence of all numbers, (2) the sequence formed from the first, third, fifth, ..., numbers, (3) the sequence formed from the first, fourth, ..., numbers, and so on. If $\alpha = 0.05$, there is a probability of 0.05 of rejecting a true hypothesis. If 10 independent sequences are examined, the probability of finding no significant autocorrelation, by chance alone, is $(0.95)^{10}$ or 0.60. Thus, 40% of the time significant autocorrelation would be detected when it does not exist. If α is 0.10 and 10 tests are conducted, there is a 65% chance of finding autocorrelation by chance alone. In conclusion, in "fishing" for autocorrelation by performing numerous tests, autocorrelation might eventually be detected, perhaps by chance alone, even when there is no autocorrelation present.

7.5 SUMMARY

This chapter described the generation of random numbers and the subsequent testing of the generated numbers for uniformity and independence. Random numbers are used to generate random variates, the subject of Chapter 8.

Of the many types of random-number generators available, ones based on the linear congruential method are the most widely used, but they are being replaced by combined linear congruential generators. Of the many types of statistical tests that are used in testing random-number generators, two different types are described: one testing for uniformity, and one testing for independence.

The simulation analyst might never work directly with a random-number generator or with the testing of random numbers from a generator. Most computers and simulation languages have routines that generate a random number, or streams of random numbers, for the asking. But even generators that have been used for years, some of which are still in use, have been found to be inadequate. So this chapter calls the simulation analyst's attention to such possibilities, with a warning to investigate and confirm that the generator has been tested thoroughly. Some researchers have attained sophisticated expertise in developing methods for generating and testing random numbers and the subsequent application of these methods. This chapter provides only a basic introduction to the subject matter; more depth and breadth are required for the reader to become a specialist in the area. The bible is Knuth [1998]; see also the reviews in Bratley, Fox, and Schrage [1996], Law and Kelton [2000], L'Ecuyer [1998], and Ripley [1987].

One final caution is due. Even if generated numbers pass all the tests (those covered in this chapter and those mentioned in the chapter), some underlying pattern might have gone undetected without the generator's having been rejected as faulty. However, the generators available in widely used simulation languages have been extensively tested and validated.

REFERENCES

BRATLEY, P., B. L. FOX, AND L. E. SCHRAGE [1996], *A Guide to Simulation,* 2d ed., Springer-Verlag, New York.

COX, D. R., AND P. A. W. LEWIS [1966], *The Statistical Analysis of Series of Events,* Methuen, London.

DURBIN, J. [1967], "Tests of Serial Independence Based on the Cumulated Periodogram," *Bulletin of the International Institute of Statistics.*

FISHMAN, G. S. [1978], *Principles of Discrete Event Simulation,* Wiley, New York.

GOOD, I. J. [1953], "The Serial Test for Sampling Numbers and Other Tests of Randomness," *Proceedings of the Cambridge Philosophical Society,* Vol. 49, pp. 276–284.

GOOD, I. J. [1967], "The Generalized Serial Test and the Binary Expansion of 4," *Journal of the Royal Statistical Society,* Ser. A, Vol. 30, No. 1, pp. 102–107.

KNUTH, D. W. [1998], *The Art of Computer Programming: Vol. 2, Semi-numerical Algorithms,* 2d ed., Addison–Wesley, Reading, MA.

LAW, A. M., AND W. D. KELTON [2000], *Simulation Modeling & Analysis,* 3d ed., McGraw–Hill, New York.

LEARMONTH, G. P., AND P. A. W. LEWIS [1973], "Statistical Tests of Some Widely Used and Recently Proposed Uniform Random Number Generators," *Proceedings of the Conference on Computer Science and Statistics: Seventh Annual Symposium on the Interface,* Western Publishing, North Hollywood, CA, pp. 163–171.

L'ECUYER, P. [1988], "Efficient and Portable Combined Random Number Generators," *Communications of the ACM,* Vol. 31, pp. 742–749, 774.

L'ECUYER, P. [1996], "Combined Multiple Recursive Random Number Generators," *Operations Research,* Vol. 44, pp. 816–822.

L'ECUYER, P. [1998], "Random Number Generation," Chapter 4 in *Handbook of Simulation,* Wiley, New York.

L'ECUYER, P. [1999], "Good Parameters and Implementations for Combined Multiple Recursive Random Number Generators," *Operations Research,* Vol. 47, pp. 159–164.

L'ECUYER, P., R. SIMARD, E. J. CHEN, AND W. D. KELTON [2002], "An Object-Oriented Random-Number Package with Many Long Streams and Substreams," *Operations Research*, Vol. 50, pp. 1073–1075.

LEHMER, D. H. [1951], *Proceedings of the Second Symposium on Large-Scale Digital Computing Machinery*, Harvard University Press, Cambridge, MA.

LEWIS, P. A. W., A. S. GOODMAN, AND J. M. MILLER [1969], "A Pseudo-Random Number Generator for the System/360," *IBM Systems Journal*, Vol. 8, pp. 136–145.

RIPLEY, B. D. [1987], *Stochastic Simulation*, Wiley, New York.

SCHMIDT, J. W., And R. E. TAYLOR [1970], *Simulation and Analysis of Industrial Systems*, Irwin, Homewood, IL.

EXERCISES

1. Describe a procedure to physically generate random numbers on the interval [0, 1] with 2-digit accuracy. (*Hint*: Consider drawing something out of a hat.)

2. List applications, other than systems simulation, for pseudo-random numbers—for example, video gambling games.

3. How could random numbers that are uniform on the interval [0, 1] be transformed into random numbers that are uniform on the interval [−11, 17]? Transformations to more general distributions are described in Chapter 8.

4. Use the linear congruential method to generate a sequence of three two-digit random integers. Let $X_0 = 27$, $a = 8$, $c = 47$, and $m = 100$.

5. Do we encounter a problem in the previous exercise if $X_0 = 0$?

6. Use the multiplicative congruential method to generate a sequence of four three-digit random integers. Let $X_0 = 117$, $a = 43$, and $m = 1000$.

7. The sequence of numbers 0.54, 0.73, 0.98, 0.11, and 0.68 has been generated. Use the Kolmogorov–Smirnov test with $\alpha = 0.05$ to learn whether the hypothesis that the numbers are uniformly distributed on the interval [0, 1] can be rejected.

8. Reverse the 100 two-digit random numbers in Example 7.7 to get a new set of random numbers. Thus, the first random number in the new set will be 0.43. Use the chi-square test, with $\alpha = 0.05$, to learn whether the hypothesis that the numbers are uniformly distributed on the interval [0, 1] can be rejected.

9. Figure out whether these linear congruential generators can achieve a maximum period; also, state restrictions on X_0 to obtain this period:

   (a) the mixed congruential method with

$$a = 2, 814, 749, 767, 109$$

$$c = 59, 482, 661, 568, 307$$

$$m = 2^{48}$$

(b) the multiplicative congruential generator with

$$a = 69,069$$
$$c = 0$$
$$m = 2^{32}$$

(c) the mixed congruential generator with

$$a = 4951$$
$$c = 247$$
$$m = 256$$

(d) the multiplicative congruential generator with

$$a = 6507$$
$$c = 0$$
$$m = 1024$$

10. Use the mixed congruential method to generate a sequence of three two-digit random numbers with $X_0 = 37$, $a = 7$, $c = 29$, and $m = 100$.

11. Use the mixed congruential method to generate a sequence of three two-digit random integers between 0 and 24 with $X_0 = 13$, $a = 9$, and $c = 35$.

12. Write a computer program that will generate four-digit random numbers, using the multiplicative congruential method. Allow the user to input values of X_0, a, c, and m.

13. If $X_0 = 3579$ in Exercise 9(c), generate the first random number in the sequence. Compute the random number to four-place accuracy.

14. Investigate the random-number generator in a spreadsheet program on a computer to which you have access. In many spreadsheets, random numbers are generated by a function called RAND or @RAND.

    (a) Check the user's manual to see whether it describes how the random numbers are generated.
    (b) Write macros to conduct each of the tests described in this chapter. Generate 100 sets of random numbers, each set containing 100 random numbers. Perform each test on each set of random numbers. Draw conclusions.

15. Consider the multiplicative congruential generator under the following circumstances:

    (a) $a = 11, m = 16, X_0 = 7$
    (b) $a = 11, m = 16, X_0 = 8$
    (c) $a = 7, m = 16, X_0 = 7$
    (d) $a = 7, m = 16, X_0 = 8$

Generate enough values in each case to complete a cycle. What inferences can be drawn? Is maximum period achieved?

16. For 16-bit computers, L'Ecuyer [1988] recommends combining three multiplicative generators, with $m_1 = 32{,}363$, $a_1 = 157$, $m_2 = 31{,}727$, $a_2 = 146$, $m_3 = 31{,}657$, and $a_3 = 142$. The period of this generator is approximately 8×10^{12}. Generate 5 random numbers with the combined generator, using the initial seeds $X_{i,0} = 100, 300, 500$, for the individual generators $i = 1, 2, 3$.

17. Apply the tests described in this chapter to the generator given in the previous exercise.

18. Use the principles described in this chapter to develop your own linear congruential random-number generator.

19. Use the principles described in this chapter to develop your own combined linear congruential random-number generator.

20. Test the following sequence of numbers for uniformity and independence, using procedures you learned in this chapter: 0.594, 0.928, 0.515, 0.055, 0.507, 0.351, 0.262, 0.797, 0.788, 0.442, 0.097, 0.798, 0.227, 0.127, 0.474, 0.825, 0.007, 0.182, 0.929, 0.852. $N = 20$

21. In some applications, it is useful to be able to quickly skip ahead in a pseudo-random number sequence without actually generating all of the intermediate values. (a) For a linear congruential generator with $c = 0$, show that $X_{i+n} = (a^n X_i) \bmod m$. (b) Next, show that $(a^n X_i) \bmod m = (a^n \bmod m) X_i \bmod m$ (this result is useful because $a^n \bmod m$ can be precomputed, making it easy to skip ahead n random numbers from any point in the sequence). (c) In Example 7.3, use this result to compute X_5, starting with $X_0 = 63$. Check your answer by computing X_5 in the usual way.

1) Use K-S test to check for uniformity in Excel

2) Autocorrelation can be done by hand but easier in Excel.

8

Random-Variate Generation

This chapter deals with procedures for sampling from a variety of widely-used continuous and discrete distributions. Previous discussions and examples indicated the usefulness of statistical distributions in modeling activities that are generally unpredictable or uncertain. For example, interarrival times and service times at queues and demands for a product are quite often unpredictable in nature, at least to a certain extent. Usually, such variables are modeled as random variables with some specified statistical distribution, and standard statistical procedures exist for estimating the parameters of the hypothesized distribution and for testing the validity of the assumed statistical model. Such procedures are discussed in Chapter 9.

In this chapter, it is assumed that a distribution has been completely specified, and ways are sought to generate samples from this distribution to be used as input to a simulation model. The purpose of the chapter is to explain and illustrate some widely-used techniques for generating random variates, not to give a state-of-the-art survey of the most efficient techniques. In practice, most simulation modelers will use either existing routines available in programming libraries or the routines built into the simulation language being used. However, some programming languages do not have built-in routines for all of the regularly used distributions, and some computer installations do not have random-variate-generation libraries; in such cases the modeler must construct an acceptable routine. Even though the chance of this happening is small, it is nevertheless worthwhile to understand how random-variate generation occurs.

This chapter discusses the inverse-transform technique and, more briefly, the acceptance–rejection technique and special properties. Another technique, the composition method, is discussed by Devroye [1986], Dagpunar [1988], Fishman [1978], and Law and Kelton [2000]. All the techniques in this chapter assume that a source of uniform [0,1] random numbers, $R_1, R_2, \ldots$

272

is readily available, where each R_i has pdf

$$f_R(x) = \begin{cases} 1, & 0 \le x \le 1 \\ 0, & \text{otherwise} \end{cases}$$

and cdf

$$F_R(x) = \begin{cases} 0, & x < 0 \\ x, & 0 \le x \le 1 \\ 1, & x > 1 \end{cases}$$

Throughout this chapter R and $R_1, R_2, \ldots$ represent random numbers uniformly distributed on [0,1] and generated by one of the techniques in Chapter 7 or taken from a random number table, such as Table A.1.

8.1 INVERSE-TRANSFORM TECHNIQUE

The inverse-transform technique can be used to sample from the exponential, the uniform, the Weibull, the triangular distributions and from empirical distributions. Additionally, it is the underlying principle for sampling from a wide variety of discrete distributions. The technique will be explained in detail for the exponential distribution and then applied to other distributions. Computationally, it is the most straightforward, but not always the most efficient, technique.

8.1.1 Exponential Distribution

The exponential distribution, discussed in Section 5.4, has the probability density function (pdf)

$$f(x) = \begin{cases} \lambda e^{-\lambda x}, & x \ge 0 \\ 0, & x < 0 \end{cases}$$

and the cumulative distribution function (cdf)

$$F(x) = \int_{-\infty}^{x} f(t)\, dt = \begin{cases} 1 - e^{-\lambda x}, & x \ge 0 \\ 0, & x < 0 \end{cases}$$

The parameter λ can be interpreted as the mean number of occurrences per time unit. For example, if interarrival times $X_1, X_2, X_3, \ldots$ had an exponential distribution with rate λ, then λ could be interpreted as the mean number of arrivals per time unit, or the arrival rate. Notice that, for any i,

$$E(X_i) = \frac{1}{\lambda}$$

and so $1/\lambda$ is the mean interarrival time. The goal here is to develop a procedure for generating values $X_1, X_2, X_3, \ldots$ that have an exponential distribution.

The inverse-transform technique can be utilized, at least in principle, for any distribution, but it is most useful when the cdf, $F(x)$, is of a form so simple that its inverse, F^{-1}, can be computed easily.[1] One step-by-step procedure for the inverse-transform technique, illustrated by the exponential distribution, consists of the following steps:

Step 1. Compute the cdf of the desired random variable X.

For the exponential distribution, the cdf is $F(x) = 1 - e^{-\lambda x}$, $x \geq 0$.

Step 2. Set $F(X) = R$ on the range of X.

For the exponential distribution, it becomes $1 - e^{-\lambda X} = R$ on the range $x \geq 0$.

X is a random variable (with the exponential distribution in this case), so $1 - e^{-\lambda X}$ is also a random variable, here called R. As will be shown later, R has a uniform distribution over the interval $[0, 1]$.

Step 3. Solve the equation $F(X) = R$ for X in terms of R.

For the exponential distribution, the solution proceeds as follows:

$$1 - e^{-\lambda X} = R$$
$$e^{-\lambda X} = 1 - R$$
$$-\lambda X = \ln(1 - R)$$
$$X = -\frac{1}{\lambda} \ln(1 - R) \tag{8.1}$$

Equation (8.1) is called a random-variate generator for the exponential distribution. In general, Equation (8.1) is written as $X = F^{-1}(R)$. Generating a sequence of values is accomplished through Step 4.

Step 4. Generate (as needed) uniform random numbers $R_1, R_2, R_3, \ldots$ and compute the desired random variates by

$$X_i = F^{-1}(R_i)$$

For the exponential case, $F^{-1}(R) = (-1/\lambda) \ln(1 - R)$ by Equation (8.1), so

$$X_i = -\frac{1}{\lambda} \ln(1 - R_i) \tag{8.2}$$

for $i = 1, 2, 3, \ldots$. One simplification that is usually employed in Equation (8.2) is to replace $1 - R_i$ by R_i to yield

$$X_i = -\frac{1}{\lambda} \ln R_i \tag{8.3}$$

This alternative is justified by the fact that both R_i and $1 - R_i$ are uniformly distributed on $[0,1]$.

---

[1] The notation F^{-1} denotes the solution of the equation $r = F(x)$ in terms of r; it does not denote $1/F$.

Table 8.1 Generation of Exponential Variates X_i with Mean 1, given Random Numbers R_i

i	1	2	3	4	5
R_i	0.1306	0.0422	0.6597	0.7965	0.7696
X_i	0.1400	0.0431	1.078	1.592	1.468

Example 8.1

Table 8.1 gives a sequence of random numbers from Table A.1 and the computed exponential variates, X_i, given by Equation (8.2) with the value $\lambda = 1$. Figure 8.1(a) is a histogram of 200 values, $R_1, R_2, \ldots, R_{200}$ from the uniform distribution, and Figure 8.1(b) is a histogram of the 200 values, $X_1, X_2, \ldots, X_{200}$, computed by Equation (8.2). Compare these empirical histograms with the theoretical density functions in Figure 8.1(c) and (d). As illustrated here, a histogram is an estimate of the underlying density function. (This fact is used in Chapter 9 as a way to identify distributions.)

Figure 8.2 gives a graphical interpretation of the inverse-transform technique. The cdf shown is $F(x) = 1 - e^{-x}$, an exponential distribution with rate $\lambda = 1$. To generate a value X_1 with cdf $F(x)$, a random number R_1 between 0 and 1 is generated, then a horizontal line is drawn from R_1 to the graph of the cdf, then a vertical line is dropped to the x axis to obtain X_1, the desired result. Notice the inverse relation between R_1 and X_1, namely

$$R_1 = 1 - e^{-X_1}$$

and

$$X_1 = -\ln(1 - R_1)$$

In general, the relation is written as

$$R_1 = F(X_1)$$

and

$$X_1 = F^{-1}(R_1)$$

Why does the random variable X_1 generated by this procedure have the desired distribution? Pick a value x_0 and compute the cumulative probability

$$P(X_1 \leq x_0) = P(R_1 \leq F(x_0)) = F(x_0) \tag{8.4}$$

To see the first equality in Equation (8.4), refer to Figure 8.2, where the fixed numbers x_0 and $F(x_0)$ are drawn on their respective axes. It can be seen that $X_1 \leq x_0$ when and only when $R_1 \leq F(x_0)$. Since $0 \leq F(x_0) \leq 1$, the second equality in Equation (8.4) follows immediately from the fact that R_1 is uniformly distributed on [0,1]. Equation (8.4) shows that the cdf of X_1 is F; hence, X_1 has the desired distribution.

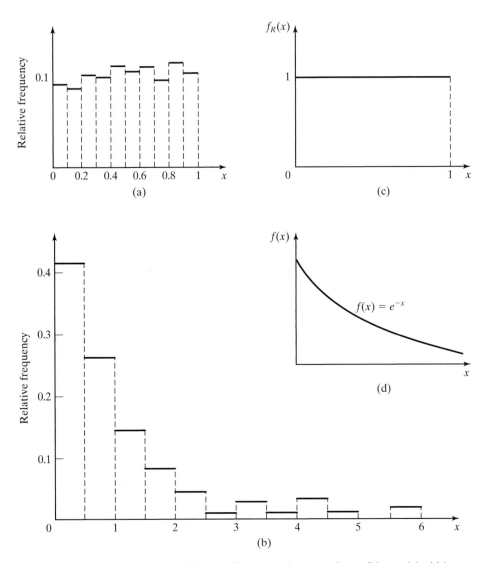

Figure 8.1   (a) Empirical histogram of 200 uniform random numbers; (b) empirical histogram of 200 exponential variates; (c) theoretical uniform density on [0,1]; (d) theoretical exponential density with mean 1.

8.1.2 Uniform Distribution

Consider a random variable X that is uniformly distributed on the interval $[a, b]$. A reasonable guess for generating X is given by

$$X = a + (b - a)R \tag{8.5}$$

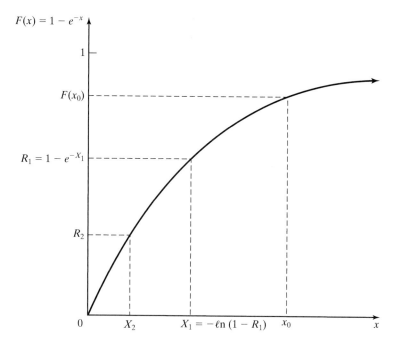

Figure 8.2   Graphical view of the inverse-transform technique.

[Recall that R is always a random number on $[0, 1]$.] The pdf of X is given by

$$f(x) = \begin{cases} \dfrac{1}{b-a}, & a \leq x \leq b \\ 0, & \text{otherwise} \end{cases}$$

The derivation of Equation (8.5) follows Steps 1 through 3 of Section 8.1.1:

Step 1. The cdf is given by

$$F(x) = \begin{cases} 0, & x < a \\ \dfrac{x-a}{b-a}, & a \leq x \leq b \\ 1, & x > b \end{cases}$$

Step 2. Set $F(X) = (X - a)/(b - a) = R$.

Step 3. Solving for X in terms of R yields $X = a + (b - a)R$, which agrees with Equation (8.5).

8.1.3 Weibull Distribution

The Weibull distribution was introduced in Section 5.4 as a model for *time to failure* for machines or electronic components. When the location parameter v is set to 0, its pdf is given by

Equation (5.47):

$$f(x) = \begin{cases} \dfrac{\beta}{\alpha^\beta} x^{\beta-1} e^{-(x/\alpha)^\beta}, & x \geq 0 \\ 0, & \text{otherwise} \end{cases}$$

where $\alpha > 0$ and $\beta > 0$ are the scale and shape parameters of the distribution. To generate a Weibull variate, follow Steps 1 through 3 of Section 8.1.1:

Step 1. The cdf is given by $F(X) = 1 - e^{-(x/\alpha)^\beta}$, $x \geq 0$.

Step 2. Set $F(X) = 1 - e^{-(X/\alpha)^\beta} = R$.

Step 3. Solving for X in terms of R yields

$$X = \alpha[-\ln(1 - R)]^{1/\beta} \tag{8.6}$$

The derivation of Equation (8.6) is left as Exercise 10 for the reader. By comparing Equations (8.6) and (8.1), it can be seen that, if X is a Weibull variate, then X^β is an exponential variate with mean α^β. Conversely, if Y is an exponential variate with mean μ, then $Y^{1/\beta}$ is a Weibull variate with shape parameter β and scale parameter $\alpha = \mu^{1/\beta}$.

8.1.4 Triangular Distribution

Consider a random variable X that has pdf

$$f(x) = \begin{cases} x, & 0 \leq x \leq 1 \\ 2 - x, & 1 < x \leq 2 \\ 0, & \text{otherwise} \end{cases}$$

as shown in Figure 8.3. This distribution is called a triangular distribution with endpoints $(0, 2)$ and mode at 1. Its cdf is given by

$$F(x) = \begin{cases} 0, & x \leq 0 \\ \dfrac{x^2}{2}, & 0 < x \leq 1 \\ 1 - \dfrac{(2-x)^2}{2}, & 1 < x \leq 2 \\ 1, & x > 2 \end{cases}$$

For $0 \leq X \leq 1$,

$$R = \frac{X^2}{2} \tag{8.7}$$

and for $1 \leq X \leq 2$,

$$R = 1 - \frac{(2 - X)^2}{2} \tag{8.8}$$

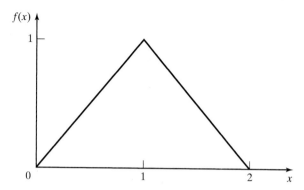

Figure 8.3 Density function for a particular triangular distribution.

By Equation (8.7), $0 \leq X \leq 1$ implies that $0 \leq R \leq \frac{1}{2}$, in which case $X = \sqrt{2R}$. By Equation (8.8), $1 \leq X \leq 2$ implies that $\frac{1}{2} \leq R \leq 1$, in which case $X = 2 - \sqrt{2(1-R)}$. Thus, X is generated by

$$X = \begin{cases} \sqrt{2R}, & 0 \leq R \leq \frac{1}{2} \\ 2 - \sqrt{2(1-R)}, & \frac{1}{2} < R \leq 1 \end{cases} \tag{8.9}$$

Exercises 2, 3, and 4 give the student practice in dealing with other triangular distributions. Notice that, if the pdf and cdf of the random variable X come in parts (i.e., require different formulas over different parts of the range of X), then the application of the inverse-transform technique for generating X will result in separate formulas over different parts of the range of R, as in Equation (8.9). A general form of the triangular distribution was discussed in Section 5.4.

8.1.5 Empirical Continuous Distributions

If the modeler has been unable to find a theoretical distribution that provides a good model for the input data, then it may be necessary to use the empirical distribution of the data. One possibility is to simply resample the observed data itself. This is known as using the *empirical distribution*, and it makes particularly good sense when the input process is known to take on a finite number of values. See Section 8.1.7 for an example of this type of situation and for a method for generating random inputs.

On the other hand, if the data are drawn from what is believed to be a continuous-valued input process, then it makes sense to interpolate between the observed data points to fill in the gaps. This section describes a method for defining and generating data from a continuous empirical distribution.

Example 8.2

Five observations of fire-crew response times (in minutes) to incoming alarms have been collected to be used in a simulation investigating possible alternative staffing and crew-scheduling policies. The data are

2.76   1.83   0.80   1.45   1.24

Table 8.2   Summary of Fire-Crew Response-Time Data

i	Interval $x_{(i-1)} < x \leq x_{(i)}$	Probability $1/n$	Cumulative Probability, i/n	Slope a_i
1	$0.0 \ \ < x \leq 0.80$	0.2	0.2	4.00
2	$0.80 < x \leq 1.24$	0.2	0.4	2.20
3	$1.24 < x \leq 1.45$	0.2	0.6	1.05
4	$1.45 < x \leq 1.83$	0.2	0.8	1.90
5	$1.83 < x \leq 2.76$	0.2	1.0	4.65

Before collecting more data, it is desired to develop a preliminary simulation model that uses a response-time distribution based on these five observations. Thus, a method for generating random variates from the response-time distribution is needed. Initially, it will be assumed that response times X have a range $0 \leq X \leq c$, where c is unknown, but will be estimated by $\hat{c} = \max\{X_i : i = 1, \ldots, n\} = 2.76$, where $\{X_i, i = 1, \ldots, n\}$ are the raw data and $n = 5$ is the number of observations.

Arrange the data from smallest to largest and let $x_{(1)} \leq x_{(2)} \leq \cdots \leq x_{(n)}$ denote these sorted values. The smallest possible value is believed to be 0, so define $x_{(0)} = 0$. Assign the probability $1/n = 1/5$ to each interval $x_{(i-1)} < x \leq x_{(i)}$, as shown in Table 8.2. The resulting empirical cdf, $\widehat{F}(x)$, is illustrated in Figure 8.4. The slope of the ith line segment is given by

$$a_i = \frac{x_{(i)} - x_{(i-1)}}{i/n - (i-1)/n} = \frac{x_{(i)} - x_{(i-1)}}{1/n}$$

The inverse cdf is calculated by

$$X = \widehat{F}^{-1}(R) = x_{(i-1)} + a_i \left(R - \frac{(i-1)}{n} \right) \tag{8.10}$$

when $(i-1)/n < R \leq i/n$.

For example, if a random number $R_1 = 0.71$ is generated, then R_1 is seen to lie in the fourth interval (between $3/5 = 0.60$ and $4/5 = 0.80$); so, by Equation (8.10),

$$X_1 = x_{(4-1)} + a_4(R_1 - (4-1)/n)$$

$$= 1.45 + 1.90(0.71 - 0.60)$$

$$= 1.66$$

The reader is referred to Figure 8.4 for a graphical view of the generation procedure.

In Example 8.2, each data point was represented in the empirical cdf. If a large sample of data is available (and sample sizes from several hundred to tens of thousands are possible with modern, automated data collection), then it might be more convenient (and computationally efficient) to first summarize the data into a frequency distribution with a much smaller number of intervals and then fit a continuous empirical cdf to the frequency distribution. Only a slight generalization

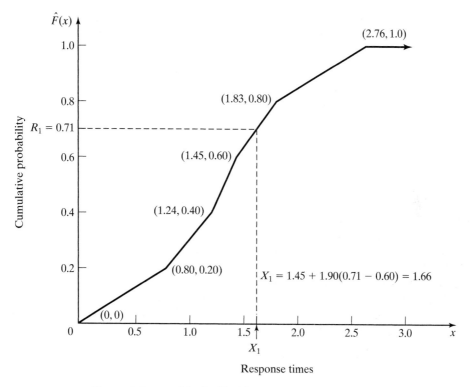

Figure 8.4 Empirical cdf of fire-crew response times.

of Equation (8.10) is required to accomplish this. Now the slope of the ith line segment is given by:

$$a_i = \frac{x_{(i)} - x_{(i-1)}}{c_i - c_{i-1}}$$

where c_i is the cumulative probability of the first i intervals of the frequency distribution and $x_{(i-1)} < x \le x_{(i)}$ is the ith interval. The inverse cdf is calculated as

$$X = \widehat{F}^{-1}(R) = x_{(i-1)} + a_i (R - c_{i-1}) \tag{8.11}$$

when $c_{i-1} < R \le c_i$.

Example 8.3
Suppose that 100 broken-widget repair times have been collected. The data are summarized in Table 8.3 in terms of the number of observations in various intervals. For example, there were 31 observations between 0 and 0.5 hour, 10 between 0.5 and 1 hour, and so on. Suppose it is known that all repairs take at least 15 minutes, so that $X \ge 0.25$ hour always. Then we set $x_{(0)} = 0.25$, as shown in Table 8.3 and Figure 8.5.

Table 8.3   Summary of Repair-Time Data

i	Interval (Hours)	Frequency	Relative Frequency	Cumulative Frequency, c_i	Slope a_i
1	$0.25 \leq x \leq 0.5$	31	0.31	0.31	0.81
2	$0.5 < x \leq 1.0$	10	0.10	0.41	5.0
3	$1.0 < x \leq 1.5$	25	0.25	0.66	2.0
4	$1.5 < x \leq 2.0$	34	0.34	1.00	1.47

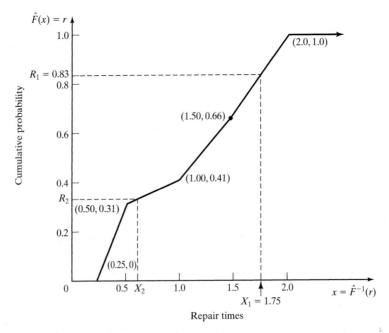

Figure 8.5   Generating variates from the empirical distribution function for repair-time data $(X \geq 0.25)$.

For example, suppose the first random number generated is $R_1 = 0.83$. Then, since R_1 is between $c_3 = 0.66$ and $c_4 = 1.00$,

$$X_1 = x_{(4-1)} + a_4(R_1 - c_{4-1}) = 1.5 + 1.47(0.83 - 0.66) = 1.75 \tag{8.12}$$

As another illustration, suppose that $R_2 = 0.33$. Since $c_1 = 0.31 < R_2 \leq 0.41 = c_2$,

$$X_2 = x_{(1)} + a_2(R_2 - c_1)$$

$$= 0.5 + 5.0(0.33 - 0.31)$$

$$= 0.6$$

The point $(R_2 = 0.33, X_2 = 0.6)$ is also shown in Figure 8.5.

Now reconsider the data of Table 8.3. The data are restricted in the range $0.25 \leq X \leq 2.0$, but the underlying distribution might have a wider range. This provides one important reason for attempting to find a theoretical statistical distribution (such as the gamma or Weibull) for the data: that these distributions do allow a wider range—namely, $0 \leq X < \infty$. On the other hand, an empirical distribution adheres closely to what is present in the data itself, and the data are often the best source of information available.

When data are summarized in terms of frequency intervals, it is recommended that relatively short intervals be used, for doing so results in a more accurate portrayal of the underlying cdf. For example, for the repair-time data of Table 8.3, for which there were $n = 100$ observations, a much more accurate estimate could have been obtained by using 10 to 20 intervals, certainly not an excessive number, rather than the four fairly wide intervals actually used here for purposes of illustration.

Several comments are in order:

1. A computerized version of the procedure will become more inefficient as the number of intervals, n, increases. A systematic computerized version is often called a table-lookup generation scheme, because, given a value of R, the computer program must search an array of c_i values to find the interval i in which R lies, namely the interval i satisfying

$$c_{i-1} < R \leq c_i$$

The more intervals there are, the longer on the average the search will take if it is imple-mented in the crude way described here. The analyst should consider this trade-off between accuracy of the estimating cdf and computational efficiency when programming the proce-dure. If a large number of observations are available, the analyst may well decide to group the observations into from 20 to 50 intervals (say) and then use the procedure of Example 8.3—or a more efficient table-lookup procedure could be used, such as the one described in Law and Kelton [2000].

2. In Example 8.2, it was assumed that response times X satisfied $0 \leq X \leq 2.76$. This assumption led to the inclusion of the points $x_{(0)} = 0$ and $x_{(5)} = 2.76$ in Figure 8.4 and Table 8.2. If it is known a priori that X falls in some other range, for example, if it is known that response times are always between 15 seconds and 3 minutes—that is,

$$0.25 \leq X \leq 3.0$$

—then the points $x_{(0)} = 0.25$ and $x_{(6)} = 3.0$ would be used to estimate the empirical cdf of response times. Notice that, because of inclusion of the new point $x_{(6)}$, there are now six intervals instead of five, and each interval is assigned probability $1/6 = 0.167$. Exercise 12 illustrates the use of these additional assumptions.

8.1.6 Continuous Distributions without a Closed-Form Inverse

A number of useful continuous distributions do not have a closed form expression for their cdf or its inverse; examples include the normal, gamma, and beta distributions. For this reason, it is often stated that the inverse-transform technique for random-variate generation is not available for these distributions. It can, in effect, become available if we are willing to *approximate* the inverse cdf, or numerically integrate and search the cdf. Although this approach sounds inaccurate, notice that even a closed-form inverse requires approximation in order to evaluate it on a computer. For example, generating exponentially distributed random variates via the inverse cdf $X = F^{-1}(R) = -\ln(1 - R)/\lambda$ requires a numerical approximation for the logarithm function. Thus, there is no essential difference between using an approximate inverse cdf and approximately evaluating a closed-form inverse. The problem with using approximate inverse cdfs is that some of them are computationally slow to evaluate.

To illustrate the idea, consider a simple approximation to the inverse cdf of the standard normal distribution, proposed by Schmeiser [1979]:

$$X = F^{-1}(R) \approx \frac{R^{0.135} - (1 - R)^{0.135}}{0.1975}$$

This approximation gives at least one-decimal-place accuracy for $0.0013499 \leq R \leq 0.9986501$. Table 8.4 compares the approximation with exact values (to four decimal places) obtained by numerical integration for several values of R. Much more accurate approximations exist that are only slightly more complicated. A good source of these approximations for a number of distributions is Bratley, Fox, and Schrage [1996].

8.1.7 Discrete Distributions

All discrete distributions can be generated via the inverse-transform technique, either numerically through a table-lookup procedure or, in some cases, algebraically, the final generation scheme being in terms of a formula. Other techniques are sometimes used for certain distributions, such as the convolution technique for the binomial distribution. Some of these methods are discussed in later sections. This subsection gives examples covering both empirical distributions and two of the standard discrete distributions, the (discrete) uniform and the geometric. Highly efficient table-lookup procedures for these and other distributions are found in Bratley, Fox, and Schrage [1996] and in Ripley [1987].

Table 8.4 Comparison of Approximate Inverse with Exact Values (To Four Decimal Places) for the Standard Normal Distribution

R	*Approximate Inverse*	*Exact Inverse*
0.01	−2.3263	−2.3373
0.10	−1.2816	−1.2813
0.25	−0.6745	−0.6713
0.50	0.0000	0.0000
0.75	0.6745	0.6713
0.90	1.2816	1.2813
0.99	2.3263	2.3373

Table 8.5 Distribution of
Number of Shipments, X

x	$p(x)$	$F(x)$
0	0.50	0.50
1	0.30	0.80
2	0.20	1.00

Example 8.4:   An Empirical Discrete Distribution

At the end of any day, the number of shipments on the loading dock of the IHW Company (whose main product is the famous "incredibly huge widget") is either 0, 1, or 2, with observed relative frequency of occurrence of 0.50, 0.30, and 0.20, respectively. Internal consultants have been asked to develop a model to improve the efficiency of the loading and hauling operations; as part of this model, they will need to be able to generate values, X, to represent the number of shipments on the loading dock at the end of each day. The consultants decide to model X as a discrete random variable with the distribution given in Table 8.5 and shown in Figure 8.6.

The probability mass function (pmf), $p(x)$, is given by

$$p(0) = P(X = 0) = 0.50$$

$$p(1) = P(X = 1) = 0.30$$

$$p(2) = P(X = 2) = 0.20$$

and the cdf, $F(x) = P(X \leq x)$, is given by

$$F(x) = \begin{cases} 0 & x < 0 \\ 0.5 & 0 \leq x < 1 \\ 0.8 & 1 \leq x < 2 \\ 1.0 & 2 \leq x \end{cases}$$

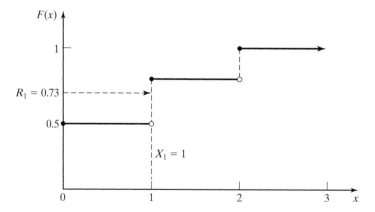

Figure 8.6 Cdf of number of shipments, X.

Table 8.6 Table for Generating the Discrete Variate X

i	Input, r_i	Output, x_i
1	0.50	0
2	0.80	1
3	1.00	2

Recall that the cdf of a discrete random variable always consists of horizontal line segments with jumps of size $p(x)$ at those points, x, that the random variable can assume. For example, in Figure 8.6, there is a jump of size $p(0) = 0.5$ at $x = 0$, of size $p(1) = 0.3$ at $x = 1$, and of size $p(2) = 0.2$ at $x = 2$.

For generating discrete random variables, the inverse transform technique becomes a table-lookup procedure, but, unlike in the case of continuous variables, interpolation is not required. To illustrate the procedure, suppose that $R_1 = 0.73$ is generated. Graphically, as illustrated in Figure 8.6, first locate $R_1 = 0.73$ on the vertical axis, next draw a horizontal line segment until it hits a "jump" in the cdf, and then drop a perpendicular to the horizontal axis to get the generated variate. Here $R_1 = 0.73$ is transformed to $X_1 = 1$. This procedure is analogous to the procedure used for empirical continuous distributions in Section 8.1.5 and illustrated in Figure 8.4, except that the final step, linear interpolation, is eliminated.

The table-lookup procedure is facilitated by construction of a table such as Table 8.6. When $R_1 = 0.73$ is generated, first find the interval in which R_1 lies. In general, for $R = R_1$, if

$$F(x_{i-1}) = r_{i-1} < R \leq r_i = F(x_i) \tag{8.13}$$

then set $X_1 = x_i$. Here, $r_0 = 0; x_0 = -\infty; x_1, x_2, \ldots, x_n$ are the possible values of the random variable; and $r_k = p(x_1) + \cdots + p(x_k), k = 1, 2, \ldots, n$. For this example, $n = 3, x_1 = 0, x_2 = 1$, and $x_3 = 2$; hence, $r_1 = 0.5, r_2 = 0.8$, and $r_3 = 1.0$. (Notice that $r_n = 1.0$ in all cases.)

Since $r_1 = 0.5 < R_1 = 0.73 \leq r_2 = 0.8$, set $X_1 = x_2 = 1$. The generation scheme is summarized as follows:

$$X = \begin{cases} 0, & R \leq 0.5 \\ 1, & 0.5 < R \leq 0.8 \\ 2, & 0.8 < R \leq 1.0 \end{cases}$$

Example 8.4 illustrated the table-lookup procedure; the next example illustrates an algebraic approach that can be used for certain distributions.

Example 8.5:   A Discrete Uniform Distribution _____

Consider the discrete uniform distribution on $\{1, 2, \ldots, k\}$ with pmf and cdf given by

$$p(x) = \frac{1}{k}, \quad x = 1, 2, \ldots, k$$

and

$$F(x) = \begin{cases} 0, & x < 1 \\ \dfrac{1}{k}, & 1 \le x < 2 \\ \dfrac{2}{k}, & 2 \le x < 3 \\ \vdots & \vdots \\ \dfrac{k-1}{k}, & k-1 \le x < k \\ 1, & k \le x \end{cases}$$

Let $x_i = i$ and $r_i = p(1) + \cdots + p(x_i) = F(x_i) = i/k$ for $i = 1, 2, \ldots, k$. Then, from Inequality (8.13), it can be seen that, if the generated random number R satisfies

$$r_{i-1} = \frac{i-1}{k} < R \le r_i = \frac{i}{k} \tag{8.14}$$

then X is generated by setting $X = i$. Now Inequality (8.14) can be solved for i:

$$i - 1 < Rk \le i$$

$$Rk \le i < Rk + 1 \tag{8.15}$$

Let $\lceil y \rceil$ denote the smallest integer $\ge y$. For example, $\lceil 7.82 \rceil = 8$, $\lceil 5.13 \rceil = 6$, and $\lceil -1.32 \rceil = -1$. For $y \ge 0$, $\lceil y \rceil$ is a function that rounds up. This notation and Inequality (8.15) yield a formula for generating X, namely

$$X = \lceil Rk \rceil \tag{8.16}$$

For example, consider the generating of a random variate X that is uniformly distributed on $\{1, 2, \ldots, 10\}$. The variate, X, might represent the number of pallets to be loaded onto a truck. Using Table A.1 as a source of random numbers R and using Equation (8.16) with $k = 10$ yields

$$R_1 = 0.78 \quad X_1 = \lceil 7.8 \rceil = 8$$
$$R_2 = 0.03 \quad X_2 = \lceil 0.3 \rceil = 1$$
$$R_3 = 0.23 \quad X_3 = \lceil 2.3 \rceil = 3$$
$$R_4 = 0.97 \quad X_4 = \lceil 9.7 \rceil = 10$$

The procedure discussed here can be modified to generate a discrete uniform random variate with any range consisting of consecutive integers. Exercise 13 asks the student to devise a procedure for one such case.

Example 8.6:   The Geometric Distribution

Consider the geometric distribution with pmf

$$p(x) = p(1-p)^x, \quad x = 0, 1, 2, \ldots$$

where $0 < p < 1$. Its cdf is given by

$$F(x) = \sum_{j=0}^{x} p(1-p)^j$$

$$= \frac{p\{1 - (1-p)^{x+1}\}}{1 - (1-p)}$$

$$= 1 - (1-p)^{x+1}$$

for $x = 0, 1, 2, \ldots$ Using the inverse-transform technique [i.e., Inequality (8.13)], recall that a geometric random variable X will assume the value x whenever

$$F(x-1) = 1 - (1-p)^x < R \le 1 - (1-p)^{x+1} = F(x) \qquad (8.17)$$

where R is a generated random number assumed $0 < R < 1$. Solving Inequality (8.17) for x proceeds as follows:

$$(1-p)^{x+1} \le 1 - R < (1-p)^x$$

$$(x+1)\ln(1-p) \le \ln(1-R) < x\ln(1-p)$$

But $1 - p < 1$ implies that $\ln(1-p) < 0$, so that

$$\frac{\ln(1-R)}{\ln(1-p)} - 1 \le x < \frac{\ln(1-R)}{\ln(1-p)} \qquad (8.18)$$

Thus, $X = x$ for that integer value of x satisfying Inequality (8.18). In brief, and using the round-up function $\lceil \cdot \rceil$, we have

$$X = \left\lceil \frac{\ln(1-R)}{\ln(1-p)} - 1 \right\rceil \qquad (8.19)$$

Since p is a fixed parameter, let $\beta = -1/\ln(1-p)$. Then $\beta > 0$ and, by Equation (8.19), $X = \lceil -\beta \ln(1-R) - 1 \rceil$. By Equation (8.1), $-\beta \ln(1-R)$ is an exponentially distributed random variable with mean β; so one way of generating a geometric variate with parameter p is to generate (by any method) an exponential variate with parameter $\beta^{-1} = -\ln(1-p)$, subtract one, and round up.

Occasionally, there is needed a geometric variate X that can assume values $\{q, q+1, q+2, \ldots\}$ with pmf $p(x) = p(1-p)^{x-q}$ $(x = q, q+1, \ldots)$. Such a variate X can be generated, via Equation (8.19), by

$$X = q + \left\lceil \frac{\ln(1-R)}{\ln(1-p)} - 1 \right\rceil \qquad (8.20)$$

One of the most common cases is $q = 1$.

Example 8.7 _____

Generate three values from a geometric distribution on the range $\{X \ge 1\}$ with mean 2. Such a geometric distribution has pmf $p(x) = p(1-p)^{x-1}$ $(x = 1, 2, \ldots)$ with mean $1/p = 2$, or $p = 1/2$.

Thus, X can be generated by Equation (8.20) with $q = 1, p = 1/2$, and $1/\ln(1 - p) = -1.443$. Using Table A.1, $R_1 = 0.932$, $R_2 = 0.105$, and $R_3 = 0.687$ yields

$$X_1 = 1 + \lceil -1.443 \ln(1 - 0.932) - 1 \rceil$$
$$= 1 + \lceil 3.878 - 1 \rceil = 4$$
$$X_2 = 1 + \lceil -1.443 \ln(1 - 0.105) - 1 \rceil = 1$$
$$X_3 = 1 + \lceil -1.443 \ln(1 - 0.687) - 1 \rceil = 2$$

Exercise 15 deals with an application of the geometric distribution.

8.2 ACCEPTANCE–REJECTION TECHNIQUE

Suppose that an analyst needed to devise a method for generating random variates, X, uniformly distributed between 1/4 and 1. One way to proceed would be to follow these steps:

Step 1. Generate a random number R.

Step 2a. If $R \geq 1/4$, accept $X = R$, then go to Step 3.

Step 2b. If $R < 1/4$, reject R, and return to Step 1.

Step 3. If another uniform random variate on [1/4, 1] is needed, repeat the procedure beginning at Step 1. If not, stop.

Each time Step 1 is executed, a new random number R must be generated. Step 2a is an "acceptance" and Step 2b is a "rejection" in this acceptance–rejection technique. To summarize the technique, random variates (R) with some distribution (here uniform on [0, 1]) are generated until some condition $(R > 1/4)$ is satisfied. When the condition is finally satisfied, the desired random variate, X (here uniform on [1/4, 1]), can be computed $(X = R)$. This procedure can be shown to be correct by recognizing that the accepted values of R are conditioned values; that is, R itself does not have the desired distribution, but R conditioned on the event $\{R \geq 1/4\}$ does have the desired distribution. To show this, take $1/4 \leq a < b \leq 1$; then

$$P(a < R \leq b|1/4 \leq R \leq 1) = \frac{P(a < R \leq b)}{P(1/4 \leq R \leq 1)} = \frac{b - a}{3/4} \tag{8.21}$$

which is the correct probability for a uniform distribution on [1/4, 1]. Equation (8.21) says that the probability distribution of R, given that R is between 1/4 and 1 (all other values of R are thrown out), is the desired distribution. Therefore, if $1/4 \leq R \leq 1$, set $X = R$.

The efficiency of an acceptance–rejection technique depends heavily on being able to minimize the number of rejections. In this example, the probability of a rejection is $P(R < 1/4) = 1/4$, so that the number of rejections is a geometrically distributed random variable with probability of "success" being $p = 3/4$ and mean number of rejections $(1/p - 1) = 4/3 - 1 = 1/3$. (Example 8.6 discussed the geometric distribution.) The mean number of random numbers R

required to generate one variate X is one more than the number of rejections; hence, it is $4/3 = 1.33$. In other words, to generate 1000 values of X would require approximately 1333 random numbers R.

In the present situation, an alternative procedure exists for generating a uniform variate on $[1/4, 1]$—namely, Equation (8.5), which reduces to $X = 1/4 + (3/4)R$. Whether the acceptance–rejection technique or an alternative procedure, such as the inverse-transform technique [Equation (8.5)], is the more efficient depends on several considerations. The computer being used, the skills of the programmer and the relative inefficiency of generating the additional (rejected) random numbers needed by acceptance–rejection should be compared to the computations required by the alternative procedure. In practice, concern with generation efficiency is left to specialists who conduct extensive tests comparing alternative methods (i.e., until a simulation model begins to require excessive computer runtime due to the generator being used).

For the uniform distribution on $[1/4, 1]$, the inverse-transform technique of Equation (8.5) is undoubtedly much easier to apply and more efficient than the acceptance–rejection technique. The main purpose of this example was to explain and motivate the basic concept of the acceptance–rejection technique. However, for some important distributions, such as the normal, gamma and beta, the inverse cdf does not exist in closed form and therefore the inverse-transform technique is difficult. These more advanced techniques are summarized by Bratley, Fox, and Schrage [1996], Fishman [1978], and Law and Kelton [2000].

In the following subsections, the acceptance–rejection technique is illustrated for the generation of random variates for the Poisson, nonstationary Poisson, and gamma distributions.

8.2.1 Poisson Distribution

A Poisson random variable, N, with mean $\alpha > 0$ has pmf

$$p(n) = P(N = n) = \frac{e^{-\alpha}\alpha^n}{n!}, \quad n = 0, 1, 2, \ldots$$

More important, however, is that N can be interpreted as the number of arrivals from a Poisson arrival process in one unit of time. Recall from Section 5.5 that the interarrival times, $A_1, A_2, \ldots$ of successive customers are exponentially distributed with rate α (i.e., α is the mean number of arrivals per unit time); in addition, an exponential variate can be generated by Equation (8.3). Thus, there is a relationship between the (discrete) Poisson distribution and the (continuous) exponential distribution:

$$N = n \tag{8.22}$$

if and only if

$$A_1 + A_2 + \cdots + A_n \leq 1 < A_1 + \cdots + A_n + A_{n+1} \tag{8.23}$$

Equation (8.22), $N = n$, says there were exactly n arrivals during one unit of time; but Relation (8.23) says that the nth arrival occurred before time 1 while the $(n + 1)$st arrival occurred after time 1. Clearly, these two statements are equivalent. Proceed now by generating exponential interarrival times until some arrival, say $n + 1$, occurs after time 1; then set $N = n$.

For efficient generation purposes, Relation (8.23) is usually simplified by first using Equation (8.3), $A_i = (-1/\alpha)\ln R_i$, to obtain

$$\sum_{i=1}^{n} -\frac{1}{\alpha}\ln R_i \leq 1 < \sum_{i=1}^{n+1} -\frac{1}{\alpha}\ln R_i$$

Next multiply through by $-\alpha$, which reverses the sign of the inequality, and use the fact that a sum of logarithms is the logarithm of a product, to get

$$\ln \prod_{i=1}^{n} R_i = \sum_{i=1}^{n}\ln R_i \geq -\alpha > \sum_{i=1}^{n+1}\ln R_i = \ln \prod_{i=1}^{n+1} R_i$$

Finally, use the relation $e^{\ln x} = x$ for any number x to obtain

$$\prod_{i=1}^{n} R_i \geq e^{-\alpha} > \prod_{i=1}^{n+1} R_i \tag{8.24}$$

which is equivalent to Relation (8.23). The procedure for generating a Poisson random variate, N, is given by the following steps:

Step 1. Set $n = 0, P = 1$.

Step 2. Generate a random number R_{n+1}, and replace P by $P \cdot R_{n+1}$.

Step 3. If $P < e^{-\alpha}$, then accept $N = n$. Otherwise, reject the current n, increase n by one, and return to step 2.

Notice that, upon completion of Step 2, P is equal to the rightmost expression in Relation (8.24). The basic idea of a rejection technique is again exhibited; if $P \geq e^{-\alpha}$ in Step 3, then n is rejected and the generation process must proceed through at least one more trial.

How many random numbers will be required, on the average, to generate one Poisson variate, N? If $N = n$, then $n + 1$ random numbers are required, so the average number is given by

$$E(N + 1) = \alpha + 1$$

which is quite large if the mean, α, of the Poisson distribution is large.

Example 8.8 _____
Generate three Poisson variates with mean $\alpha = 0.2$. First, compute $e^{-\alpha} = e^{-0.2} = 0.8187$. Next, get a sequence of random numbers R from Table A.1 and follow the previously described Steps 1 to 3:

Step 1. Set $n = 0, P = 1$.

Step 2. $R_1 = 0.4357, P = 1 \cdot R_1 = 0.4357$.

Step 3. Since $P = 0.4357 < e^{-\alpha} = 0.8187$, accept $N = 0$.

Step 1–3. ($R_1 = 0.4146$ leads to $N = 0$.)

Step 1. Set $n = 0, P = 1$.

Step 2. $R_1 = 0.8353$, $P = 1 \cdot R_1 = 0.8353$.

Step 3. Since $P \geq e^{-\alpha}$, reject $n = 0$ and return to Step 2 with $n = 1$.

Step 2. $R_2 = 0.9952$, $P = R_1 R_2 = 0.8313$.

Step 3. Since $P \geq e^{-\alpha}$, reject $n = 1$ and return to Step 2 with $n = 2$.

Step 2. $R_3 = 0.8004$, $P = R_1 R_2 R_3 = 0.6654$.

Step 3. Since $P < e^{-\alpha}$, accept $N = 2$.

The calculations required for the generation of these three Poisson random variates are summarized as follows:

n	R_{n+1}	P	Accept/Reject	Result
0	0.4357	0.4357	$P < e^{-\alpha}$ (accept)	$N = 0$
0	0.4146	0.4146	$P < e^{-\alpha}$ (accept)	$N = 0$
0	0.8353	0.8353	$P \geq e^{-\alpha}$ (reject)	
1	0.9952	0.8313	$P \geq e^{-\alpha}$ (reject)	
2	0.8004	0.6654	$P < e^{-\alpha}$ (accept)	$N = 2$

It took five random numbers, R, to generate three Poisson variates here ($N = 0, N = 0$, and $N = 2$), but in the long run, to generate, say, 1000 Poisson variates with mean $\alpha = 0.2$, it would require approximately $1000(\alpha + 1)$ or 1200 random numbers.

Example 8.9

Buses arrive at the bus stop at Peachtree and North Avenue according to a Poisson process with a mean of one bus per 15 minutes. Generate a random variate, N, which represents the number of arriving buses during a 1-hour time slot. Now, N is Poisson distributed with a mean of four buses per hour. First compute $e^{-\alpha} = e^{-4} = 0.0183$. Using a sequence of 12 random numbers from Table A.1 yields the following summarized results:

n	R_{n+1}	P	Accept/Reject	Result
0	0.4357	0.4357	$P \geq e^{-\alpha}$ (reject)	
1	0.4146	0.1806	$P \geq e^{-\alpha}$ (reject)	
2	0.8353	0.1508	$P \geq e^{-\alpha}$ (reject)	
3	0.9952	0.1502	$P \geq e^{-\alpha}$ (reject)	
4	0.8004	0.1202	$P \geq e^{-\alpha}$ (reject)	
5	0.7945	0.0955	$P \geq e^{-\alpha}$ (reject)	
6	0.1530	0.0146	$P < e^{-\alpha}$ (accept)	$N = 6$

It is immediately seen that a larger value of α (here $\alpha = 4$) usually requires more random numbers; if 1000 Poisson variates were desired, approximately $1000(\alpha + 1) = 5000$ random numbers would be required.

When α is large, say $\alpha \geq 15$, the rejection technique outlined here becomes quite expensive, but fortunately an approximate technique based on the normal distribution works quite well. When the mean, α, is large, then

$$Z = \frac{N - \alpha}{\sqrt{\alpha}}$$

is approximately normally distributed with mean zero and variance 1; this observation suggests an approximate technique. First, generate a standard normal variate Z, by Equation (8.28) in Section 8.3.1, then generate the desired Poisson variate, N, by

$$N = \lceil \alpha + \sqrt{\alpha}Z - 0.5 \rceil \tag{8.25}$$

where $\lceil \cdot \rceil$ is the round-up function described in Section 8.1.7. (If $\alpha + \sqrt{\alpha}Z - 0.5 < 0$, then set $N = 0$.) The "0.5" used in the formula makes the round-up function become a "round to the nearest integer" function. Equation (8.25) is not an acceptance–rejection technique, but, when used as an alternative to the acceptance–rejection method, it provides a fairly efficient and accurate method for generating Poisson variates with a large mean.

8.2.2 Nonstationary Poisson Process

Another type of acceptance–rejection method (which is also called "thinning") can be used to generate interarrival times from a nonstationary Poisson process (NSPP) with arrival rate $\lambda(t), 0 \leq t \leq T$. A NSPP is an arrival process with an arrival rate that varies with time; see Section 5.5.2.

Consider, for instance, the arrival-rate function given in Table 8.7 that changes every hour. The idea behind thinning is to generate a stationary Poisson arrival process at the fastest rate (1/5 customer per minute in the example), but "accept" or admit only a portion of the arrivals, thinning out just enough to get the desired time-varying rate. Next we give the generic algorithm, which generates T_i as the time of the ith arrival. Remember that, in a stationary Poisson arrival process, the times between arrivals are exponentially distributed.

Step 1. Let $\lambda^* = \max_{0 \leq t \leq T} \lambda(t)$ be the maximum of the arrival rate function and set $t = 0$ and $i = 1$.

Step 2. Generate E from the exponential distribution with rate λ^* and let $t = t + E$ (this is the arrival time of the stationary Poisson process).

Step 3. Generate random number R from the $U(0, 1)$ distribution. If $R \leq \lambda(t)/\lambda^*$ then $T_i = t$ and $i = i + 1$.

Step 4. Go to Step 2.

The thinning algorithm can be inefficient if there are large differences between the typical and the maximum arrival rate. However, thinning has the advantage that it works for any integrable arrival rate function, not just a piecewise-constant function as in this example.

Table 8.7 Arrival Rate for NSPP Example

t (min)	*Mean Time between Arrivals (min)*	*Arrival Rate $\lambda(t)$ (arrivals/min)*
0	15	1/15
60	12	1/12
120	7	1/7
180	5	1/5
240	8	1/8
300	10	1/10
360	15	1/15
420	20	1/20
480	20	1/20

Example 8.10

For the arrival-rate function in Table 8.7, generate the first two arrival times.

Step 1. $\lambda^* = \max_{0 \leq t \leq T} \lambda(t) = 1/5$, $t = 0$ and $i = 1$.

Step 2. For random number $R = 0.2130$, $E = -5\ln(0.213) = 13.13$ and $t = 0 + 13.13 = 13.13$.

Step 3. Generate $R = 0.8830$. Since $R = 0.8830 \not\leq \lambda(13.13)/\lambda^* = (1/15)/(1/5) = 1/3$, do not generate the arrival.

Step 4. Go to Step 2.

Step 2. For random number $R = 0.5530$, $E = -5\ln(0.553) = 2.96$, and $t = 13.13 + 2.96 = 16.09$.

Step 3. Generate $R = 0.0240$. Since $R = 0.0240 \leq \lambda(16.09)/\lambda^* = (1/15)/(1/5) = 1/3$, set $T_1 = t = 16.09$ and $i = i + 1 = 2$.

Step 4. Go to Step 2.

Step 2. For random number $R = 0.0001$, $E = -5\ln(0.0001) = 46.05$ and $t = 16.09 + 46.05 = 62.14$.

Step 3. Generate $R = 0.1443$. Since $R = 0.1443 \leq \lambda(62.14)/\lambda^* = (1/12)/(1/5) = 5/12$, set $T_2 = t = 62.14$ and $i = i + 1 = 3$.

Step 4. Go to Step 2.

8.2.3 Gamma Distribution

Several acceptance–rejection techniques for generating gamma random variates have been developed. (See Bratley, Fox, and Schrage [1996]; Fishman [1978]; and Law and Kelton [2000].) One of the more efficient is by Cheng [1977]; the mean number of trials is between 1.13 and 1.47 for any value of the shape parameter $\beta \geq 1$.

If the shape parameter β is an integer, say $\beta = k$, one possibility is to use the convolution technique in Example 8.12, because the Erlang distribution is a special case of the more general gamma distribution. On the other hand, the acceptance–rejection technique described here would be a highly efficient method for the Erlang distribution especially if $\beta = k$ were large. The routine generates gamma random variates with scale parameter θ and shape parameter β—that is, with mean $1/\theta$ and variance $1/\beta\theta^2$. The steps are as follows:

Step 1. Compute $a = 1/(2\beta - 1)^{1/2}$, $b = \beta - \ln 4$.

Step 2. Generate R_1 and R_2. Set $V = R_1/(1 - R_1)$.

Step 3. Compute $X = \beta V^a$.

Step 4a. If $X > b + (\beta a + 1)\ln(V) - \ln(R_1^2 R_2)$, reject X and return to Step 2.

Step 4b. If $X \leq b + (\beta a + 1)\ln(V) - \ln(R_1^2 R_2)$, use X as the desired variate.

The generated variates from Step 4b will have mean and variance both equal to β. If it is desired to have mean $1/\theta$ and variance $1/\beta\theta^2$ as in Section 5.4, then include Step 5.

(Step 5. Replace X by $X/(\beta\theta)$.)

The basic idea of all acceptance–rejection methods is again illustrated here, but the proof of this example is beyond the scope of this book. In Step 3, $X = \beta V^a = \beta[R_1/(1 - R_1)]^a$ is not gamma distributed, but rejection of certain values of X in Step 4a guarantees that the accepted values in Step 4b do have the gamma distribution.

Example 8.11 _____

Downtimes for a high-production candy-making machine have been found to be gamma distributed with mean 2.2 minutes and variance 2.10 minutes^2. Thus, $1/\theta = 2.2$ and $1/\beta\theta^2 = 2.10$, which together imply that $\beta = 2.30$ and $\theta = 0.4545$.

Step 1. $a = 0.53$, $b = 0.91$.

Step 2. Generate $R_1 = 0.832$, $R_2 = 0.021$. Set $V = 0.832/(1 - 0.832) = 4.952$.

Step 3. Compute $X = 2.3(4.952)^{0.53} = 5.37$.

Step 4. $X = 5.37 > 0.91 + [2.3(0.53) + 1]\ln(4.952) - \ln[(0.832)^2 0.021] = 8.68$, so reject X and return to Step 2.

Step 2. Generate $R_1 = 0.434$, $R_2 = 0.716$. Set $V = 0.434/(1 - 0.434) = 0.767$.

Step 3. Compute $X = 2.3(0.767)^{0.53} = 2.00$.

Step 4. Since $X = 2.00 \leq 0.91 + [2.3(0.53) + 1]\ln(0.767) - \ln[(0.434)^2 0.716] = 2.32$, accept X.

Step 5. Divide X by $\beta\theta = 1.045$ to get $X = 1.91$.

This example took two trials (i.e., one rejection) to generate an acceptable gamma-distributed random variate, but, on the average, to generate, say, 1000 gamma variates, the method will require between 1130 and 1470 trials, or equivalently, between 2260 and 2940 random numbers. The method is somewhat cumbersome for hand calculations, but is easy to program on the computer and is one of the most efficient gamma generators known.

8.3 SPECIAL PROPERTIES

"Special properties" are just as the name implies: They are variate-generation techniques that are based on features of a particular family of probability distributions, rather than being general-purpose techniques like the inverse-transform or acceptance–rejection techniques.

8.3.1 Direct Transformation for the Normal and Lognormal Distributions

Many methods have been developed for generating normally distributed random variates. The inverse-transform technique cannot easily be applied, however, because the inverse cdf cannot be written in closed form. The standard normal cdf is given by

$$\Phi(x) = \int_{-\infty}^{x} \frac{1}{\sqrt{2\pi}} e^{-t^2/2} \, dt, \quad -\infty < x < \infty$$

This section describes an intuitively appealing direct transformation that produces an independent pair of standard normal variates with mean zero and variance 1. The method is due to Box and Muller [1958]. Although not as efficient as many more modern techniques, it is easy to program in a scientific language, such as FORTRAN, C, C++, Visual Basic, or Java. We then show how to transform a standard normal variate into a normal variate with mean μ and variance σ^2. Once we have a method (this or any other) for generating X from a $N(\mu, \sigma^2)$ distribution, then we can generate a lognormal random variate Y with parameters μ and σ^2 by using the direct transformation $Y = e^X$. (Recall that μ and σ^2 are *not* the mean and variance of the lognormal; see Equations (5.58) and (5.59).)

Consider two standard normal random variables, Z_1 and Z_2, plotted as a point in the plane as shown in Figure 8.7 and represented in polar coordinates as

$$Z_1 = B \cos \theta$$
$$Z_2 = B \sin \theta \tag{8.26}$$

It is known that $B^2 = Z_1^2 + Z_2^2$ has the chi-square distribution with 2 degrees of freedom, which is equivalent to an exponential distribution with mean 2. Thus, the radius, B, can be generated by use of Equation (8.3):

$$B = (-2 \ln R)^{1/2} \tag{8.27}$$

By the symmetry of the normal distribution, it seems reasonable to suppose, and indeed it is the case, that the angle is uniformly distributed between 0 and 2π radians. In addition, the radius, B,

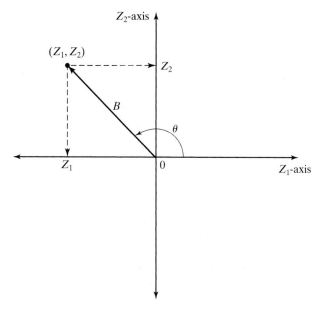

Figure 8.7 Polar representation of a pair of standard normal variables.

and the angle, θ, are mutually independent. Combining Equations (8.26) and (8.27) gives a direct method for generating two independent standard normal variates, Z_1 and Z_2, from two independent random numbers, R_1 and R_2:

$$\begin{aligned} Z_1 &= (-2\ln R_1)^{1/2}\cos(2\pi R_2) \\ Z_2 &= (-2\ln R_1)^{1/2}\sin(2\pi R_2) \end{aligned} \tag{8.28}$$

To illustrate the generation scheme, consider Equation (8.28) with $R_1 = 0.1758$ and $R_2 = 0.1489$. Two standard normal random variates are generated as follows:

$$Z_1 = [-2\ln(0.1758)]^{1/2}\cos(2\pi\,0.1489) = 1.11$$

$$Z_2 = [-2\ln(0.1758)]^{1/2}\sin(2\pi\,0.1489) = 1.50$$

To obtain normal variates X_i with mean μ and variance σ^2, we then apply the transformation

$$X_i = \mu + \sigma Z_i \tag{8.29}$$

to the standard normal variates. For example, to transform the two standard normal variates into normal variates with mean $\mu = 10$ and variance $\sigma^2 = 4$, we compute

$$X_1 = 10 + 2(1.11) = 12.22$$

$$X_2 = 10 + 2(1.50) = 13.00$$

8.3.2 Convolution Method

The probability distribution of a sum of two or more independent random variables is called a convolution of the distributions of the original variables. The convolution method thus refers to adding together two or more random variables to obtain a new random variable with the desired distribution. This technique can be applied to obtain Erlang variates and binomial variates. What is important is not the cdf of the desired random variable, but rather its relation to other variates more easily generated.

Example 8.12:   Erlang Distribution ────────────────────────────────────

As was discussed in Section 5.4, an Erlang random variable X with parameters (k, θ) can be shown to be the sum of k independent exponential random variables, $X_i, i = 1, \ldots, k$, each having mean $1/k\theta$—that is,

$$X = \sum_{i=1}^{k} X_i$$

The convolution approach is to generate $X_1, X_2, \ldots, X_k$, then sum them to get X. In the case of the Erlang, each X_i can be generated by Equation (8.3) with $1/\lambda = 1/k\theta$. Therefore, an Erlang variate can be generated by

$$X = \sum_{i=1}^{k} -\frac{1}{k\theta} \ln R_i$$

$$= -\frac{1}{k\theta} \ln \left(\prod_{i=1}^{k} R_i \right) \tag{8.30}$$

It is more efficient computationally to multiply all the random numbers first and then to compute only one logarithm.

Example 8.13 ────────────────────────────────────

Trucks arrive at a large warehouse in a completely random fashion that is modeled as a Poisson process with arrival rate $\lambda = 10$ trucks per hour. The guard at the entrance sends trucks alternately to the north and south docks. An analyst has developed a model to study the loading/unloading process at the south docks and needs a model of the arrival process at the south docks alone. An interarrival time X between successive truck arrivals at the south docks is equal to the sum of two interarrival times at the entrance and thus it is the sum of two exponential random variables, each having mean 0.1 hour, or 6 minutes. Thus, X has the Erlang distribution with $K = 2$ and mean $1/\theta = 2/\lambda = 0.2$ hour. To generate the variate X, first obtain $K = 2$ random numbers from Table A.1, say $R_1 = 0.937$ and $R_2 = 0.217$. Then, by Equation (8.30),

$$X = -0.1 \ln[0.937(0.217)]$$

$$= 0.159 \text{ hour} = 9.56 \text{ minutes}$$

In general, Equation (8.30) implies that K uniform random numbers are needed for each Erlang variate generated. If K is large, it is more efficient to generate Erlang variates by other techniques, such as one of the many acceptance–rejection techniques for the gamma distribution given in Section 8.2.3, or by Bratley, Fox and Schrage [1996], Fishman [1978], and Law and Kelton [2000].

8.3.3 More Special Properties

There are many relationships among probability distributions that can be exploited for random-variate generation. The convolution method in the Section 8.3.2 is one example. Another particularly useful example is the relationship between the beta distribution and the gamma distribution.

Suppose that X_1 has a gamma distribution with shape parameter β_1 and scale parameter $\theta_1 = 1/\beta_1$, while X_2 has a gamma distribution with shape parameter β_2 and scale parameter $\theta_2 = 1/\beta_2$, and that these two random variables are independent. Then

$$Y = \frac{X_1}{X_1 + X_2}$$

has a beta distribution with parameters β_1 and β_2 on the interval $(0, 1)$. If, instead, we want Y to be defined on the interval (a, b), then set

$$Y = a + (b - a)\left(\frac{X_1}{X_1 + X_2}\right)$$

Thus, using the acceptance–rejection technique for gamma variates defined in the previous section, we can generate beta variates, with two gamma variates required for each beta.

Although this method of beta generation is convenient, there are faster methods based on acceptance–rejection ideas. See, for instance, Devroye [1986] or Dagpunar [1988].

8.4 SUMMARY

The basic principles of random-variate generation via the inverse-transform technique, the acceptance–rejection technique, and special properties have been introduced and illustrated by examples. Methods for generating many of the important continuous and discrete distributions, plus all empirical distributions, have been given. See Schmeiser [1980] for an excellent survey; for a state-of-the-art treatment, the reader is referred to Devroye [1986] or Dagpunar [1988].

REFERENCES

BRATLEY, P., B. L. FOX, AND L. E. SCHRAGE [1996], *A Guide to Simulation*, 2d ed., Springer-Verlag, New York.

BOX, G. E. P., AND M. F. MULLER [1958], "A Note on the Generation of Random Normal Deviates," *Annals of Mathematical Statistics*, Vol. 29, pp. 610–611.

CHENG, R. C. H. [1977], "The Generation of Gamma Variables," *Applied Statistician,* Vol. 26, No. 1, pp. 71–75.

DAGPUNAR, J. [1988], *Principles of Random Variate Generation,* Clarendon Press, Oxford, New York.

DEVROYE, L. [1986], *Non-Uniform Random Variate Generation,* Springer-Verlag, New York.

FISHMAN, G. S. [1978], *Principles of Discrete Event Simulation,* Wiley, New York.

LAW, A. M., AND W. D. KELTON [2000], *Simulation Modeling & Analysis,* 3d ed., McGraw–Hill, New York.

RIPLEY, B. D. [1987], *Stochastic Simulation,* Wiley, New York.

SCHMEISER, BRUCE W. [1979], "Approximations to the Inverse Cumulative Normal Function for Use on Hand Calculators," *Applied Statistics,* Vol. 28, pp. 175–176.

SCHMEISER, B. W. [1980], "Random Variate Generation: A Survey," in *Simulation with Discrete Models: A State of the Art View,* T. I. Ören, C. M. Shub, and P. F. Roth, eds., IEEE, New York.

EXERCISES

1. Develop a random-variate generator for a random variable X with the pdf

$$f(x) = \begin{cases} e^{2x}, & -\infty < x \leq 0 \\ e^{-2x}, & 0 < x < \infty \end{cases}$$

2. Develop a generation scheme for the triangular distribution with pdf

$$F(x) = \begin{cases} \frac{1}{4}x^2 - x, & 2 \leq x \leq 3 \\ x - \frac{1}{12} \cdot x^2, & 3 \leq x \leq 6 \\ 0, & o/w \end{cases}$$

$$f(x) = \begin{cases} \frac{1}{2}(x-2), & 2 \leq x \leq 3 \\ \frac{1}{2}\left(2 - \frac{x}{3}\right), & 3 < x \leq 6 \\ 0, & \text{otherwise} \end{cases}$$

 Lookup cdf.

   Generate 10 values of the random variate, compute the sample mean, and compare it to the true mean of the distribution.

3. Develop a generator for a triangular distribution with range (1, 10) and mode at $x = 4$.

4. Develop a generator for a triangular distribution with range (1, 10) and a mean of 4.

5. Given the following cdf for a continuous variable with range from -3 to 4, develop a generator for the variable.

$$F(x) = \begin{cases} 0, & x \leq -3 \\ \frac{1}{2} + \frac{x}{6}, & -3 < x \leq 0 \\ \frac{1}{2} + \frac{x^2}{32}, & 0 < x \leq 4 \\ 1, & x > 4 \end{cases}$$

6. Given the cdf $F(x) = x^4/16$ on $0 \leq x \leq 2$, develop a generator for this distribution.

7. Given the pdf $f(x) = x^2/9$ on $0 \le x \le 3$, develop a generator for this distribution.

8. Develop a generator for a random variable whose pdf is

$$
f(x) = \begin{cases} \dfrac{1}{3}, & 0 \le x \le 2 \\[2mm] \dfrac{1}{24}, & 2 < x \le 10 \\[2mm] 0, & \text{otherwise} \end{cases}
$$

9. The cdf of a discrete random variable X is given by

$$
F(x) = \frac{x(x+1)(2x+1)}{n(n+1)(2n+1)}, \quad x = 1, 2, \dots, n
$$

When $n = 4$, generate three values of X, using $R_1 = 0.83$, $R_2 = 0.24$, and $R_3 = 0.57$.

10. Times to failure for an automated production process have been found to be randomly distributed with a Weibull distribution with parameters $\beta = 2$ and $\alpha = 10$. Derive Equation (8.6), and then use it to generate five values from this Weibull distribution, using five random numbers taken from Table A.1.

11. Data have been collected on service times at a drive-in bank window at the Shady Lane National Bank. This data are summarized into intervals as follows:

Interval (Seconds)	Frequency
15–30	10
30–45	20
45–60	25
60–90	35
90–120	30
120–180	20
180–300	10

Set up a table like Table 8.3 for generating service times by the table-lookup method, and generate five values of service time, using four-digit random numbers.

12. In Example 8.2, assume that fire-crew response times satisfy $0.25 \le x \le 3$. Modify Table 8.2 to accommodate this assumption. Then generate five values of response time, using four-digit uniform random numbers from Table A.1.

13. For a preliminary version of a simulation model, the number of pallets, X, to be loaded onto a truck at a loading dock was assumed to be uniformly distributed between 8 and 24. Devise a method for generating X, assuming that the loads on successive trucks are independent. Use the technique of Example 8.5 for discrete uniform distributions. Finally, generate loads for 10 successive trucks by using four-digit random numbers.

14. Develop a method for generating values from a negative binomial distribution with parameters p and k, as described in Section 5.3. Generate 3 values when $p = 0.8$ and $k = 2$. [*Hint*: Think about the definition of the negative binomial as the number of Bernoulli trials until the kth success.]

15. The weekly demand, X, for a slow-moving item has been found to be approximated well by a geometric distribution on the range $\{0, 1, 2, \ldots\}$ with mean weekly demand of 2.5 items. Generate 10 values of X, demand per week, using random numbers from Table A.1. (*Hint*: For a geometric distribution on the range $\{q, q+1, \ldots\}$ with parameter p, the mean is $1/p+q-1$.)

16. In Exercise 15, suppose that the demand has been found to have a Poisson distribution with mean 2.5 items per week. Generate 10 values of X, demand per week, using random numbers from Table A.1. Discuss the differences between the geometric and the Poisson distributions.

17. Lead times have been found to be exponentially distributed with mean 3.7 days. Generate five random lead times from this distribution.

18. Regular maintenance of a production routine has been found to vary and has been modeled as a normally distributed random variable with mean 33 minutes and variance 4 minutes^2. Generate five random maintenance times from the given distribution.

19. A machine is taken out of production either if it fails or after 5 hours, whichever comes first. By running similar machines until failure, it has been found that time to failure, X, has the Weibull distribution with $\alpha = 8$, $\beta = 0.75$, and $\nu = 0$ (refer to Sections 5.4 and 8.1.3). Thus, the time until the machine is taken out of production can be represented as $Y = \min(X, 5)$. Develop a step-by-step procedure for generating Y.

20. The time until a component is taken out of service is uniformly distributed on from 0 to 8 hours. Two such independent components are put in series, and the whole system goes down when one of the components goes down. If X_i $(i = 1, 2)$ represents the component runtimes, then $Y = \min(X_1, X_2)$ represents the system lifetime. Devise two distinct ways to generate Y. [*Hint*: One way is relatively straightforward. For a second method, first compute the cdf of Y : $F_Y(y) = P(Y \leq y) = 1 - P(Y > y)$, for $0 \leq y \leq 8$. Use the equivalence $\{Y > y\} = \{X_1 > y$ and $X_2 > y\}$ and the independence of X_1, and X_2. After finding $F_Y(y)$, proceed with the inverse-transform technique.]

21. In Exercise 20, component lifetimes are exponentially distributed, one with mean 2 hours and the other with mean 6 hours. Rework Exercise 20 under this new assumption. Discuss the relative efficiency of the two generation schemes devised.

22. Develop a technique for generating a binomial random variable, X, via the convolution technique. [*Hint*: X can be represented as the number of successes in n independent Bernoulli

trials, each success having probability p. Thus, $X = \sum_{i=1}^{n} X_i$, where $P(X_i = 1) = p$ and $P(X_i = 0) = 1 - p$.]

23. Develop an acceptance–rejection technique for generating a geometric random variable, X, with parameter p on the range $\{0, 1, 2, \ldots\}$. (*Hint*: X can be thought of as the number of trials before the first success occurs in a sequence of independent Bernoulli trials.)

24. Write a computer routine to generate standard normal variates by the exact method discussed in this chapter. Use it to generate 1000 values. Compare the true probability, $\Phi(z)$, that a value lies in $(-\infty, z)$ to the actual observed relative frequency that values were $\leq z$, for $z = -4, -3, -2, -1, 0, 1, 2, 3$, and 4.

25. Write a computer routine to generate gamma variates with shape parameter β and scale parameter θ. Generate 1000 values with $\beta = 2.5$ and $\theta = 0.2$ and compare the true mean, $1/\theta = 5$, to the sample mean.

26. Write a computer routine to generate 200 values from one of the variates in Exercises 1 to 23. Make a histogram of the 200 values and compare it to the theoretical density function (or probability mass function, for discrete random variables).

27. Many spreadsheet, symbolic-calculation, and statistical-analysis programs have built-in routines for generating random variates from standard distributions. Try to find out what variate-generation methods are used in one of these packages by looking at the documentation. Should you trust a variate generator if the method is not documented?

28. Suppose that, somehow, we have available a source of exponentially distributed random variates with mean 1. Write an algorithm to generate random variates with a triangular distribution by transforming the exponentially distributed random variates. [*Hint*: First transform to obtain uniformly distributed random variates.]

29. Let time $t = 0$ correspond to 6 A.M., and suppose that the arrival rate (in arrivals per hour) of customers to a breakfast restaurant that is open 6–9 A.M. is

$$\lambda(t) = \begin{cases} 30, & 0 \leq t < 1 \\ 45, & 1 \leq t < 2 \\ 20, & 2 \leq t \leq 4 \end{cases}$$

Derive a thinning algorithm for generating arrivals from this NSPP.

30. Generate 10 values from a beta distribution on the interval $[0, 1]$ with parameters $\beta_1 = 1.47$ and $\beta_2 = 2.16$. Next transform them to be on the interval $[-10, 20]$.

Part IV

Analysis of Simulation Data

9

Input Modeling

Input models provide the driving force for a simulation model. In the simulation of a queueing system, typical input models are the distributions of time between arrivals and of service times. For an inventory-system simulation, input models include the distributions of demand and of lead time. For the simulation of a reliability system, the distribution of time to failure of a component is an example of an input model.

In the examples and exercises in Chapters 2 and 3, the appropriate distributions were specified for you. In real-world simulation applications, however, coming up with appropriate distributions for input data is a major task from the standpoint of time and resource requirements. Regardless of the sophistication of the analyst, faulty models of the inputs will lead to outputs whose interpretation could give rise to misleading recommendations.

There are four steps in the development of a useful model of input data:

1. Collect data from the real system of interest. This often requires a substantial time and resource commitment. Unfortunately, in some situations it is not possible to collect data (for example, when time is extremely limited, when the input process does not yet exist, or when laws or rules prohibit the collection of data). When data are not available, expert opinion and knowledge of the process must be used to make educated guesses.

2. Identify a probability distribution to represent the input process. When data are available, this step typically begins with the development of a frequency distribution, or histogram, of the data. Given the frequency distribution and a structural knowledge of the process, a family of distributions is chosen. Fortunately, as was described in Chapter 5, several well-known distributions often provide good approximations in practice.

3. Choose parameters that determine a specific instance of the distribution family. When data are available, these parameters may be estimated from the data.

4. Evaluate the chosen distribution and the associated parameters for goodness of fit. Goodness of fit may be evaluated informally, via graphical methods, or formally, via statistical tests. The chi-square and the Kolmogorov–Smirnov tests are standard goodness-of-fit tests. If not satisfied that the chosen distribution is a good approximation of the data, then the analyst returns to the second step, chooses a different family of distributions, and repeats the procedure. If several iterations of this procedure fail to yield a fit between an assumed distributional form and the collected data, the empirical form of the distribution may be used, as was described in Section 8.1.5.

Each of these steps is discussed in this chapter. Although software is now widely available to accomplish Steps 2, 3, and 4—including such stand-alone programs as ExpertFit® and Stat::Fit® and such integrated programs as Arena's Input Processor and @Risk's BestFit®—it is still important to understand what the software does, so that it can be used appropriately. Unfortunately, software is not as readily available for input modeling when there is a relationship between two or more variables of interest or when no data are available. These two topics are discussed toward the end of the chapter.

9.1 DATA COLLECTION

Problems are found at the end of each chapter, as exercises for the reader, in textbooks about mathematics, physics, chemistry, and other technical subjects. Years and years of working these problems could give the reader the impression that data are readily available. Nothing could be further from the truth. Data collection is one of the biggest tasks in solving a real problem. It is one of the most important and difficult problems in simulation. And, even when data are available, they have rarely been recorded in a form that is directly useful for simulation input modeling.

"GIGO," or "garbage-in–garbage-out," is a basic concept in computer science, and it applies equally in the area of discrete-system simulation. Even when the model structure is valid, if the input data are inaccurately collected, inappropriately analyzed, or not representative of the environment, the simulation output data will be misleading and possibly damaging or costly when used for policy or decision making.

Example 9.1:   The Laundromat ————————————————————————————————
As budding simulation students, the first two authors had assignments to simulate the operation of an ongoing system. One of these systems, which seemed to be a rather simple operation, was a self-service laundromat with 10 washing machines and six dryers.

However, the data-collection aspect of the problem rapidly became rather enormous. The interarrival-time distribution was not homogeneous; it changed by time of day and by day of week. The laundromat was open 7 days a week for 16 hours per day, or 112 hours per week. It would have been impossible to cover the operation of the laundromat with the limited resources available (two students who were also taking four other courses) and with a tight time constraint (the simulation was to be completed in a 4-week period). Additionally, the distribution of time between arrivals during one week might not have been followed during the next week. As a compromise, a sample of times was selected, and the interarrival-time distributions were classified according to arrival rate (perhaps inappropriately) as "high," "medium," and "low."

Service-time distributions also presented a difficult problem from many perspectives. The proportion of customers demanding the various service combinations had to be observed and recorded. The simplest case was the customer desiring one washer followed by one dryer. However, a customer might choose two washing machines followed by one dryer, one dryer only, and so on. The customers used numbered machines, and it was possible to follow the customers via that reference, rather than remembering them by personal characteristics. Because of the dependence between washer demand and dryer demand for an individual customer, it would have been inappropriate to treat the service times for washers and dryers separately as independent variables.

Some customers waited patiently for their clothes to complete the washing or drying cycle, and then they removed their clothes promptly. Others left the premises and returned after their clothes had finished their cycle on the machine being used. In a very busy period, the manager would remove a customer's clothes after the cycle and set them aside in a basket. It was decided that service termination would be measured as that point in time at which the machine was emptied of its contents.

Also, machines would break down from time to time. The length of the breakdown varied from a few moments, when the manager repaired the machine, to several days (a breakdown on Friday night, requiring a part not in the laundromat storeroom, would not be fixed until the following Monday). The short-term repair times were recorded by the student team. The long-term repair completion times were estimated by the manager. Breakdowns then became part of the simulation.

Many lessons can be learned from an actual experience at data collection. The first five exercises at the end of this chapter suggest some situations in which the student can gain such experience.

The following suggestions might enhance and facilitate data collection, although they are not all inclusive.

1. A useful expenditure of time is in planning. This could begin by a practice or preobserving session. Try to collect data while preobserving. Devise forms for this purpose. It is very likely that these forms will have to be modified several times before the actual data collection begins. Watch for unusual circumstances, and consider how they will be handled. When possible, videotape the system and extract the data later by viewing the tape. Planning is important, even if data will be collected automatically (e.g., via computer data collection), to ensure that the appropriate data are available. When data have already been collected by someone else, be sure to allow plenty of time for converting the data into a usable format.

2. Try to analyze the data as they are being collected. Figure out whether the data being collected are adequate to provide the distributions needed as input to the simulation. Find out whether any data being collected are useless to the simulation. There is no need to collect superfluous data.

3. Try to combine homogeneous data sets. Check data for homogeneity in successive time periods and during the same time period on successive days. For example, check for homogeneity of data from 2:00 P.M. to 3:00 P.M. and 3:00 P.M. to 4:00 P.M., and check

to see whether the data are homogeneous for 2:00 P.M. to 3:00 P.M. on Thursday and Friday. When checking for homogeneity, an initial test is to see whether the means of the distributions (the average interarrival times, for example) are the same. The two-sample t test can be used for this purpose. A more thorough analysis would require a test of the equivalence of the distributions, perhaps via a quantile–quantile plot (described later).

4. Be aware of the possibility of data censoring, in which a quantity of interest is not observed in its entirety. This problem most often occurs when the analyst is interested in the time required to complete some process (for example, produce a part, treat a patient, or have a component fail), but the process begins prior to, or finishes after the completion of, the observation period. Censoring can result in especially long process times being left out of the data sample.

5. To discover whether there is a relationship between two variables, build a scatter diagram. Sometimes an eyeball scan of the scatter diagram will indicate whether there is a relationship between two variables of interest. Section 9.7 describes models for statistically dependent input data.

6. Consider the possibility that a sequence of observations that appear to be independent actually has autocorrelation. Autocorrelation can exist in successive time periods or for successive customers. For example, the service time for the ith customer could be related to the service time for the $(i + n)$th customer. A brief introduction to autocorrelation was provided in Section 7.4.2, and some input models that account for autocorrelation are presented in Section 9.7.

7. Keep in mind the difference between input data and output or performance data, and be sure to collect input data. Input data typically represent the uncertain quantities that are largely beyond the control of the system and will not be altered by changes made to improve the system. Output data, on the other hand, represent the performance of the system when subjected to the inputs, performance that we might be trying to improve. In a queueing simulation, the customer arrival times are usually inputs, whereas the customer delay is an output. Performance data are useful for model validation, however—see Chapter 10.

Again, these are just a few suggestions. As a rule, data collection and analysis must be approached with great care.

9.2 IDENTIFYING THE DISTRIBUTION WITH DATA

In this section, we discuss methods for selecting families of input distributions when data are available. The specific distribution within a family is specified by estimating its parameters, as described in Section 9.3. Section 9.6 takes up the case in which data are unavailable.

9.2.1 Histograms

A frequency distribution or histogram is useful in identifying the shape of a distribution. A histogram is constructed as follows:

1. Divide the range of the data into intervals. (Intervals are usually of equal width; however, unequal widths may be used if the heights of the frequencies are adjusted.)

2. Label the horizontal axis to conform to the intervals selected.
3. Find the frequency of occurrences within each interval.
4. Label the vertical axis so that the total occurrences can be plotted for each interval.
5. Plot the frequencies on the vertical axis.

The number of class intervals depends on the number of observations and on the amount of scatter or dispersion in the data. Hines, Montgomery, Goldsman, and Borrow [2002] state that choosing the number of class intervals approximately equal to the square root of the sample size often works well in practice. If the intervals are too wide, the histogram will be coarse, or blocky, and its shape and other details will not show well. If the intervals are too narrow, the histogram will be ragged and will not smooth the data. Examples of ragged, coarse, and appropriate histograms of the same data are shown in Figure 9.1. Modern data-analysis software often allows the interval sizes to be changed easily and interactively until a good choice is found.

The histogram for continuous data corresponds to the probability density function of a theoretical distribution. If continuous, a line drawn through the center point of each class interval frequency should result in a shape like that of a pdf.

Histograms for discrete data, where there are a large number of data points, should have a cell for each value in the range of the data. However, if there are few data points, it could be necessary to combine adjacent cells to eliminate the ragged appearance of the histogram. If the histogram is associated with discrete data, it should look like a probability mass function.

Example 9.2:   Discrete Data _____

The number of vehicles arriving at the northwest corner of an intersection in a 5-minute period between 7:00 A.M. and 7:05 A.M. was monitored for five workdays over a 20-week period. Table 9.1 shows the resulting data. The first entry in the table indicates that there were 12 5-minute periods during which zero vehicles arrived, 10 periods during which one vehicle arrived, and so on.

The number of automobiles is a discrete variable, and there are ample data, so the histogram may have a cell for each possible value in the range of the data. The resulting histogram is shown in Figure 9.2.

Example 9.3:   Continuous Data _____

Life tests were performed on a random sample of electronic components at 1.5 times the nominal voltage, and their lifetime (or time to failure), in days, was recorded:

79.919	3.081	0.062	1.961	5.845
3.027	6.505	0.021	0.013	0.123
6.769	59.899	1.192	34.760	5.009
18.387	0.141	43.565	24.420	0.433
144.695	2.663	17.967	0.091	9.003
0.941	0.878	3.371	2.157	7.579
0.624	5.380	3.148	7.078	23.960
0.590	1.928	0.300	0.002	0.543
7.004	31.764	1.005	1.147	0.219
3.217	14.382	1.008	2.336	4.562

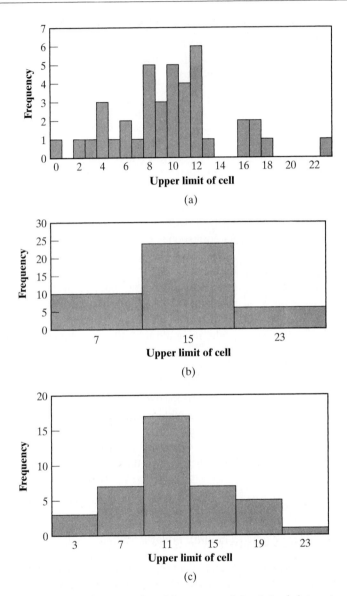

(a)

(b)

(c)

Figure 9.1   Ragged, coarse, and appropriate histograms: (a) original data---too ragged; (b) combining adjacent cells—too coarse; (c) combining adjacent cells—appropriate.

Lifetime, usually considered a continuous variable, is recorded here to three-decimal-place accuracy. The histogram is prepared by placing the data in class intervals. The range of the data is rather large, from 0.002 day to 144.695 days. However, most of the values (30 of 50) are in the zero-to-5-day range. Using intervals of width three results in Table 9.2. The data of Table 9.2 are then used to prepare the histogram shown in Figure 9.3.

Table 9.1 Number of Arrivals in a 5-Minute Period

Arrivals per Period	Frequency	Arrivals per Period	Frequency
0	12	6	7
1	10	7	5
2	19	8	5
3	17	9	3
4	10	10	3
5	8	11	1

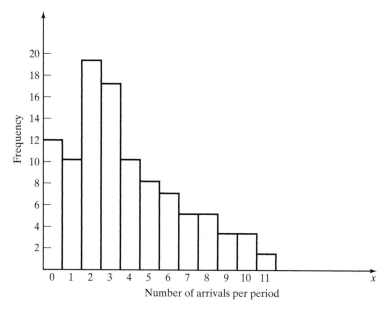

Figure 9.2 Histogram of number of arrivals per period.

9.2.2 Selecting the Family of Distributions

In Chapter 5, some distributions that arise often in simulation were described. Additionally, the shapes of these distributions were displayed. The purpose of preparing a histogram is to infer a known pdf or pmf. A family of distributions is selected on the basis of what might arise in the context being investigated along with the shape of the histogram. Thus, if interarrival-time data have been collected, and the histogram has a shape similar to the pdf in Figure 5.9, the assumption of an exponential distribution would be warranted. Similarly, if measurements of the weights of pallets of freight are being made, and the histogram appears symmetric about the mean with a shape like that shown in Figure 5.11, the assumption of a normal distribution would be warranted.

The exponential, normal, and Poisson distributions are frequently encountered and are not difficult to analyze from a computational standpoint. Although more difficult to analyze, the beta,

Table 9.2 Electronic Component
Data

Component Life (Days)	Frequency
$0 \leq x_j < 3$	23
$3 \leq x_j < 6$	10
$6 \leq x_j < 9$	5
$9 \leq x_j < 12$	1
$12 \leq x_j < 15$	1
$15 \leq x_j < 18$	2
$18 \leq x_j < 21$	0
$21 \leq x_j < 24$	1
$24 \leq x_j < 27$	1
$27 \leq x_j < 30$	0
$30 \leq x_j < 33$	1
$33 \leq x_j < 36$	1
.	.
.	.
.	.
$42 \leq x_j < 45$	1
.	.
.	.
$57 \leq x_j < 60$	1
.	.
.	.
$78 \leq x_j < 81$	1
.	.
.	.
$144 \leq x_j < 147$	1

gamma, and Weibull distributions provide a wide array of shapes and should not be overlooked during modeling of an underlying probabilistic process. Perhaps an exponential distribution was assumed, but it was found not to fit the data. The next step would be to examine where the lack of fit occurred. If the lack of fit was in one of the tails of the distribution, perhaps a gamma or Weibull distribution would fit the data more adequately.

There are literally hundreds of probability distributions that have been created; many were created with some specific physical process in mind. One aid to selecting distributions is to use the physical basis of the distributions as a guide. Here are some examples:

Binomial: Models the number of successes in n trials, when the trials are independent with common success probability, p; for example, the number of defective computer chips found in a lot of n chips.

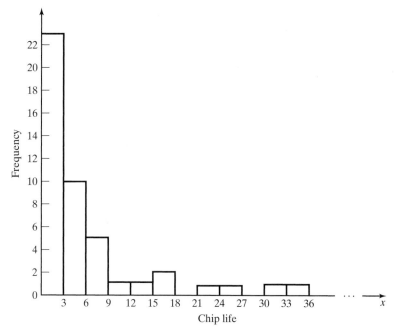

Figure 9.3   Histogram of component life.

Negative Binomial (includes the geometric distribution):   Models the number of trials required to achieve k successes; for example, the number of computer chips that we must inspect to find 4 defective chips.

Poisson: Models the number of independent events that occur in a fixed amount of time or space; for example, the number of customers that arrive to a store during 1 hour, or the number of defects found in 30 square meters of sheet metal.

Normal: Models the distribution of a process that can be thought of as the sum of a number of component processes; for example, a time to assemble a product that is the sum of the times required for each assembly operation. Notice that the normal distribution admits negative values, which could be impossible for process times.

Lognormal: Models the distribution of a process that can be thought of as the product of (meaning to multiply together) a number of component processes—for example, the rate on an investment, when interest is compounded, is the product of the returns for a number of periods.

Exponential: Models the time between independent events, or a process time that is memoryless (knowing how much time has passed gives no information about how much additional time will pass before the process is complete)—for example, the times between the arrivals from a large population of potential customers who act independently of each other. The exponential is a highly variable distribution; it is sometimes overused, because it often leads to mathematically tractable models. Recall that, if the time between events is exponentially distributed, then the number of events in a fixed period of time is Poisson.

Gamma: An extremely flexible distribution used to model nonnegative random variables. The gamma can be shifted away from 0 by adding a constant.

Beta: An extremely flexible distribution used to model bounded (fixed upper and lower limits) random variables. The beta can be shifted away from 0 by adding a constant and can be given a range larger than [0, 1] by multiplying by a constant.

Erlang: Models processes that can be viewed as the sum of several exponentially distributed processes—for example, a computer network fails when a computer and two backup computers fail, and each has a time to failure that is exponentially distributed. The Erlang is a special case of the gamma.

Weibull: Models the time to failure for components—for example, the time to failure for a disk drive. The exponential is a special case of the Weibull.

Discrete or Continuous Uniform: Models complete uncertainty: All outcomes are equally likely. This distribution often is used inappropriately, when there are no data.

Triangular: Models a process for which only the minimum, most likely, and maximum values of the distribution are known; for example, the minimum, most likely, and maximum time required to test a product. This model is often a marked improvement over a uniform distribution.

Empirical: Resamples from the actual data collected; often used when no theoretical distribution seems appropriate.

Do not ignore physical characteristics of the process when selecting distributions. Is the process naturally discrete or continuous valued? Is it bounded, or is there no natural bound? This knowledge, which does not depend on data, can help narrow the family of distributions from which to choose. And keep in mind that there is no "true" distribution for any stochastic input process. An input model is an approximation of reality, so the goal is to obtain an approximation that yields useful results from the simulation experiment.

The reader is encouraged to complete Exercises 6 through 11 to learn more about the shapes of the distributions mentioned in this section. Examining the variations in shape as the parameters change is very instructive.

9.2.3 Quantile–Quantile Plots

The construction of histograms, as discussed in Section 9.2.1, and the recognition of a distributional shape, as discussed in Section 9.2.2, are necessary ingredients for selecting a family of distributions to represent a sample of data. However, a histogram is not as useful for evaluating the *fit* of the chosen distribution. When there is a small number of data points, say 30 or fewer, a histogram can be rather ragged. Further, our perception of the fit depends on the widths of the histogram intervals. But, even if the intervals are chosen well, grouping data into cells makes it difficult to compare a histogram to a continuous probability density function. A quantile–quantile $(q-q)$ plot is a useful tool for evaluating distribution fit, one that does not suffer from these problems.

If X is a random variable with cdf F, then the q-quantile of X is that value γ such that $F(\gamma) = P(X \leq \gamma) = q$, for $0 < q < 1$. When F has an inverse, we write $\gamma = F^{-1}(q)$.

Now let $\{x_i, i = 1, 2, \ldots, n\}$ be a sample of data from X. Order the observations from the smallest to the largest, and denote these as $\{y_j, j = 1, 2, \ldots n\}$, where $y_1 \leq y_2 \leq \cdots \leq y_n$. Let j

denote the ranking or order number. Therefore, $j = 1$ for the smallest and $j = n$ for the largest. The q–q plot is based on the fact that y_j is an estimate of the $(j - 1/2)/n$ quantile of X. In other words,

$$y_j \text{ is approximately } F^{-1}\left(\frac{j - \frac{1}{2}}{n}\right)$$

Now suppose that we have chosen a distribution with cdf F as a possible representation of the distribution of X. If F is a member of an appropriate family of distributions, then a plot of y_j versus $F^{-1}((j - 1/2)/n)$ will be *approximately a straight line*. If F is from an appropriate family of distributions and also has appropriate parameter values, then the line will have slope 1. On the other hand, if the assumed distribution is inappropriate, the points will deviate from a straight line, usually in a systematic manner. The decision about whether to reject some hypothesized model is subjective.

Example 9.4:   Normal Q–Q Plot

A robot is used to install the doors on automobiles along an assembly line. It was thought that the installation times followed a normal distribution. The robot is capable of measuring installation times accurately. A sample of 20 installation times was automatically taken by the robot, with the following results, where the values are in seconds:

99.79	99.56	100.17	100.33
100.26	100.41	99.98	99.83
100.23	100.27	100.02	100.47
99.55	99.62	99.65	99.82
99.96	99.90	100.06	99.85

The sample mean is 99.99 seconds, and the sample variance is $(0.2832)^2$ seconds^2. These values can serve as the parameter estimates for the mean and variance of the normal distribution. The observations are now ordered from smallest to largest as follows:

j	Value	j	Value	j	Value	j	Value
1	99.55	6	99.82	11	99.98	16	100.26
2	99.56	7	99.83	12	100.02	17	100.27
3	99.62	8	99.85	13	100.06	18	100.33
4	99.65	9	99.90	14	100.17	19	100.41
5	99.79	10	99.96	15	100.23	20	100.47

The ordered observations are then plotted versus $F^{-1}((j - 1/2)/20)$, for $j = 1, 2, \ldots, 20$, where F is the cdf of the normal distribution with mean 99.99 and variance $(0.2832)^2$, to obtain a q–q plot. The plotted values are shown in Figure 9.4, along with a histogram of the data that has the density function of the normal distribution superimposed. Notice that it is difficult to tell whether the data are well represented by a normal distribution from looking at the histogram, but the

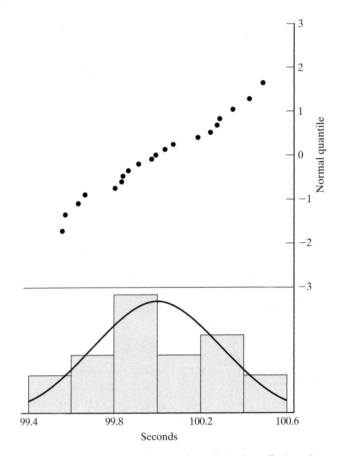

Figure 9.4   Histogram and $q-q$ plot of the installation times.

general perception of a straight line is quite clear in the $q-q$ plot and supports the hypothesis of a normal distribution.

In the evaluation of the linearity of a $q-q$ plot, the following should be considered:

1. The observed values will never fall exactly on a straight line.
2. The ordered values are not independent; they have been ranked. Hence, if one point is above a straight line, it is likely that the next point will also lie above the line. And it is unlikely that the points will be scattered about the line.
3. The variances of the extremes (largest and smallest values) are much higher than the variances in the middle of the plot. Greater discrepancies can be accepted at the extremes. The linearity of the points in the middle of the plot is more important than the linearity at the extremes.

Modern data-analysis software often includes tools for generating $q-q$ plots, especially for the normal distribution. The $q-q$ plot can also be used to compare two samples of data to see whether they can be represented by the same distribution (that is, that they are homogeneous). If $x_1, x_2, \ldots, x_n$ are a sample of the random variable X, and $z_1, z_2, \ldots, z_n$ are a sample of the random variable Z, then plotting the ordered values of X versus the ordered values of Z will reveal approximately a straight line if both samples are well represented by the same distribution (Chambers, Cleveland, and Tukey [1983]).

9.3 PARAMETER ESTIMATION

After a family of distributions has been selected, the next step is to estimate the parameters of the distribution. Estimators for many useful distributions are described in this section. In addition, many software packages—some of them integrated into simulation languages—are now available to compute these estimates.

9.3.1 Preliminary Statistics: Sample Mean and Sample Variance

In a number of instances, the sample mean, or the sample mean and sample variance, are used to estimate the parameters of a hypothesized distribution; see Example 9.4. In the following paragraphs, three sets of equations are given for computing the sample mean and sample variance. Equations (9.1) and (9.2) can be used when discrete or continuous raw data are available. Equations (9.3) and (9.4) are used when the data are discrete and have been grouped in a frequency distribution. Equations (9.5) and (9.6) are used when the data are discrete or continuous and have been placed in class intervals. Equations (9.5) and (9.6) are approximations and should be used only when the raw data are unavailable.

If the observations in a sample of size n are $X_1, X_2, \ldots, X_n$, the sample mean $(\bar{X})$ is defined by

$$\bar{X} = \frac{\sum_{i=1}^{n} X_i}{n} \tag{9.1}$$

and the sample variance, S^2, is defined by

$$S^2 = \frac{\sum_{i=1}^{n} X_i^2 - n\bar{X}^2}{n - 1} \tag{9.2}$$

If the data are discrete and have been grouped in a frequency distribution, Equations (9.1) and (9.2) can be modified to provide for much greater computational efficiency. The sample mean can be computed as

$$\bar{X} = \frac{\sum_{j=1}^{k} f_j X_j}{n} \tag{9.3}$$

and the sample variance as

$$S^2 = \frac{\sum_{j=1}^{k} f_j X_j^2 - n\bar{X}^2}{n - 1} \tag{9.4}$$

where k is the number of distinct values of X and f_j is the observed frequency of the value X_j of X.

Example 9.5:   Grouped Data ————————————————————————————————

The data in Table 9.1 can be analyzed to obtain $n = 100$, $f_1 = 12$, $X_1 = 0$, $f_2 = 10$, $X_2 = 1, \dots$, $\sum_{j=1}^{k} f_j X_j = 364$, and $\sum_{j=1}^{k} = f_j X_j^2 = 2080$. From Equation (9.3),

$$\bar{X} = \frac{364}{100} = 3.64$$

and, from Equation (9.4),

$$S^2 = \frac{2080 - 100(3.64)^2}{99} = 7.63$$

The sample standard deviation, S, is just the square root of the sample variance. In this case, $S = \sqrt{7.63} = 2.76$. Equations (9.1) and (9.2) would have yielded exactly the same results for $\bar{X}$ and S^2.

It is preferable to use the raw data, if possible, when the values are continuous. However, data sometimes are received after having been placed in class intervals. Then it is no longer possible to obtain the exact sample mean and variance. In such cases, the sample mean and sample variance are approximated from the following equations:

$$\bar{X} \doteq \frac{\sum_{j=1}^{c} f_j m_j}{n} \tag{9.5}$$

and

$$S^2 \doteq \frac{\sum_{j=1}^{c} f_j m_j^2 - n\bar{X}^2}{n - 1} \tag{9.6}$$

where f_j is the observed frequency in the jth class interval, m_j is the midpoint of the jth interval, and c is the number of class intervals.

Example 9.6:   Continuous Data in Class Intervals ————————————————————

Assume that the raw data on component life shown in Example 9.3 either was discarded or was lost. However, the data shown in Table 9.2 are still available. To approximate values for $\bar{X}$ and S^2, Equations (9.5) and (9.6) are used. The following values are created: $f_1 = 23$, $m_1 = 1.5$, $f_2 = 10$, $m_2 = 4.5, \dots$, $\sum_{j=1}^{49} f_j m_j = 614$ and $\sum_{j=1}^{49} f_j m_j^2 = 37{,}226.5$. With $n = 50$, $\bar{X}$ is approximated from Equation (9.5) as

$$\bar{X} \doteq \frac{614}{50} = 12.28$$

Then, S^2 is approximated from Equation (9.6) as

$$S^2 \doteq \frac{37{,}226.5 - 50(12.28)^2}{49} = 605.849$$

and

$$S \doteq 24.614$$

Applying Equations (9.1) and (9.2) to the original data in Example 9.3 results in $\bar{X} = 11.894$ and $S = 24.953$. Thus, when the raw data are either discarded or lost, inaccuracies could result.

9.3.2 Suggested Estimators

Numerical estimates of the distribution parameters are needed to reduce the family of distributions to a specific distribution and to test the resulting hypothesis. Table 9.3 contains suggested estimators for distributions often used in simulation, all of which were described in Chapter 5. Except for an adjustment to remove bias in the estimate of σ^2 for the normal distribution, these estimators

Table 9.3 Suggested Estimators for Distributions Often Used in Simulation

Distribution	Parameter(s)	Suggested Estimator(s)
Poisson	α	$\widehat{\alpha} = \bar{X}$
Exponential	λ	$\widehat{\lambda} = \dfrac{1}{\bar{X}}$
Gamma	β, θ	$\widehat{\beta}$ (see Table A.9)
		$\widehat{\theta} = \dfrac{1}{\bar{X}}$
Normal	μ, σ^2	$\widehat{\mu} = \bar{X}$
		$\widehat{\sigma}^2 = S^2$ (unbiased)
Lognormal	μ, σ^2	$\widehat{\mu} = \bar{X}$ (after taking ln of the data)
		$\widehat{\sigma}^2 = S^2$ (after taking ln of the data)
Weibull with $\nu = 0$	α, β	$\widehat{\beta}_0 = \dfrac{\bar{X}}{S}$
		$\widehat{\beta}_j = \widehat{\beta}_{j-1} - \dfrac{f(\widehat{\beta}_{j-1})}{f'(\widehat{\beta}_{j-1})}$
		See Equations (9.12) and (9.15) for $f(\widehat{\beta})$ and $f'(\widehat{\beta})$
		Iterate until convergence
		$\widehat{\alpha} = \left(\dfrac{1}{n} \sum_{i=1}^{n} X_i^{\widehat{\beta}} \right)^{1/\widehat{\beta}}$
Beta	β_1, β_2	$\Psi(\widehat{\beta}_1) + \Psi(\widehat{\beta}_1 - \widehat{\beta}_2) = \ln(G_1)$
		$\Psi(\widehat{\beta}_2) + \Psi(\widehat{\beta}_1 - \widehat{\beta}_2) = \ln(G_2)$
		where Ψ is the digamma function,
		$G_1 = \left(\prod_{i=1}^{n} X_i \right)^{1/n}$ and
		$G_2 = \left(\prod_{i=1}^{n} (1 - X_i) \right)^{1/n}$

are the maximum-likelihood estimators based on the raw data. (If the data are in class intervals, these estimators must be modified.) The reader is referred to Fishman [1973] and Law and Kelton [2000] for parameter estimates for the uniform, binomial, and negative binomial distributions. The triangular distribution is usually employed when no data are available, with the parameters obtained from educated guesses for the minimum, most likely, and maximum possible values; the uniform distribution may also be used in this way if only minimum and maximum values are available.

Examples of the use of the estimators are given in the following paragraphs. The reader should keep in mind that a parameter is an unknown constant, but the estimator is a statistic (or random variable), because it depends on the sample values. To distinguish the two clearly here, if, say, a parameter is denoted by α, the estimator will be denoted by $\widehat{\alpha}$.

Example 9.7:   Poisson Distribution

Assume that the arrival data in Table 9.1 require analysis. By comparison with Figure 5.7, an examination of Figure 9.2 suggests a Poisson distributional assumption with unknown parameter α. From Table 9.3, the estimator of α is $\bar{X}$, which was found in Example 9.5. Thus, $\widehat{\alpha} = 3.64$. Recall that the true mean and variance are equal for the Poisson distribution. In Example 9.5, the sample variance was estimated as $S^2 = 7.63$. However, it should never be expected that the sample mean and the sample variance will be precisely equal, because each is a random variable.

Example 9.8:   Lognormal Distribution

The rates of return on 10 investments in a portfolio are 18.8, 27.9, 21.0, 6.1, 37.4, 5.0, 22.9, 1.0, 3.1 and 8.3 percent. To estimate the parameters of a lognormal model of these data, we first take the natural log of the data and obtain 2.9, 3.3, 3.0, 1.8, 3.6, 1.6, 3.1, 0, 1.1, and 2.1. Then we set $\widehat{\mu} = \bar{X} = 2.3$ and $\widehat{\sigma}^2 = S^2 = 1.3$.

Example 9.9:   Normal Distribution

The parameters of the normal distribution, μ and σ^2, are estimated by $\bar{X}$ and S^2, as shown in Table 9.3. The $q-q$ plot in Example 9.4 leads to a distributional assumption that the installation times are normal. From Equations (9.1) and (9.2), the data in Example 9.4 yield $\widehat{\mu} = \bar{X} = 99.9865$ and $\widehat{\sigma}^2 = S^2 = (0.2832)^2$ second^2.

Example 9.10:   Gamma Distribution

The estimator $\widehat{\beta}$ for the gamma distribution is chosen by the use of Table A.9, from Choi and Wette [1969]. Table A.9 requires the computation of the quantity $1/M$, where

$$M = \ln \bar{X} - \frac{1}{n} \sum_{i=1}^{n} \ln X_i \tag{9.7}$$

Also, it can be seen in Table 9.3 that $\widehat{\theta}$ is given by

$$\widehat{\theta} = \frac{1}{\bar{X}} \tag{9.8}$$

In Chapter 5, it was stated that lead time is often gamma distributed. Suppose that the lead times (in days) associated with 20 orders have been accurately measured as follows:

Order	Lead Time (Days)	Order	Lead Time (Days)
1	70.292	11	30.215
2	10.107	12	17.137
3	48.386	13	44.024
4	20.480	14	10.552
5	13.053	15	37.298
6	25.292	16	16.314
7	14.713	17	28.073
8	39.166	18	39.019
9	17.421	19	32.330
10	13.905	20	36.547

To estimate $\widehat{\beta}$ and $\widehat{\theta}$, it is first necessary to compute M from Equation (9.7). Here, $\bar{X}$ is found, from Equation (9.1), to be

$$\bar{X} = \frac{564.32}{20} = 28.22$$

Then,

$$\ln \bar{X} = 3.34$$

Next,

$$\sum_{i=1}^{20} \ln X_i = 63.99$$

Then,

$$M = 3.34 - \frac{63.99}{20} = 0.14$$

and

$$1/M = 7.14$$

By interpolation in Table A.9, $\widehat{\beta} = 3.728$. Finally, Equation (9.8) results in

$$\widehat{\theta} = \frac{1}{28.22} = 0.035$$

Example 9.11:   **Exponential Distribution**

Assuming that the data in Example 9.3 come from an exponential distribution, the parameter estimate, $\widehat{\lambda}$, can be determined. In Table 9.3, $\widehat{\lambda}$ is obtained from $\bar{X}$ as follows:

$$\widehat{\lambda} = \frac{1}{\bar{X}} = \frac{1}{11.894} = 0.084 \text{ per day}$$

Example 9.12:   **Weibull Distribution**

Suppose that a random sample of size n, $X_1, X_2, \ldots, X_n$, has been taken and that the observations are assumed to come from a Weibull distribution. The likelihood function derived by using the pdf given by Equation (5.47) can be shown to be

$$L(\alpha, \beta) = \frac{\beta^n}{\alpha^{\beta n}} \left[\prod_{i=1}^{n} X_i^{(\beta-1)} \right] \exp\left[-\sum_{i=1}^{n} \left(\frac{X_i}{\alpha} \right)^{\beta} \right] \tag{9.9}$$

The maximum-likelihood estimates are those values of $\widehat{\alpha}$ and $\widehat{\beta}$ that maximize $L(\alpha, \beta)$ or, equivalently, maximize $\ln L(\alpha, \beta)$, denoted by $l(\alpha, \beta)$. The maximum value of $l(\alpha, \beta)$ is obtained by taking the partial derivatives $\partial l(\alpha, \beta)/\partial \alpha$ and $\partial l(\alpha, \beta)/\partial \beta$, setting each to zero, and solving the resulting equations, which after substitution become

$$f(\beta) = 0 \tag{9.10}$$

and

$$\alpha = \left(\frac{1}{n} \sum_{i=1}^{n} X_i^{\beta} \right)^{1/\beta} \tag{9.11}$$

where

$$f(\beta) = \frac{n}{\beta} + \sum_{i=1}^{n} \ln X_i - \frac{n \sum_{i=1}^{n} X_i^{\beta} \ln X_i}{\sum_{i=1}^{n} X_i^{\beta}} \tag{9.12}$$

The maximum-likelihood estimates, $\widehat{\alpha}$ and $\widehat{\beta}$, are the solutions of Equations (9.10) and (9.11). First, $\widehat{\beta}$ is found via the iterative procedure explained shortly. Then $\widehat{\alpha}$ is found from Equation (9.11), with $\beta = \widehat{\beta}$.

Equation (9.10) is nonlinear, so it is necessary to use a numerical-analysis technique to solve it. In Table 9.3, an iterative method for computing $\widehat{\beta}$ is given as

$$\widehat{\beta}_j = \widehat{\beta}_{j-1} - \frac{f(\widehat{\beta}_{j-1})}{f'(\widehat{\beta}_{j-1})} \tag{9.13}$$

Equation (9.13) employs Newton's method in reaching $\widehat{\beta}$, where $\widehat{\beta}_j$ is the jth iteration, beginning with an initial estimate for $\widehat{\beta}_0$, given in Table 9.3, as follows:

$$\widehat{\beta}_0 = \frac{\bar{X}}{S} \tag{9.14}$$

If the initial estimate, $\widehat{\beta}_0$, is sufficiently close to the solution $\widehat{\beta}$, then $\widehat{\beta}_j$ approaches $\widehat{\beta}$ as $j \longrightarrow \infty$. In Newton's method, $\widehat{\beta}$ is approached through increments of size $f(\widehat{\beta}_{j-1})/f'(\widehat{\beta}_{j-1})$.

Equation (9.12) is used to compute $f(\widehat{\beta}_{j-1})$ and Equation (9.15) is used to compute $f'(\widehat{\beta}_{j-1})$, as follows:

$$f'(\beta) = -\frac{n}{\beta^2} - \frac{n \sum_{i=1}^{n} X_i^{\beta} (\ln X_i)^2}{\sum_{i=1}^{n} X_i^{\beta}} + \frac{n \left(\sum_{i=1}^{n} X_i^{\beta} \ln X_i \right)^2}{\left(\sum_{i=1}^{n} X_i^{\beta} \right)^2} \tag{9.15}$$

Equation (9.15) can be derived from Equation (9.12) by differentiating $f(\beta)$ with respect to β. The iterative process continues until $f(\widehat{\beta}_j) \doteq 0$, for example, until $|f(\widehat{\beta}_j)| \leq 0.001$.

Consider the data given in Example 9.3. These data concern the failure of electronic components and looks to come from an exponential distribution. In Example 9.11, the parameter $\widehat{\lambda}$ was estimated on the hypothesis that the data were from an exponential distribution. If the hypothesis that the data came from an exponential distribution is rejected, an alternative hypothesis is that the data come from a Weibull distribution. The Weibull distribution is suspected because the data pertain to electronic component failures, which occur suddenly.

Equation (9.14) is used to compute $\widehat{\beta}_0$. For the data in Example 9.3, $n = 50$, $\bar{X} = 11.894$, $\bar{X}^2 = 141.467$, and $\sum_{i=1}^{50} X_i^2 = 37,575.850$; so S^2 is found by Equation (9.2) to be

$$S^2 = \frac{37,578.850 - 50(141.467)}{49} = 622.650$$

and $S = 24.953$. Thus,

$$\widehat{\beta}_0 = \frac{11.894}{24.953} = 0.477$$

To compute $\widehat{\beta}_1$ by using Equation (9.13) requires the calculation of $f(\widehat{\beta}_0)$ and $f'(\widehat{\beta}_0)$ from Equations (9.12) and (9.15). The following additional values are needed: $\sum_{i=1}^{50} X_i^{\widehat{\beta}_0} = 115.125$, $\sum_{i=1}^{50} \ln X_i = 38.294$, $\sum_{i=1}^{50} X_i^{\widehat{\beta}_0} \ln X_i = 292.629$, and $\sum_{i=1}^{50} X_i^{\widehat{\beta}_0} (\ln X_i)^2 = 1057.781$. Thus,

$$f(\widehat{\beta}_0) = \frac{50}{0.477} + 38.294 - \frac{50(292.629)}{115.125} = 16.024$$

and

$$f'(\widehat{\beta}_0) = \frac{-50}{(0.477)^2} - \frac{50(1057.781)}{115.125} + \frac{50(292.629)^2}{(115.125)^2} = -356.110$$

Then, by Equation (9.13),

$$\widehat{\beta}_1 = 0.477 - \frac{16.024}{-356.110} = 0.522$$

After four iterations, $|f(\widehat{\beta}_3)| \leq 0.001$, at which point $\widehat{\beta} \doteq \widehat{\beta}_4 = 0.525$ is the approximate solution to Equation (9.10). Table 9.4 contains the values needed to complete each iteration.

Now, $\widehat{\alpha}$ can be computed from Equation (9.11) with $\beta = \widehat{\beta} = 0.525$, as follows:

$$\widehat{\alpha} = \left[\frac{130.608}{50} \right]^{1/0.525} = 6.227$$

Table 9.4   Iterative Estimation of Parameters of the Weibull Distribution

j	$\widehat{\beta}_j$	$\sum\limits_{i=1}^{50} X_i^{\widehat{\beta}_j}$	$\sum\limits_{i=1}^{50} X_i^{\widehat{\beta}_j} \ln X_i$	$\sum\limits_{i=1}^{50} X_i^{\widehat{\beta}_j} (\ln X_i)^2$	$f(\widehat{\beta}_j)$	$f'(\widehat{\beta}_j)$	$\widehat{\beta}_{j+1}$
0	0.477	115.125	292.629	1057.781	16.024	-356.110	0.522
1	0.522	129.489	344.713	1254.111	1.008	-313.540	0.525
2	0.525	130.603	348.769	1269.547	0.004	-310.853	0.525
3	0.525	130.608	348.786	1269.614	0.000	-310.841	0.525

```
betaMLE := proc(X, n)
       local G1, G2, beta1, beta2, eqns, solns;
       G1 := product(X[i], i=1..n)^(1/n);
       G2 := product(1-X[i],i=1..n)^(1/n);
       eqns := {Psi(beta1) - Psi(beta1 + beta2) = ln(G1),
                Psi(beta2) - Psi(beta1 + beta2) = ln(G2)};
       solns := fsolve(eqns, {beta1=0..infinity, beta2=0..infinity});
       RETURN(solns);
       end;
```

Figure 9.5   Maple procedure to compute the maximum likelihood estimates for the beta distribution parameters.

If $\widehat{\beta}_0$ is sufficiently close to $\widehat{\beta}$, the procedure converges quickly, usually in four to five iterations. However, if the procedure appears to be diverging, try other initial guesses for $\widehat{\beta}_0$—for example, one-half the initial estimate or twice the initial estimate.

The difficult task of estimating parameters for the Weibull distribution by hand emphasizes the value of having software support for input modeling.

Example 9.13:   Beta Distribution
The percentage of customers each month who bring in store coupons must be between 0 and 100 percent. Observations at a store for eight months gave the values 25%, 74%, 20%, 32%, 81%, 47%, 31%, and 8%. To fit a beta distribution to these data, we first need to rescale it to the interval $(0, 1)$ by dividing all the values by 100, to get 0.25, 0.74, 0.20, 0.32, 0.81, 0.47, 0.31, 0.08.

The maximum-likelihood estimators of the parameters β_1, β_2 solve the system of equations shown in Table 9.3. Such equations can be solved by modern symbolic/numerical calculation programs, such as Maple; a Maple procedure for the beta parameters is shown in Figure 9.5. In this case, the solutions are $\widehat{\beta}_1 = 1.47$ and $\widehat{\beta}_2 = 2.16$.

9.4 GOODNESS-OF-FIT TESTS

Hypothesis testing was discussed in Section 7.4 with respect to testing random numbers. In Section 7.4.1, the Kolmogorov–Smirnov test and the chi-square test were introduced. These two tests are applied in this section to hypotheses about distributional forms of input data.

Goodness-of-fit tests provide helpful guidance for evaluating the suitability of a potential input model; however, there is no single correct distribution in a real application, so you should not be a slave to the verdict of such a test. It is especially important to understand the effect of sample size. If very little data are available, then a goodness-of-fit test is unlikely to reject *any* candidate distribution; but if a lot of data are available, then a goodness-of-fit test will likely reject *all* candidate distributions. Therefore, failing to reject a candidate distribution should be taken as one piece of evidence in favor of that choice, and rejecting an input model as only one piece of evidence against the choice.

9.4.1 Chi-Square Test

One procedure for testing the hypothesis that a random sample of size n of the random variable X follows a specific distributional form is the chi-square goodness-of-fit test. This test formalizes the intuitive idea of comparing the histogram of the data to the shape of the candidate density or mass function. The test is valid for large sample sizes and for both discrete and continuous distributional assumptions when parameters are estimated by maximum likelihood. The test procedure begins by arranging the n observations into a set of k class intervals or cells. The test statistic is given by

$$\chi_0^2 = \sum_{i=1}^{k} \frac{(O_i - E_i)^2}{E_i} \tag{9.16}$$

where O_i is the observed frequency in the ith class interval and E_i is the expected frequency in that class interval. The expected frequency for each class interval is computed as $E_i = np_i$, where p_i is the theoretical, hypothesized probability associated with the ith class interval.

It can be shown that χ_0^2 approximately follows the chi-square distribution with $k - s - 1$ degrees of freedom, where s represents the number of parameters of the hypothesized distribution estimated by the sample statistics. The hypotheses are the following:

H_0: The random variable, X, conforms to the distributional assumption with the parameter(s) given by the parameter estimate(s).

H_1: The random variable X does not conform.

The critical value $\chi_{\alpha,k-s-1}^2$ is found in Table A.6. The null hypothesis, H_0, is rejected if $\chi_0^2 > \chi_{\alpha,k-s-1}^2$.

When applying the test, if expected frequencies are too small, χ_0^2 will reflect not only the departure of the observed from the expected frequency, but also the smallness of the expected frequency as well. Although there is no general agreement regarding the minimum size of E_i, values of 3, 4, and 5 have been widely used. In Section 7.4.1, when the chi-square test was discussed, the minimum expected frequency five was suggested. If an E_i value is too small, it can be combined with expected frequencies in adjacent class intervals. The corresponding O_i values should also be combined, and k should be reduced by one for each cell that is combined.

If the distribution being tested is discrete, each value of the random variable should be a class interval, unless it is necessary to combine adjacent class intervals to meet the minimum-expected-cell-frequency requirement. For the discrete case, if combining adjacent cells is not required,

$$p_i = p(x_i) = P(X = x_i)$$

Otherwise, p_i is found by summing the probabilities of appropriate adjacent cells.

If the distribution being tested is continuous, the class intervals are given by $[a_{i-1}, a_i)$, where a_{i-1} and a_i are the endpoints of the ith class interval. For the continuous case with assumed pdf $f(x)$, or assumed cdf $F(x)$, p_i can be computed as

$$p_i = \int_{a_{i-1}}^{a_i} f(x)\, dx = F(a_i) - F(a_{i-1})$$

For the discrete case, the number of class intervals is determined by the number of cells resulting after combining adjacent cells as necessary. However, for the continuous case, the number of class intervals must be specified. Although there are no general rules to be followed, the recommendations in Table 9.5 are made to aid in determining the number of class intervals for continuous data.

Example 9.14: Chi-Square Test Applied to Poisson Assumption

In Example 9.7, the vehicle-arrival data presented in Example 9.2 were analyzed. The histogram of the data, shown in Figure 9.2, appeared to follow a Poisson distribution; hence the parameter, $\widehat{\alpha} = 3.64$, was found. Thus, the following hypotheses are formed:

H_0: the random variable is Poisson distributed.
H_1: the random variable is not Poisson distributed.

The pmf for the Poisson distribution was given in Equation (5.19):

$$p(x) = \begin{cases} \dfrac{e^{-\alpha}\alpha^x}{x!}, & x = 0, 1, 2, \ldots \\ 0, & \text{otherwise} \end{cases} \qquad (9.17)$$

For $\alpha = 3.64$, the probabilities associated with various values of x are obtained from Equation (9.17):

p(0) = 0.026	p(6) = 0.085
p(1) = 0.096	p(7) = 0.044
p(2) = 0.174	p(8) = 0.020
p(3) = 0.211	p(9) = 0.008
p(4) = 0.192	p(10) = 0.003
p(5) = 0.140	p(≥ 11) = 0.001

From this information, Table 9.6 is constructed. The value of E_1 is given by $np_0 = 100(0.026) = 2.6$. In a similar manner, the remaining E_i values are computed. Since $E_1 = 2.6 < 5$, E_1 and E_2 are combined. In that case, O_1 and O_2 are also combined, and k is reduced by one. The last five class intervals are also combined, for the same reason, and k is further reduced by four.

The calculated χ_0^2 is 27.68. The degrees of freedom for the tabulated value of χ^2 is $k - s - 1 = 7 - 1 - 1 = 5$. Here, $s = 1$, since one parameter, $\widehat{\alpha}$ was estimated from the data. At the 0.05

Table 9.5 Recommendations for Number of Class Intervals for Continuous Data

Sample Size, n	Number of Class Intervals, k
20	Do not use the chi-square test
50	5 to 10
100	10 to 20
>100	$\sqrt{n}$ to $n/5$

Table 9.6   Chi-Square Goodness-of-Fit Test for Example 9.14

x_i	Observed Frequency, O_i		Expected Frequency, E_i		$\dfrac{(O_i - E_i)^2}{E_i}$
0	12 } 22		2.6 } 12.2		} 7.87
1	10		9.6		
2	19		17.4		0.15
3	17		21.1		0.80
4	10		19.2		4.41
5	8		14.0		2.57
6	7		8.5		0.26
7	5		4.4		
8	5		2.0		
9	3 } 17		0.8 } 7.6		} 11.62
10	3		0.3		
≥11	1		0.1		
	100		100.0		27.68

level of significance, the critical value $\chi^2_{0.05,5}$ is 11.1. Thus, H_0 would be rejected at level of significance 0.05. The analyst, therefore, might want to search for a better-fitting model or use the empirical distribution of the data.

9.4.2 Chi-Square Test with Equal Probabilities

If a continuous distributional assumption is being tested, class intervals that are equal in probability rather than equal in width of interval should be used. This has been recommended by a number of authors [Mann and Wald, 1942; Gumbel, 1943; Law and Kelton, 2000; Stuart, Ord, and Arnold, 1998]. It should be noted that the procedure is not applicable to data collected in class intervals, where the raw data have been discarded or lost.

Unfortunately, there is as yet no method for figuring out the probability associated with each interval that maximizes the power for a test of a given size. (The power of a test is defined as the probability of rejecting a false hypothesis.) However, if using equal probabilities, then $p_i = 1/k$.

We recommend

$$E_i = np_i \geq 5$$

so substituting for p_i yields

$$\frac{n}{k} \geq 5$$

and solving for k yields

$$k \leq \frac{n}{5} \tag{9.18}$$

Equation (9.18) was used in coming up with the recommendations for maximum number of class intervals in Table 9.5.

If the assumed distribution is normal, exponential, or Weibull, the method described in this section is straightforward. Example 9.15 indicates how the procedure is accomplished for the exponential distribution. If the assumed distribution is gamma (but not Erlang) or certain other distributions, then the computation of endpoints for class intervals is complex and could require numerical integration of the density function. Statistical-analysis software is very helpful in such cases.

Example 9.15:   Chi-Square Test for Exponential Distribution ——————————————
In Example 9.11, the failure data presented in Example 9.3 were analyzed. The histogram of the data, shown in Figure 9.3, appeared to follow an exponential distribution, so the parameter $\hat{\lambda} = 1/\bar{X} = 0.084$ was computed. Thus, the following hypotheses are formed:

H_0: the random variable is exponentially distributed.
H_1: the random variable is not exponentially distributed.

In order to perform the chi-square test with intervals of equal probability, the endpoints of the class intervals must be found. Equation (9.18) indicates that the number of intervals should be less than or equal to $n/5$. Here, $n = 50$, and so $k \leq 10$. In Table 9.5, it is recommended that 7 to 10 class intervals be used. Let $k = 8$; then each interval will have probability $p = 0.125$. The endpoints for each interval are computed from the cdf for the exponential distribution, given in Equation (5.28), as follows:

$$F(a_i) = 1 - e^{-\lambda a_i} \tag{9.19}$$

where a_i represents the endpoint of the ith interval, $i = 1, 2, \ldots, k$. Since $F(a_i)$ is the cumulative area from zero to a_i, $F(a_i) = ip$, so Equation (9.19) can be written as

$$ip = 1 - e^{-\lambda a_i}$$

or

$$e^{-\lambda a_i} = 1 - ip$$

Table 9.7   Chi-Square Goodness-of-Fit Test for Example 9.15

Class Interval	Observed Frequency, O_i	Expected Frequency, E_i	$\dfrac{(O_i - E_i)^2}{E_i}$
[0, 1.590)	19	6.25	26.01
[1.590, 3.425)	10	6.25	2.25
[3.425, 5.595)	3	6.25	0.81
[5.595, 8.252)	6	6.25	0.01
[8.252, 11.677)	1	6.25	4.41
[11.677, 16.503)	1	6.25	4.41
[16.503, 24.755)	4	6.25	0.81
[24.755, ∞)	6	6.25	0.01
	50	50	39.6

Taking the logarithm of both sides and solving for a_i gives a general result for the endpoints of k equiprobable intervals for the exponential distribution:

$$a_i = -\frac{1}{\lambda} \ln(1 - ip), \quad i = 0, 1, \dots, k \tag{9.20}$$

Regardless of the value of λ, Equation (9.20) will always result in $a_0 = 0$ and $a_k = \infty$. With $\widehat{\lambda} = 0.084$ and $k = 8$, a_1 is computed from Equation (9.20) as

$$a_1 = -\frac{1}{0.084} \ln(1 - 0.125) = 1.590$$

Continued application of Equation (9.20) for $i = 2, 3, \dots, 7$ results in $a_2, \dots, a_7$ as 3.425, 5.595, 8.252, 11.677, 16.503, and 24.755. Since $k = 8$, $a_8 = \infty$. The first interval is [0, 1.590), the second interval is [1.590, 3.425), and so on. The expectation is that 0.125 of the observations will fall in each interval. The observations, the expectations, and the contributions to the calculated value of χ_0^2 are shown in Table 9.7. The calculated value of χ_0^2 is 39.6. The degrees of freedom are given by $k - s - 1 = 8 - 1 - 1 = 6$. At $\alpha = 0.05$, the tabulated value of $\chi_{0.05,6}^2$ is 12.6. Since $\chi_0^2 > \chi_{0.05,6}^2$, the null hypothesis is rejected. (The value of $\chi_{0.01,6}^2$ is 16.8, so the null hypothesis would also be rejected at level of significance $\alpha = 0.01$.)

9.4.3 Kolmogorov–Smirnov Goodness-of-Fit Test

The chi-square goodness-of-fit test can accommodate the estimation of parameters from the data with a resultant decrease in the degrees of freedom (one for each parameter estimated). The chi-square test requires that the data be placed in class intervals; in the case of a continuous distributional assumption, this grouping is arbitrary. Changing the number of classes and the interval width affects the value of the calculated and tabulated chi-square. A hypothesis could be accepted when the data are grouped one way, but rejected when they are grouped another way. Also, the distribution of the chi-square test statistic is known only approximately, and the power of the test is sometimes rather low. As a result of these considerations, goodness-of-fit tests other than the chi-square, are desired. The Kolmogorov–Smirnov test formalizes the idea behind examining a q–q plot.

The Kolmogorov–Smirnov test was presented in Section 7.4.1 to test for the uniformity of numbers. Both of these uses fall into the category of testing for goodness of fit. Any continuous distributional assumption can be tested for goodness of fit by using the method of Section 7.4.1.

The Kolmogorov–Smirnov test is particularly useful when sample sizes are small and when no parameters have been estimated from the data. When parameter estimates have been made, the critical values in Table A.8 are biased; in particular, they are too conservative. In this context, "conservative" means that the critical values will be too large, resulting in smaller Type I (α) errors than those specified. The exact value of α can be worked out in some instances, as is discussed at the end of this section.

The Kolmogorov–Smirnov test does not take any special tables when an exponential distribution is assumed. The following example indicates how the test is applied in this instance. (Notice that it is not necessary to estimate the parameter of the distribution in this example, so we may use Table A.8.)

Example 9.16:   Kolmogorov–Smirnov Test for Exponential Distribution _____

Suppose that 50 interarrival times (in minutes) are collected over the following 100-minute interval (arranged in order of occurrence):

0.44	0.53	2.04	2.74	2.00	0.30	2.54	0.52	2.02	1.89	1.53	0.21
2.80	0.04	1.35	8.32	2.34	1.95	0.10	1.42	0.46	0.07	1.09	0.76
5.55	3.93	1.07	2.26	2.88	0.67	1.12	0.26	4.57	5.37	0.12	3.19
1.63	1.46	1.08	2.06	0.85	0.83	2.44	1.02	2.24	2.11	3.15	2.90
6.58	0.64										

The null hypothesis and its alternate are formed as follows:

H_0: the interarrival times are exponentially distributed.

H_1: the interarrival times are not exponentially distributed.

The data were collected over the interval from 0 to $T = 100$ minutes. It can be shown that, if the underlying distribution of interarrival times $\{T_1, T_2, \ldots\}$ is exponential, the arrival times are uniformly distributed on the interval $(0, T)$. The arrival times $T_1, T_1 + T_2, T_1 + T_2 + T_3, \ldots,$ $T_1 + \cdots + T_{50}$ are obtained by adding interarrival times. The arrival times are then normalized to a $(0, 1)$ interval so that the Kolmogorov-Smirnov test, as presented in Section 7.4.1, can be applied. On a $(0, 1)$ interval, the points will be $[T_1/T, (T_1 + T_2)/T, \ldots, (T_1 + \cdots + T_{50})/T]$. The resulting 50 data points are as follows:

0.0044	0.0097	0.0301	0.0575	0.0775	0.0805	0.1059	0.1111	0.1313	0.1502
0.1655	0.1676	0.1956	0.1960	0.2095	0.2927	0.3161	0.3356	0.3366	0.3508
0.3553	0.3561	0.3670	0.3746	0.4300	0.4694	0.4796	0.5027	0.5315	0.5382
0.5494	0.5520	0.5977	0.6514	0.6526	0.6845	0.7008	0.7154	0.7262	0.7468
0.7553	0.7636	0.7880	0.7982	0.8206	0.8417	0.8732	0.9022	0.9680	0.9744

Following the procedure in Example 7.6 produces a D^+ of 0.1054 and a D^- of 0.0080. Therefore, the Kolmogorov–Smirnov statistic is $D = \max(0.1054, 0.0080) = 0.1054$. The critical value of D obtained from Table A.8 for a level of significance of $\alpha = 0.05$ and $n = 50$ is $D_{0.05} = 1.36/\sqrt{n} = 0.1923$; but $D = 0.1054$, so the hypothesis that the interarrival times are exponentially distributed cannot be rejected.

The Kolmogorov–Smirnov test has been modified so that it can be used in several situations where the parameters are estimated from the data. The computation of the test statistic is the same, but different tables of critical values are used. Different tables of critical values are required for different distributional assumptions. Lilliefors [1967] developed a test for normality. The null hypothesis states that the population is one of the family of normal distributions, without specifying the parameters of the distribution. The interested reader might wish to study Lilliefors' original work; he describes how simulation was used to develop the critical values.

Lilliefors [1969] also modified the critical values of the Kolmogorov–Smirnov test for the exponential distribution. Lilliefors again used random sampling to obtain approximate critical values, but Durbin [1975] subsequently obtained the exact distribution. Connover [1998] gives examples of Kolmogorov–Smirnov tests for the normal and exponential distributions. He also refers to several other Kolmogorov–Smirnov-type tests that might be of interest to the reader.

A test that is similar in spirit to the Kolmogorov–Smirnov test is the *Anderson–Darling test*. Like the Kolmogorov–Smirnov test, the Anderson–Darling test is based on the difference between the empirical cdf and the fitted cdf; unlike the Kolmogorov–Smirnov test, the Anderson–Darling test is based on a more comprehensive measure of difference (not just the maximum difference) and is more sensitive to discrepancies in the tails of the distributions. The critical values for the Anderson–Darling test also depend on the candidate distribution and on whether parameters have been estimated. Fortunately, this test and the Kolmogorov–Smirnov test have been implemented in a number of software packages that support simulation-input modeling.

9.4.4 *p*-Values and "Best Fits"

To apply a goodness-of-fit test, a significance level must be chosen. Recall that the significance level is the probability of falsely rejecting H_0: the random variable conforms to the distributional assumption. The traditional significance levels are $0.1, 0.05$ and 0.01. Prior to the availability of high-speed computing, having a small set of standard values made it possible to produce tables of useful critical values. Now most statistical software computes critical values as needed, rather than storing them in tables. Thus, the analyst can employ a different level of significance—say, 0.07.

However, rather than require a prespecified significance level, many software packages compute a *p-value* for the test statistic. The *p*-value is the significance level at which one would *just reject* H_0 for the given value of the test statistic. Therefore, a large *p*-value tends to indicate a good fit (we would have to accept a large chance of error in order to reject), while a small *p*-value suggests a poor fit (to accept we would have to insist on almost no risk).

Recall Example 9.14, in which a chi-square test was used to check the Poisson assumption for the vehicle-arrival data. The value of the test statistic was $\chi_0^2 = 27.68$, with 5 degrees of freedom. The *p*-value for this test statistic is 0.00004, meaning that we would reject the hypothesis that the data are Poisson at the 0.00004 significance level. (Recall that we rejected the hypothesis at the 0.05 level; now we know that we would also to reject it at even lower levels.)

The *p*-value can be viewed as a measure of fit, with larger values being better. This suggests that we could fit every distribution at our disposal, compute a test statistic for each fit, and then choose the distribution that yields the largest *p*-value. We know of no input modeling software that implements this specific algorithm, but many such packages do include a "best fit" option, in which the software recommends an input model to the user after evaluating all feasible models. The software might also take into account other factors—such as whether the data are discrete or continuous, bounded or unbounded—but, in the end, some summary measure of fit, like the *p*-value, is used to rank the distributions. There is nothing wrong with this, but there are several things to keep in mind:

1. The software might know nothing about the physical basis of the data, whereas that information can suggest distribution families that are appropriate. (See the list in Section 9.2.2.) Remember that the goal of input modeling is often to fill in gaps or smooth the data, rather than find an input model that conforms as closely as possible to the given sample.
2. Recall that both the Erlang and the exponential distributions are special cases of the gamma and that the exponential is also a special case of the more flexible Weibull. Automated best-fit procedures tend to choose the more flexible distributions (gamma and Weibull over Erlang and exponential), because the extra flexibility allows closer conformance to the data and a better summary measure of fit. But again, close conformance to the data does not always lead to the most appropriate input model.
3. A summary statistic, like the *p*-value, is just that, a summary measure. It says little or nothing about where the lack of fit occurs (in the body of the distribution, in the right tail, or in the left tail). A human, using graphical tools, can see where the lack of fit occurs and decide whether or not it is important for the application at hand.

Our recommendation is that automated distribution selection be used as one of several ways to suggest candidate distributions. Always inspect the automatic selection, using graphical methods, and remember that the final choice is yours.

9.5 FITTING A NONSTATIONARY POISSON PROCESS

Fitting a nonstationary Poisson process (NSPP) to arrival data is a difficult problem, in general, because we seldom have knowledge about the appropriate form of the arrival rate function $\lambda(t)$. (See Chapter 5, Section 5.5 for the definition of a NSPP). One approach is to choose a very flexible model with lots of parameters and fit it with a method such as maximum likelihood; see Johnson, Lee, and Wilson [1994] for an example of this approach. A second method, and the one we consider here, is to approximate the arrival rate as being constant over some basic interval of time, such as an hour, or a day, or a month, but varying from time interval to time interval. The problem then becomes choosing the basic time interval and estimating the arrival rate within each interval.

Suppose we need to model arrivals over a time period, say $[0, T]$. The approach that we describe is most appropriate when it is possible to observe the time period $[0, T]$ repeatedly and count arrivals. For instance, if the problem involves modeling the arrival of e-mail throughout the

Table 9.8 Monday E-mail Arrival Data for NSPP Example

Time Period	Number of Arrivals			Estimated Arrival Rate (arrivals/hour)
	Day 1	Day 2	Day 3	
8:00–8:30	12	14	10	24
8:30–9:00	23	26	32	54
9:00–9:30	27	19	32	52
9:30–10:00	20	13	12	30

business day (8 A.M. to 6 P.M.), and we believe that the arrival rate is approximately constant over half-hour intervals, then we need to be able to count arrivals during half-hour intervals for several days. If it is possible to record actual arrival times, rather than counts, then actual arrival times are clearly better since they can later be grouped into any interval lengths we desire. However, we will assume from here on that only counts are available.

Divide the time period $[0, T]$ into k equal intervals of length $\Delta t = T/k$. For instance, if we are considering a 10-hour business day from 8 A.M. to 6 P.M. and if we allow the rate to change every half hour, then $T = 10$, $k = 20$, and $\Delta t = 1/2$. Over n periods of observation (e.g., n days), let C_{ij} be the number of arrivals that occurred during the ith time interval on the jth period of observation. In our example, C_{23} would be the number of arrivals from 8:30 A.M. to 9 A.M. (second half-hour period) on the third day of observation.

The estimated arrival rate during the ith time period, $(i - 1)\Delta t < t \leq i \Delta t$, is then just the average number of arrivals scaled by the length of the time interval:

$$\widehat{\lambda}(t) = \frac{1}{n \Delta t} \sum_{j=1}^{n} C_{ij} \tag{9.21}$$

After the arrival rates for each time interval have been estimated, adjacent intervals whose rates appear to be the same can be combined.

For instance, consider the e-mail arrival counts during the first two hours of the business day on three Mondays, shown in Table 9.8. The estimated arrival rate for 8:30–9:00 is

$$\frac{1}{3(1/2)}(23 + 26 + 32) = 54 \text{ arrivals/hour}$$

After seeing these results we might consider combining the interval 8:30–9:00 with the interval 9:00–9:30, because the rates are so similar. Note also that the goodness-of-fit tests described in the previous section can be applied to the data from each time interval individually, to check the Poisson approximation.

9.6 SELECTING INPUT MODELS WITHOUT DATA

Unfortunately, it is often necessary in practice to develop a simulation model—perhaps for demonstration purposes or a preliminary study—before any process data are available. In this case, the modeler must be resourceful in choosing input models and must carefully check the sensitivity of results to the chosen models.

There are a number of ways to obtain information about a process even if data are not available:

Engineering data: Often a product or process has performance ratings provided by the manufacturer (for example, the mean time to failure of a disk drive is 10000 hours; a laser printer can produce 8 pages/minute; the cutting speed of a tool is 1 cm/second; etc.). Company rules might specify time or production standards. These values provide a starting point for input modeling by fixing a central value.

Expert option: Talk to people who are experienced with the process or similar processes. Often, they can provide optimistic, pessimistic, and most-likely times. They might also be able to say whether the process is nearly constant or highly variable, and they might be able to define the source of variability.

Physical or conventional limitations: Most real processes have physical limits on performance—for example, computer data entry cannot be faster than a person can type. Because of company policies, there could be upper limits on how long a process may take. Do not ignore obvious limits or bounds that narrow the range of the input process.

The nature of the process: The description of the distributions in Section 9.2.2 can be used to justify a particular choice even when no data are available.

When data are not available, the uniform, triangular, and beta distributions are often used as input models. The uniform can be a poor choice, because the upper and lower bounds are rarely just as likely as the central values in real processes. If, in addition to upper and lower bounds, a most-likely value can be given, then the triangular distribution can be used. The triangular distribution places much of its probability near the most-likely value, and much less near the extremes. (See Section 5.4.) If a beta distribution is used, then be sure to plot the density function of the selected distribution; the beta can take unusual shapes.

A useful refinement is obtained when a minimum, a maximum, and one or more "breakpoints" can be given. A breakpoint is an intermediate value together with a probability of being less than or equal to that value. The following example illustrates how breakpoints are used.

Example 9.17 _____

For a production-planning simulation, the sales volume of various products is required. The salesperson responsible for product XYZ-123 says that no fewer than 1000 units will be sold (because of existing contracts) and no more than 5000 units will be sold (because that is the entire market for the product). Given her experience, she believes that there is a 90% chance of selling more than 2000 units, a 25% chance of selling more than 3500 units, and only a 1% chance of selling more than 4500 units.

Table 9.9 summarizes this information. Notice that the chances of exceeding certain sales goals have been translated into the cumulative probability of being less than or equal to those goals. With the information in this form, the method of Section 8.1.5 can be employed to generate simulation-input data.

When input models have been selected without data, it is especially important to test the sensitivity of simulation results to the distribution chosen. Check sensitivity not only to the center of the distribution, but also to the variability or limits. Extreme sensitivity of output results to

Table 9.9 Summary of Sales Information

i	Interval (Sales)	Cumulative Frequency, c_i
1	$1000 \leq x \leq 2000$	0.10
2	$2000 < x \leq 3500$	0.75
3	$3500 < x \leq 4500$	0.99
4	$4500 < x \leq 5000$	1.00

the input model provides a convincing argument against making critical decisions based on the results and in favor of undertaking data collection.

For additional discussion of input modeling in the absence of data, see Pegden, Shannon, and Sadowski [1995].

9.7 MULTIVARIATE AND TIME-SERIES INPUT MODELS

In Sections 9.1–9.4, the random variables presented were considered to be independent of any other variables within the context of the problem. However, variables may be related, and, if the variables appear in a simulation model as inputs, the relationship should be investigated and taken into consideration.

Example 9.18

An inventory simulation includes the lead time and annual demand for industrial robots. An increase in demand results in an increase in lead time: The final assembly of the robots must be made according to the specifications of the purchaser. Therefore, rather than treat lead time and demand as independent random variables, a multivariate input model should be developed.

Example 9.19

A simulation of the web-based trading site of a stock broker includes the time between arrivals of orders to buy and sell. Investors tend to react to what other investors are doing, so these buy and sell orders arrive in bursts. Therefore, rather than treat the time between arrivals as independent random variables, a time-series model should be developed.

We distinguish *multivariate input models* of a fixed, finite number of random variables (such as the two random variables *lead time* and *annual demand* in Example 9.18) from *time-series input models* of a (conceptually infinite) sequence of related random variables (such as the successive times between orders in Example 9.19). We will describe input models appropriate for these examples after reviewing two measures of dependence, the covariance and the correlation.

9.7.1 Covariance and Correlation

Let X_1 and X_2 be two random variables, and let $\mu_i = E(X_i)$ and $\sigma_i^2 = V(X_i)$ be the mean and variance of X_i, respectively. The *covariance* and *correlation* are measures of the linear dependence between X_1 and X_2. In other words, the covariance and correlation indicate how well the relationship between X_1 and X_2 is described by the model

$$(X_1 - \mu_1) = \beta(X_2 - \mu_2) + \epsilon$$

where ϵ is a random variable with mean 0 that is independent of X_2. If, in fact, $(X_1 - \mu_1) = \beta(X_2 - \mu_2)$, then this model is perfect. On the other hand, if X_1 and X_2 are statistically independent, then $\beta = 0$ and the model is of no value. In general, a positive value of β indicates that X_1 and X_2 tend to be above or below their means together; a negative value of β indicates that they tend to be on opposite sides of their means.

The covariance between X_1 and X_2 is defined to be

$$\text{cov}(X_1, X_2) = E[(X_1 - \mu_1)(X_2 - \mu_2)] = E(X_1 X_2) - \mu_1 \mu_2 \tag{9.22}$$

The value $\text{cov}(X_1, X_2) = 0$ implies $\beta = 0$ in our model of dependence, and $\text{cov}(X_1, X_2) < 0$ (>0) implies $\beta < 0$ (>0).

The covariance can take any value between $-\infty$ and ∞. The correlation standardizes the covariance to be between -1 and 1:

$$\rho = \text{corr}(X_1, X_2) = \frac{\text{cov}(X_1, X_2)}{\sigma_1 \sigma_2} \tag{9.23}$$

Again, the value $\text{corr}(X_1, X_2) = 0$ implies $\beta = 0$ in our model, and $\text{corr}(X_1, X_2) < 0$ (>0) implies $\beta < 0$ (>0). The closer ρ is to -1 or 1, the stronger the linear relationship is between X_1 and X_2.

Now suppose that we have a sequence of random variables $X_1, X_2, X_3, \ldots$ that are identically distributed (implying that they all have the same mean and variance), but could be dependent. We refer to such a sequence as a *time series* and to $\text{cov}(X_t, X_{t+h})$ and $\text{corr}(X_t, X_{t+h})$ as the *lag-h autocovariance* and *lag-h autocorrelation*, respectively. If the value of the autocovariance depends only on h and not on t, then we say that the time series is *covariance stationary*; this concept is discussed further in Chapter 11. For a covariance-stationary time series, we use the shorthand notation

$$\rho_h = \text{corr}(X_t, X_{t+h})$$

for the *lag-h* autocorrelation. Notice that autocorrelation measures the dependence between random variables that are separated by $h - 1$ others in the time series.

9.7.2 Multivariate Input Models

If X_1 and X_2 each are normally distributed, then dependence between them can be modeled by the bivariate normal distribution with parameters $\mu_1, \mu_2, \sigma_1^2, \sigma_2^2$, and $\rho = \text{corr}(X_1, X_2)$. Estimation of μ_1, μ_2, σ_1^2, and σ_2^2 was described in Section 9.3.2. To estimate ρ, suppose that we have n independent and identically distributed pairs $(X_{11}, X_{21}), (X_{12}, X_{22}), \ldots, (X_{1n}, X_{2n})$. Then the sample covariance is

$$\widehat{\text{cov}}(X_1, X_2) = \frac{1}{n-1} \sum_{j=1}^{n} \left(X_{1j} - \bar{X}_1\right)(X_{2j} - \bar{X}_2)$$

$$= \frac{1}{n-1} \left(\sum_{j=1}^{n} X_{1j} X_{2j} - n \bar{X}_1 \bar{X}_2 \right) \tag{9.24}$$

where $\bar{X}_1$ and $\bar{X}_2$ are the sample means. The correlation is estimated by

$$\widehat{\rho} = \frac{\widehat{\text{cov}}(X_1, X_2)}{\widehat{\sigma}_1 \widehat{\sigma}_2} \tag{9.25}$$

where $\widehat{\sigma}_1$ and $\widehat{\sigma}_2$ are the sample variances.

Example 9.20:   Example 9.18 Continued

Let X_1 represent the average lead time to deliver (in months), and X_2 the annual demand, for industrial robots. The following data are available on demand and lead time for the last ten years:

lead time	demand
6.5	103
4.3	83
6.9	116
6.0	97
6.9	112
6.9	104
5.8	106
7.3	109
4.5	92
6.3	96

Standard calculations give $\bar{X}_1 = 6.14$, $\widehat{\sigma}_1 = 1.02$, $\bar{X}_2 = 101.80$, and $\widehat{\sigma}_2 = 9.93$ as estimates of μ_1, σ_1, μ_2, and σ_2, respectively. To estimate the correlation, we need

$$\sum_{j=1}^{10} X_{1j} X_{2j} = 6328.5$$

Therefore, $\widehat{\text{cov}} = [6328.5 - (10)(6.14)(101.80)]/(10 - 1) = 8.66$, and

$$\widehat{\rho} = \frac{8.66}{(1.02)(9.93)} = 0.86$$

Clearly, lead time and demand are strongly dependent. Before we accept this model, however, lead time and demand should be checked individually to see whether they are represented well by normal distributions. In particular, demand is a discrete-valued quantity, so the continuous normal distribution is certainly at best an approximation.

The following simple algorithm can be used to generate bivariate normal random variables:

Step 1. Generate Z_1 and Z_2, independent standard normal random variables (see Section 8.3.1).

Step 2. Set $X_1 = \mu_1 + \sigma_1 Z_1$

Step 3. Set $X_2 = \mu_2 + \sigma_2 \left(\rho Z_1 + \sqrt{1 - \rho^2} Z_2 \right)$

Obviously, the bivariate normal distribution will not be appropriate for all multivariate-input modeling problems. It can be generalized to the k-variate normal distribution to model the dependence among more than two random variables, but, in many instances, a normal distribution is not appropriate in any form. We provide one method for handling nonnormal distributions in Section 9.7.4. Good references for other models are Johnson [1987] and Nelson and Yamnitsky [1998].

9.7.3 Time-Series Input Models

If $X_1, X_2, X_3, \ldots$ is a sequence of identically distributed, but dependent and covariance-stationary random variables, then there are a number of time series models that can be used to represent the process. We will describe two models that have the characteristic that the autocorrelations take the form

$$\rho_h = \text{corr}(X_t, X_{t+h}) = \rho^h$$

for $h = 1, 2, \ldots$. Notice that the *lag-h* autocorrelation decreases geometrically as the lag increases, so that observations far apart in time are nearly independent. For one model to be shown shortly, each X_t is normally distributed; for the other model, each X_t is exponentially distributed. More general time-series input models are described in Section 9.7.4 and in Nelson and Yamnitsky [1998].

AR(1) MODEL. Consider the time-series model

$$X_t = \mu + \phi(X_{t-1} - \mu) + \varepsilon_t \tag{9.26}$$

for $t = 2, 3, \ldots$, where $\varepsilon_2, \varepsilon_3, \ldots$ are independent and identically (normally) distributed with mean 0 and variance σ_ε^2, and $-1 < \phi < 1$. If the initial value X_1 is chosen appropriately (see shortly), then $X_1, X_2, \ldots$ are all normally distributed with mean μ, variance $\sigma_\varepsilon^2/(1 - \phi^2)$, and

$$\rho_h = \phi^h$$

for $h = 1, 2, \ldots$. This time-series model is called the autoregressive order-1 model, or AR(1) for short.

Estimation of the parameter ϕ can be obtained from the fact that

$$\phi = \rho^1 = \text{corr}(X_t, X_{t+1})$$

the lag-1 autocorrelation. Therefore, to estimate ϕ, we first estimate the lag-1 autocovariance by

$$\widehat{\text{cov}}(X_t, X_{t+1}) = \frac{1}{n-1} \sum_{t=1}^{n-1} (X_t - \bar{X})(X_{t+1} - \bar{X})$$

$$\doteq \frac{1}{n-1} \left(\sum_{t=1}^{n-1} X_t X_{t+1} - (n-1)\bar{X}^2 \right) \tag{9.27}$$

and the variance $\sigma^2 = \text{var}(X)$ by the usual estimator $\widehat{\sigma}^2$. Then

$$\widehat{\phi} = \frac{\widehat{\text{cov}}(X_t, X_{t+1})}{\widehat{\sigma}^2}$$

Finally, estimate μ and σ_ε^2 by $\widehat{\mu} = \bar{X}$ and

$$\widehat{\sigma}_\varepsilon^2 = \widehat{\sigma}^2 (1 - \widehat{\phi}^2)$$

respectively.

The following algorithm generates a stationary AR(1) time series, given values of the parameters ϕ, μ, and σ_ε^2:

Step 1. Generate X_1 from the normal distribution with mean μ and variance $\sigma_\varepsilon^2/(1 - \phi^2)$. Set $t = 2$.

Step 2. Generate ε_t from the normal distribution with mean 0 and variance σ_ε^2.

Step 3. Set $X_t = \mu + \phi(X_{t-1} - \mu) + \varepsilon_t$.

Step 4. Set $t = t + 1$ and go to Step 2.

EAR(1) MODEL.   Consider the time-series model

$$X_t = \begin{cases} \phi X_{t-1}, & \text{with probability } \phi \\ \phi X_{t-1} + \varepsilon_t, & \text{with probability } 1 - \phi \end{cases} \tag{9.28}$$

for $t = 2, 3, \ldots$, where $\varepsilon_2, \varepsilon_3, \ldots$ are independent and identically (exponentially) distributed with mean $1/\lambda$ and $0 \le \phi < 1$. If the initial value X_1 is chosen appropriately (see shortly), then $X_1, X_2, \ldots$ are all exponentially distributed with mean $1/\lambda$ and

$$\rho_h = \phi^h$$

for $h = 1, 2, \ldots$. This time-series model is called the exponential autoregressive order-1 model, or EAR(1) for short. Only autocorrelations greater than 0 can be represented by this model.

Estimation of the parameters proceeds as for the AR(1) by setting $\widehat{\phi} = \widehat{\rho}$, the estimated lag-1 autocorrelation, and setting $\widehat{\lambda} = 1/\bar{X}$.

The following algorithm generates a stationary EAR(1) time series, given values of the parameters ϕ and λ:

Step 1. Generate X_1 from the exponential distribution with mean $1/\lambda$. Set $t = 2$.

Step 2. Generate U from the uniform distribution on $[0, 1]$. If $U \leq \phi$, then set

$$X_t = \phi X_{t-1}$$

Otherwise, generate ε_t from the exponential distribution with mean $1/\lambda$ and set

$$X_t = \phi X_{t-1} + \varepsilon_t$$

Step 3. Set $t = t + 1$ and go to Step 2.

Example 9.21:   Example 9.19 Continued _____

The stock broker would typically have a large sample of data, but, for the sake of illustration, suppose that the following twenty time gaps between customer buy and sell orders had been recorded (in seconds): 1.95, 1.75, 1.58, 1.42, 1.28, 1.15, 1.04, 0.93, 0.84, 0.75, 0.68, 0.61, 11.98, 10.79, 9.71, 14.02, 12.62, 11.36, 10.22, 9.20. Standard calculations give $\bar{X} = 5.2$ and $\widehat{\sigma}^2 = 26.7$. To estimate the lag-1 autocorrelation, we need

$$\sum_{j=1}^{19} X_t X_{t+1} = 924.1$$

Thus, $\widehat{\text{cov}} = [924.1 - (20 - 1)(5.2)^2]/(20 - 1) = 21.6$, and

$$\widehat{\rho} = \frac{21.6}{26.7} = 0.8$$

Therefore, we could model the interarrival times as an EAR(1) process with $\widehat{\lambda} = 1/5.2 = 0.192$ and $\widehat{\phi} = 0.8$, provided that an exponential distribution is a good model for the individual gaps.

_____

9.7.4 The Normal-to-Anything Transformation

The bivariate normal distribution and the AR(1) and EAR(1) time-series models are useful input models that are easy to fit and simulate. However, the marginal distribution is either normal or exponential, which is certainly not the best choice for many applications. Fortunately, we can start with a bivariate normal or AR(1) model and *transform* it to have any marginal distributions we want (including exponential).

Suppose we want to simulate a random variable X with cdf $F(x)$. Let Z be a standard normal random variable (mean 0 and variance 1), and let $\Phi(z)$ be its cdf. Then it can be shown that

$$R = \Phi(Z)$$

is a $U(0, 1)$ random variable. As we learned in Chapter 8, if we have a $U(0, 1)$ random variable, we can get X by using the inverse cdf transformation

$$X = F^{-1}[R] = F^{-1}[\Phi(Z)]$$

We refer this as the *normal to anything transformation*, or NORTA for short.

Of course, if all we want is X, then there is no reason to go to this trouble; we can just generate R directly, using the methods in Chapter 8. But suppose we want a bivariate random vector (X_1, X_2) such that X_1 and X_2 are correlated but their distributions are not normal. Then we can start with a bivariate normal random vector (Z_1, Z_2) and apply the NORTA transformation to obtain

$$X_1 = F_1^{-1}[\Phi(Z_1)] \text{ and } X_2 = F_2^{-1}[\Phi(Z_2)]$$

There is not even a requirement that F_1 and F_2 be from the same distribution family; for instance, F_1 could be an exponential distribution and F_2 a beta distribution.

The same idea applies for time series. If Z_t is generated by an AR(1) with $N(0, 1)$ marginals, then

$$X_t = F^{-1}[\Phi(Z_t)]$$

will be a time-series model with marginal distribution $F(x)$. To insure that Z_t is $N(0, 1)$, we set $\mu = 0$ and $\sigma_\varepsilon^2 = 1 - \phi^2$ in the AR(1) model.

Although the NORTA method is very general, there are two technical issues that must be addressed to implement it:

1. The NORTA approach requires being able to evaluate that standard normal cdf, $\Phi(z)$, and the inverse cdf of the distributions of interest, $F^{-1}(u)$. There is no closed-form expression for $\Phi(z)$ and no closed-form expression for $F^{-1}(u)$ for many distributions. Therefore, numerical approximations are required. Fortunately, these functions are built into many symbolic calculation and spreadsheet programs, and we give one example next. In addition, Bratley, Fox, and Schrage [1987] contains algorithms for many distributions.
2. The correlation between the standard normal random variables (Z_1, Z_2) is distorted when it passes through the NORTA transformation. To be more specific, if (Z_1, Z_2) have correlation ρ, then in general $X_1 = F_1^{-1}[\Phi(Z_1)]$ and $X_2 = F_2^{-1}[\Phi(Z_2)]$ will have a correlation $\rho_X \neq \rho$. The difference is often small, but not always.

The second issue is more critical, because in input-modeling problems we want to specify the bivariate or lag-1 correlation. Thus, we need to find the bivariate normal correlation ρ that gives us the input correlation ρ_X that we want (recall that we specify the time series model via the lag-1 correlation, $\rho_X = \text{corr}(X_t, X_{t+1})$). There has been much research on this problem, including Cario and Nelson [1996, 1998] and Biller and Nelson [2003]. Fortunately, it has been shown that ρ_X is a nondecreasing function of ρ, and ρ and ρ_X will always have the same sign. Thus, we can do a relatively simple search based on the following algorithm:

Step 1. Set $\rho = \rho_X$ to start.

Step 2. Generate a large number of bivariate normal pairs (Z_1, Z_2) with correlation ρ, and transform them into (X_1, X_2)'s, using the NORTA transformation.

Step 3. Compute the sample correlation between (X_1, X_2), using Equation (9.24), and call it $\widehat{\rho}_T$. If $\widehat{\rho}_T > \rho_X$, then reduce ρ and go to Step 2; if $\widehat{\rho}_T < \rho_X$, then increase ρ and go to Step 2. If $\widehat{\rho}_T \approx \rho_X$ then stop.

Example 9.22 _____

Suppose we needed X_1 to have an exponential distribution with mean 1, X_2 to have a beta distribution with $\beta_1 = 1$, $\beta_2 = 1/2$, and the two of them to have correlation $\rho_X = 0.45$. Figure 9.5 shows a procedure in Maple that will estimate the required value of ρ. In the procedure, n is the number of sample pairs used to estimate the correlation. Running this procedure with n set to 1000 gives $\rho = 0.52$.

9.8 SUMMARY

Input-data collection and analysis require major time and resource commitments in a discrete-event simulation project. However, regardless of the validity or sophistication of the simulation model, unreliable inputs can lead to outputs whose subsequent interpretation could result in faulty recommendations.

This chapter discussed four steps in the development of models of input data: collecting the raw data, identifying the underlying statistical distribution, estimating the parameters, and testing for goodness of fit.

Some suggestions were given for facilitating the data-collection step. However, experience, such as that obtained by completing any of Exercises 1 through 5, will increase awareness of the difficulty of problems that can arise in data collection and of the need for planning.

Once the data have been collected, a statistical model should be hypothesized. Constructing a histogram is very useful at this point if sufficient data are available. A distribution based on the underlying process and on the shape of the histogram can usually be selected for further investigation.

The investigation proceeds with the estimation of parameters for the hypothesized distribution. Suggested estimators were given for distributions used often in simulation. In a number of instances, these are functions of the sample mean and sample variance.

The last step in the process is the testing of the distributional hypothesis. The $q-q$ plot is a useful graphical method for assessing fit. The Kolmogorov–Smirnov, chi-square, and Anderson–Darling goodness-of-fit tests can be applied to many distributional assumptions. When a distributional assumption is rejected, another distribution is tried. When all else fails, the empirical distribution could be used in the model.

Unfortunately, in some situations, a simulation study must be undertaken when there is not time or resources to collect data on which to base input models. When this happens, the analyst must use any available information—such as manufacturer specifications and expert opinion—to construct the input models. When input models are derived without the benefit of data, it is particularly important to examine the sensitivity of the results to the models chosen.

Many, but not all, input processes can be represented as sequences of independent and identically distributed random variables. When inputs should exhibit dependence, then multivariate-input models are required. The bivariate normal distribution (and more generally the multivariate normal distribution) is often used to represent a finite number of dependent random variables.

```
NORTARho := proc(rhoX, n)
local Z1, Z2, Ztemp, X1, X2, R1, R2, rho, rhoT, lower, upper;
randomize(123456);
Z1 := [random[normald[0,1]](n)]:
ZTemp := [random[normald[0,1]](n)]:
Z2 := [0]:
# set up bisection search
rho := rhoX:
if (rhoX < 0) then
   lower := -1:
   upper := 0:
else
   lower := 0:
   upper := 1:
fi:
Z2 := rho*Z1 + sqrt(1-rho^2)*ZTemp:
R1 := statevalf[cdf,normald[0,1]](Z1):
R2 := statevalf[cdf,normald[0,1]](Z2):
X1 := statevalf[icdf,exponential[1,0]](R1):
X2 := statevalf[icdf,beta[1,2]](R2):
rhoT := describe[linearcorrelation](X1, X2);
# do bisection search until 5% relative error
while abs(rhoT - rhoX)/abs(rhoX) > 0.05 do
   if (rhoT > rhoX) then
      upper := rho:
   else
      lower := rho:
   fi:
   rho := evalf((lower + upper)/2):
   Z2 := rho*Z1 + sqrt(1-rho^2)*ZTemp:
   R1 := statevalf[cdf,normald[0,1]](Z1):
   R2 := statevalf[cdf,normald[0,1]](Z2):
   X1 := statevalf[icdf,exponential[1,0]](R1):
   X2 := statevalf[icdf,beta[1,2]](R2):
   rhoT := describe[linearcorrelation](X1, X2);
end do;
RETURN(rho);
end;
```

Figure 9.5 Maple procedure to estimate the bivariate normal correlation required for the NORTA method.

Time-series models are useful for representing a (conceptually infinite) sequence of dependent inputs. The NORTA transformation facilitates developing multivariate-input models with marginal distributions that are not normal.

REFERENCES

BILLER, B., AND B. L. NELSON [2003], "Modeling and Generating Multivariate Time Series with Arbitrary Marginals Using an Autoregressive Technique," *ACM Transactions on Modeling and Computer Simulation*, Vol. 13, pp. 211–237.

BRATLEY, P., B. L. FOX, AND L. E. SCHRAGE [1987], *A Guide to Simulation*, 2d ed., Springer-Verlag, New York.

CARIO, M. C., AND B. L. NELSON [1996], "Autoregressive to Anything: Time-Series Input Processes for Simulation," *Operations Research Letters*, Vol. 19, pp. 51–58.

CARIO, M. C., AND B. L. NELSON [1998],"Numerical Methods for Fitting and Simulating Autoregressive-to-Anything Processes," *INFORMS Journal on Computing*, Vol. 10, pp. 72–81.

CHOI, S. C., AND R. WETTE [1969], "Maximum Likelihood Estimation of the Parameters of the Gamma Distribution and Their Bias," *Technometrics*, Vol. 11, No. 4, pp. 683–890.

CHAMBERS, J. M., CLEVELAND, W. S., AND TUKEY, P. A. [1983], *Graphical Methods for Data Analysis*, CRC Press, Boca Raton, FL.

CONNOVER, W. J. [1998], *Practical Nonparametric Statistics*, 3d ed., Wiley, New York.

DURBIN, J. [1975], "Kolmogorov–Smirnov Tests When Parameters Are Estimated with Applications to Tests of Exponentiality and Tests on Spacings," *Biometrika*, Vol. 65, pp. 5–22.

FISHMAN, G. S. [1973], *Concepts and Methods in Discrete Event Digital Simulation*, Wiley, New York.

GUMBEL, E. J. [1943], "On the Reliability of the Classical Chi-squared Test," *Annals of Mathematical Statistics*, Vol. 14, pp. 253ff.

HINES, W. W., D. C. MONTGOMERY, D. M. GOLDSMAN, AND C. M. BORROR [2002], *Probability and Statistics in Engineering and Management Science*, 4th ed., Wiley, New York.

JOHNSON, M. A., S. LEE, AND J. R. WILSON [1994], "NPPMLE and NPPSIM: Software for Estimating and Simulating Nonhomogeneous Poisson Processes Having Cyclic Behavior," *Operations Research Letters*, Vol. 15, pp. 273–282.

JOHNSON, M. E. [1987], *Multivariate Statistical Simulation*, Wiley, New York.

LAW, A. M., AND W. D. KELTON [2000], *Simulation Modeling & Analysis*, 3d ed., McGraw–Hill, New York.

LILLIEFORS, H. W. [1967], "On the Kolmogorov–Smirnov Test for Normality with Mean and Variance Unknown," *Journal of the American Statistical Association*, Vol. 62, pp. 339–402.

LILLIEFORS, H. W. [1969], "On the Kolmogorov–Smirnov Test for the Exponential Distribution with Mean Unknown," *Journal of the American Statistical Association*, Vol. 64, pp. 387–389.

MANN, H. B., AND A. WALD [1942], "On the Choice of the Number of Intervals in the Application of the Chi-squared Test," *Annals of Mathematical Statistics*, Vol. 18, p. 50ff.

NELSON, B. L., AND M. YAMNITSKY [1998], "Input Modeling Tools for Complex Problems," in *Proceedings of the 1998 Winter Simulation Conference*, eds. D. Medeiros, E. Watson, J. Carson, and M. Manivannan, pp. 105–112, The Institute for Electrical and Electronics Engineers, Piscataway, NJ.

PEGDEN, C. D., R. E. SHANNON, AND R. P. SADOWSKI [1995], *Introduction to Simulation Using SIMAN*, 2d ed. McGraw–Hill, New York.

STUART, A., J. K. ORD, AND E. ARNOLD [1998], *Kendall's Advanced Theory of Statistics*, 6th ed., Vol. 2, Oxford University Press, Oxford, NY.

EXERCISES

1. Go to a small store, and record the interarrival- and service-time distributions. If there are several workers, how do the service-time distributions compare to each other? Do service-time distributions need to be constructed for each type of appliance? (Make sure that the management gives permission to perform this study.)

2. Go to a cafeteria, and collect data on the distributions of interarrival and service times. The distribution of interarrival times is probably different for each of the three daily meals and

might also vary during the meal—that is, the interarrival time distribution for 11:00 A.M. to 12:00 noon could be different from that for 12:00 noon to 1:00 P.M. Define service time as the time from when the customer reaches the point at which the first selection could be made until the time of exiting from the cafeteria line. (Any reasonable modification of this definition is acceptable.) The service-time distribution probably changes for each meal. Can times of the day or days of the week for either distribution be grouped to suit the homogeneity of the data? (Make sure that the management gives permission to perform this study.)

3. Go to a major traffic intersection, and record the interarrival-time distributions from each direction. Some arrivals want to go straight, some turn left, some turn right. The interarrival-time distribution varies during the day and by day of the week. Every now and then an accident occurs.

4. Go to a grocery store, and construct the interarrival and service distributions at the checkout counters. These distributions might vary by time of day and by day of week. Record, also, the number of service channels available at all times. (Make sure that the management gives permission to perform this study.)

5. Go to a laundromat, and "relive" the authors' data-collection experience discussed in Example 9.1. (Make sure that the management gives permission to perform this study.)

6. Prepare four theoretical normal density functions, all on the same figure, each distribution having mean zero, but let the standard deviations be 1/4, 1/2, 1, and 2.

7. On one figure, draw the pdfs of the Erlang distribution where $\theta = 1/2$ and $k = 1, 2, 4$, and 8.

8. On one figure, draw the pdfs of the Erlang distribution where $\theta = 2$ and $k = 1, 2, 4$, and 8.

9. Draw the pmf of the Poisson distribution that results when the parameter α is equal to the following:

   (a) $\alpha = 1/2$
   (b) $\alpha = 1$
   (c) $\alpha = 2$
   (d) $\alpha = 4$

10. On one figure, draw the two exponential pdf's that result when the parameter, λ, equals 0.6 and 1.2.

11. On one figure, draw the three Weibull pdf's that result when $\nu = 0$, $\alpha = 1/2$, and $\beta = 1, 2$, and 4.

12. The following data are generated randomly from a gamma distribution:

1.691	1.437	8.221	5.976
1.116	4.435	2.345	1.782
3.810	4.589	5.313	10.90
2.649	2.432	1.581	2.432
1.843	2.466	2.833	2.361

Compute the maximum-likelihood estimators $\widehat{\beta}$ and $\widehat{\theta}$.

13. The following data are generated randomly from a Weibull distribution where $v = 0$:

$$
\begin{array}{llll}
7.936 & 5.224 & 3.937 & 6.513 \\
4.599 & 7.563 & 7.172 & 5.132 \\
5.259 & 2.759 & 4.278 & 2.696 \\
6.212 & 2.407 & 1.857 & 5.002 \\
4.612 & 2.003 & 6.908 & 3.326
\end{array}
$$

Compute the maximum-likelihood estimators $\widehat{\alpha}$ and $\widehat{\beta}$. (This exercise requires a programmable calculator, a computer, or a lot of patience.)

14. The highway between Atlanta, Georgia and Athens, Georgia has a high incidence of accidents along its 100 kilometers. Public safety officers say that the occurrence of accidents along the highway is randomly (uniformly) distributed, but the news media say otherwise. The Georgia Department of Public Safety published records for the month of September. These records indicated the point at which 30 accidents involving an injury or death occurred, as follows (the data points representing the distance from the city limits of Atlanta):

$$
\begin{array}{lllll}
88.3 & 40.7 & 36.3 & 27.3 & 36.8 \\
91.7 & 67.3 & 7.0 & 45.2 & 23.3 \\
98.8 & 90.1 & 17.2 & 23.7 & 97.4 \\
32.4 & 87.8 & 69.8 & 62.6 & 99.7 \\
20.6 & 73.1 & 21.6 & 6.0 & 45.3 \\
76.6 & 73.2 & 27.3 & 87.6 & 87.2
\end{array}
$$

Use the Kolmogorov–Smirnov test to discover whether the distribution of location of accidents is uniformly distributed for the month of September.

15. Show that the Kolmogorov–Smirnov test statistic for Example 9.16 is $D = 0.1054$.

16. Records pertaining to the monthly number of job-related injuries at an underground coalmine were being studied by a federal agency. The values for the past 100 months were as follows:

Injuries per Month	Frequency of Occurrence
0	35
1	40
2	13
3	6
4	4
5	1
6	1

(a) Apply the chi-square test to these data to test the hypothesis that the underlying distribution is Poisson. Use the level of significance $\alpha = 0.05$.

(b) Apply the chi-square test to these data to test the hypothesis that the distribution is Poisson with mean 1.0. Again let $\alpha = 0.05$.

(c) What are the differences between parts (a) and (b), and when might each case arise?

17. The time required for 50 different employees to compute and record the number of hours worked during the week was measured, with the following results in minutes:

Employee	Time (Minutes)	Employee	Time (Minutes)
1	1.88	26	0.04
2	0.54	27	1.49
3	1.90	28	0.66
4	0.15	29	2.03
5	0.02	30	1.00
6	2.81	31	0.39
7	1.50	32	0.34
8	0.53	33	0.01
9	2.62	34	0.10
10	2.67	35	1.10
11	3.53	36	0.24
12	0.53	37	0.26
13	1.80	38	0.45
14	0.79	39	0.17
15	0.21	40	4.29
16	0.80	41	0.80
17	0.26	42	5.50
18	0.63	43	4.91
19	0.36	44	0.35
20	2.03	45	0.36
21	1.42	46	0.90
22	1.28	47	1.03
23	0.82	48	1.73
24	2.16	49	0.38
25	0.05	50	0.48

Use the chi-square test (as in Example 9.15) to test the hypothesis that these service times are exponentially distributed. Let the number of class intervals be $k = 6$. Use the level of significance $\alpha = 0.05$.

18. Studentwiser Beer Company is trying to find out the distribution of the breaking strength of their glass bottles. Fifty bottles are selected at random and tested for breaking strength, with

the following results (in pounds per square inch):

218.95	232.75	212.80	231.10	215.95
237.55	235.45	228.25	218.65	212.80
230.35	228.55	216.10	229.75	229.00
199.75	225.10	208.15	213.85	205.45
219.40	208.15	198.40	238.60	219.55
243.10	198.85	224.95	212.20	222.90
218.80	203.35	223.45	213.40	206.05
229.30	239.20	201.25	216.85	207.25
204.85	219.85	226.15	230.35	211.45
227.95	229.30	225.25	201.25	216.10

Using input-modeling software, apply as many tests for normality as are available in the software. If the chi-square test is available, apply it with at least two different choices for the number of intervals. Do all of the tests reach the same conclusion?

19. The Crosstowner is a bus that cuts a diagonal path from northeast Atlanta to southwest Atlanta. The time required to complete the route is recorded by the bus operator. The bus runs from Monday through Friday. The times of the last fifty 8:00 A.M. runs, in minutes, are as follows:

92.3	92.8	106.8	108.9	106.6
115.2	94.8	106.4	110.0	90.9
104.6	72.0	86.0	102.4	99.8
87.5	111.4	105.9	90.7	99.2
97.8	88.3	97.5	97.4	93.7
99.7	122.7	100.2	106.5	105.5
80.7	107.9	103.2	116.4	101.7
84.8	101.9	99.1	102.2	102.5
111.7	101.5	95.1	92.8	88.5
74.4	98.9	111.9	96.5	95.9

How are these run times distributed? Develop and test a suitable model.

20. The time required for the transmission of a message (in minutes) is sampled electronically at a communications center. The last 50 values in the sample are as follows:

7.936	4.612	2.407	4.278	5.132
4.599	5.224	2.003	1.857	2.696
5.259	7.563	3.937	6.908	5.002
6.212	2.759	7.172	6.513	3.326
8.761	4.502	6.188	2.566	5.515
3.785	3.742	4.682	4.346	5.359
3.535	5.061	4.629	5.298	6.492
3.502	4.266	3.129	1.298	3.454
5.289	6.805	3.827	3.912	2.969
4.646	5.963	3.829	4.404	4.924

How are the transmission times distributed? Develop and test an appropriate model.

21. The time (in minutes) between requests for the hookup of electric service was accurately recorded at the Gotwatts Flash and Flicker Company, with the following results for the last 50 requests:

Can use Input Analyzer.

0.661	4.910	8.989	12.801	20.249
5.124	15.033	58.091	1.543	3.624
13.509	5.745	0.651	0.965	62.146
15.512	2.758	17.602	6.675	11.209
2.731	6.892	16.713	5.692	6.636
2.420	2.984	10.613	3.827	10.244
6.255	27.969	12.107	4.636	7.093
6.892	13.243	12.711	3.411	7.897
12.413	2.169	0.921	1.900	0.315
4.370	0.377	9.063	1.875	0.790

How are the times between requests for service distributed? Develop and test a suitable model.

22. Daily demands for transmission overhaul kits for the D-3 dragline were maintained by Earth Moving Tractor Company, with the following results:

0	2	0	0	0
1	0	1	1	1
0	1	0	0	0
2	0	1	0	1
0	1	0	0	2
1	0	1	0	0
0	0	0	0	0
1	0	1	0	1
0	0	3	0	1
1	0	0	0	0

How are the daily demands distributed? Develop and test an appropriate model.

23. A simulation is to be conducted of a job shop that performs two operations: milling and planing, in that order. It would be possible to collect data about processing times for each operation, then generate random occurrences from each distribution. However, the shop manager says that the times might be related; large milling jobs take lots of planing. Data are collected for the next 25 orders, with the following results in minutes:

Order	Milling Time (Minutes)	Planing Time (Minutes)	Order	Milling Time (Minutes)	Planing Time (Minutes)
1	12.3	10.6	14	24.6	16.6
2	20.4	13.9	15	28.5	21.2
3	18.9	14.1	16	11.3	9.9
4	16.5	10.1	17	13.3	10.7

5	8.3	8.4	18	21.0	14.0
6	6.5	8.1	19	19.5	13.0
7	25.2	16.9	20	15.0	11.5
8	17.7	13.7	21	12.6	9.9
9	10.6	10.2	22	14.3	13.2
10	13.7	12.1	23	17.0	12.5
11	26.2	16.0	24	21.2	14.2
12	30.4	18.9	25	28.4	19.1
13	9.9	7.7			

(a) Plot milling time on the horizontal axis and planing time on the vertical axis. Do these data seem dependent?

(b) Compute the sample correlation between milling time and planing time.

(c) Fit a bivariate normal distribution to these data.

24. Write a computer program to compute the maximum-likelihood estimators $(\hat{\alpha}, \hat{\beta})$ of the Weibull distribution. Inputs to the program should include the sample size, n; the observations, $x_1, x_2, \ldots, x_n$; a stopping criterion, ϵ (stop when $|f(\hat{\beta_j})| \leq \epsilon$); and a print option, OPT (usually set = 0). Output would be the estimates $\hat{\alpha}$ and $\hat{\beta}$. If OPT = 1, additional output would be printed, as in Table 9.4, showing convergence. Make the program as "user friendly" as possible.

25. Examine a computer-software library or simulation-support environment to which you have access. Obtain documentation on data-analysis software that would be useful in solving exercises 7 through 24. Use the software as an aid in solving selected problems.

26. The numbers of patrons staying at a small hotel on 20 successive nights was observed to be 20, 14, 21, 19, 14, 18, 21, 25, 27, 26, 22, 18, 13, 18, 18, 18, 25, 23, 20, 21. Fit both an AR(1) and an EAR(1) model to these data. Decide which model provides a better fit by looking at a histogram of the data.

27. The following data represent the time to perform transactions in a bank, measured in minutes: 0.740, 1.28, 1.46, 2.36, 0.354, 0.750, 0.912, 4.44, 0.114, 3.08, 3.24, 1.10, 1.59, 1.47, 1.17, 1.27, 9.12, 11.5, 2.42, 1.77. Develop an input model for these data.

28. Two types of jobs (A and B) are released to the input buffer of a job shop as orders arrive, and the arrival of orders is uncertain. The following data are available from the last week of production:

Day	Number of Jobs	Number of A's
1	83	53
2	93	62
3	112	66
4	65	41
5	78	55

Develop an input model for the number of new arrivals of each type each day.

29. The following data are available on the processing time at a machine (in minutes): 0.64, 0.59, 1.1, 3.3, 0.54, 0.04, 0.45, 0.25, 4.4, 2.7, 2.4, 1.1, 3.6, 0.61, 0.20, 1.0, 0.27, 1.7, 0.04, 0.34. Develop an input model for the processing time.

30. For the arrival-count data shown here, estimate the arrival-rate function, $\lambda(t)$, in arrivals per hour:

Time Period	Number of Arrivals			
	Day 1	Day 2	Day 3	Day 4
8:00–10:00	22	24	20	28
10:00–12:00	23	26	32	30
12:00–2:00	40	33	32	38

31. Using the web, research some of the input-modeling software packages mentioned in this chapter. What are their features? What distributions do they include?

10

Verification and Validation of Simulation Models

One of the most important and difficult tasks facing a model developer is the verification and validation of the simulation model. The engineers and analysts who use the model outputs to aid in making design recommendations and the managers who make decisions based on these recommendations—justifiably look upon a model with some degree of skepticism about its validity. It is the job of the model developer to work closely with the end users throughout the period of development and validation to reduce this skepticism and to increase the model's credibility.

The goal of the validation process is twofold: (1) to produce a model that represents true system behavior closely enough for the model to be used as a substitute for the actual system for the purpose of experimenting with the system, analyzing system behavior, and predicting system performance; and (2) to increase to an acceptable level the credibility of the model, so that the model will be used by managers and other decision makers.

Validation should not be seen as an isolated set of procedures that follows model development, but rather as an integral part of model development. Conceptually, however, the verification and validation process consists of the following components:

1. Verification is concerned with building the model correctly. It proceeds by the comparison of the conceptual model to the computer representation that implements that conception. It asks the questions: Is the model implemented correctly in the simulation software? Are the input parameters and logical structure of the model represented correctly?

2. Validation is concerned with building the correct model. It attempts to confirm that a model is an accurate representation of the real system. Validation is usually achieved through the calibration of the model, an iterative process of comparing the model to actual system behavior and using the discrepancies between the two, and the insights gained, to improve the model. This process is repeated until model accuracy is judged to be acceptable.

This chapter describes methods that have been recommended and used in the verification and validation process. Most of the methods are informal subjective comparisons; a few are formal statistical procedures. The use of the latter procedures involves issues related to output analysis, the subject of Chapters 11 and 12. Output analysis refers to analysis of the data produced by a simulation and to drawing inferences from these data about the behavior of the real system. To summarize their relationship, validation is the process by which model users gain confidence that output analysis is making valid inferences about the real system under study.

Many articles and chapters in texts have been written on verification and validation. For discussion of the main issues, the reader is referred to Balci [1994, 1998, 2003], Carson [1986, 2002], Gass [1983], Kleijnen [1995], Law and Kelton [2000], Naylor and Finger [1967], Oren [1981], Sargent [2003], Shannon [1975], and van Horn [1969, 1971]. For statistical techniques relevant to various aspects of validation, the reader can obtain the foregoing references plus those by Balci and Sargent [1982a,b; 1984a], Kleijnen [1987], and Schruben [1980]. For case studies in which validation is emphasized, the reader is referred to Carson *et al.* [1981a,b], Gafarian and Walsh [1970], Kleijnen [1993], and Shechter and Lucas [1980]. Bibliographies on validation have been published by Balci and Sargent [1984b] and by Youngblood [1993].

10.1 MODEL BUILDING, VERIFICATION, AND VALIDATION

The first step in model building consists of observing the real system and the interactions among their various components and of collecting data on their behavior. But observation alone seldom yields sufficient understanding of system behavior. Persons familiar with the system, or any subsystem, should be questioned to take advantage of their special knowledge. Operators, technicians, repair and maintenance personnel, engineers, supervisors, and managers understand certain aspects of the system that might be unfamiliar to others. As model development proceeds, new questions may arise, and the model developers will return to this step of learning true system structure and behavior.

The second step in model building is the construction of a conceptual model—a collection of assumptions about the components and the structure of the system, plus hypotheses about the values of model input parameters. As is illustrated by Figure 10.1, conceptual validation is the comparison of the real system to the conceptual model.

The third step is the implementation of an operational model, usually by using simulation software and incorporating the assumptions of the conceptual model into the worldview and concepts of the simulation software. In actuality, model building is not a linear process with three steps. Instead, the model builder will return to each of these steps many times while building, verifying, and validating the model. Figure 10.1 depicts the ongoing model building process, in

Figure 10.1   Model building, verification, and validation.

which the need for verification and validation causes continual comparison of the real system to the conceptual model and to the operational model and induces repeated modification of the model to improve its accuracy.

10.2 VERIFICATION OF SIMULATION MODELS

The purpose of model verification is to assure that the conceptual model is reflected accurately in the operational model. The conceptual model quite often involves some degree of abstraction about system operations or some amount of simplification of actual operations. Verification asks the following question: Is the conceptual model (assumptions about system components and system structure, parameter values, abstractions, and simplifications) accurately represented by the operational model?

Many common-sense suggestions can be given for use in the verification process:

1. Have the operational model checked by someone other than its developer, preferably an expert in the simulation software being used.
2. Make a flow diagram that includes each logically possible action a system can take when an event occurs, and follow the model logic for each action for each event type. (An example of a logic flow diagram is given in Figures 2.2 and 2.3 for the model of a single-server queue.)
3. Closely examine the model output for reasonableness under a variety of settings of the input parameters. Have the implemented model display a wide variety of output statistics, and examine all of them closely.

4. Have the operational model print the input parameters at the end of the simulation, to be sure that these parameter values have not been changed inadvertently.

5. Make the operational model as self-documenting as possible. Give a precise definition of every variable used and a general description of the purpose of each submodel, procedure (or major section of code), component, or other model subdivision.

6. If the operational model is animated, verify that what is seen in the animation imitates the actual system. Examples of errors that can be observed through animation are automated guided vehicles (AGVs) that pass through one another on a unidirectional path or at an intersection and entities that disappear (unintentionally) during a simulation.

7. The Interactive Run Controller (IRC) or debugger is an essential component of successful simulation model building. Even the best of simulation analysts makes mistakes or commits logical errors when building a model. The IRC assists in finding and correcting those errors in the following ways:

    (a) The simulation can be monitored as it progresses. This can be accomplished by advancing the simulation until a desired time has elapsed, then displaying model information at that time. Another possibility is to advance the simulation until a particular condition is in effect, and then display information.

    (b) Attention can be focused on a particular entity line, of code, or procedure. For instance, every time that an entity enters a specified procedure, the simulation will pause so that information can be gathered. As another example, every time that a specified entity becomes active, the simulation will pause.

    (c) Values of selected model components can be observed. When the simulation has paused, the current value or status of variables, attributes, queues, resources, counters, and so on can be observed.

    (d) The simulation can be temporarily suspended, or paused, not only to view information, but also to reassign values or redirect entities.

8. Graphical interfaces are recommended for accomplishing verification and validation [Bortscheller and Saulnier, 1992]. The graphical representation of the model is essentially a form of self-documentation. It simplifies the task of understanding the model.

These suggestions are basically the same ones any software engineer would follow.

Among these common-sense suggestions, one that is very easily implemented, but quite often overlooked, especially by students who are learning simulation, is a close and thorough examination of model output for reasonableness (suggestion 3). For example, consider a model of a complex network of queues consisting of many service centers in series and parallel configurations. Suppose that the modeler is interested mainly in the response time, defined as the time required for a customer to pass through a designated part of the network. During the verification (and calibration) phase of model development, it is recommended that the program collect and print out many statistics in addition to response times, such as utilizations of servers and time-average number of customers in various subsystems. Examination of the utilization of a server, for example, might reveal that it is unreasonably low (or high), a possible error that could be caused by wrong specification of mean service time, or by a mistake in model logic that sends too few (or too many) customers to this particular server, or by any number of other possible parameter misspecifications or errors in logic.

In a simulation language that automatically collects many standard statistics (average queue lengths, average waiting times, etc.), it takes little or no extra programming effort to display almost all statistics of interest. The effort required can be considerably greater in a general-purpose language such as Java, C, or C++, which do not have statistics-gathering capabilities to aid the programmer.

Two sets of statistics that can give a quick indication of model reasonableness are *current contents* and *total count*. These statistics apply to any system having items of some kind flowing through it, whether these items be called customers, transactions, inventory, or vehicles. "Current contents" refers to the number of items in each component of the system at a given time. "Total count" refers to the total number of items that have entered each component of the system by a given time. In some simulation software, these statistics are kept automatically and can be displayed at any point in simulation time. In other simulation software, simple counters might have to be added to the operational model and displayed at appropriate times. If the current contents in some portion of the system are high, this condition indicates that a large number of entities are delayed. If the output is displayed for successively longer simulation run times and the current contents tend to grow in a more or less linear fashion, it is highly likely that a queue is unstable and that the server(s) will fall further behind as time continues. This indicates possibly that the number of servers is too small or that a service time is misspecified. (Unstable queues were discussed in Chapter 6.) On the other hand, if the total count for some subsystem is zero, this indicates that no items entered that subsystem—again, a highly suspect occurrence. Another possibility is that the current count and total count are equal to one. This could indicate that an entity has captured a resource, but never freed that resource. Careful evaluation of these statistics for various run lengths can aid in the detection of mistakes in model logic and data misspecifications. Checking for output reasonableness will usually fail to detect the more subtle errors, but it is one of the quickest ways to discover gross errors. To aid in error detection, it is best for the model developer to forecast a reasonable range for the value of selected output statistics before making a run of the model. Such a forecast reduces the possibility of rationalizing a discrepancy and failing to investigate the cause of unusual output.

For certain models, it is possible to consider more than whether a particular statistic is reasonable. It is possible to compute certain long-run measures of performance. For example, as seen in Chapter 6, the analyst can compute the long-run server utilization for a large number of queueing systems without any special assumptions regarding interarrival or service-time distributions. Typically, the only information needed is the network configuration, plus arrival and service rates. Any measure of performance that can be computed analytically and then compared to its simulated counterpart provides another valuable tool for verification. Presumably, the objective of the simulation is to estimate some measure of performance, such as mean response time, that cannot be computed analytically; but, as illustrated by the formulas in Chapter 6 for a number of special queues ($M/M/1$, $M/G/1$, etc.), all the measures of performance in a queueing system are interrelated. Thus, if a simulation model is predicting one measure (such as utilization) correctly, then confidence in the model's predictive ability for other related measures (such as response time) is increased (even though the exact relation between the two measures is, of course, unknown in general and varies from model to model). Conversely, if a model incorrectly predicts utilization, its prediction of other quantities, such as mean response time, is highly suspect.

.

Another important way to aid the verification process is the oft-neglected documentation phase. If a model builder writes brief comments in the operational model, plus definitions of all variables and parameters, plus descriptions of each major section of the operational model, it becomes much simpler for someone else, or the model builder at a later date, to verify the model logic. Documentation is also important as a means of clarifying the logic of a model and verifying its completeness.

A more sophisticated technique is the use of a trace. In general, a trace is a detailed computer printout which gives the value of every variable (in a specified set of variables) in a computer program, every time that one of these variables changes in value. A trace designed specifically for use in a simulation program would give the value of selected variables each time the simulation clock was incremented (i.e., each time an event occurred). Thus, a simulation trace is nothing more than a detailed printout of the state of the simulation model as it changes over time.

Example 10.1

When verifying the operational model (in a general purpose language such as FORTRAN, Pascal, C or C++, or most simulation languages) of the single-server queue model of Example 2.1, an analyst made a run over 16 units of time and observed that the time-average length of the waiting line was $\hat{L}_Q = 0.4375$ customer, which is certainly reasonable for a short run of only 16 time units. Nevertheless, the analyst decided that a more detailed verification would be of value.

The trace in Figure 10.2 gives the hypothetical printout from simulation time CLOCK = 0 to CLOCK = 16 for the simple single-server queue of Example 2.1. This example illustrates how an error can be found with a trace, when no error was apparent from the examination of the summary output statistics (such as $\hat{L}_Q$). Note that, at simulation time CLOCK = 3, the number of customers in the system is NCUST = 1, but the server is idle (STATUS = 0). The source of this error could be incorrect logic, or simply not setting the attribute STATUS to the value 1 (when coding in a general purpose language or most simulation languages).

In any case, the error must be found and corrected. Note that the less sophisticated practice of examining the summary measures, or output, did not detect the error. By using equation (6.1), the reader can verify that $\hat{L}_Q$ was computed correctly from the data ($\hat{L}_Q$ is the time-average value of NCUST minus STATUS):

$$\hat{L}_Q = \frac{(0-0)3 + (1-0)2 + (0-0)6 + (1-0)1 + (2-1)4}{3+2+6+1+4}$$
$$= \frac{7}{16} = 0.4375$$

as previously mentioned. Thus, the output measure, $\hat{L}_Q$, had a reasonable value and was computed correctly from the data, but its value was indeed wrong because the attribute STATUS was not assuming correct values. As is seen from Figure 10.2, a trace yields information on the actual history of the model that is more detailed and informative than the summary measures alone.

Most simulation software has a built-in capability to conduct a trace without the programmer having to do any extensive programming. In addition, a 'print' or 'write' statement can be used to implement a tracing capability in a general-purpose language.

As can be easily imagined, a trace over a large span of simulation time can quickly produce an extremely large amount of computer printout, which would be extremely cumbersome to check

Definition of Variables:
CLOCK = Simulation clock
EVTYP = Event type (start, arrival, departure, or stop)
NCUST = Number of customers in system at time 'CLOCK'
STATUS = Status of server (1–busy, 0–idle)

State of System Just After the Named Event Occurs:

CLOCK = 0	EVTYP = 'Start'	NCUST = 0	STATUS = 0
CLOCK = 3	EVTYP = 'Arrival'	NCUST = 1	STATUS = 0
CLOCK = 5	EVTYP = 'Depart'	NCUST = 0	STATUS = 0
CLOCK = 11	EVTYP = 'Arrival'	NCUST = 1	STATUS = 0
CLOCK = 12	EVTYP = 'Arrival'	NCUST = 2	STATUS = 1
CLOCK = 16	EVTYP = 'Depart'	NCUST = 1	STATUS = 1

Figure 10.2   Simulation Trace of Example 2.1.

in detail for correctness. The purpose of the trace is to verify the correctness of the computer program by making detailed paper-and-pencil calculations. To make this practical, a simulation with a trace is usually restricted to a very short period of time. It is desirable, of course, to ensure that each type of event (such as ARRIVAL) occurs at least once, so that its consequences and effect on the model can be checked for accuracy. If an event is especially rare in occurrence, it may be necessary to use artificial data to force it to occur during a simulation of short duration. This is legitimate, as the purpose is to verify that the effect on the system of the rare event is as intended.

Some software allows a selective trace. For example, a trace could be set for specific locations in the model or could be triggered to begin at a specified simulation time. Whenever an entity goes through the designated locations, the simulation software writes a time-stamped message to a trace file. Some simulation software allows tracing a selected entity; any time the designated entity becomes active, the trace is activated and time-stamped messages are written. This trace is very useful in following one entity through the entire model. Another example of a selective trace is to set it for the occurrence of a particular condition. For example, whenever the queue before a certain resource reaches five or more, turn on the trace. This allows running the simulation until something unusual occurs, then examining the behavior from that point forward in time. Different simulation software packages support tracing to various extents. In practice, it is often implemented by the model developer by adding printed messages at appropriate points into a model.

Of the three classes of techniques—the common-sense techniques, thorough documentation, and traces—it is recommended that the first two always be carried out. Close examination of model output for reasonableness is especially valuable and informative. A generalized trace may provide voluminous data, far more than can be used or examined carefully. A selective trace can provide useful information on key model components and keep the amount of data to a manageable level.

10.3 CALIBRATION AND VALIDATION OF MODELS

Verification and validation, although conceptually distinct, usually are conducted simultaneously by the modeler. Validation is the overall process of comparing the model and its behavior to the real system and its behavior. Calibration is the iterative process of comparing the model to the real system, making adjustments (or even major changes) to the model, comparing the revised model to reality, making additional adjustments, comparing again, and so on. Figure 10.3 shows the relationship of model calibration to the overall validation process. The comparison of the model to reality is carried out by a variety of tests—some subjective, others objective. Subjective tests usually involve people, who are knowledgeable about one or more aspects of the system, making judgments about the model and its output. Objective tests always require data on the system's behavior, plus the corresponding data produced by the model. Then one or more statistical tests are performed to compare some aspect of the system data set with the same aspect of the model data set. This iterative process of comparing model with system and then revising both the conceptual and operational models to accommodate any perceived model deficiencies is continued until the model is judged to be sufficiently accurate.

A possible criticism of the calibration phase, were it to stop at this point, is that the model has been validated only for the one data set used—that is, the model has been "fitted" to one data set. One way to alleviate this criticism is to collect a new set of system data (or to reserve a portion of the original system data) to be used at this final stage of validation. That is, after the model has been calibrated by using the original system data set, a "final" validation is conducted, using the second system data set. If unacceptable discrepancies between the model and the real system are discovered in the "final" validation effort, the modeler must return to the calibration phase and modify the model until it becomes acceptable.

Validation is not an either/or proposition—no model is ever totally representative of the system under study. In addition, each revision of the model, as pictured in Figure 10.3, involves

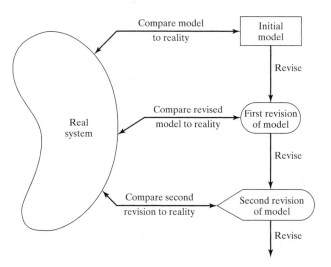

Figure 10.3 Iterative process of calibrating a model.

some cost, time, and effort. The modeler must weigh the possible, but not guaranteed, increase in model accuracy versus the cost of increased validation effort. Usually, the modeler (and model users) have some maximum discrepancy between model predictions and system behavior that would be acceptable. If this level of accuracy cannot be obtained within the budget constraints, either expectations of model accuracy must be lowered, or the model must be abandoned.

As an aid in the validation process, Naylor and Finger [1967] formulated a three-step approach that has been widely followed:

1. Build a model that has high face validity.
2. Validate model assumptions.
3. Compare the model input–output transformations to corresponding input–output transformations for the real system.

The next five subsections investigate these three steps in detail.

10.3.1 Face Validity

The first goal of the simulation modeler is to construct a model that appears reasonable on its face to model users and others who are knowledgeable about the real system being simulated. The potential users of a model should be involved in model construction from its conceptualization to its implementation, to ensure that a high degree of realism is built into the model through reasonable assumptions regarding system structure and through reliable data. Potential users and knowledgeable persons can also evaluate model output for reasonableness and can aid in identifying model deficiencies. Thus, the users can be involved in the calibration process as the model is improved iteratively by the insights gained from identification of the initial model deficiencies. Another advantage of user involvement is the increase in the model's perceived validity, or credibility, without which a manager would not be willing to trust simulation results as a basis for decision making.

Sensitivity analysis can also be used to check a model's face validity. The model user is asked whether the model behaves in the expected way when one or more input variables is changed. For example, in most queueing systems, if the arrival rate of customers (or demands for service) were to increase, it would be expected that utilizations of servers, lengths of lines, and delays would tend to increase (although by how much might well be unknown). From experience and from observations on the real system (or similar related systems), the model user and model builder would probably have some notion at least of the direction of change in model output when an input variable is increased or decreased. For most large-scale simulation models, there are many input variables and thus many possible sensitivity tests. The model builder must attempt to choose the most critical input variables for testing if it is too expensive or time consuming to vary all input variables. If real system data are available for at least two settings of the input parameters, objective scientific sensitivity tests can be conducted via appropriate statistical techniques.

10.3.2 Validation of Model Assumptions

Model assumptions fall into two general classes: structural assumptions and data assumptions. Structural assumptions involve questions of how the system operates and usually involve simplifications and abstractions of reality. For example, consider the customer queueing and service facility in a bank. Customers can form one line, or there can be an individual line for each teller. If there are many lines, customers could be served strictly on a first-come–first-served basis, or some customers could change lines if one line is moving faster. The number of tellers could be fixed or variable. These structural assumptions should be verified by actual observation during appropriate time periods and by discussions with managers and tellers regarding bank policies and actual implementation of these policies.

Data assumptions should be based on the collection of reliable data and correct statistical analysis of the data. (Example 9.1 discussed similar issues for a model of a laundromat.) For example, in the bank study previously mentioned, data were collected on

1. interarrival times of customers during several 2-hour periods of peak loading ("rush-hour" traffic);
2. interarrival times during a slack period;
3. service times for commercial accounts;
4. service times for personal accounts.

The reliability of the data was verified by consultation with bank managers, who identified typical rush hours and typical slack times. When combining two or more data sets collected at different times, data reliability can be further enhanced by objective statistical tests for homogeneity of data. (Do two data sets $\{X_i\}$ and $\{Y_i\}$ on service times for personal accounts, collected at two different times, come from the same parent population? If so, the two sets can be combined.) Additional tests might be required, to test for correlation in the data. As soon as the analyst is assured of dealing with a random sample (i.e., correlation is not present), the statistical analysis can begin.

The procedures for analyzing input data from a random sample were discussed in detail in Chapter 9. Whether done manually or by special-purpose software, the analysis consists of three steps:

1. Identify an appropriate probability distribution.
2. Estimate the parameters of the hypothesized distribution.
3. Validate the assumed statistical model by a goodness-of-fit test, such as the chi-square or Kolmogorov–Smirnov test, and by graphical methods.

The use of goodness-of-fit tests is an important part of the validation of data assumptions.

10.3.3 Validating Input–Output Transformations

The ultimate test of a model, and in fact the only objective test of the model as a whole, is the model's ability to predict the future behavior of the real system when the model input data match the real inputs and when a policy implemented in the model is implemented at some point in the system. Furthermore, if the level of some input variables (e.g., the arrival rate of customers to a service facility) were to increase or decrease, the model should accurately predict what would happen in the real system under similar circumstances. In other words, the structure of the model should be accurate enough for the model to make good predictions, not just for one input data set, but for the range of input data sets that are of interest.

In this phase of the validation process, the model is viewed as an input–output transformation—that is, the model accepts values of the input parameters and transforms these inputs into output measures of performance. It is this correspondence that is being validated.

Instead of validating the model input–output transformations by predicting the future, the modeler could use historical data that have been reserved for validation purposes only—that is, if one data set has been used to develop and calibrate the model, it is recommended that a separate data set be used as the final validation test. Thus, accurate "prediction of the past" can replace prediction of the future for the purpose of validating the model.

A model is usually developed with primary interest in a specific set of system responses to be measured under some range of input conditions. For example, in a queueing system, the responses may be server utilization and customer delay, and the range of input conditions (or input variables) may include two or three servers at some station and a choice of scheduling rules. In a production system, the response may be throughput (i.e., production per hour), and the input conditions may be a choice of several machines that run at different speeds, with each machine having its own breakdown and maintenance characteristics.

In any case, the modeler should use the main responses of interest as the primary criteria for validating a model. If the model is used later for a purpose different from its original purpose, the model should be revalidated in terms of the new responses of interest and under the possibly new input conditions.

A necessary condition for the validation of input–output transformations is that some version of the system under study exist, so that system data under at least one set of input conditions can be collected to compare to model predictions. If the system is in the planning stages and no system operating data can be collected, complete input–output validation is not possible. Other types of validation should be conducted, to the extent possible. In some cases, subsystems of the planned system may exist, and a partial input–output validation can be conducted.

Presumably, the model will be used to compare alternative system designs or to investigate system behavior under a range of new input conditions. Assume for now that some version of the system is operating and that the model of the existing system has been validated. What, then, can be said about the validity of the model when different inputs are used?—that is, if model inputs are being changed to represent a new system design, or a new way to operate the system, or even hypothesized future conditions, what can be said about the validity of the model with respect to this new but nonexistent proposed system or to the system under new input conditions?

First, the responses of the two models under similar input conditions will be used as the criteria for comparison of the existing system to the proposed system. Validation increases the modeler's confidence that the model of the existing system is accurate. Second, in many cases, the proposed system is a modification of the existing system, and the modeler hopes that confidence in the model of the existing system can be transferred to the model of the new system. This transfer of confidence usually can be justified if the new model is a relatively minor modification of the old model in terms of changes to the operational model (it may be a major change for the actual system). Changes in the operational model ranging from relatively minor to relatively major include the following:

1. minor changes of single numerical parameters, such as the speed of a machine, the arrival rate of customers (with no change in distributional form of interarrival times), the number

of servers in a parallel service center, or the mean time to failure or mean time to repair of a machine;

2. minor changes of the form of a statistical distribution, such as the distribution of a service time or a time to failure of a machine;

3. major changes in the logical structure of a subsystem, such as a change in queue discipline for a waiting-line model or a change in the scheduling rule for a job-shop model;

4. major changes involving a different design for the new system, such as a computerized inventory control system replacing an older noncomputerized system, or an automated storage-and-retrieval system replacing a warehouse system in which workers pick items manually using fork trucks.

If the change to the operational model is minor, such as in items 1 or 2, these changes can be carefully verified and output from the new model accepted with considerable confidence. If a sufficiently similar subsystem exists elsewhere, it might be possible to validate the submodel that represents the subsystem and then to integrate this submodel with other validated submodels to build a complete model. In this way, partial validation of the substantial model changes in items 3 and 4 might be possible. Unfortunately, there is no way to validate the input–output transformations of a model of a nonexisting system completely. In any case, within time and budget constraints, the modeler should use as many validation techniques as possible, including input–output validation of subsystem models if operating data can be collected on such subsystems.

Example 10.2 will illustrate some of the techniques that are possible for input–output validation and will discuss the concepts of an input variable, uncontrollable variable, decision variable, output or response variable, and input–output transformation in more detail.

Example 10.2: The Fifth National Bank of Jaspar
The Fifth National Bank of Jaspar, as shown in Figure 10.4, is planning to expand its drive-in service at the corner of Main Street. Currently, there is one drive-in window serviced by one teller. Only one or two transactions are allowed at the drive-in window, so it was assumed that each service time was a random sample from some underlying population. Service times $\{S_i, i = 1, 2, \ldots, 90\}$ and interarrival times $\{A_i, i = 1, 2, \ldots, 90\}$ were collected for the 90 customers who arrived between 11:00 A.M. and 1:00 P.M. on a Friday. This time slot was selected for data collection after consultation with management and the teller because it was felt to be representative of a typical rush hour.

Data analysis (as outlined in Chapter 9) led to the conclusion that arrivals could be modeled as a Poisson process at a rate of 45 customers per hour and that service times were approximately normally distributed, with mean 1.1 minutes and standard deviation 0.2 minute. Thus, the model has two input variables:

1. interarrival times, exponentially distributed (i.e., a Poisson arrival process) at rate $\lambda = 45$ per hour;

2. service times, assumed to be $N(1.1, (0.2)^2)$.

Each input variable has a level: the rate ($\lambda = 45$ per hour) for the interarrival times, and the mean 1.1 minutes and standard deviation 0.2 minute for the service times. The interarrival times are

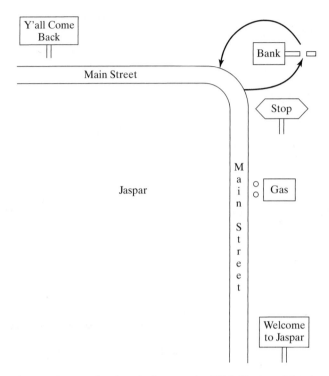

Figure 10.4   Drive-in window at the Fifth National Bank.

examples of uncontrollable variables (i.e., uncontrollable by management in the real system). The service times are also treated as uncontrollable variables, although the level of the service times might be partially controllable. If the mean service time could be decreased to 0.9 minute by installing a computer terminal, the level of the service-time variable becomes a decision variable or controllable parameter. Setting all decision variables at some level constitutes a policy. For example, the current bank policy is one teller ($D_1 = 1$), mean service time $D_2 = 1.1$ minutes, and one line for waiting cars ($D_3 = 1$). ($D_1, D_2, \ldots$ are used to denote decision variables.) Decision variables are under management's control; the uncontrollable variables, such as arrival rate and actual arrival times, are not under management's control. The arrival rate might change from time to time, but such change is treated as being due to external factors not under management control.

A model of current bank operations was developed and verified in close consultation with bank management and employees. Model assumptions were validated, as discussed in Section 10.3.2. The resulting model is now viewed as a "black box" that takes all input-variable specifications and transforms them into a set of output or response variables. The output variables consist of all statistics of interest generated by the simulation about the model's behavior. For example, management is interested in the teller's utilization at the drive-in window (percent of time the teller is busy at the window), average delay in minutes of a customer from arrival to beginning of service, and the maximum length of the line during the rush hour. These input and output variables are shown in Figure 10.5 and are listed in Table 10.1, together with some additional

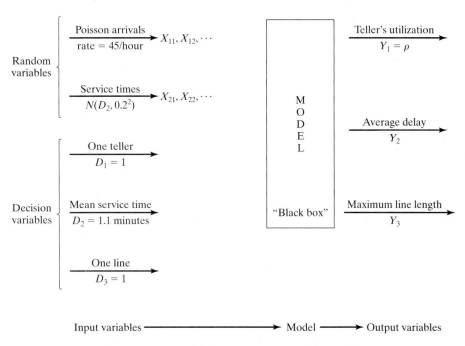

Figure 10.5 Model input–output transformation.

Table 10.1 Input and Output Variables for Model of Current Bank Operations

Input Variables	Model Output Variables, Y
D = decision variables	Variables of primary interest
X = other variables	to management (Y_1, Y_2, Y_3)
	Y_1 = teller's utilization
Poisson arrivals at rate = 45/hour	Y_2 = average delay
$X_{11}, X_{12}, \ldots$	Y_3 = maximum line length
Service times, $N(D_2, 0.2^2)$	Other output variables of
$X_{21}, X_{22}, \ldots$	secondary interest
	Y_4 = observed arrival rate
$D_1 = 1$ (one teller)	Y_5 = average service time
$D_2 = 1.1$ minutes (mean service time)	Y_6 = sample standard deviation of service
$D_3 = 1$ (one line)	times
	Y_7 = average length of waiting line

output variables. The uncontrollable input variables are denoted by X, the decision variables by D, and the output variables by Y. From the "black box" point of view, the model takes the inputs X and D and produces the outputs Y, namely

$$(X, D) \xrightarrow{f} Y$$

or

$$f(X, D) = Y$$

Here f denotes the transformation that is due to the structure of the model. For the Fifth National Bank study, the exponentially distributed interarrival time generated in the model (by the methods of Chapter 8) between customer $n - 1$ and customer n is denoted by X_{1n}. (Do not confuse X_{1n} with A_n; the latter was an observation made on the real system.) The normally distributed service time generated in the model for customer n is denoted by X_{2n}. The set of decision variables, or policy, is $D = (D_1, D_2, D_3) = (1, 1.1, 1)$ for current operations. The output, or response, variables are denoted by $Y = (Y_1, Y_2, \ldots, Y_7)$ and are defined in Table 10.1.

For validation of the input–output transformations of the bank model to be possible, real system data must be available, comparable to at least some of the model output Y of Table 10.1. The system responses should have been collected during the same time period (from 11:00 A.M. to 1:00 P.M. on the same Friday) in which the input data $\{A_i, S_i\}$ were collected. This is important because, if system response data were collected on a slower day (say, an arrival rate of 40 per hour), the system responses such as teller utilization (Z_1), average delay (Z_2), and maximum line length (Z_3) would be expected to be lower than the same variables during a time slot when the arrival rate was 45 per hour, as observed. Suppose that the delay of successive customers was measured on the same Friday between 11:00 A.M. and 1:00 P.M. and that the average delay was found to be $Z_2 = 4.3$ minutes. For the purpose of validation, we will consider this to be the true mean value $\mu_0 = 4.3$.

When the model is run with generated random variates X_{1n} and X_{2n}, it is expected that observed values of average delay, Y_2, should be close to $Z_2 = 4.3$ minutes. The generated input values (X_{1n} and X_{2n}) cannot be expected to replicate the actual input values (A_n and S_n) of the real system exactly, but they are expected to replicate the statistical pattern of the actual inputs. Hence, simulation-generated values of Y_2 are expected to be consistent with the observed system variable, $Z_2 = 4.3$ minutes. Now consider how the modeler might test this consistency.

The modeler makes a small number of statistically independent replications of the model. Statistical independence is guaranteed by using nonoverlapping sets of random numbers produced by the random-number generator or by choosing seeds for each replication independently (from a random number table). The results of six independent replications, each of 2 hours duration, are given in Table 10.2.

Observed arrival rate Y_4 and sample average service time Y_5 for each replication of the model are also noted, to be compared with the specified values of 45/hour and 1.1 minutes, respectively. The validation test consists of comparing the system response, namely average delay $Z_2 = 4.3$ minutes, to the model responses, Y_2. Formally, a statistical test of the null hypothesis

$$H_0 : E(Y_2) = 4.3 \text{ minutes}$$

versus $\qquad\qquad\qquad\qquad\qquad\qquad\qquad\qquad\qquad\qquad\qquad\qquad$ (10.1)

$$H_1 : E(Y_2) \neq 4.3 \text{ minutes}$$

is conducted. If H_0 is not rejected, then, on the basis of this test, there is no reason to consider the model invalid. If H_0 is rejected, the current version of the model is rejected, and the modeler is forced to seek ways to improve the model, as illustrated by Figure 10.3. As formulated here, the appropriate statistical test is the t test, which is conducted in the following manner:

Table 10.2   Results of Six Replications of the First Bank Model

Replication	Y_4 (Arrivals/Hour)	Y_5 (Minutes)	Y_2 = Average Delay (Minutes)
1	51	1.07	2.79
2	40	1.12	1.12
3	45.5	1.06	2.24
4	50.5	1.10	3.45
5	53	1.09	3.13
6	49	1.07	2.38
Sample mean			2.51
Standard deviation			0.82

Choose a level of significance, α, and a sample size, n. For the bank model, choose

$$\alpha = 0.05, \quad n = 6$$

Compute the sample mean, $\bar{Y}_2$, and the sample standard deviation, S, over the n replications, by using Equations (9.1) and (9.2):

$$\bar{Y}_2 = \frac{1}{n} \sum_{i=1}^{n} Y_{2i} = 2.51 \text{ minutes}$$

and

$$S = \left[\frac{\sum_{i=1}^{n} (Y_{2i} - \bar{Y}_2)^2}{n-1} \right]^{1/2} = 0.82 \text{ minute}$$

where $Y_{2i}, i = 1, \ldots, 6$, are as shown in Table 10.2.

Get the critical value of t from Table A.5. For a two-sided test, such as that in equation (10.1), use $t_{\alpha/2,n-1}$; for a one-sided test, use $t_{\alpha,n-1}$ or $-t_{\alpha,n-1}$, as appropriate ($n - 1$ being the degrees of freedom). From Table A.5, $t_{0.025,5} = 2.571$ for a two-sided test.

Compute the test statistic

$$t_0 = \frac{\bar{Y}_2 - \mu_0}{S/\sqrt{n}} \tag{10.2}$$

where μ_0 is the specified value in the null hypothesis, H_0. Here $\mu_0 = 4.3$ minutes, so that

$$t_0 = \frac{2.51 - 4.3}{0.82/\sqrt{6}} = -5.34$$

For the two-sided test, if $|t_0| > t_{\alpha/2,n-1}$, reject H_0. Otherwise, do not reject H_0. [For the one-sided test with $H_1 : E(Y_2) > \mu_0$, reject H_0 if $t > t_{\alpha,n-1}$; with $H_1 : E(Y_2) < \mu_0$, reject H_0 if $t < -t_{\alpha,n-1}$.]

Since $|t| = 5.34 > t_{0.025,5} = 2.571$, reject H_0, and conclude that the model is inadequate in its prediction of average customer delay.

Recall that, in the testing of hypotheses, rejection of the null hypothesis H_0 is a strong conclusion, because

$$P(H_0 \text{ rejected} \mid H_0 \text{ is true}) = \alpha \qquad (10.3)$$

and the level of significance α is chosen small, say $\alpha = 0.05$, as was done here. Equation (10.3) says that the probability of making the error of rejecting H_0 when H_0 is in fact true is low ($\alpha = 0.05$)—that is, the probability is small of declaring the model invalid when it is valid (with respect to the variable being tested). The assumptions justifying a t test are that the observations (Y_{2i}) are normally and independently distributed. Are these assumptions met in the present case?

1. The ith observation Y_{2i} is the average delay of all drive-in customers who began service during the ith simulation run of 2 hours; thus, by a Central Limit Theorem effect, it is reasonable to assume that each observation Y_{2i} is approximately normally distributed, provided that the number of customers it is based on is not too small.
2. The observations $Y_{2i}, i = 1, \ldots, 6$, are statistically independent by design—that is, by choice of the random-number seeds independently for each replication or by use of nonoverlapping streams.
3. The t statistic computed by Equation (10.2) is a robust statistic—that is, it is distributed approximately as the t distribution with $n - 1$ degrees of freedom, even when $Y_{21}, Y_{22}, \ldots$ are not exactly normally distributed, and thus the critical values in Table A.5 can reliably be used.

Now that the model of the Fifth National Bank of Jaspar has been found lacking, what should the modeler do? Upon further investigation, the modeler realized that the model contained two unstated assumptions:

1. When a car arrived to find the window immediately available, the teller began service immediately.
2. There is no delay between one service ending and the next beginning, when a car is waiting.

Assumption 2 was found to be approximately correct, because a service time was considered to begin when the teller actually began service but was not considered to have ended until the car had exited the drive-in window and the next car, if any, had begun service, or the teller saw that the line was empty. On the other hand, assumption 1 was found to be incorrect because the teller had other duties—mainly, serving walk-in customers if no cars were present—and tellers always finished with a previous customer before beginning service on a car. It was found that walk-in customers were always present during rush hour; that the transactions were mostly commercial in nature, taking a considerably longer time than the time required to service drive-up customers; and that, when an arriving car found no other cars at the window, it had to wait until the teller finished with the present walk-in customer. To correct this model inadequacy, the structure of the model was changed to include the additional demand on the teller's time, and data were collected

Table 10.3 Results of Six Replications of the Revised Bank Model

Obs. Avr. Rate   Avg. Svc Time

Replication	Y_4 (Arrivals/Hour)	Y_5 (Minutes)	Y_2 = Average Delay (Minutes)
1	51	1.07	5.37
2	40	1.11	1.98
3	45.5	1.06	5.29
4	50.5	1.09	3.82
5	53	1.08	6.74
6	49	1.08	5.49
Sample mean			4.78
Standard deviation			1.66

on service times of walk-in customers. Analysis of these data found that they were approximately exponentially distributed with a mean of 3 minutes.

The revised model was run, yielding the results in Table 10.3. A test of the null hypothesis $H_0 : E(Y_2) = 4.3$ minutes [as in equation (10.1)] was again conducted, according to the procedure previously outlined.

Choose $\alpha = 0.05$ and $n = 6$ (sample size).

Compute $\bar{Y}_2 = 4.78$ minutes, $S = 1.66$ minutes.

Look up, in Table A.5, the critical value $t_{0.25,5} = 2.571$.

Compute the test statistic $t_0 = (\bar{Y}_2 - \mu_0)/(S/\sqrt{n}) = 0.710$.

Since $|t_0| < t_{0.025,5} = 2.571$, do not reject H_0, and thus tentatively accept the model as valid.

Failure to reject H_0 must be considered as a weak conclusion unless the power of the test has been estimated and found to be high (close to 1)—that is, it can be concluded only that the data at hand ($Y_{21}, \ldots, Y_{26}$) were not sufficient to reject the hypothesis $H_0 : \mu_0 = 4.3$ minutes. In other words, this test detects no inconsistency between the sample data ($Y_{21}, \ldots, Y_{26}$) and the specified mean μ_0.

The power of a test is the probability of detecting a departure from $H_0 : \mu = \mu_0$ when in fact such a departure exists. In the validation context, the power of the test is the probability of detecting an invalid model. The power may also be expressed as 1 minus the probability of a Type II, or β, error, where $\beta = P(\text{Type II error}) = P(\text{failing to reject } H_0|H_1 \text{ is true})$ is the probability of accepting the model as valid when it is not valid.

To consider failure to reject H_0 as a strong conclusion, the modeler would want β to be small. Now, β depends on the sample size n and on the true difference between $E(Y_2)$ and $\mu_0 = 4.3$ minutes—that is, on

$$\delta = \frac{|E(Y_2) - \mu_0|}{\sigma}$$

where σ, the population standard deviation of an individual Y_{2i}, is estimated by S. Tables A.10 and A.11 are typical operating-characteristic (OC) curves, which are graphs of the probability of a Type II error $\beta(\delta)$ versus δ for given sample size n. Table A.10 is for a two-sided t test; Table A.11 is for a one-sided t test. Suppose that the modeler would like to reject H_0 (model validity) with probability at least 0.90 if the true mean delay of the model, $E(Y_2)$, differed from the average delay in the system, $\mu_0 = 4.3$ minutes, by 1 minute. Then δ is estimated by

$$\hat{\delta} = \frac{|E(Y_2) - \mu_0|}{S} = \frac{1}{1.66} = 0.60$$

For the two-sided test with $\alpha = 0.05$, use of Table A.10 results in

$$\beta(\hat{\delta}) = \beta(0.6) = 0.75 \text{ for } n = 6$$

To guarantee that $\beta(\hat{\delta}) \leq 0.10$, as was desired by the modeler, Table A.10 reveals that a sample size of approximately $n = 30$ independent replications would be required—that is, for a sample size $n = 6$ and assuming that the population standard deviation is 1.66, the probability of accepting H_0 (model validity), when in fact the model is invalid ($|E(Y_2) - \mu_0| = 1$ minute), is $\beta = 0.75$, which is quite high. If a 1-minute difference is critical, and if the modeler wants to control the risk of declaring the model valid when model predictions are as much as 1 minute off, a sample size of $n = 30$ replications is required to achieve a power of 0.9. If this sample size is too high, either a higher β risk (lower power) or a larger difference δ must be considered.

In general, it is always best to control the Type II error, or β error, by specifying a critical difference δ and choosing a sample size by making use of an appropriate OC curve. (Computation of power and use of OC curves for a wide range of tests is discussed in Hines, Montgomery, Goldsman, and Borror [2002].) In summary, in the context of model validation, the Type I error is the rejection of a valid model and is easily controlled by specifying a small level of significance α (say $\alpha = 0.1, 0.05,$ or 0.01). The Type II error is the acceptance of a model as valid when it is invalid. For a fixed sample size n, increasing α will decrease β, the probability of a Type II error. Once α is set, and the critical difference to be detected is selected, the only way to decrease β is to increase the sample size. A Type II error is the more serious of the two types of errors; thus, it is important to design the simulation experiments to control the risk of accepting an invalid model. The two types of error are summarized in Table 10.4, which compares statistical terminology to modeling terminology.

Note that validation is not to be viewed as an either/or proposition, but rather should be viewed in the context of calibrating a model, as conceptually exhibited in Figure 10.3. If the current version of the bank model produces estimates of average delay (Y_2) that are not close enough to real system behavior ($\mu_0 = 4.3$ minutes), the source of the discrepancy is sought, and the model is revised in light of this new knowledge. This iterative scheme is repeated until model accuracy is judged adequate.

Table 10.4   Types of Error in Model Validation

Statistical Terminology	Modeling Terminology	Associated Risk
Type I: rejecting H_0 when H_0 is true	Rejecting a valid model	α
Type II: failure to reject H_0 when H_1 is true	Failure to reject an invalid model	β

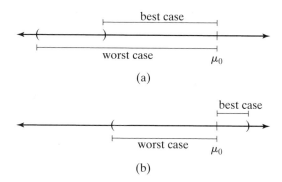

Figure 10.6 Validation of the input–output transformation (a) when the true value falls outside, (b) when the true value falls inside, the confidence interval.

Philosophically, the hypothesis-testing approach tries to evaluate whether the simulation and the real system are *the same* with respect to some output performance measure or measures. A different, but closely related, approach is to attempt to evaluate whether the simulation and the real-system performance measures are *close enough* by using confidence intervals.

We continue to assume that there is a known output performance measure for the existing system, denoted by μ_0, and an unknown performance measure of the simulation, μ, that we hope is close. The hypothesis-testing formulation tested whether $\mu = \mu_0$; the confidence-interval formulation tries to bound the difference $|\mu - \mu_0|$ to see whether it is $\leq \varepsilon$, a difference that is small enough to allow valid decisions to be based on the simulation. The value of ε is set by the analyst.

Specifically, if Y is the simulation output, and $\mu = E(Y)$, then we execute the simulation and form a confidence interval for μ, such as $\bar{Y} \pm t_{\alpha/2,n-1}S/\sqrt{n}$. The determination of whether to accept the model as valid or to refine the model depends on the best-case and worst-case error implied by the confidence interval.

1. Suppose the confidence interval does not contain μ_0. (See Figure 10.6(a).)
   (a) If the best-case error is $> \varepsilon$, then the difference in performance is large enough, even in the best case, to indicate that we need to refine the simulation model.
   (b) If the worst-case error is $\leq \varepsilon$, then we can accept the simulation model as close enough to be considered valid.
   (c) If the best-case error is $\leq \varepsilon$, but the worst-case error is $> \varepsilon$, then additional simulation replications are necessary to shrink the confidence interval until a conclusion can be reached.
2. Suppose the confidence interval does contain μ_0. (See Figure 10.6(b).)
   (a) If either the best-case or worst-case error is $> \varepsilon$, then additional simulation replications are necessary to shrink the confidence interval until a conclusion can be reached.
   (b) If the worst-case error is $\leq \varepsilon$, then we can accept the simulation model as close enough to be considered valid.

In Example 10.2, $\mu_0 = 4.3$ minutes, and "close enough" was $\varepsilon = 1$ minute of expected customer delay. A 95% confidence interval, based on the 6 replications in Table 10.2, is

$$\bar{Y} \pm t_{0.025,5} S / \sqrt{n}$$

$$2.51 \pm 2.571(0.82/\sqrt{6})$$

yielding the interval [1.65, 3.37]. As in Figure 10.6(a), $\mu_0 = 4.3$ falls outside the confidence interval. Since in the best case $|3.37 - 4.3| = 0.93 < 1$, but in the worst case $|1.65 - 4.3| = 2.65 > 1$, additional replications are needed to reach a decision.

10.3.4 Input–Output Validation: Using Historical Input Data

When using artificially generated data as input data, as was done to test the validity of the bank models in Section 10.3.3, the modeler expects the model to produce event patterns that are compatible with, but not identical to, the event patterns that occurred in the real system during the period of data collection. Thus, in the bank model, artificial input data $\{X_{1n}, X_{2n}, n = 1, 2, \ldots\}$ for interarrival and service times were generated, and replicates of the output data Y_2 were compared to what was observed in the real system by means of the hypothesis test stated in equation (10.1). An alternative to generating input data is to use the actual historical record, $\{A_n, S_n, n = 1, 2, \ldots\}$, to drive the simulation model and then to compare model output with system data.

To implement this technique for the bank model, the data $A_1, A_2, \ldots$ and $S_1, S_2, \ldots$ would have to be entered into the model into arrays, or stored in a file to be read as the need arose. Just after customer n arrived at time $t_n = \sum_{i=1}^{n} A_i$, customer $n + 1$ would be scheduled on the future event list to arrive at future time $t_n + A_{n+1}$ (without any random numbers being generated). If customer n were to begin service at time t'_n, a service completion would be scheduled to occur at time $t'_n + S_n$ This event scheduling without random-number generation could be implemented quite easily in a general-purpose programming language or most simulation languages by using arrays to store the data or reading the data from a file.

When using this technique, the modeler hopes that the simulation will duplicate as closely as possible the important events that occurred in the real system. In the model of the Fifth National Bank of Jaspar, the arrival times and service durations will exactly duplicate what happened in the real system on that Friday between 11:00 A.M. and 1:00 P.M. If the model is sufficiently accurate, then the delays of customers, lengths of lines, utilizations of servers, and departure times of customers predicted by the model will be close to what actually happened in the real system. It is, of course, the model-builder's and model-user's judgment that determines the level of accuracy required.

To conduct a validation test using historical input data, it is important that all the input data $(A_n, S_n, \ldots)$ and all the system response data, such as average delay (Z_2), be collected during the same time period. Otherwise, the comparison of model responses to system responses, such as the comparison of average delay in the model (Y_2) to that in the system (Z_2), could be misleading. The responses $(Y_2$ and $Z_2)$ depend both on the inputs $(A_n$ and $S_n)$ and on the structure of the system (or model). Implementation of this technique could be difficult for a large system, because of the need

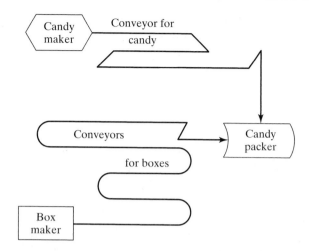

Figure 10.7 Production line at the candy factory.

for simultaneous data collection of all input variables and those response variables of primary interest. In some systems, electronic counters and devices are used to ease the data-collection task by automatically recording certain types of data. The following example was based on two simulation models reported in Carson *et al.* [1981a, b], in which simultaneous data collection and the subsequent validation were both completed successfully.

Example 10.3: The Candy Factory

The production line at the Sweet Lil' Things Candy Factory in Decatur consists of three machines that make, package, and box their famous candy. One machine (the candy maker) makes and wraps individual pieces of candy and sends them by conveyor to the packer. The second machine (the packer) packs the individual pieces into a box. A third machine (the box maker) forms the boxes and supplies them by conveyor to the packer. The system is illustrated in Figure 10.7.

Each machine is subject to random breakdowns due to jams and other causes. These breakdowns cause the conveyor to begin to empty or fill. The conveyors between the two makers and the packer are used as a temporary storage buffer for in-process inventory. In addition to the randomly occurring breakdowns, if the candy conveyor empties, a packer runtime is interrupted and the packer remains idle until more candy is produced. If the box conveyor empties because of a long random breakdown of the box machine, an operator manually places racks of boxes onto the packing machine. If a conveyor fills, the corresponding maker becomes idle. The purpose of the model is to investigate the frequency of those operator interventions that require manual loading of racks of boxes as a function of various combinations of individual machines and lengths of conveyor. Different machines have different production speeds and breakdown characteristics, and longer conveyors can hold more in-process inventory. The goal is to hold operator interventions to an acceptable level while maximizing production. Machine stoppages (whether due to a full or an empty conveyor) cause damage to the product, so this is also a factor in production.

A simulation model of the Candy Factory was developed, and a validation effort using historical inputs was conducted. Engineers in the Candy Factory set aside a 4-hour time slot from

7:00 A.M. to 11:00 A.M. to collect data on an existing production line. For each machine—say, machine i—time to failure and downtime duration

$$T_{i1}, D_{i1}, T_{i2}, D_{i2} \ldots$$

were collected. For machine $i (i = 1, 2, 3)$, T_{ij} is the jth runtime (or time to failure), and D_{ij} is the successive downtime. A runtime, T_{ij}, can be interrupted by a full or empty conveyor (as appropriate), but resumes when conditions are right. Initial system conditions at 7:00 A.M. were recorded so that they could be duplicated in the model as initial conditions at time 0. Additionally, system responses of primary interest—the production level (Z_1), and the number (Z_2) and time of occurrence (Z_3) of operator interventions—were recorded for comparison with model predictions.

The system input data, T_{ij} and D_{ij}, were fed into the model and used as runtimes and random downtimes. The structure of the model determined the occurrence of shutdowns due to a full or empty conveyor and the occurrence of operator interventions. Model response variables ($Y_i, i = 1, 2, 3$) were collected for comparison to the corresponding system response variables ($Z_i, i = 1, 2, 3$).

The closeness of model predictions to system performance aided the engineering staff considerably in convincing management of the validity of the model. These results are shown in Table 10.5. A simple display such as Table 10.5 can be quite effective in convincing skeptical engineers and managers of a model's validity—perhaps more effectively than the most sophisticated statistical methods!

With only one set of historical input and output data, only one set of simulated output data can be obtained, and thus no simple statistical tests are possible that are based on summary measures; but, if K historical input data sets are collected, and K observations $Z_{i1}, Z_{i2}, \ldots, Z_{iK}$ of some system response variable, Z_i, are collected, such that the output measure Z_{ij} corresponds to the jth input set, an objective statistical test becomes possible. For example, Z_{ij} could be the average delay of all customers who were served during the time the jth input data set was collected. With the K input data sets in hand, the modeler now runs the model K times, once for each input set, and observes the simulated results $W_{i1}, W_{i2}, \ldots, W_{iK}$ corresponding to $Z_{ij}, j = 1, \ldots, K$. Continuing the same example, W_{ij} would be the average delay predicted by the model for the jth input set. The data available for comparison appears as in Table 10.6.

If the K input data sets are fairly homogeneous, it is reasonable to assume that the K observed differences $d_j = Z_{ij} - W_{ij}, j = 1, \ldots, K$, are identically distributed. Furthermore, if the collection of the K sets of input data was separated in time—say, on different days—it is reasonable to assume that the K differences $d_1, \ldots, d_K$ are statistically independent and, hence, that the differences $d_1, \ldots, d_K$ constitute a random sample. In many cases, each Z_i and W_i is a sample

Table 10.5 Validation of the Candy-Factory Model

Response, i	System, Z_i	Model, Y_i
1. Production level	897,208	883,150
2. Number of operator interventions	3	3
3. Time of occurrence	7:22, 8:41, 10:10	7:24, 8:42, 10:14

Table 10.6 Comparison of System and Model Output Measures for Identical Historical Inputs

Input Data Set	System Output, Z_{ij}	Model Output, W_{ij}	Observed Difference, d_j	Squared Deviation from Mean, $(d_j - \bar{d})^2$
1	Z_{i1}	W_{i1}	$d_1 = Z_{i1} - W_{i1}$	$(d_1 - \bar{d})^2$
2	Z_{i2}	W_{i2}	$d_2 = Z_{i2} - W_{i2}$	$(d_2 - \bar{d})^2$
3	Z_{i3}	W_{i3}	$d_3 = Z_{i3} - W_{i3}$	$(d_3 - \bar{d})^2$
.	.	.	.	.
.	.	.	.	.
.	.	.	.	.
K	Z_{iK}	W_{iK}	$d_K = Z_{iK} - W_{iK}$	$(d_K - \bar{d})^2$

$$\bar{d} = \frac{1}{K} \sum_{j=1}^{K} d_j \qquad S_d^2 = \frac{1}{K-1} \sum_{j=1}^{K} (d_j - \bar{d})^2$$

average over customers, and so (by the Central Limit Theorem) the differences $d_j = Z_{ij} - W_{ij}$ are approximately normally distributed with some mean μ_d and variance σ_d^2. The appropriate statistical test is then a t test of the null hypothesis of no mean difference:

$$H_0 : \mu_d = 0$$

versus the alternative of significant difference:

$$H_1 : \mu_d \neq 0$$

The proper test is a paired t test (Z_{i1} is paired with W_{i1}, each having been produced by the first input data set, and so on). First, compute the sample mean difference, $\bar{d}$, and the sample variance, S_d^2, by the formulas given in Table 10.6. Then, compute the t statistic as

$$t_0 = \frac{\bar{d} - \mu_d}{S_d / \sqrt{K}} \tag{10.4}$$

(with $\mu_d = 0$), and get the critical value $t_{\alpha/2, K-1}$ from Table A.5, where α is the prespecified significance level and $K - 1$ is the number of degrees of freedom. If $|t_0| > t_{\alpha/2, K-1}$, reject the hypothesis H_0 of no mean difference, and conclude that the model is inadequate. If $|t_0| \leq t_{\alpha/2, K-1}$, do not reject H_0, and hence conclude that this test provides no evidence of model inadequacy.

Example 10.4: The Candy Factory, Continued _____

Engineers at the Sweet Lil' Things Candy Factory decided to expand the initial validation effort reported in Example 10.3. Electronic devices were installed that could automatically monitor one of the production lines, and the validation effort of Example 10.3 was repeated with $K = 5$ sets of input data. The system and the model were compared on the basis of production level. The results are shown in Table 10.7.

Table 10.7   Validation of the Candy-Factory Model (Continued)

Input Data Set, j	System Production, Z_{1j}	Model Production, W_{1j}	Observed Difference, d_j	Squared Deviation from Mean, $(d_j - \bar{d})^2$
1	897,208	883,150	14,058	7.594×10^7
2	629,126	630,550	−1,424	4.580×10^7
3	735,229	741,420	−6,191	1.330×10^8
4	797,263	788,230	9,033	1.362×10^7
5	825,430	814,190	11,240	3.4772×10^7
			$\bar{d} = 5,343.2$	$S_d^2 = 7.580 \times 10^7$

A paired t test was conducted to test $H_0 : \mu_d = 0$, or equivalently, $H_0 : E(Z_1) = E(W_1)$, where Z_1 is the system production level and W_1 is the production level predicted by the simulated model. Let the level of significance be $\alpha = 0.05$. Using the results in Table 10.7, the test statistic, as given by equation (10.4), is

$$t_0 = \frac{\bar{d}}{S_d/\sqrt{K}} = \frac{5343.2}{8705.85/\sqrt{5}} = 1.37$$

From Table A.5, the critical value is $t_{\alpha/2,K-1} = t_{0.025,4} = 2.78$. Since $|t_0| = 1.37 < t_{0.025,4} = 2.78$, the null hypothesis cannot be rejected on the basis of this test—that is, no inconsistency is detected between system response and model predictions in terms of mean production level. If H_0 had been rejected, the modeler would have searched for the cause of the discrepancy and revised the model, in the spirit of Figure 10.3.

10.3.5 Input–Output Validation: Using a Turing Test

In addition to statistical tests, or when no statistical test is readily applicable, persons knowledgeable about system behavior can be used to compare model output to system output. For example, suppose that five reports of system performance over five different days are prepared, and simulation output data are used to produce five "fake" reports. The 10 reports should all be in exactly the same format and should contain information of the type that managers and engineers have previously seen on the system. The 10 reports are randomly shuffled and given to the engineer, who is asked to decide which reports are fake and which are real. If the engineer identifies a substantial number of the fake reports, the model builder questions the engineer and uses the information gained to improve the model. If the engineer cannot distinguish between fake and real reports with any consistency, the modeler will conclude that this test provides no evidence of model inadequacy. For further discussion and an application to a real simulation, the reader is referred to Schruben [1980]. This type of validation test is commonly called a Turing test. Its use as model development proceeds can be a valuable tool in detecting model inadequacies and, eventually, in increasing model credibility as the model is improved and refined.

10.4 SUMMARY

Validation of simulation models is of great importance. Decisions are made on the basis of simulation results; thus, the accuracy of these results should be subject to question and investigation.

Quite often, simulations appear realistic on the surface because simulation models, unlike analytic models, can incorporate any level of detail about the real system. To avoid being "fooled" by this apparent realism, it is best to compare system data to model data and to make the comparison by using a wide variety of techniques, including an objective statistical test, if at all possible.

As discussed by Van Horn [1969, 1971], some of the possible validation techniques, in order of increasing cost-to-value ratios, include

1. Develop models with high face validity by consulting persons knowledgeable about system behavior on both model structure, model input, and model output. Use any existing knowledge in the form of previous research and studies, observation, and experience.
2. Conduct simple statistical tests of input data for homogeneity, for randomness, and for goodness of fit to assumed distributional forms.
3. Conduct a Turing test. Have knowledgeable people (engineers, managers) compare model output to system output and attempt to detect the difference.
4. Compare model output to system output by means of statistical tests.
5. After model development, collect new system data and repeat techniques 2 to 4.
6. Build the new system (or redesign the old one) conforming to the simulation results, collect data on the new system, and use the data to validate the model (not recommended if this is the only technique used).
7. Do little or no validation. Implement simulation results without validating. (Not recommended.)

It is usually too difficult, too expensive, or too time consuming to use all possible validation techniques for every model that is developed. It is an important part of the model-builder's task to choose those validation techniques most appropriate, both to assure model accuracy and to promote model credibility.

REFERENCES

BALCI, O. [1994], "Validation, Verification and Testing Techniques throughout the Life Cycle of a Simulation Study," *Annals of Operations Research*, Vol. 53, pp. 121–174.

BALCI, O. [1998] "Verification, Validation, and Testing," in *Handbook of Simulation*, J. Banks, ed., John Wiley, New York.

BALCI, O. [2003], "Verification, Validation, and Certification of Modeling and Simulation Applications," in *Proceedings of the 2003 Winter Simulation Conference*, ed. by S. Chick, P. J. Sánchez, D. Ferrin, and D. J. Morrice, pp. 150–158, Association for Computing Machinery, New York.

BALCI, O., AND R. G. SARGENT [1982a], "Some Examples of Simulation Model Validation Using Hypothesis Testing," in *Proceedings of the Winter Simulation Conference*, ed. by H. J. Highland, Y. W. Chao, and O. S. Madrigal, pp. 620–629, Association for Computing Machinery, New York.

BALCI, O., AND R. G. SARGENT [1982b], "Validation of Multivariate Response Models Using Hotelling's Two-Sample T^2 Test," *Simulation*, Vol. 39, No. 6 (Dec), pp. 185–192.

BALCI, O., AND R. G. SARGENT [1984a], "Validation of Simulation Models via Simultaneous Confidence Intervals," *American Journal of Mathematical Management Sciences*, Vol. 4, Nos. 3 & 4, pp. 375–406.

BALCI, O., AND R. G. SARGENT [1984b], "A Bibliography on the Credibility Assessment and Validation of Simulation and Mathematical Models," *Simuletter*, Vol. 15, No. 3, pp. 15–27.

BORTSCHELLER, B. J., AND E. T. SAULNIER [1992], "Model Reusability in a Graphical Simulation Package," in *Proceedings of the Winter Simulation Conference*, ed. by J. J. Swain, D. Goldsman, R. C. Crain, and J. R. Wilson, pp. 764–772, Association for Computing Machinery, New York.

CARSON, J. S. [2002], "Model Verification and Validation," in *Proceedings of the Winter Simulation Conference*, ed. by E. Yücesan, C.-H. Chen, J. L. Snowdon, and J. M. Charnes, pp. 52–58, Association for Computing Machinery, New York.

CARSON, J. S., N. WILSON, D. CARROLL, AND C. H. WYSOWSKI [1981a], "A Discrete Simulation Model of a Cigarette Fabrication Process," *Proceedings of the Twelfth Modeling and Simulation Conference*, University of Pittsburgh, PA.

CARSON, J. S., N. WILSON, D. CARROLL, AND C. H. WYSOWSKI [1981b], "Simulation of a Filter Rod Manufacturing Process," *Proceedings of the 1981 Winter Simulation Conference*, ed. by T. I. Oren, C. M. Delfosse, and C. M. Shub, pp. 535–541, Association for Computing Machinery, New York.

CARSON, J. S., [1986], "Convincing Users of Model's Validity is Challenging Aspect of Modeler's Job," *Industrial Engineering*, June, pp. 76–85.

GAFARIAN, A. V., AND J. E. WALSH [1970], "Methods for Statistical Validation of a Simulation Model for Freeway Traffic near an On-Ramp," *Transportation Research*, Vol. 4, p. 379–384.

GASS, S. I. [1983], "Decision-Aiding Models: Validation, Assessment, and Related Issues for Policy Analysis," *Operations Research*, Vol. 31, No. 4, pp. 601–663.

HINES, W. W., D. C. MONTGOMERY, D. M. GOLDSMAN, AND C. M. BORROR [2002], *Probability and Statistics in Engineering*, 4th ed., Wiley, New York.

KLEIJNEN, J. P. C. [1987], *Statistical Tools for Simulation Practitioners*, Marcel Dekker, New York.

KLEIJNEN, J. P. C. [1993], "Simulation and Optimization in Production Planning: A Case Study," *Decision Support Systems*, Vol. 9, pp. 269–280.

KLEIJNEN, J. P. C. [1995], "Theory and Methodology: Verification and Validation of Simulation Models," *European Journal of Operational Research*, Vol. 82, No. 1, pp. 145–162.

LAW, A. M., AND W. D. KELTON [2000], *Simulation Modeling and Analysis*, 3d ed., McGraw–Hill, New York.

NAYLOR, T. H., AND J. M. FINGER [1967], "Verification of Computer Simulation Models," *Management Science*, Vol. 2, pp. B92–B101.

OREN, T. [1981], "Concepts and Criteria to Assess Acceptability of Simulation Studies: A Frame of Reference," *Communications of the Association for Computing Machinery*, Vol. 24, No. 4, pp. 180–89.

SARGENT, R. G. [2003], "Verification and Validation of Simulation Models," in *Proceedings of the 2003 Winter Simulation Conference*, ed. by S. Chick, P. J. Sánchez, D. Ferrin, and D. J. Morrice, pp. 37–48, Association for Computing Machinery, New York.

SCHECTER, M., AND R. C. LUCAS [1980], "Validating a Large Scale Simulation Model of Wilderness Recreation Travel," *Interfaces*, Vol. 10, pp. 11–18.

SCHRUBEN, L. W. [1980], "Establishing the Credibility of Simulations," *Simulation*, Vol. 34, pp. 101–105.

SHANNON, R. E. [1975], *Systems Simulation: The Art and Science*. Prentice–Hall, Englewood Cliffs, N.J.

VAN HORN, R. L. [1969], "Validation," in *The Design of Computer Simulation Experiments*, ed. by T. H. Naylor, Duke University Press, Durham, N.C.

VAN HORN, R. L. [1971], "Validation of Simulation Results," *Management Science*, Vol. 17, pp. 247–258.

YOUNGBLOOD, S. M. [1993] "Literature Review and Commentary on the Verification, Validation and Accreditation of Models," in *Proceedings of the Summer Computer Simulation Conference*, Laurel, MD.

EXERCISES

1. A simulation model of a job shop was developed to investigate different scheduling rules. To validate the model, the scheduling rule currently used was incorporated into the model and the resulting output was compared against observed system behavior. By searching the previous year's database records, it was estimated that the average number of jobs in the shop was 22.5 on a given day. Seven independent replications of the model were run, each of 30 days' duration, with the following results for average number of jobs in the shop:

$$18.9 \quad 22.0 \quad 19.4 \quad 22.1 \quad 19.8 \quad 21.9 \quad 20.2$$

   (a) Develop and conduct a statistical test to evaluate whether model output is consistent with system behavior. Use the level of significance $\alpha = 0.05$.
   (b) What is the power of this test if a difference of two jobs is viewed as critical? What sample size is needed to guarantee a power of 0.8 or higher? (Use $\alpha = 0.05$.)

2. System data for the job shop of Exercise 1 revealed that the average time spent by a job in the shop was approximately 4 working days. The model made the following predictions, on seven independent replications, for average time spent in the shop:

$$3.70 \quad 4.21 \quad 4.35 \quad 4.13 \quad 3.83 \quad 4.32 \quad 4.05$$

   (a) Is model output consistent with system behavior? Conduct a statistical test, using the level of significance $\alpha = 0.01$.
   (b) If it is important to detect a difference of 0.5 day, what sample size is needed to have a power of 0.90? Interpret your results in terms of model validity or invalidity. (Use $\alpha = 0.01$.)

3. For the job shop of Exercise 1, four sets of input data were collected over four different 10-day periods, together with the average number of jobs in the shop (Z_i) for each period. The input data were used to drive the simulation model for four runs of 10 days each, and model predictions of average number of jobs in the shop (Y_i) were collected, with these results:

i	1	2	3	4
Z_i	21.7	19.2	22.8	19.4
Y_i	24.6	21.1	19.7	24.9

   (a) Conduct a statistical test to check the consistency of system output and model output. Use the level of significance $\alpha = 0.05$.
   (b) If a difference of two jobs is viewed as important to detect, what sample size is required to guarantee a probability of at least 0.80 of detecting this difference if it indeed exists? (Use $\alpha = 0.05$.)

4. Obtain at least two of the papers or reports listed in the References dealing with validation and verification. Write a short essay comparing and contrasting the various philosophies and approaches to the topic of verification and validation.

5. Find several examples of actual simulations reported in the literature in which the authors discuss validation of their model. Is enough detail given to judge the adequacy of the validation effort? If so, compare the reported validation with the criteria set forth in this chapter. Did the authors use any validation technique not discussed in this chapter? [Several potential sources of articles on simulation applications include the journal *Interfaces* and *Simulation*, and the *Winter Simulation Conference Proceedings* at www.informs-cs.org.]

6. Compare and contrast the various simulation-software packages in their capability to aid the modeler in the often arduous task of debugging and verification. (Articles discussing the nature of simulation software packages may be found in the *Winter Simulation Conference Proceedings* at www.informs-cs.org.)

7. (a) Compare validation in simulation to the validation of theories in the physical sciences.

(b) Compare the issues involved and the techniques available for validation of models of physical systems versus models of social systems.

(c) Contrast the difficulties, and compare the techniques, in validating a model of a manually operated warehouse with fork trucks and other manually operated vehicles, versus a model of a facility with automated guided vehicles, conveyors, and an automated storage-and-retrieval system.

(d) Repeat (c) for a model of a production system involving considerable manual labor and human decision making, versus a model of the same production system after it has been automated.

8. Use the confidence-interval approach to assess the validity of the revised bank model in Example 10.2, using the output data in Table 10.3.

11

Output Analysis for a Single Model

Output analysis is the examination of data generated by a simulation. Its purpose is either to predict the performance of a system or to compare the performance of two or more alternative system designs. This chapter deals with the analysis of a single system; Chapter 12 deals with the comparison of two or more systems. The need for statistical output analysis is based on the observation that the output data from a simulation exhibits random variability when random-number generators are used to produce the values of the input variables—that is, two different streams or sequences of random numbers will produce two sets of outputs, which (probably) will differ. If the performance of the system is measured by a parameter θ, the result of a set of simulation experiments will be an estimator $\widehat{\theta}$ of θ. The precision of the estimator $\widehat{\theta}$ can be measured by the standard error of $\widehat{\theta}$ or by the width of a confidence interval for θ. The purpose of the statistical analysis is either to estimate this standard error or confidence interval or to figure out the number of observations required to achieve a standard error or confidence interval of a given size—or both.

Consider a typical output variable, Y, the total cost per week of an inventory system; Y should be treated as a random variable with an unknown distribution. A simulation run of length 1 week provides a single sample observation from the population of all possible observations on Y. By increasing the run length, the sample size can be increased to n observations, $Y_1, Y_2, \ldots, Y_n$, based on a run length of n weeks. However, these observations do not constitute a random sample, in the classical sense, because they are not statistically independent. In this case, the inventory on hand at the end of one week is the beginning inventory on hand for the next week, and thus the value of Y_i has some influence on the value of Y_{i+1}. Thus, the sequence of random variables $Y_1, Y_2, \ldots, Y_n$ could be autocorrelated (i.e., correlated with itself). This autocorrelation, which is a measure of a lack of statistical independence, means that classical methods of statistics, which assume independence, are not directly applicable to the analysis of these output data. The methods must be properly modified and the simulation experiments properly designed for valid inferences to be made.

In addition to the autocorrelation present in most simulation output data, the specification of the initial conditions of the system at time 0 can pose a problem for the simulation analyst and could influence the output data. For example, the inventory on hand and the number of backorders at time 0 would most likely influence the value of Y_1, the total cost for week 1. Because of the autocorrelation, these initial conditions would also influence the costs $(Y_2, \ldots, Y_n)$ for subsequent weeks. The specified initial conditions, if not chosen well, can have an especially deleterious effect on attempting to estimate the steady-state (long-run) performance of a simulation model. For purposes of statistical analysis, the effect of the initial conditions is that the output observations might not be identically distributed and that the initial observations might not be representative of the steady-state behavior of the system.

Section 11.1 distinguishes between two types of simulation—transient versus steady state—and defines commonly used measures of system performance for each type of simulation. Section 11.2 illustrates by example the inherent variability in a stochastic (i.e., probabilistic) discrete-event simulation and thereby demonstrates the need for a statistical analysis of the output. Section 11.3 covers the statistical estimation of performance measures. Section 11.4 discusses the analysis of transient simulations, Section 11.5 the analysis of steady-state simulations.

11.1 TYPES OF SIMULATIONS WITH RESPECT TO OUTPUT ANALYSIS

In the analyzing of simulation output data, a distinction is made between terminating or transient simulations and steady-state simulations. A *terminating* simulation is one that runs for some duration of time T_E, where E is a specified event (or set of events) that stops the simulation. Such a simulated system "opens" at time 0 under well-specified *initial conditions* and "closes" at the *stopping time T_E*. The next four examples are terminating simulations.

Example 11.1

The Shady Grove Bank opens at 8:30 A.M. (time 0) with no customers present and 8 of the 11 tellers working (initial conditions) and closes at 4:30 P.M. (time $T_E = 480$ minutes). Here, the event E is merely the fact that the bank has been open for 480 minutes. The simulation analyst is interested in modeling the interaction between customers and tellers over the entire day, including the effect of starting up and of closing down at the end of the day.

Example 11.2

Consider the Shady Grove Bank of Example 11.1, but restricted to the period from 11:30 A.M. (time 0) to 1:30 P.M., when it is especially busy. The simulation run length is $T_E = 120$ minutes. The initial conditions at time 0 (11:30 A.M.) could be specified in essentially two ways: (1) the real system could be observed at 11:30 on a number of different days and a distribution of number of customers in system (at 11:30 A.M.) could be estimated, then these data could be used to load

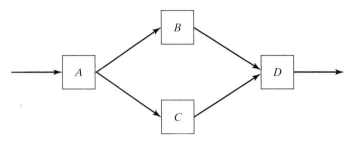

Figure 11.1 Example of a communications system.

the simulation model with customers at time 0; or (2) the model could be simulated from 8:30 A.M. to 11:30 A.M. without collecting output statistics, and the ending conditions at 11:30 A.M. used as initial conditions for the 11:30 A.M. to 1:30 P.M. simulation.

Example 11.3

A communications system consists of several components plus several backup components. It is represented schematically in Figure 11.1. Consider the system over a period of time, T_E, until the system fails. The stopping event E is defined by $E = \{A$ fails, or D fails, or $(B$ and C both fail)$\}$. Initial conditions are that all components are new at time 0.

Notice that, in the bank model of Example 11.1, the stopping time $T_E = 480$ minutes is known, but in Example 11.3, the stopping time T_E is generally unpredictable in advance; in fact, T_E is probably the output variable of interest, as it represents the total time until the system breaks down. One goal of the simulation might be to estimate $E(T_E)$, the mean time to system failure.

Example 11.4

A widget-manufacturing process runs continuously from Monday mornings until Saturday mornings. The first shift of each workweek is used to load inventory buffers and chemical tanks with the components and catalysts needed to make the final product (28 varieties of widget). These components and catalysts are made continually throughout the week, except for the last shift Friday night, which is used for cleanup and maintenance. Thus, most inventory buffers are near empty at the end of the week. During the first shift on Monday, a buffer stock is built up to cover the eventuality of breakdown in some part of the process. It is desired to simulate this system during the first shift (time 0 to time $T_E = 8$ hours) to study various scheduling policies for loading inventory buffers.

In the simulating of a terminating system, the initial conditions of the system at time 0 must be specified, and the stopping time T_E—or, alternatively, the stopping event E—must be well defined. Although it is certainly true that the Shady Grove Bank in Example 11.1 will open again the next day, the simulation analyst has chosen to consider it a terminating system because the object of interest is one day's operation, including start up and close down. On the other hand, if the simulation analyst were interested in some other aspect of the bank's operations, such as the flow of money or operation of automated teller machines, then the system might be considered as a nonterminating one. Similar comments apply to the communications system of Example 11.3. If the failed component were replaced and the system continued to operate, and, most important, if the simulation analyst were interested in studying its long-run behavior, it might be considered as a nonterminating system. In Example 11.3, however, interest is in its short-run behavior, from time 0 until the first system failure at time T_E. *Therefore, whether a simulation is considered to be terminating depends on both the objectives of the simulation study and the nature of the system.*

Example 11.4 is a terminating system, too. It is also an example of a *transient* (or nonstationary) simulation: the variables of interest are the in-process inventory levels, which are increasing from zero or near zero (at time 0) to full or near full (at time 8 hours).

A *nonterminating system* is a system that runs continuously, or at least over a very long period of time. Examples include assembly lines that shut down infrequently, continuous production systems of many different types, telephone systems and other communications systems such as the Internet, hospital emergency rooms, police dispatching and patrolling operations, fire departments, and continuously operating computer networks.

A simulation of a nonterminating system starts at simulation time 0 under initial conditions defined by the analyst and runs for some analyst-specified period of time T_E. (Significant problems arise concerning the specification of these initial and stopping conditions, problems that we discuss later.) Usually, the analyst wants to study steady-state, or long-run, properties of the system—that is, properties that are not influenced by the initial conditions of the model at time 0. *A steady-state simulation is a simulation whose objective is to study long-run, or steady-state, behavior of a nonterminating system.* The next two examples are steady-state simulations.

Example 11.5

Consider the widget-manufacturing process of Example 11.4, beginning with the second shift when the complete production process is under way. It is desired to estimate long-run production levels and production efficiencies. For the relatively long period of 13 shifts, this may be considered as a steady-state simulation. To obtain sufficiently precise estimates of production efficiency and other response variables, the analyst could decide to simulate for any length of time, T_E (even longer than 13 shifts)—that is, T_E is not determined by the nature of the problem (as it was in terminating simulations); rather, it is set by the analyst as one parameter in the design of the simulation experiment.

Example 11.6

HAL Inc., a large computer-service bureau, has many customers worldwide. Thus, its large computer system with many servers, workstations, and peripherals runs continuously, 24 hours per day. To handle an increased work load, HAL is considering additional CPUs, memory, and storage devices in various configurations. Although the load on HAL's computers varies throughout the day, management wants the system to be able to accommodate sustained periods of peak load. Furthermore, the time frame in which HAL's business will change in any substantial way is unknown, so there is no fixed planning horizon. Thus, a steady-state simulation at peak-load conditions is appropriate. HAL systems staff develops a simulation model of the existing system with the current peak work load and then explores several possibilities for expanding capacity. HAL is interested in long-run average throughput and utilization of each computer. The stopping time, T_E, is determined not by the nature of the problem, but rather by the simulation analyst, either arbitrarily or with a certain statistical precision in mind.

11.2 STOCHASTIC NATURE OF OUTPUT DATA

Consider one run of a simulation model over a period of time $[0, T_E]$. Since the model is an input-output transformation, as illustrated by Figure 10.5, and since some of the model input variables are random variables, it follows that the model output variables are random variables. Three examples are now given to illustrate the nature of the output data from stochastic simulations and to give a preliminary discussion of several important properties of these data. Do not be concerned if some of these properties and the associated terminology are not entirely clear on a first reading. They will be explained carefully later in the chapter.

Example 11.7:   Able and Baker, Revisited _____

Consider the Able–Baker technical-support call center problem (Example 2.2) which involved customers arriving according to the distribution of Table 2.11 and being served either by Able, whose service-time distribution is given in Table 2.12, or by Baker, whose service-time distribution is given in Table 2.13. The purpose of the simulation is to estimate Able's utilization, ρ, and the mean time spent in the system per customer, w, over the first 2 hours of the workday. Therefore, each run of the model is for a 2-hour period, with the system being empty and idle at time 0. Four statistically independent runs were made by using four distinct streams of random numbers to generate the interarrival and service times. Table 11.1 presents the results. The estimated utilization for run r is given by $\widehat{\rho}_r$ and the estimated average system time by $\widehat{w}_r$ (i.e., $\widehat{w}_r$ is the sample average time in system for all customers served during run r). Notice that, in this sample, the observed utilization ranges from 0.708 to 0.875 and the observed average system time ranges from 3.74 minutes to 4.53 minutes. The stochastic nature of the output data $\{\widehat{\rho}_1, \widehat{\rho}_2, \widehat{\rho}_3, \widehat{\rho}_4\}$ and $\{\widehat{w}_1, \widehat{w}_2, \widehat{w}_3, \widehat{w}_4\}$ is demonstrated by the results in Table 11.1.

There are two general questions that we will address by a statistical analysis—say, of the observed utilizations $\widehat{\rho}_r$, $r = 1, \ldots, 4$:

1. estimation of the true utilization $\rho = E(\widehat{\rho}_r)$ by a single number, called a point estimate;
2. estimation of the error in our point estimate, in the form either of a standard error or of a confidence interval.

These questions are addressed in Section 11.4 for terminating simulations, such as Example 11.7. Classical methods of statistics may be used because $\widehat{\rho}_1$, $\widehat{\rho}_2$, $\widehat{\rho}_3$, and $\widehat{\rho}_4$ constitute a random sample—that is, they are independent and identically distributed. In addition, $\rho = E(\widehat{\rho}_r)$ is the parameter

Table 11.1   Results of Four Independent Runs of 2-Hour Duration of the Able–Baker Queueing Problem

Run, r	Able's Utilization $\hat{\rho}_r$	Average System Time, $\widehat{w}_r$ (Minutes)
1	0.808	3.74
2	0.875	4.53
3	0.708	3.84
4	0.842	3.98

being estimated, so each $\widehat{\rho}_r$ is an unbiased estimate of the true mean utilization ρ. The analysis of Example 11.7 is considered in Example 11.10 of Section 11.4. A survey of statistical methods applicable to terminating simulations is given by Law [1980]. Additional guidance may be found in Alexopoulos and Seila [1998], Kleijnen [1987], Law and Kelton [2000], and Nelson [2001].

The next example illustrates the effects of correlation and initial conditions on the estimation of long-run mean measures of performance of a system.

Example 11.8

Consider a single-server queue with Poisson arrivals at an average rate of one every 10 minutes ($\lambda = 0.1$ per minute) and service times that are normally distributed, with mean 9.5 minutes and standard deviation 1.75 minutes.[1] This is an $M/G/1$ queue, which was described and analyzed in Section 6.4.1. By Equation (6.11), the true long-run server utilization is $\rho = \lambda E(S) = (0.1)(9.5) = 0.95$. We typically would not need to simulate such a system, because we can analyze it mathematically; but we simulate it here to illustrate difficulties that occur in trying to estimate the long-run mean queue length, L_Q, defined by Equation (6.4).

Suppose we run a single simulation for a total of 5000 minutes and observe the output process $\{L_Q(t), 0 \le t \le 5000\}$, where $L_Q(t)$ is the number of customers in the waiting line at time t minutes. To make this continuous-time process a little easier to analyze, we divide the time interval [0, 5000) into five equal subintervals of 1000 minutes and compute the average number of customers in queue for each interval individually. Specifically, the average number of customers in the queue from time $(j-1)1000$ to $j(1000)$ is

$$Y_j = \frac{1}{1000} \int_{(j-1)1000}^{j(1000)} L_Q(t)\, dt, \quad j = 1, \ldots, 5 \tag{11.1}$$

Thus, $Y_1 = \int_0^{1000} L_Q(t)\, dt / 1000$ is the time-weighted average number of customers in the queue from time 0 to time 1000, $Y_2 = \int_{1000}^{2000} L_Q(t)\, dt / 1000$ is the same average over [1000, 2000), and so on. Equation (11.1) is a special case of Equation (6.4). The observations $\{Y_1, Y_2, Y_3, Y_4, Y_5\}$ provide an example of "batching" of raw simulation data—in this case, $\{L_Q(t), 0 \le t \le 5000\}$—and the Y_j are called *batch means*. The use of batch means in analyzing output data is discussed in Section 11.5.5. For now, simply notice that batching transforms the continuous-time queue-length process, $\{L_Q(t), 0 \le t \le 5000\}$, into a discrete-time batch-means process $\{Y_i, i = 1, 2, 3, 4, 5\}$, where each Y_i is an estimator of L_Q.

The simulation results of three statistically independent replications are shown in Table 11.2. Each replication, or run, uses a distinct stream of random numbers. For replication 1, Y_{1j} is the batch mean for batch j (the jth interval), as defined by Equation (11.1); similarly, Y_{2j} and Y_{3j} are defined for batch j for replications 2 and 3, respectively. Table 11.2 also gives the sample mean over each replication, $\bar{Y}_{r.}$, for replications $r = 1, 2, 3$.[2] That is,

$$\bar{Y}_{r.} = \frac{1}{5} \sum_{j=1}^{5} Y_{rj}, \quad r = 1, 2, 3 \tag{11.2}$$

---

[1] The range of a service time is restricted to ± 5 standard deviations, to exclude the possibility of a negative service time; that range covers well over 99.999% of the normal distribution.

[2] The dot, as in the subscript $r\cdot$, indicates summation over the second subscript; the bar, as in $\bar{Y}_{r.}$, indicates an average.

Table 11.2 Batched Average Queue Length for Three Independent
Replications

Batching Interval (Minutes)	Batch, j	Replication 1, Y_{1j}	Replication 2, Y_{2j}	Replication 3, Y_{3j}
[0, 1000)	1	3.61	2.91	7.67
[1000, 2000)	2	3.21	9.00	19.53
[2000, 3000)	3	2.18	16.15	20.36
[3000, 4000)	4	6.92	24.53	8.11
[4000, 5000)	5	2.82	25.19	12.62
[0, 5000)		$\bar{Y}_{1.} = 3.75$	$\bar{Y}_{2.} = 15.56$	$\bar{Y}_{3.} = 13.66$

It probably will not surprise you that, if we take batch averages first, then average the batch means, or just average everything together, we get the same thing. In other words, each $\bar{Y}_{r.}$ is equivalent to the time average over the entire interval [0, 5000) for replication r, as given by Equation (6.4).

Table 11.2 illustrates the inherent variability in stochastic simulations both *within* a single replication and *across* different replications. Consider the variability within replication 3, in which the average queue length over the batching intervals varies from a low of $Y_{31} = 7.67$ customers during the first 1000 minutes to a high of $Y_{33} = 20.36$ customers during the third subinterval of 1000 minutes. Table 11.2 also shows the variability across replications. Compare Y_{15} to Y_{25} to Y_{35}, the average queue lengths over the intervals 4000 to 5000 minutes across all three replications.

Suppose, for the moment, that a simulation analyst makes only one replication of this model and gets the result $\bar{Y}_{1.} = 3.75$ customers as an estimate of mean queue length, L_Q. How precise is the estimate? This question is usually answered by attempting to estimate the standard error of $\bar{Y}_{1.}$ or by forming a confidence interval. The simulation analyst might think that the five batch means $Y_{11}, Y_{12}, \dots, Y_{15}$ could be regarded as a random sample; however, the terms in the sequence are not independent, and in fact they are autocorrelated, because all of the data are obtained from within one replication. If $Y_{11}, \dots, Y_{15}$ were mistakenly assumed to be independent observations, and their autocorrelation were ignored, the usual classical methods of statistics might severely underestimate the standard error of $\bar{Y}_{1.}$, possibly resulting in the simulation analyst's thinking that a high degree of precision had been achieved. On the other hand, the averages across the three replications, $\bar{Y}_{1.}$, $\bar{Y}_{2.}$, and $\bar{Y}_{3.}$, can be regarded as independent observations, because they are derived from three different replications.

Intuitively, Y_{11} and Y_{12} are correlated because in replication 1 the queue length at the end of the time interval [0, 1000) is the queue length at the beginning of the interval [1000, 2000)—similarly for any two adjacent batches within a given replication. If the system is congested at the end of one interval, it will be congested for a while at the beginning of the next time interval. Similarly, periods of low congestion tend to follow each other. Within a replication, say for $Y_{r1}, Y_{r2}, \dots, Y_{r5}$, high values of a batch mean tend to be followed by high values and low values by low. This tendency of adjacent observations to have like values is known as positive autocorrelation. The effect of ignoring autocorrelation when it is present is discussed in more detail in Section 11.5.2.

Now suppose that the purpose of the $M/G/1$ queueing simulation of Example 11.8 is to estimate "steady-state" mean queue length, that is, mean queue length under "typical operating conditions over the long run." However, each of the three replications was begun in the empty and idle state (no customers in the queue and the server available). The empty and idle initial state means that, within a given replication, there will be a higher-than- "typical" probability that the system will be uncongested for times close to 0. The practical effect is that an estimator of L_Q —say, $\bar{Y}_r$. for replication r—will be biased low [i.e., $E(\bar{Y}_r.) < L_Q$]. The extent of the bias decreases as the run length increases, but, for short-run-length simulations with atypical initial conditions, this initialization bias can produce misleading results. The problem of initialization bias is discussed further in Section 11.5.1.

11.3 MEASURES OF PERFORMANCE AND THEIR ESTIMATION

Consider the estimation of a performance parameter, θ (or ϕ), of a simulated system. It is desired to have a point estimate and an interval estimate of θ (or ϕ). The length of the interval estimate is a measure of the error in the point estimate. The simulation output data are of the form $\{Y_1, Y_2, \dots, Y_n\}$ for estimating θ; we refer to such output data as *discrete-time data*, because the index n is discrete valued. The simulation output data are of the form $\{Y(t), 0 \leq t \leq T_E\}$ for estimating ϕ; we refer to such output data as *continuous-time data*, because the index t is continuous valued. For example, Y_i might be the delay of customer i, or the total cost in week i; $Y(t)$ might be the queue length at time t, or the number of backlogged orders at time t. The parameter θ is an ordinary mean; ϕ will be referred to as a time-weighted mean. Whether we call the performance parameter θ or ϕ does not really matter; we use two different symbols here simply to provide a distinction between ordinary means and time-weighted means.

11.3.1 Point Estimation

The point estimator of θ based on the data $\{Y_1, \dots, Y_n\}$ is defined by

$$\widehat{\theta} = \frac{1}{n} \sum_{i=1}^{n} Y_i \tag{11.3}$$

where $\widehat{\theta}$ is a sample mean based on a sample of size n. Computer simulation languages may refer to this as a "discrete-time," "collect," "tally" or "observational" statistic.

The point estimator $\widehat{\theta}$ is said to be unbiased for θ if its expected value is θ—that is, if

$$E(\widehat{\theta}) = \theta \tag{11.4}$$

In general, however,

$$E(\widehat{\theta}) \neq \theta \tag{11.5}$$

and $E(\widehat{\theta}) - \theta$ is called the *bias* in the point estimator $\widehat{\theta}$. It is desirable to have estimators that are unbiased, or, if this is not possible, have a small bias relative to the magnitude of θ. Examples of estimators of the form of Equation (11.3) include $\widehat{w}$ and $\widehat{w}_Q$ of Equations (6.5) and (6.7), in which case Y_i is the time spent in the (sub)system by customer i.

The point estimator of ϕ based on the data $\{Y(t), 0 \leq t \leq T_E\}$, where T_E is the simulation run length, is defined by

$$\widehat{\phi} = \frac{1}{T_E} \int_0^{T_E} Y(t)\, dt \tag{11.6}$$

and is called a time average of $Y(t)$ over $[0, T_E]$. Simulation languages may refer to this as a "continuous-time," "discrete-change" or "time-persistent" statistic. In general,

$$E(\widehat{\phi}) \neq \phi \tag{11.7}$$

and $\widehat{\phi}$ is said to be biased for ϕ. Again, we would like to obtain unbiased or low-bias estimators. Examples of time averages include $\widehat{L}$ and $\widehat{L}_Q$ of Equations (6.3) and (6.4) and Y_j of Equation (11.1).

Generally, θ and ϕ are regarded as mean measures of performance of the system being simulated. Other measures usually can be put into this common framework. For example, consider estimation of the proportion of days on which sales are lost through an out-of-stock situation. In the simulation, let

$$Y_i = \begin{cases} 1, & \text{if out of stock on day } i \\ 0, & \text{otherwise} \end{cases}$$

With n equal to the total number of days, $\widehat{\theta}$ defined by Equation (11.3) is a point estimator of θ, the proportion of out-of-stock days. For a second example, consider estimation of the proportion of time queue length is greater than k_0 customers (for example, $k_0 = 10$). If $L_Q(t)$ represents simulated queue length at time t, then (in the simulation) define

$$Y(t) = \begin{cases} 1, & \text{if } L_Q(t) > k_0 \\ 0, & \text{otherwise} \end{cases}$$

Then $\widehat{\phi}$, as defined by Equation (11.6), is a point estimator of ϕ, the proportion of time that the queue length is greater than k_0 customers. Thus, estimation of proportions or probabilities is a special case of the estimation of means.

A performance measure that does not fit this common framework is a quantile or percentile. Quantiles describe the level of performance that can be delivered with a given probability, p. For instance, suppose that Y represents the delay in queue that a customer experiences in a service system, measured in minutes. Then the 0.85 quantile of Y is the value θ such that

$$\Pr\{Y \leq \theta\} = p \tag{11.8}$$

where $p = 0.85$ in this case. As a percentage, θ is the $100p$th or 85th percentile of customer delay. Therefore, 85% of all customers will experience a delay of θ minutes or less. Stated differently, a customer has only a 0.15 probability of experiencing a delay of longer than θ minutes. A widely used performance measure is the median, which is the 0.5 quantile or 50th percentile.

The problem of estimating a quantile is the inverse of the problem of estimating a proportion or probability. Consider Equation (11.8). In estimating a proportion, θ is given and p is to be estimated; but, in estimating a quantile, p is given and θ is to be estimated.

The most intuitive method for estimating a quantile is to form a histogram of the observed values of Y, then find a value $\widehat{\theta}$ such that $100p\%$ of the histogram is to the left of (smaller than) $\widehat{\theta}$. For instance, if we observe $n = 250$ customer delays $\{Y_1, \ldots, Y_{250}\}$, then an estimate of the 85th percentile of delay is a value $\widehat{\theta}$ such that $(0.85)(250) = 212.5 \approx 213$ of the observed values are less than or equal to θ. An obvious estimate is, therefore, to set $\widehat{\theta}$ equal to the 213th smallest value in the sample (this requires sorting the data). When the output is a continuous-time process, such as the queue-length process $\{L_Q(t), 0 \leq t \leq T_E\}$, then a histogram gives the *fraction of time* that the process spent at each possible level (queue length in this example). However, the method for quantile estimation remains the same: Find a value $\widehat{\theta}$ such that $100p\%$ of the histogram is to the left of $\widehat{\theta}$.

11.3.2 Confidence-Interval Estimation

To understand confidence intervals fully, it is important to understand the difference between a *measure of error* and a *measure of risk*. One way to make the difference clear is to contrast a confidence interval with a *prediction interval* (which is another useful output-analysis tool).

Both confidence and prediction intervals are based on the premise that the data being produced by the simulation is represented well by a probability model. Suppose that model is the normal distribution with mean θ and variance σ^2, both unknown. To make the example concrete, let $\bar{Y}_{i\cdot}$ be the average cycle time for parts produced on the ith replication (representing a day of production) of the simulation. Therefore, θ is the mathematical expectation of $\bar{Y}_{i\cdot}$, and σ is represents the day-to-day variation of the average cycle time.

Suppose our goal is to estimate θ. If we are planning to be in business for a long time, producing parts day after day, then θ is a relevant parameter, because it is the long-run mean daily cycle time. Our average cycle time will vary from day to day, but over the long run the *average of the averages* will be close to θ.

The natural estimator for θ is the overall sample mean of R independent replications, $\bar{Y}_{\cdot\cdot} = \sum_{i=1}^{R} \bar{Y}_{i\cdot}/R$. But $\bar{Y}_{\cdot\cdot}$ is not θ, it is an estimate, based on a sample, and it has error. A confidence interval is a measure of that error. Let

$$S^2 = \frac{1}{R-1} \sum_{i=1}^{R} \left(Y_{i\cdot} - \bar{Y}_{\cdot\cdot} \right)^2$$

be the sample variance across the R replications. The usual confidence interval, which assumes the $Y_{i\cdot}$ are normally distributed, is

$$\bar{Y}_{\cdot\cdot} \pm t_{\alpha/2, R-1} \frac{S}{\sqrt{R}}$$

where $t_{\alpha/2, R-1}$ is the quantile of the t distribution with $R - 1$ degrees of freedom that cuts off $\alpha/2$ of the area of each tail. (See Table A.5.) We cannot know for certain exactly how far $\bar{Y}_{\cdot\cdot}$ is from θ, but the confidence interval attempts to bound that error. Unfortunately, the confidence interval itself may be wrong. A confidence level, such as 95%, tells us how much we can trust the interval to actually bound the error between $\bar{Y}_{..}$ and θ. The more replications we make, the less error there is in $\bar{Y}_{..}$, and our confidence interval reflects that because $t_{\alpha/2,R-1}S/\sqrt{R}$ will tend to get smaller as R increases, converging to 0 as R goes to infinity.

Now suppose we need to make a promise about what the average cycle time will be on a particular day. A good guess is our estimator $\bar{Y}_{..}$, but it is unlikely to be exactly right. Even θ itself, which is the center of the distribution, is not likely to be the *actual average cycle time* on any particular day, because the daily average cycle time varies. A prediction interval, on the other hand, is designed to be wide enough to contain the actual average cycle time on any particular day with high probability. A prediction interval is a measure of risk; a confidence interval is a measure of error.

The normal-theory prediction interval is

$$\bar{Y}_{..} \pm t_{\alpha/2,R-1}S\sqrt{1 + \frac{1}{R}}$$

The length of this interval will not go to 0 as R increases. In fact, in the limit it becomes

$$\theta \pm z_{\alpha/2}\sigma$$

to reflect the fact that, no matter how much we simulate, our daily average cycle time still varies.

In summary, a prediction interval is a measure of risk, and a confidence interval is a measure of error. We can simulate away error by making more and more replications, but we can never simulate away risk, which is an inherent part of the system. We can, however, do a better job of evaluating risk by making more replications.

Example 11.9
Suppose that the overall average of the average cycle time on 120 replications of a manufacturing simulation is 5.80 hours, with a sample standard deviation of 1.60 hours. Since $t_{0.025,119} = 1.98$, a 95% confidence interval for the long-run expected daily average cycle time is $5.80 \pm 1.98(1.60/\sqrt{120})$ or 5.80 ± 0.29 hours. Thus, our best guess of the long-run average of the daily average cycle times is 5.80 hours, but there could be as much as ± 0.29 hours error in this estimate.

On any particular day, we are 95% confident that the average cycle time for all parts produced on that day will be

$$5.80 \pm 1.98(1.60)\sqrt{1 + \frac{1}{120}}$$

or 5.80 ± 3.18 hours. The ± 3.18 hours reflects the inherent variability in the daily average cycle times and the fact that we want to be 95% confident of covering the actual average cycle time on a particular day (rather than simply covering the long-run average).

11.4 OUTPUT ANALYSIS FOR TERMINATING SIMULATIONS

Consider a terminating simulation that runs over a simulated time interval $[0, T_E]$ and results in observations $Y_1, \ldots, Y_n$. The sample size, n, may be a fixed number, or it may be a random variable (say, the number of observations that occur during time T_E). A common goal in simulation is to estimate

$$\theta = E\left(\frac{1}{n}\sum_{i=1}^{n}Y_i\right)$$

When the output data are of the form $\{Y(t), 0 \leq t \leq T_E\}$, the goal is to estimate

$$\phi = E\left(\frac{1}{T_E}\int_0^{T_E}Y(t)\,dt\right)$$

The method used in each case is the method of *independent replications*. The simulation is repeated a total of R times, each run using a different random number stream and independently chosen initial conditions (which includes the case that all runs have identical initial conditions). We now address this problem.

11.4.1 Statistical Background

Perhaps the most confusing aspect of simulation output analysis is distinguishing *within-replication* data from *across-replication* data, and understanding the properties and uses of each. The issue can be further confused by the fact that simulation languages often provide only summary measures, like sample means, sample variances, and confidence intervals, rather than all of the raw data. Sometimes these summary measures are *all* the simulation language provides without a lot of extra work.

To illustrate the key ideas, think in terms of the simulation of a manufacturing system and two performance measures of that system, the cycle time for parts (time from release into the factory until completion) and the work in process (WIP, the total number of parts in the factory at any time). In computer applications, these two measures could correspond to the response time and the length of the task queue at the CPU; in a service application, they could be the time to fulfill a customer's request and the number of requests on the "to do" list; in a supply-chain application, they could be the order fill time and the inventory level. Similar measures appear in many systems.

Here is the usual set up for something like cycle time: Let Y_{ij} be the cycle time for the jth part produced in the ith replication. If each replication represents two shifts of production, then the number of parts produced in each replication might differ. Table 11.3 shows, symbolically, the results of R replications.

Table 11.3 Within and Across-Replication Cycle-Time Data

Within-Rep Data				Across-Rep Data
Y_{11}	Y_{12}	$\cdots$	Y_{1n_1}	$\bar{Y}_{1\cdot}, S_1^2, H_1$
Y_{21}	Y_{22}	$\cdots$	Y_{2n_2}	$\bar{Y}_{2\cdot}, S_2^2, H_2$
$\vdots$	$\vdots$	$\cdots$	$\vdots$	$\vdots$
Y_{R1}	Y_{R2}	$\cdots$	Y_{Rn_R}	$\bar{Y}_{R\cdot}, S_R^2, H_R$
				$\bar{Y}_{\cdot\cdot}, S^2, H$

The across-replication data are formed by summarizing within-replication data: $\bar{Y}_{i\cdot}$ is the sample mean of the n_i cycle times from the ith replication, S_i^2 is the sample variance of the same data, and

$$H_i = t_{\alpha/2, n_i - 1} \frac{S_i}{\sqrt{n_i}} \qquad (11.9)$$

is a confidence-interval half-width based on this dataset.

From the across-replication data, we compute overall statistics, the average of the daily cycle time averages

$$\bar{Y}_{\cdot\cdot} = \frac{1}{R} \sum_{i=1}^{R} \bar{Y}_{i\cdot} \qquad (11.10)$$

the sample variance of the daily cycle time averages,

$$S^2 = \frac{1}{R-1} \sum_{i=1}^{R} (\bar{Y}_{i\cdot} - \bar{Y}_{\cdot\cdot})^2 \qquad (11.11)$$

and finally, the confidence-interval half-width

$$H = t_{\alpha/2, R-1} \frac{S}{\sqrt{R}} \qquad (11.12)$$

The quantity $S/\sqrt{R}$ is the standard error, which is sometimes interpreted as the average error in $\bar{Y}_{\cdot\cdot}$ as an estimator of θ. Notice that S^2 is *not* the average of the within-replication sample variances, S_i^2; rather, it is the sample variance of the within-replication averages $\bar{Y}_{1\cdot}, \bar{Y}_{2\cdot}, \ldots, \bar{Y}_{R\cdot}$.

Within a replication, work in process (WIP) is a continuous-time output, denoted $Y_i(t)$. The stopping time for the ith replication, T_{E_i}, could be a random variable, in general; in this example, it is the end of the second shift. Table 11.4 is an abstract representation of the data produced.

The within-replication sample mean and variance are defined appropriately for continuous-time data:

$$\bar{Y}_{i\cdot} = \frac{1}{T_{E_i}} \int_0^{T_{E_i}} Y_i(t)\,dt \qquad (11.13)$$

Table 11.4   Within and Across-Replication WIP Data

Within-Rep Data	Across-Rep Data
$Y_1(t), 0 \leq t \leq T_{E_1}$ $Y_2(t), 0 \leq t \leq T_{E_2}$	$\bar{Y}_{1\cdot}, S_1^2, H_1$ $\bar{Y}_{2\cdot}, S_2^2, H_2$
$\vdots$	$\vdots$
$Y_R(t), 0 \leq t \leq T_{E_R}$	$\bar{Y}_{R\cdot}, S_R^2, H_R$ $\bar{Y}_{\cdot\cdot}, S^2, H$

and

$$S_i^2 = \frac{1}{T_{E_i}} \int_0^{T_{E_i}} \left(Y_i(t) - \bar{Y}_{i\cdot}\right)^2 dt \tag{11.14}$$

A definition for H_i is more problematic, but, to be concrete, take it to be

$$H_i = z_{\alpha/2} \frac{S_i}{\sqrt{T_{E_i}}}. \tag{11.15}$$

Frankly, it is difficult to conceive of a situation in which H_i is relevant, a topic we discuss later. Although the definitions of the within-replication data change for continuous-time data, the across-replication statistics are unchanged, and this is a critical observation.

Here are the key points that must be understood:

- The overall sample average, $\bar{Y}_{\cdot\cdot}$, and the individual replication sample averages, $\bar{Y}_{i\cdot}$, are always unbiased estimators of the expected daily average cycle time or daily average WIP.
- Across-replication data are independent (since they are based on different random numbers), are identically distributed (since we are running the same model on each replication), and tend to be normally distributed if they are averages of within-replication data, as they are here. This implies that the confidence interval $\bar{Y}_{\cdot\cdot} \pm H$ is often pretty good.
- Within-replication data, on the other hand, might have none of these properties. The individual cycle times may not be identically distributed (if the first few parts of the day find the system empty); they are almost certainly not independent (because one part follows another); and whether they are normally distributed is difficult to know in advance. For this reason, S_i^2 and H_i, which are computed under the assumption of independent and identically distributed (i.i.d.) data, tend not to be useful (although there are exceptions).
- There are situations in which $\bar{Y}_{\cdot\cdot}$ and $\bar{Y}_{i\cdot}$ are valid estimators of the expected cycle time for an *individual part* or the *expected WIP at any point in time*, rather than the daily average. (See Section 11.5 on steady-state simulations.) Even when this is the case, the confidence interval $\bar{Y}_{\cdot\cdot} \pm H$ is valid, and $\bar{Y}_{i\cdot} \pm H_i$ is not. The difficulty occurs because S_i^2 is a reasonable estimator of the variance of the cycle time, but S_i^2/n_i and S_i^2/T_{E_i} are not good estimators of the $\text{Var}[\bar{Y}_{i\cdot}]$—more on this in Section 11.5.2.

Example 11.10:   The Able–Baker Problem, Continued

Consider Example 11.7, the Able–Baker technical-support call center problem, with the data for $R = 4$ replications given in Table 11.1. The four utilization estimates, $\hat{\rho}_r$, are time averages of the form of Equation (11.13). The simulation produces output data of the form

$$Y_r(t) = \begin{cases} 1, & \text{if Able is busy at time } t \\ 0, & \text{otherwise} \end{cases}$$

and $\hat{\rho}_r = \bar{Y}_r$. as computed by Equation (11.13) with $T_E = 2$ hours. Similarly, the four average system times, $\hat{w}_1, \ldots, \hat{w}_4$, are analogous to $\bar{Y}_r$. of Table 11.3, where Y_{ri} is the actual time spent in system by customer i on replication r.

First, suppose that the analyst desires a 95% confidence interval for Able's true utilization, ρ. Using Equation (11.10), compute an overall point estimator

$$\bar{Y}_{..} = \hat{\rho} = \frac{0.808 + 0.875 + 0.708 + 0.842}{4} = 0.808$$

Using Equation (11.11), compute its estimated variance:

$$S^2 = \frac{(0.808 - 0.808)^2 + \cdots + (0.842 - 0.808)^2}{4 - 1} = (0.072)^2$$

Thus, the standard error of $\hat{\rho} = 0.808$ is estimated by s.e.$(\hat{\rho}) = S/\sqrt{4} = 0.036$. Obtain $t_{0.025,3} = 3.18$ from Table A.5, and compute the 95% confidence interval half-width by (11.12) as

$$H = t_{0.025,3} \frac{S}{\sqrt{4}} = (3.18)(0.036) = 0.114$$

giving 0.808 ± 0.114 or, with 95% confidence,

$$0.694 \leq \rho \leq 0.922$$

In a similar fashion, compute a 95% confidence interval for mean time in system w:

$$\hat{w} = \frac{3.74 + 4.53 + 3.84 + 3.98}{4} = 4.02 \text{ minutes}$$

$$S^2 = \frac{(3.74 - 4.02)^2 + \cdots + (3.98 - 4.02)^2}{3 - 1} = (0.352)^2$$

so that

$$H = t_{0.025,3} \frac{S}{\sqrt{4}} = (3.18)(0.176) = 0.560$$

or

$$4.02 - 0.56 \leq w \leq 4.02 + 0.56$$

Thus, the 95% confidence interval for w is $3.46 \leq w \leq 4.58$.

11.4.2 Confidence Intervals with Specified Precision

By Expression (11.12), the half-length H of a $100(1 - \alpha)\%$ confidence interval for a mean θ, based on the t distribution, is given by

$$H = t_{\alpha/2, R-1} \frac{S}{\sqrt{R}}$$

where S^2 is the sample variance and R is the number of replications. Suppose that an error criterion ϵ is specified; in other words, it is desired to estimate θ by $\bar{Y}_{..}$ to within $\pm\epsilon$ with high probability—say, at least $1 - \alpha$. Thus, it is desired that a sufficiently large sample size, R, be taken to satisfy

$$P(|\bar{Y}.. - \theta| < \epsilon) \geq 1 - \alpha$$

When the sample size, R, is fixed, no guarantee can be given for the resulting error. But if the sample size can be increased, an error criterion can be specified.

Assume that an initial sample of size R_0 replications has been observed—that is, the simulation analyst initially makes R_0 independent replications. We must have $R_0 \geq 2$, with 10 or more being desirable. The R_0 replications will be used to obtain an initial estimate S_0^2 of the population variance σ^2. To meet the half-length criterion, a sample size R must be chosen such that $R \geq R_0$ and

$$H = t_{\alpha/2, R-1} \frac{S_0}{\sqrt{R}} \leq \epsilon \tag{11.16}$$

Solving for R in Inequality (11.16) shows that R is the smallest integer satisfying $R \geq R_0$ and

$$R \geq \left(\frac{t_{\alpha/2, R-1} S_0}{\epsilon} \right)^2 \tag{11.17}$$

Since $t_{\alpha/2, R-1} \geq z_{\alpha/2}$, an initial estimate for R is given by

$$R \geq \left(\frac{z_{\alpha/2} S_0}{\epsilon} \right)^2 \tag{11.18}$$

where $z_{\alpha/2}$ is the $100(1 - \alpha/2)$ percentage point of the standard normal distribution from Table A.3. And since $t_{\alpha/2, R-1} \approx z_{\alpha/2}$ for large R (say, $R \geq 50$), the second inequality for R is adequate when R is large. After determining the final sample size, R, collect $R - R_0$ additional observations (i.e., make $R - R_0$ additional replications, or start over and make R total replications) and form the $100(1 - \alpha)\%$ confidence interval for θ by

$$\bar{Y}.. - t_{\alpha/2, R-1} \frac{S}{\sqrt{R}} \leq \theta \leq \bar{Y}.. + t_{\alpha/2, R-1} \frac{S}{\sqrt{R}} \tag{11.19}$$

where $\bar{Y}..$ and S^2 are computed on the basis of all R replications, $\bar{Y}..$ by Equation (11.10), and S^2 by Equation (11.11). The half-length of the confidence interval given by Inequality (11.19) should be approximately ϵ or smaller; however, with the additional $R - R_0$ observations, the variance estimator S^2 could differ somewhat from the initial estimate S_0^2, possibly causing the half-length to be greater than desired. If the confidence interval (11.19) is too large, the procedure may be repeated, using Inequality (11.17), to determine an even larger sample size.

Example 11.11

Suppose that it is desired to estimate Able's utilization in Example 11.7 to within ± 0.04 with probability 0.95. An initial sample of size $R_0 = 4$ is taken, with the results given in Table 11.1. An initial estimate of the population variance is $S_0^2 = (0.072)^2 = 0.00518$. (See Example 11.10 for the relevant data.) The error criterion is $\epsilon = 0.04$, and the confidence coefficient is 1 - From Inequality (11.18), the final sample size must be at least as large as

$$\frac{z_{0.025}^2 S_0^2}{\epsilon^2} = \frac{(1.96)^2(0.00518)}{(0.04)^2} = 12.44$$

Next, Inequality (11.17) can be used to test possible candidates $(R = 13, 14, \ldots)$ for final sample size, as follows:

R	13	14	15
$t_{0.025,R-1}$	2.18	2.16	2.14
$\dfrac{t_{0.025,R-1}^2 S_0^2}{\epsilon^2}$	15.39	15.10	14.83

Thus, $R = 15$ is the smallest integer satisfying Inequality (11.17), so $R - R_0 = 15 - 4 = 11$ additional replications are needed. After obtaining the additional outputs, we would again need to compute the half-width H to ensure that it is as small as is desired.

11.4.3 Quantiles

To present the interval estimator for quantiles, it is helpful to review the interval estimator for a mean in the special case when the mean represents a proportion or probability, p. In this book, we have chosen to treat a proportion or probability as just a special case of a mean. However, in many statistics texts, probabilities are treated separately.

When the number of independent replications $Y_1, \ldots, Y_R$ is large enough that $t_{\alpha/2,n-1} \doteq z_{\alpha/2}$, the confidence interval for a probability p is often written as

$$\widehat{p} \pm z_{\alpha/2}\sqrt{\frac{\widehat{p}(1-\widehat{p})}{R-1}}$$

where $\widehat{p}$ is the sample proportion (tedious algebra shows that this formula for the half-width is precisely equivalent to Equation (11.12) when used in estimating a proportion).

As mentioned in Section 11.3, the quantile-estimation problem is the inverse of the probability-estimation problem: Find θ such that $\Pr\{Y \leq \theta\} = p$. Thus, to estimate the p quantile, we find that value $\widehat{\theta}$ such that $100p\%$ of the data in a histogram of Y is to the left of $\widehat{\theta}$ (or stated differently, the npth smallest value of $Y_1, \ldots, Y_R$).

Extending this idea, an approximate $(1-\alpha)100\%$ confidence interval for θ can be obtained by finding two values: θ_ℓ that cuts of $100p_\ell\%$ of the histogram and θ_u that cuts off $100p_u\%$ of the histogram, where

$$p_\ell = p - z_{\alpha/2}\sqrt{\frac{p(1-p)}{R-1}}$$

$$p_u = p + z_{\alpha/2}\sqrt{\frac{p(1-p)}{R-1}} \tag{11.20}$$

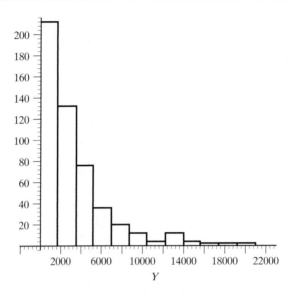

Figure 11.2   Failure data in hours for 500 replications of the communications system.

(Recall that we know p.) In terms of sorted values, $\widehat{\theta}_\ell$ is the Rp_ℓ smallest value (rounded down) and $\widehat{\theta}_u$ is the Rp_u smallest value (rounded up), of $Y_1, \ldots, Y_R$.

Example 11.12

Suppose that we want to estimate the 0.8 quantile of the time to failure (in hours) for the communications system in Example 11.3 and form a 95% confidence interval for it. A histogram of $R = 500$ independent replications is shown in Figure 11.2.

The point estimator is $\widehat{\theta} = 4644$ hours, because 80% of the data in the histogram is to the left of 4644. Equivalently, it is the $500 \times 0.8 = 400$th smallest value of the sorted data.

To obtain the confidence interval, we first compute

$$p_\ell = p - z_{\alpha/2}\sqrt{\frac{p(1-p)}{R-1}} = 0.8 - 1.96\sqrt{\frac{0.8(0.2)}{499}} = 0.765$$

$$p_u = p + z_{\alpha/2}\sqrt{\frac{p(1-p)}{R-1}} = 0.8 + 1.96\sqrt{\frac{0.8(0.2)}{499}} = 0.835$$

The lower bound of the confidence interval is $\widehat{\theta}_\ell = 4173$ (the $500 \times p_\ell = 382$nd smallest value, rounding down); the upper bound of the confidence interval is $\widehat{\theta}_u = 5119$ hours (the $500 \times p_u = 418$th smallest value, rounding up).

11.4.4 Estimating Probabilities and Quantiles from Summary Data

Knowing the equation for the confidence interval half-width is important if all the simulation software provides is $\bar{Y}_{..}$ and H and you need to work out the number of replications required to get a prespecified precision, or if you need to estimate a probability or quantile. You know the number of replications, so the sample standard deviation can be extracted from H by using the formula

$$S = \frac{H\sqrt{R}}{t_{\alpha/2,R-1}}$$

With this information, the method in Section 11.4.2 can be employed.

The more difficult problem is estimating a probability or quantile from summary data. When all we have available is the sample mean and confidence-interval halfwidth (which gives us the sample standard deviation), then one approach is to use a normal-theory approximation for the probabilities or quantiles we desire, specifically

$$\Pr\{\bar{Y}_{i\cdot} \le c\} \approx \Pr\left\{Z \le \frac{c - \bar{Y}_{\cdot\cdot}}{S}\right\}$$

and

$$\widehat{\theta} \approx \bar{Y}_{\cdot\cdot} + z_p S$$

The following example illustrates how this is done.

Example 11.13

From 25 replications of the manufacturing simulation, a 90% confidence interval for the daily average WIP is 218 ± 32. What is the probability that the daily average WIP is less than 350? What is the 85th percentile of daily average WIP?

First, we extract the standard deviation:

$$S = \frac{H\sqrt{R}}{t_{0.05,24}} = \frac{32\sqrt{25}}{1.71} = 93$$

Then, we use the normal approximations and Table A.3 to get

$$\Pr\{\bar{Y}_{i\cdot} \le 350\} \approx \Pr\left\{Z \le \frac{350 - 218}{93}\right\} = \Pr\{Z \le 1.42\} = 0.92$$

and

$$\widehat{\theta} \approx \bar{Y}_{\cdot\cdot} + z_{0.85}S = 218 + 1.04(93) = 315 \text{ parts}$$

There are shortcomings to obtaining our probabilities and quantiles this way. The approximation depends heavily on whether the output variable of interest is normally distributed. If the output variable itself is not an average, then this approximation is suspect. Therefore, we expect the approximation to work well for statements about the average daily cycle time, for instance, but very poorly for the cycle time of an individual part.

11.5 OUTPUT ANALYSIS FOR STEADY-STATE SIMULATIONS

Consider a single run of a simulation model whose purpose is to estimate a *steady-state*, or *long-run*, characteristic of the system. Suppose that the single run produces observations $Y_1, Y_2, \ldots$, which, generally, are samples of an autocorrelated time series. The steady-state (or long-run) measure of performance, θ, is defined by

$$\theta = \lim_{n \longrightarrow \infty} \frac{1}{n} \sum_{i=1}^{n} Y_i \tag{11.21}$$

with probability 1, where the value of θ is independent of the initial conditions. (The phrase "with probability 1" means that essentially all simulations of the model, using different random numbers, will produce series $Y_i, i = 1, 2, \ldots$ whose sample average converges to θ.) For example, if Y_i was the time customer i spent talking to an operator, then θ would be the long-run average time a customer spends talking to an operator; and, because θ is defined as a limit, it is independent of the call center's conditions at time 0. Similarly, the steady-state performance for a continuous-time output measure $\{Y(t), t \geq 0\}$, such as the number of customers in the call center's hold queue, is defined as

$$\phi = \lim_{T_E \longrightarrow \infty} \frac{1}{T_E} \int_0^{T_E} Y(t)\, dt$$

with probability 1.

Of course, the simulation analyst could decide to stop the simulation after some number of observations—say, n—have been collected; or the simulation analyst could decide to simulate for some length of time T_E that determines n (although n may vary from run to run). The sample size n (or T_E) is a *design* choice; it is not inherently determined by the nature of the problem. The simulation analyst will choose simulation run length (n or T_E) with several considerations in mind:

1. Any bias in the point estimator that is due to artificial or arbitrary initial conditions. (The bias can be severe if run length is too short, but generally it decreases as run length increases.)
2. The desired precision of the point estimator, as measured by the standard error or confidence interval half-width.
3. Budget constraints on computer resources.

The next subsection discusses initialization bias and the following subsections outline two methods of estimating point-estimator variability. For clarity of presentation, we discuss only estimation of θ from a discrete-time output process. Thus, when discussing one replication (or run), the notation

$$Y_1, Y_2, Y_3, \ldots$$

will be used; if several replications have been made, the output data for replication r will be denoted by

$$Y_{r1}, Y_{r2}, Y_{r3}, \ldots \tag{11.22}$$

11.5.1 Initialization Bias in Steady-State Simulations

There are several methods of reducing the point-estimator bias caused by using artificial and unrealistic initial conditions in a steady-state simulation. The first method is to initialize the simulation in a state that is more representative of long-run conditions. This method is sometimes called intelligent initialization. Examples include

1. setting the inventory levels, number of backorders, and number of items on order and their arrival dates in an inventory simulation;
2. placing customers in queue and in service in a queueing simulation;
3. having some components failed or degraded in a reliability simulation.

There are at least two ways to specify the initial conditions intelligently. If the system exists, collect data on it and use these data to specify more nearly typical initial conditions. This method sometimes requires a large data-collection effort. In addition, if the system being modeled does not exist—for example, if it is a variant of an existing system—this method is impossible to implement. Nevertheless, it is recommended that simulation analysts use any available data on existing systems to help initialize the simulation, as this will usually be better than assuming the system to be "completely stocked," "empty and idle," or "brand new" at time 0.

A related idea is to obtain initial conditions from a second model of the system that has been simplified enough to make it mathematically solvable. The queueing models in Chapter 6 are very useful for this purpose. The simplified model can be solved to find long-run expected or most likely conditions—such as the expected number of customers in the queue—and these conditions can be used to initialize the simulation.

A second method to reduce the impact of initial conditions, possibly used in conjunction with the first, is to divide each simulation run into two phases: first, an initialization phase, from time 0 to time T_0, followed by a data-collection phase from time T_0 to the stopping time $T_0 + T_E$—that is, the simulation begins at time 0 under specified initial conditions I_0 and runs for a specified period of time T_0. Data collection on the response variables of interest does not begin until time T_0 and continues until time $T_0 + T_E$. The choice of T_0 is quite important, because the system state at time T_0, denoted by I, should be more nearly representative of steady-state behavior than are the original initial conditions at time 0, I_0. In addition, the length T_E of the data-collection phase should be long enough to guarantee sufficiently precise estimates of steady-state behavior. Notice that the system state, I, at time T_0 is a random variable and to say that the system has reached an approximate steady state is to say that the probability distribution of the system state at time T_0 is sufficiently close to the steady-state probability distribution as to make the bias in point estimates of response variables negligible. Figure 11.3 illustrates the two phases of a steady-state simulation. The effect of starting a simulation run of a queueing model in the empty and idle state, as well as several useful plots to aid the simulation analyst in choosing an appropriate value of T_0, are given in the following example.

Example 11.14

Consider the $M/G/1$ queue discussed in Example 11.8. Suppose that a total of 10 independent replications were made ($R = 10$), each replication beginning in the empty and idle state. The total simulation run length on each replication was $T_0 + T_E = 15,000$ minutes. The response variable

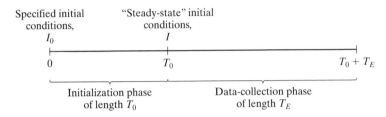

Figure 11.3   Initialization and data collection phases of a steady-state simulation run.

was queue length, $L_Q(t, r)$, at time t, where the second argument, r, denotes the replication ($r = 1, \ldots, 10$). The raw output data were batched, as in Example 11.8, Equation (11.1), in batching intervals of 1000 minutes, to produce the following batch means:

$$Y_{rj} = \frac{1}{1000} \int_{(j-1)1000}^{j(1000)} L_Q(t, r)dt \qquad (11.23)$$

for replication $r = 1, \ldots, 10$ and for batch $j = 1, 2, \ldots, 15$. The estimator in Equation (11.23) is simply the time-weighted-average queue length over the time interval $[(j - 1)1000, j(1000))$, similar to that in Equation (6.4). The 15 batch means for the 10 replications are given in Table 11.5.

Normally we average all the batch means *within* each replication to obtain a replication average. However, our goal at this stage is to identify the trend in the data due to initialization bias and find out when it dissipates. To do this, we will average corresponding batch means *across* replications and plot them (this idea is usually attributed to Welch [1983]). Such averages are known as *ensemble averages*. Specifically, for each batch j, define the ensemble average across all R replications to be

$$\bar{Y}_{\cdot j} = \frac{1}{R} \sum_{r=1}^{R} Y_{rj} \qquad (11.24)$$

($R = 10$ here). The ensemble averages $\bar{Y}_{\cdot j}, j = 1, \ldots, 15$ are displayed in the third column of Table 11.6. Notice that $\bar{Y}_{\cdot 1} = 4.03$ and $\bar{Y}_{\cdot 2} = 5.45$ are estimates of mean queue length over the time periods $[0, 1000)$ and $[1000, 2000)$, respectively, and they are less than all other ensemble averages $\bar{Y}_{\cdot j}$ ($j = 3, \ldots, 15$). The simulation analyst may suspect that this is due to the downward bias in these estimators, which in turn is due to the queue being empty and idle at time 0. This downward bias is further illustrated in the plots that follow.

Figure 11.4 is a plot of the ensemble averages, $\bar{Y}_{\cdot j}$, versus $1000j$, for $j = 1, 2, \ldots, 15$. The actual values, $\bar{Y}_{\cdot j}$, are the discrete set of points in circles, which have been connected by straight lines as a visual aid. Figure 11.4 illustrates the downward bias of the initial observations. As time becomes larger, the effect of the initial conditions on later observations lessens and the observations appear to vary around a common mean. When the simulation analyst feels that this point has been reached, then the data-collection phase begins.

Table 11.6 also gives the cumulative average sample mean after deleting zero, one, and two batch means from the beginning—that is, using the ensemble average batch means $\bar{Y}_{\cdot j}$, when

Table 11.5 Individual Batch Means (Y_{rj}) for $M/G/1$ Simulation with Empty and Idle Initial State

Replication	Batch														
	1	2	3	4	5	6	7	8	9	10	11	12	13	14	15
1	3.61	3.21	2.18	6.92	2.82	1.59	3.55	5.60	3.04	2.57	1.41	3.07	4.03	2.70	2.71
2	2.91	9.00	16.15	24.53	25.19	21.63	24.47	8.45	8.53	14.84	23.65	27.58	24.19	8.58	4.06
3	7.67	19.53	20.36	8.11	12.62	22.15	14.10	9.87	23.96	24.50	14.56	6.08	4.83	16.04	23.41
4	6.62	1.75	12.87	8.77	1.25	1.16	1.92	6.29	4.74	17.43	18.24	18.59	4.62	2.76	1.57
5	2.18	1.32	2.14	2.18	2.59	1.20	4.11	6.21	7.31	1.58	2.16	3.08	2.32	2.21	3.32
6	0.93	3.54	4.80	0.72	2.95	5.56	1.96	2.07	2.74	3.45	14.24	13.39	7.87	0.94	3.19
7	1.12	2.59	5.05	1.16	2.72	5.12	5.03	4.14	4.98	15.81	9.29	2.14	8.72	29.80	28.94
8	1.54	5.94	5.33	2.91	2.69	1.91	3.27	3.61	10.35	9.66	4.13	6.14	7.90	2.61	7.95
9	8.93	4.78	0.74	2.56	9.43	18.63	8.14	1.49	4.51	1.69	12.62	11.28	3.32	3.42	3.35
10	4.78	2.84	10.39	5.87	1.01	2.59	16.77	27.25	26.81	20.96	7.26	2.32	5.04	8.50	9.11

Table 11.6   Summary of Data for *M/G/1* Simulation: Ensemble Batch Means and Cumulative Means, Averaged Over 10 Replications

Run Length T	*Batch j*	*Average Batch Mean,* $\bar{Y}_{\cdot j}$	*Cumulative Average (No Deletion),* $\bar{Y}..(j, 0)$	*Cumulative Average (Delete 1),* $\bar{Y}..(j, 1)$	*Cumulative Average (Delete 2),* $\bar{Y}..(j, 2)$
1,000	1	4.03	4.03	—	—
2,000	2	5.45	4.74	5.45	—
3,000	3	8.00	5.83	6.72	8.00
4,000	4	6.37	5.96	6.61	7.18
5,000	5	6.33	6.04	6.54	6.90
6,000	6	8.15	6.39	6.86	7.21
7,000	7	8.33	6.67	7.11	7.44
8,000	8	7.50	6.77	7.16	7.45
9,000	9	9.70	7.10	7.48	7.77
10,000	10	11.25	7.51	7.90	8.20
11,000	11	10.76	7.81	8.18	8.49
12,000	12	9.37	7.94	8.29	8.58
13,000	13	7.28	7.89	8.21	8.46
14,000	14	7.76	7.88	8.17	8.40
15,000	15	8.76	7.94	8.21	8.43

Figure 11.4   Ensemble averages $\bar{Y}_{\cdot j}$ for *M/G/1* queue.

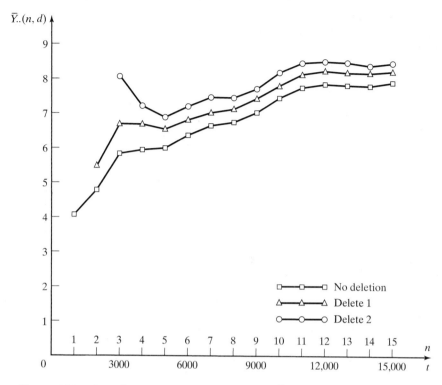

Figure 11.5   Cumulative average queue length $\bar{Y}_{..}(n, d)$ versus time $1000n$.

deleting d observations out of a total of n observations, compute

$$\bar{Y}_{..}(n, d) = \frac{1}{n - d} \sum_{j=d+1}^{n} \bar{Y}_{.j} \tag{11.25}$$

The results in Table 11.6 for the $M/G/1$ simulation are for $d = 0, 1$, and 2, and $n = d+1, \dots, 15$. These cumulative averages with deletion, namely $\bar{Y}_{..}(n, d)$, are plotted for comparison purposes in Figure 11.5. We do not recommend using cumulative averages to determine the initialization phase, for reasons given next.

From Figures 11.4 and 11.5, it is apparent that downward bias is present, and this initialization bias in the point estimator can be reduced by deletion of one or more observations. For the 15 ensemble average batch means, it appears that the first two observations have considerably more bias than any of the remaining ones. The effect of deleting first one and then two batch means is also illustrated in Table 11.6 and Figure 11.5. As expected, the estimators increase in value as more data are deleted; that is, $\bar{Y}_{..}(15, 2) = 8.43$ and $\bar{Y}_{..}(15, 1) = 8.21$ are larger than $\bar{Y}_{..}(15, 0) = 7.94$. It also appears from Figure 11.5 that $\bar{Y}_{..}(n, d)$ is increasing for $n = 5, 6, \dots, 11$ (and all $d = 0, 1, 2$), and thus there may still be some initialization bias. It seems, however, that deletion of the first two batches removes most of the bias.

Unfortunately, there is no widely accepted, objective, and proven technique to guide how much data to delete to reduce initialization bias to a negligible level. Plots can, at times, be misleading, but they are still recommended. Several points should be kept in mind:

1. Ensemble averages, such as Figure 11.4, will reveal a smoother and more precise trend as the number of replications, R, is increased. Since each ensemble average is the sample mean of i.i.d. observations, a confidence interval based on the t distribution can be placed around each point, as shown in Figure 11.6, and these intervals can be used to judge whether or not the plot is precise enough to judge that bias has diminished. *This is the preferred method to determine a deletion point.*

2. Ensemble averages can be smoothed further by plotting a moving average, rather than the original ensemble averages. In a moving average, each plotted point is actually the average of several adjacent ensemble averages. Specifically, the jth plot point would be

$$\tilde{Y}_j = \frac{1}{2m+1} \sum_{i=j-m}^{j+m} \bar{Y}_{\cdot i}$$

for some $m \geq 1$, rather than the original ensemble average $\bar{Y}_{\cdot j}$. The value of m is typically chosen by trial and error until a smooth plot is obtained. See Law and Kelton [2000] or Welch [1983] for further discussion of smoothing.

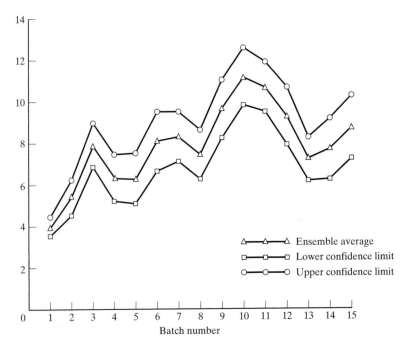

Figure 11.6   Ensemble averages $\bar{Y}_{\cdot j}$ for $M/G/1$ queue with 95% confidence intervals.

3. Cumulative averages, such as in Figure 11.5, become less variable as more data are averaged. Therefore, it is expected that the left side of the curve will always be less smooth than the right side. More importantly, cumulative averages tend to converge more slowly to long-run performance than do ensemble averages, because cumulative averages contain all observations, including the most biased ones from the beginning of the run. *For this reason, cumulative averages should be used only if it is not feasible to compute ensemble averages*, such as when only a single replication is possible.

4. Simulation data, especially from queueing models, usually exhibit positive autocorrelation. The more correlation present, the longer it takes for $\bar{Y}_{.j}$ to approach steady state. The positive correlation between successive observations (i.e., batch means) $\bar{Y}_{.1}, \bar{Y}_{.2}, \ldots$ can be seen in Figure 11.4.

5. In most simulation studies, the analyst is interested in several different output performance measures at once, such as the number in queue, customer waiting time, and utilization of the servers. Unfortunately, different performance measures could approach steady state at different rates. Thus, it is important to examine each performance measure individually for initialization bias and use a deletion point that is adequate for all of them.

There has been no shortage of solutions to the initialization-bias problem. Unfortunately, for every "solution" that works well in some situations, there are other situations in which either it is not applicable or it performs poorly. Important ideas include testing for bias (e.g., Kelton and Law [1983], Schruben [1980], Goldsman, Schruben, and Swain [1994]); modeling the bias (e.g., Snell and Schruben [1985]); and randomly sampling the initial conditions on multiple replications (e.g., Kelton [1989]).

11.5.2 Error Estimation for Steady-State Simulation

If $\{Y_1, \ldots, Y_n\}$ are not statistically independent, then S^2/n, given by Equation (11.11), is a biased estimator of the true variance, $V(\hat{\theta})$. This is almost always the case when $\{Y_1, \ldots, Y_n\}$ is a sequence of output observations from within a single replication. In this situation, $Y_1, Y_2, \ldots$ is an autocorrelated sequence, sometimes called a time series. Example 11.8 (the $M/G/1$ queue) provides an illustration of this situation.

Suppose that our point estimator for θ is the sample mean $\bar{Y} = \sum_{i=1}^{n} Y_i/n$. A general result from mathematical statistics is that the variance of $\bar{Y}$ is[3]

$$V(\bar{Y}) = \frac{1}{n^2} \sum_{i=1}^{n} \sum_{j=1}^{n} \text{cov}(Y_i, Y_j) \qquad (11.26)$$

where $\text{cov}(Y_i, Y_i) = V(Y_i)$. To construct a confidence interval for θ, an estimate of $V(\bar{Y})$ is required. But obtaining an estimate of (11.26) is pretty much hopeless, because each term $\text{cov}(Y_i, Y_j)$ could be different, in general. Fortunately, systems that have a steady state will, if simulated long enough to pass the transient phase (such as the production-line startup in Example 11.4), produce an output process that is approximately *covariance stationary*. Intuitively, stationarity implies that Y_{i+k} depends on Y_{i+1} in the same manner as Y_k depends on Y_1.

---

[3]This general result can be derived from the fact that, for two random variables Y_1 and Y_2, $V(Y_1 \pm Y_2) = V(Y_1) + V(Y_2) \pm 2\text{cov}(Y_1, Y_2)$.

In particular, the covariance between two random variables in the time series depends only on the number of observations between them, called the *lag*.

For a covariance-stationary time series, $\{Y_1, Y_2, \ldots\}$, define the lag-k autocovariance by

$$\gamma_k = \text{cov}(Y_1, Y_{1+k}) = \text{cov}(Y_i, Y_{i+k}) \tag{11.27}$$

which, by definition of covariance stationarity, is not a function of i. For $k = 0$, γ_0 becomes the population variance σ^2—that is,

$$\gamma_0 = \text{cov}(Y_i, Y_{i+0}) = \text{V}(Y_i) = \sigma^2 \tag{11.28}$$

The lag-k autocorrelation is the correlation between any two observations k apart. It is defined by

$$\rho_k = \frac{\gamma_k}{\sigma^2} \tag{11.29}$$

and has the property that

$$-1 \le \rho_k \le 1, \quad k = 1, 2, \ldots$$

If a time series is covariance stationary, then Equation (11.26) can be simplified substantially. Tedious algebra shows that

$$\text{V}(\bar{Y}) = \frac{\sigma^2}{n} \left[1 + 2 \sum_{k=1}^{n-1} \left(1 - \frac{k}{n} \right) \rho_k \right] \tag{11.30}$$

where ρ_k is the lag-k autocorrelation given by Equation (11.29).

When $\rho_k > 0$ for all k (or most k), the time series is said to be positively autocorrelated. In this case, large observations tend to be followed by large observations, small observations by small ones. Such a series will tend to drift slowly above and then below its mean. Figure 11.7(a) is an example of a stationary time series exhibiting positive autocorrelation. The output data from most queueing simulations are positively autocorrelated.

On the other hand, if some of the $\rho_k < 0$, the series $Y_1, Y_2, \ldots$ will display the characteristics of negative autocorrelation. In this case, large observations tend to be followed by small observations, and vice versa. Figure 11.7(b) is an example of a stationary time series exhibiting negative autocorrelation. The output of certain inventory simulations might be negatively autocorrelated.

Figure 11.7(c) also shows an example of a time series with an upward trend. Such a time series is not stationary; the probability distribution of Y_i is changing with the index i.

Why does autocorrelation make it difficult to estimate $\text{V}(\bar{Y})$? Recall that the standard estimator for the variance of a sample mean is S^2/n. By using Equation (11.30), it can be shown [Law, 1977] that the expected value of the variance estimator S^2/n is

$$E\left(\frac{S^2}{n} \right) = B\text{V}(\bar{Y}) \tag{11.31}$$

where

$$B = \frac{n/c - 1}{n - 1} \tag{11.32}$$

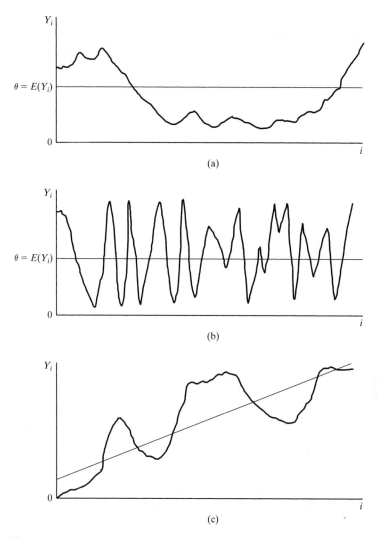

Figure 11.7   (a) Stationary time series Y_i exhibiting positive autocorrelation; (b) stationary time series Y_i exhibiting negative autocorrelation; (c) nonstationary time series with an upward trend.

and c is the quantity in brackets in Equation (11.30). The effect of the autocorrelation on the estimator S^2/n is derived by an examination of Equations (11.30) and (11.32). There are essentially three possibilities:

Case 1

If the Y_i are independent, then $\rho_k = 0$ for $k = 1, 2, 3, \ldots$. Therefore, $c = 1 + 2\sum_{k=1}^{n-1}(1 - k/n)\rho_k = 1$ and Equation (11.30) reduces to the familiar σ^2/n. Notice also that $B = 1$, so S^2/n

is an unbiased estimator of $V(\bar{Y})$. The Y_i will always be independent when they are obtained from different replications; that independence is the primary reason that we prefer experiment designs calling for multiple replications.

Case 2

If the autocorrelations ρ_k are primarily positive, then $c = 1 + 2\sum_{k=1}^{n-1}(1 - k/n)\rho_k > 1$, so that $n/c < n$, and hence $B < 1$. Therefore, S^2/n is biased low as an estimator of $V(\bar{Y})$. If this correlation were ignored, the nominal $100(1-\alpha)\%$ confidence interval given by Expression (11.12) would be too short, and its true confidence coefficient would be less than $1 - \alpha$. The practical effect would be that the simulation analyst would have unjustified confidence (in the apparent precision of the point estimator) due to the shortness of the confidence interval. If the correlations ρ_k are large, B could be quite small, implying a significant underestimation.

Case 3

If the autocorrelations ρ_k are substantially negative, then $0 \leq c < 1$, and it follows that $B > 1$ and S^2/n is biased high for $V(\bar{Y})$. In other words, the true precision of the point estimator $\bar{Y}$ would be greater than what is indicated by its variance estimator S^2/n, because

$$V(\bar{Y}) < E\left(\frac{S^2}{n}\right)$$

As a result, the nominal $100(1 - \alpha)\%$ confidence interval of Expression (11.12) would have true confidence coefficient greater than $1 - \alpha$. This error is less serious than Case 2, because we are unlikely to make incorrect decisions if our estimate is actually more precise than we think it is.

A simple example demonstrates why we are especially concerned about positive correlation: Suppose you want to know how students on a university campus will vote in an upcoming election. To estimate their preferences, you plan to solicit 100 responses. The standard experiment is to randomly select 100 students to poll; call this experiment A. An alternative is to randomly select 20 students and ask each of them to state their preference 5 times in the same day; call this experiment B. Both experiments obtain 100 responses, but clearly an estimate based on experiment B will be less precise (will have larger variance) than an estimate based on experiment A. Experiment A obtains 100 independent responses, whereas experiment B obtains only 20 independent responses and 80 dependent ones. The five opinions from any one student are perfectly positively correlated (assuming a student names the same candidate all five times). Although this is an extreme example, it illustrates that estimates based on positively correlated data are more variable than estimates based on independent data. Therefore, a confidence interval or other measure of error should account correctly for dependent data, but S^2/n does not.

Two methods for eliminating or reducing the deleterious effects of autocorrelation upon estimation of a mean are given in the following sections. Unfortunately, some simulation languages either use or facilitate the use of S^2/n as an estimator of $V(\bar{Y})$, the variance of the sample mean, in all situations. If used uncritically in a simulation with positively autocorrelated output data, the downward bias in S^2/n and the resulting shortness of a confidence interval for θ will convey the impression of much greater precision than actually exists. When such positive autocorrelation is present in the output data, the true variance of the point estimator, $\bar{Y}$, can be many times greater than is indicated by S^2/n.

11.5.3 Replication Method for Steady-State Simulations

If initialization bias in the point estimator has been reduced to a negligible level (through some combination of intelligent initialization and deletion), then the method of independent replications can be used to estimate point-estimator variability and to construct a confidence interval. The basic idea is simple: Make R replications, initializing and deleting from each one the same way.

If, however, significant bias remains in the point estimator and a large number of replications are used to reduce point-estimator variability, the resulting confidence interval can be misleading. This happens because *bias is not affected by the number of replications* (R); it is affected only by deleting more data (i.e., increasing T_0) or extending the length of each run (i.e., increasing T_E). Thus, increasing the number of replications (R) could produce shorter confidence intervals around the "wrong point." Therefore, it is important to do a thorough job of investigating the initial-condition bias.

If the simulation analyst decides to delete d observations of the total of n observations in a replication, then the point estimator of θ is $\bar{Y}_{..}(n, d)$, defined by Equation (11.25)—that is, the point estimator is the average of the remaining data. The basic raw output data, $\{Y_{rj}, r = 1, \dots, R; j = 1, \dots, n\}$, are exhibited in Table 11.7. Each Y_{rj} is derived in one of the following ways:

Case 1

Y_{rj} is an individual observation from within replication r; for example, Y_{rj} could be the delay of customer j in a queue, or the response time to job j in a job shop.

Case 2

Y_{rj} is a batch mean from within replication r of some number of discrete-time observations. (Batch means are discussed further in Section 11.5.5.)

Table 11.7   Raw Output Data from a Steady-State Simulation

Replication	\multicolumn{5}{c}{Observations}	Replication Averages
	1	$\cdots$	*d*	*d+1*	$\cdots$	*n*	
1	$Y_{1,1}$	$\cdots$	$Y_{1,d}$	$Y_{1,d+1}$	$\cdots$	$Y_{1,n}$	$\bar{Y}_{1.}(n, d)$
2	$Y_{2,1}$	$\cdots$	$Y_{2,d}$	$Y_{2,d+1}$	$\cdots$	$Y_{2,n}$	$\bar{Y}_{2.}(n, d)$
.	.		.	.		.	.
.	.		.	.		.	.
.	.		.	.		.	.
R	$Y_{R,1}$	$\cdots$	$Y_{R,d}$	$Y_{R,d+1}$	$\cdots$	$Y_{R,n}$	$\bar{Y}_{R.}(n, d)$
	$\bar{Y}_{.1}$	$\cdots$	$\bar{Y}_{.d}$	$\bar{Y}_{.,d+1}$	$\cdots$	$\bar{Y}_{.n}$	$\bar{Y}_{..}(n, d)$

Case 3

Y_{rj} is a batch mean of a continuous-time process over time interval j; for instance, as in Example 11.14, Equation (11.23) defines Y_{rj} as the time-average (batch mean) number in queue over the interval $[1000(j-1), 1000j)$.

In Case 1, the number d of deleted observations and the total number of observations n might vary from one replication to the next, in which case replace d by d_r and n by n_r. For simplicity, assume that d and n are constant over replications. In Cases 2 and 3, d and n will be constant.

When using the replication method, each replication is regarded as a single sample for the purpose of estimating θ. For replication r, define

$$\bar{Y}_{r \cdot}(n, d) = \frac{1}{n-d} \sum_{j=d+1}^{n} Y_{rj} \tag{11.33}$$

as the sample mean of all (nondeleted) observations in replication r. Because all replications use different random-number streams and all are initialized at time 0 by the same set of initial conditions (I_0), the replication averages

$$\bar{Y}_{1 \cdot}(n, d), \dots, \bar{Y}_{R \cdot}(n, d)$$

are independent and identically distributed random variables—that is, they constitute a random sample from some underlying population having unknown mean

$$\theta_{n,d} = \mathrm{E}[\bar{Y}_{r \cdot}(n, d)] \tag{11.34}$$

The overall point estimator, given in Equation (11.25), is also given by

$$\bar{Y}_{\cdot \cdot}(n, d) = \frac{1}{R} \sum_{r=1}^{R} \bar{Y}_{r \cdot}(n, d) \tag{11.35}$$

as can be seen from Table 11.7 or from using Equation (11.24). Thus, it follows that

$$\mathrm{E}[\bar{Y}_{\cdot \cdot}(n, d)] = \theta_{n,d}$$

also. If d and n are chosen sufficiently large, then $\theta_{n,d} \approx \theta$, and $\bar{Y}_{\cdot \cdot}(n, d)$ is an approximately unbiased estimator of θ. The bias in $\bar{Y}_{\cdot \cdot}(n, d)$ is $\theta_{n,d} - \theta$.

For convenience, when the value of n and d are understood, abbreviate $\bar{Y}_{r \cdot}(n, d)$ (the mean of the undeleted observations from the rth replication) and $\bar{Y}_{\cdot \cdot}(n, d)$ (the mean of $\bar{Y}_{1 \cdot}(n, d), \dots,$ $\bar{Y}_{R \cdot}(n, d)$) by $\bar{Y}_{r \cdot}$ and $\bar{Y}_{\cdot \cdot}$, respectively. To estimate the standard error of $\bar{Y}_{\cdot \cdot}$, first compute the sample variance,

$$S^2 = \frac{1}{R-1} \sum_{r=1}^{R} (\bar{Y}_{r \cdot} - \bar{Y}_{\cdot \cdot})^2 = \frac{1}{R-1} \left(\sum_{r=1}^{R} \bar{Y}_{r \cdot}^2 - R \bar{Y}_{\cdot \cdot}^2 \right) \tag{11.36}$$

The standard error of $\bar{Y}_{\cdot \cdot}$ is given by

$$\text{s.e.}(\bar{Y}_{\cdot \cdot}) = \frac{S}{\sqrt{R}} \tag{11.37}$$

A $100(1 - \alpha)\%$ confidence interval for θ, based on the t distribution, is given by

$$\bar{Y}_{..} - t_{\alpha/2, R-1} \frac{S}{\sqrt{R}} \leq \theta \leq \bar{Y}_{..} + t_{\alpha/2, R-1} \frac{S}{\sqrt{R}} \qquad (11.38)$$

where $t_{\alpha/2, R-1}$ is the $100(1 - \alpha/2)$ percentage point of a t distribution with $R - 1$ degrees of freedom. This confidence interval is valid only if the bias of $\bar{Y}_{..}$ is approximately zero.

As a rough rule, the length of each replication, beyond the deletion point, should be at least ten times the amount of data deleted. In other words, $(n - d)$ should at least $10d$ (or more generally, T_E should be at least $10T_0$). Given this run length, the number of replications should be as many as time permits, up to about 25 replications. Kelton [1986] established that there is little value in dividing the available time into more than 25 replications, so, if time permits making more than 25 replications of length $T_0 + 10T_0$, then make 25 replications of longer than $T_0 + 10T_0$, instead. Again, these are rough rules that need not be followed slavishly.

Example 11.15

Consider again the $M/G/1$ queueing simulation of Examples 11.8 and 11.14. Suppose that the simulation analyst decides to make $R = 10$ replications, each of length $T_E = 15,000$ minutes, each starting at time 0 in the empty and idle state, and each initialized for $T_0 = 2000$ minutes before data collection begins. The raw output data consist of the batch means defined by Equation (11.23); recall that each batch mean is simply the average number of customers in queue for a 1000-minute interval. The first two batch means are deleted ($d = 2$). The purpose of the simulation is to estimate, by a 95% confidence interval, the long-run time-average queue length, denoted by L_Q.

The replication averages $\bar{Y}_{r.}(15, 2)$, $r = 1, 2, \ldots, 10$, are shown in Table 11.8 in the rightmost column. The point estimator is computed by Equation (11.35) as

$$\bar{Y}_{..}(15, 2) = 8.43$$

Its standard error is given by Equation (11.37) as

$$\text{s.e.}(\bar{Y}_{..}(15, 2)) = 1.59$$

and using $\alpha = 0.05$ and $t_{0.025, 9} = 2.26$, the 95% confidence interval for long-run mean queue length is given by Inequality (11.38) as

$$8.43 - 2.26(1.59) \leq L_Q \leq 8.43 + 2.26(1.59)$$

or

$$4.84 \leq L_Q \leq 12.02$$

The simulation analyst may conclude with a high degree of confidence that the long-run mean queue length is between 4.84 and 12.02 customers. The confidence interval computed here as given by Inequality (11.38) should be used with caution, because a key assumption behind its validity is that enough data have been deleted to remove any significant bias due to initial conditions—that is, that d and n are sufficiently large that the bias $\theta_{n,d} - \theta$ is negligible.

Table 11.8   Data Summary for *M/G/1* Simulation by Replication

	Sample Mean for Replication *r*		
Replication, *r*	*(No Deletion)* $\bar{Y}_r.(15, 0)$	*(Delete 1)* $\bar{Y}_r.(15, 1)$	*(Delete 2)* $\bar{Y}_r.(15, 2)$
1	3.27	3.24	3.25
2	16.25	17.20	17.83
3	15.19	15.72	15.43
4	7.24	7.28	7.71
5	2.93	2.98	3.11
6	4.56	4.82	4.91
7	8.44	8.96	9.45
8	5.06	5.32	5.27
9	6.33	6.14	6.24
10	10.10	10.48	11.07
$\bar{Y}..(15, d)$	7.94	8.21	8.43
$\sum_{r=1}^{R} \bar{Y}_{r.}^2$	826.20	894.68	938.34
S^2	21.75	24.52	25.30
S	4.66	4.95	5.03
$S/\sqrt{10} = $ s.e.$(\bar{Y}..)$	1.47	1.57	1.59

Example 11.16 ———

Suppose that, in Example 11.15, the simulation analyst had decided to delete one batch ($d = 1$) or no batches ($d = 0$). The quantities needed to compute 95% confidence intervals are shown in Table 11.8. The resulting 95% confidence intervals are computed by Inequality (11.38) as follows:

$$(d = 1) \quad 4.66 = 8.21 - 2.26(1.57) \leq L_Q \leq 8.21 + 2.26(1.57) = 11.76$$
$$(d = 0) \quad 4.62 = 7.94 - 2.26(1.47) \leq L_Q \leq 7.94 + 2.26(1.47) = 11.26$$

Notice that, for a fixed total sample size, *n*, two things happen as fewer data are deleted:

1. The confidence interval shifts downward, reflecting the greater downward bias in $\bar{Y}..(15, d)$ as *d* decreases.
2. The standard error of $\bar{Y}..(n, d)$, namely $S/\sqrt{R}$, decreases as *d* decreases.

In this example, $\bar{Y}..(n, d)$ is based on a run length of $T_E = 1000(n - d) = 15{,}000 - 1000d$ minutes. Thus, as *d* decreases, T_E increases, and, in effect, the sample mean $\bar{Y}..$ is based on a larger "sample size" (i.e., longer run length). In general, the larger the sample size, the smaller the standard error of the point estimator. This larger sample size can be due to a longer run length (T_E) per replication, or to more replications (R).

Therefore, there is a trade-off between reducing bias and increasing the variance of a point estimator, when the total sample size (R and $T_0 + T_E$) is fixed. The more deletion (i.e., the larger T_0 is and the smaller T_E is, keeping $T_0 + T_E$ fixed), the less bias but greater variance there is in the point estimator.

Recall that each batch in Examples 11.15 and 11.16 consists of 1000 minutes of simulated time. Therefore, discarding $d = 2$ batches really means discarding 2000 minutes of data, a substantial amount. It is not uncommon for very large deletions to be necessary to overcome the initial conditions.

11.5.4 Sample Size in Steady-State Simulations

Suppose it is desired to estimate a long-run performance measure, θ, within $\pm\epsilon$, with confidence $100(1 - \alpha)\%$. In a steady-state simulation, a specified precision may be achieved either by increasing the number of replications (R) or by increasing the run length (T_E). The first solution, controlling R, is carried out as given in Section 11.4.2 for terminating simulations.

Example 11.17 _____

Consider the data in Table 11.8 for the $M/G/1$ queueing simulation as an initial sample of size $R_0 = 10$. Assuming that $d = 2$ observations were deleted, the initial estimate of variance is $S_0^2 = 25.30$. Suppose that it is desired to estimate long-run mean queue length, L_Q, within $\epsilon = 2$ customers with 90% confidence. The final sample size needed must satisfy Inequality (11.17). Using $\alpha = 0.10$ in Inequality (11.18) yields an initial estimate:

$$R \geq \left(\frac{z_{0.05}S_0}{\epsilon}\right)^2 = \frac{1.645^2(25.30)}{2^2} = 17.1$$

Thus, at least 18 replications will be needed. Proceeding as in Example 11.11, next try $R = 18$, $R = 19, \ldots$ as follows:

R	18	19
$t_{0.05,R-1}$	1.74	1.73
$\left(\dfrac{t_{0.05,R-1}S_0}{\epsilon}\right)^2$	19.15	18.93

$R = 19 \geq (t_{0.05,18}\, S_0/\epsilon)^2 = 18.93$ is the smallest integer R satisfying Inequality (11.17), so a total sample size of $R = 19$ replications is needed to estimate L_Q to within ± 2 customers. Therefore, $R - R_0 = 19 - 10 = 9$ additional replications are needed to achieve the specified error.

An alternative to increasing R is to increase total run length $T_0 + T_E$ within each replication. If the calculations in Section 11.4.2, as illustrated in Example 11.17, indicate that $R - R_0$ additional replications are needed beyond the initial number, R_0, then an alternative is to increase run length $(T_0 + T_E)$ in the same proportion (R/R_0) to a new run length $(R/R_0)(T_0 + T_E)$. Thus, additional data will be deleted, from time 0 to time $(R/R_0)T_0$, and more data will be used to compute the point estimates, as illustrated by Figure 11.8. However, the total amount of simulation effort is the same as if we had simply increased the number of replications but maintained the same run length. The advantage of increasing total run length per replication and deleting a fixed proportion $[T_0/(T_0 + T_E)]$ of the total run length is that any residual bias in the point estimator

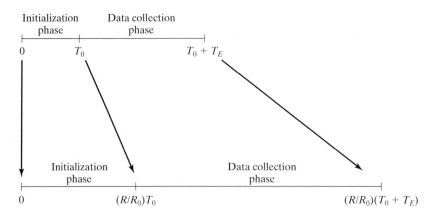

Figure 11.8   Increasing runlength to achieve specified accuracy.

should be further reduced by the additional deletion of data at the beginning of the run. A possible disadvantage of the method is that, in order to continue the simulation of all R replications [from time $T_0 + T_E$ to time $(R/R_0)(T_0 + T_E)$], it is necessary to have saved the state of the model at time $T_0 + T_E$ and to be able to restart the model and run it for the additional required time. Otherwise, the simulations would have to be rerun from time 0, which could be time consuming for a complex model. Some simulation languages have the capability to save enough information that a replication can be continued from time T_E onward, rather than having to start over from time 0.

Example 11.18
In Example 11.17, suppose that run length was to be increased to achieve the desired error, ± 2 customers. Since $R/R_0 = 19/10 = 1.9$, the run length should be almost doubled to $(R/R_0)(T_0 + T_E) = 1.9(15,000) = 28,500$ minutes. The data collected from time 0 to time $(R/R_0)T_0 = 1.9(2000) = 3800$ minutes would be deleted, and the data from time 3800 to time 28,500 used to compute new point estimates and confidence intervals.

11.5.5 Batch Means for Interval Estimation in Steady-State Simulations

One disadvantage of the replication method is that data must be deleted on each replication and, in one sense, deleted data are wasted data, or at least lost information. This suggests that there might be merit in using an experiment design that is based on a single, long replication. The disadvantage of a single-replication design arises when we try to compute the standard error of the sample mean. Since we only have data from within one replication, the data are dependent, and the usual estimator is biased.

The method of *batch means* attempts to solve this problem by dividing the output data from one replication (after appropriate deletion) into a few large batches and then treating the means of these batches as if they were independent. When the raw output data after deletion form a continuous-time process, $\{Y(t), T_0 \leq t \leq T_0 + T_E\}$, such as the length of a queue or the level of inventory, then we form k batches of size $m = T_E/k$ and compute the batch means as

$$\bar{Y}_j = \frac{1}{m} \int_{(j-1)m}^{jm} Y(t + T_0)\, dt$$

for $j = 1, 2, \ldots, k$. In other words, the jth batch mean is just the time-weighted average of the process over the time interval $[T_0 + (j-1)m, T_0 + jm)$, exactly as in Example 11.8.

When the raw output data after deletion form a discrete-time process, $\{Y_i, i = d + 1, d + 2, \ldots, n\}$, such as the customer delays in a queue or the cost per period of an inventory system, then we form k batches of size $m = (n - d)/k$ and compute the batch means as

$$\bar{Y}_j = \frac{1}{m} \sum_{i=(j-1)m+1}^{jm} Y_{i+d}$$

for $j = 1, 2, \ldots, k$ (assuming k divides $n - d$ evenly, otherwise round down to the nearest integer). That is, the batch means are formed as shown here:

$$\underbrace{Y_1, \ldots, Y_d,}_{\text{deleted}} \underbrace{Y_{d+1}, \ldots, Y_{d+m},}_{\bar{Y}_1} \underbrace{Y_{d+m+1}, \ldots, Y_{d+2m},}_{\bar{Y}_2} \ldots, \underbrace{Y_{d+(k-1)m+1}, \ldots, Y_{d+km}}_{\bar{Y}_k}$$

Starting with either continuous-time or discrete-time data, the variance of the sample mean is estimated by

$$\frac{S^2}{k} = \frac{1}{k} \sum_{j=1}^{k} \frac{(\bar{Y}_j - \bar{Y})^2}{k - 1} = \frac{\sum_{j=1}^{k} \bar{Y}_j^2 - k\bar{Y}^2}{k(k - 1)} \tag{11.39}$$

where $\bar{Y}$ is the overall sample mean of the data after deletion. As was discussed in Section 11.2, the batch means $\bar{Y}_1, \bar{Y}_2, \ldots, \bar{Y}_k$ are not independent; however, if the batch size is sufficiently large, successive batch means will be approximately independent, and the variance estimator will be approximately unbiased.

Unfortunately, there is no widely accepted and relatively simple method for choosing an acceptable batch size m (or equivalently choosing a number of batches k). But there are some general guidelines that can be culled from the research literature:

- Schmeiser [1982] found that for a *fixed total sample size* there is little benefit from dividing it into more than $k = 30$ batches, even if we could do so and still retain independence between the batch means. Therefore, there is no reason to consider numbers of batches much greater than 30, no matter how much raw data are available. He also found that the performance of the confidence interval, in terms of its width and the variability of its width, is poor for fewer than 10 batches. Therefore, a number of batches between 10 and 30 should be used in most applications.
- Although there is typically autocorrelation between batch means at all lags, the lag-1 autocorrelation $\rho_1 = \text{corr}(\bar{Y}_j, \bar{Y}_{j+1})$ is usually studied to assess the dependence between

batch means. When the lag-1 autocorrelation is nearly 0, then the batch means are treated as independent. This approach is based on the observation that the autocorrelation in many stochastic processes decreases as the lag increases. Therefore, all lag autocorrelations should be smaller (in absolute value) than the lag-1 autocorrelation.

- The lag-1 autocorrelation between batch means can be estimated as described shortly. However, the autocorrelation should not be estimated from a small number of batch means (such as the $10 \leq k \leq 30$ recommended above); there is bias in the autocorrelation estimator. Law and Carson [1979] suggest estimating the lag-1 autocorrelation from a large number of batch means based on a smaller batch size (perhaps $100 \leq k \leq 400$). When the autocorrelation between these batch means is approximately 0, then the autocorrelation will be even smaller if we rebatch the data to between 10 and 30 batch means based on a larger batch size. Hypothesis tests for 0 autocorrelation are available, as described next.

- If the *total sample size is to be chosen sequentially*, say to attain a specified precision, then it is helpful to allow the batch size and number of batches to grow as the run length increases. It can be shown that a good strategy is to allow the number of batches to increase as the square root of the sample size after first finding a batch size at which the lag-1 autocorrelation is approximately 0. Although we will not discuss this point further, an algorithm based on it can be found in Fishman and Yarberry [1997]; see also Steiger and Wilson [2002].

Given these insights, we recommend the following general strategy:

1. Obtain output data from a single replication and delete as appropriate. Recall our guideline: collecting at least 10 times as much data as is deleted.
2. Form up to $k = 400$ batches (but at least 100 batches) with the retained data, and compute the batch means. Estimate the sample lag-1 autocorrelation of the batch means as

$$\widehat{\rho}_1 = \frac{\sum_{j=1}^{k-1}(\bar{Y}_j - \bar{Y})(\bar{Y}_{j+1} - \bar{Y})}{\sum_{j=1}^{k}(\bar{Y}_j - \bar{Y})^2}$$

3. Check the correlation to see whether it is sufficiently small.
    (a) If $\widehat{\rho}_1 \leq 0.2$, then rebatch the data into $30 \leq k \leq 40$ batches, and form a confidence interval using $k - 1$ degrees of freedom for the t distribution and Equation (11.39) to estimate the variance of $\bar{Y}$.
    (b) If $\widehat{\rho}_1 > 0.2$, then extend the replication by 50% to 100% and go to Step 2. If it is not possible to extend the replication, then rebatch the data into approximately $k = 10$ batches, and form the confidence interval, using $k - 1$ degrees of freedom for the t distribution and Equation (11.39) to estimate the variance of $\bar{Y}$.
4. As an additional check on the confidence interval, examine the batch means (at the larger or smaller batch size) for independence, using the following test. (See, for instance, Alexopoulos and Seila [1998].) Compute the test statistic

$$C = \sqrt{\frac{k^2 - 1}{k - 2}} \left(\widehat{\rho}_1 + \frac{(\bar{Y}_1 - \bar{Y})^2 + (\bar{Y}_k - \bar{Y})^2}{2\sum_{j=1}^{k}(\bar{Y}_j - \bar{Y})^2} \right)$$

If $C < z_\beta$ then accept the independence of the batch means, where β is the Type I error level of the test (such as 0.1, 0.05, 0.01). Otherwise, extend the replication by 50% to 100% and go to Step 2. If it is not possible to extend the replication, then rebatch the data into approximately $k = 10$ batches, and form the confidence interval, using $k - 1$ degrees of freedom for the t distribution and Equation (11.39) to estimate the variance of $\bar{Y}$.

This procedure, including the final check, is conservative in several respects. First, if the lag-1 autocorrelation is substantially negative then we proceed to form the confidence interval anyway. A dominant negative correlation tends to make the confidence interval wider than necessary, which is an error, but not one that will cause us to make incorrect decisions. The requirement that $\widehat{\rho}_1 < 0.2$ at $100 \le k \le 400$ batches is pretty stringent and will tend to force us to get more data (and therefore create larger batches) if there is any hint of positive dependence. And finally, the hypothesis test at the end has a probability of β of forcing us to get more data when none are really needed. But this conservatism is by design; the cost of an incorrect decision is typically much greater than the cost of some additional computer run time.

The batch-means approach to confidence-interval estimation is illustrated in the next example.

Example 11.19

Reconsider the $M/G/1$ simulation of Example 11.8, except that the mean service time is changed from 9.5 minutes to 7 minutes (implying a long-run server utilization of 0.7). Suppose that we want to estimate the steady-state expected delay in queue, w_Q, by a 95% confidence interval. To illustrate the method of batch means, assume that one run of the model has been made, simulating 3000 customers after the deletion point. We then form batch means from $k = 100$ batches of size $m = 30$ and estimate the lag-1 autocorrelation to be $\widehat{\rho}_1 = 0.346 > 0.2$. Thus, we decide to extend the simulation to 6000 customers after the deletion point, and again we estimate the lag-1 autocorrelation. This estimate, based on $k = 100$ batches of size $m = 60$, is $\widehat{\rho}_1 = 0.004 < 0.2$.

Having passed the correlation check, we rebatch the data into $k = 30$ batches of size $m = 200$. The point estimate is the overall mean

$$\bar{Y} = \frac{1}{6000} \sum_{j=1}^{6000} \bar{Y}_j = 9.04$$

minutes. The variance of $\bar{Y}$, computed from the 30 batch means, is

$$\frac{S^2}{k} = \frac{\sum_{j=1}^{30} \bar{Y}_j^2 - 30\bar{Y}^2}{30(29)} = 0.604$$

Thus, a 95% confidence interval is given by

$$\bar{Y} - t_{0.025,29}\sqrt{0.604} \le w_Q \le \bar{Y} + t_{0.025,29}\sqrt{0.604}$$

or

$$7.45 = 9.04 - 2.04(0.777) \le w_Q \le 9.04 + 2.04(0.777) = 10.63$$

Thus, we assert with 95% confidence that true mean delay in queue, w_Q, is between 7.45 and 10.63 minutes. If these results are not sufficiently precise for practical use, the run length should be increased to achieve greater precision.

As a further check on the validity of the confidence interval, we can apply the correlation hypothesis test. To do so, we compute the test statistic from the $k = 30$ batches of size $m = 200$ used to form the confidence interval. This gives

$$C = -0.31 < 1.96 = z_{0.05}$$

confirming the lack of correlation at the 0.05 significance level. Notice that, at this small number of batches, the estimated lag-1 autocorrelation appears to be slightly negative, illustrating our point about the difficulty of estimating correlation with small numbers of observations.

11.5.6 Quantiles

Constructing confidence intervals for quantile estimates in a steady-state simulation can be tricky, especially if the output process of interest is a continuous-time process, such as $L_Q(t)$, the number of customers in queue at time t. In this section, we outline the main issues.

Taking the easier case first, suppose that the output process from a single replication (after appropriate deletion of initial data) is $Y_{d+1}, \ldots, Y_n$. To be concrete, Y_i might be the delay in queue of the ith customer. Then the point estimate of the pth quantile can be obtained as before, either from the histogram of the data or from the sorted values. Of course, only the data after the deletion point are used. Suppose we make R replications and let $\widehat{\theta}_r$ be the quantile estimate from the rth. Then the R quantile estimates, $\widehat{\theta}_1, \ldots, \widehat{\theta}_R$, are independent and identically distributed. Their average is

$$\widehat{\theta}_. = \frac{1}{R} \sum_{i=1}^{R} \widehat{\theta}_i$$

It can be used as the point estimator of θ; and an approximate confidence interval is

$$\widehat{\theta}_. \pm t_{\alpha/2, R-1} \frac{S}{\sqrt{R}}$$

where S^2 is the usual sample variance of $\widehat{\theta}_1, \ldots, \widehat{\theta}_R$.

What if only a single replication is obtained? Then the same reasoning applies if we let $\widehat{\theta}_i$ be the quantile estimate from *within* the ith batch of data. This requires sorting the data, or forming a histogram, within each batch. If the batches are large enough, then these within-batch quantile estimates will also be approximately i.i.d.

When we have a continuous-time output process, then, in principle, the same methods apply. However, we must be careful not to transform the data in a way that changes the problem. In particular, we cannot first form batch means—as we have done throughout this chapter—and then estimate the quantile from these batch means. The p quantile of the *batch means* of $L_Q(t)$ is *not* the same as the p quantile of $L_Q(t)$ itself. Thus, the quantile point estimate must be formed from the histogram of the raw data—either from each run, if we make replications, or within each batch, if we make a single replication.

11.6 SUMMARY

This chapter emphasized the idea that a stochastic discrete-event simulation is a statistical experiment. Therefore, before sound conclusions can be drawn on the basis of the simulation-generated output data, a proper statistical analysis is required. The purpose of the simulation experiment is to obtain estimates of the performance measures of the system under study. The purpose of the statistical analysis is to acquire some assurance that these estimates are sufficiently precise for the proposed use of the model.

A distinction was made between terminating simulations and steady-state simulations. Steady-state simulation output data are more difficult to analyze, because the simulation analyst must address the problem of initial conditions and the choice of run length. Some suggestions were given regarding these problems, but unfortunately no simple, complete, and satisfactory solution exists. Nevertheless, simulation analysts should be aware of the potential problems, and of the possible solutions—namely, deletion of data and increasing of the run length. More advanced statistical techniques (not discussed in this text) are given in Alexopoulos and Seila [1998], Bratley, Fox, and Schrage [1996], and Law and Kelton [2000].

The statistical precision of point estimators can be measured by a standard-error estimate or by a confidence interval. The method of independent replications was emphasized. With this method, the simulation analyst generates statistically independent observations, and thus standard statistical methods can be employed. For steady-state simulations, the method of batch means was also discussed.

The main point is that simulation output data contain some amount of random variability; without some assessment of its size, the point estimates cannot be used with any degree of reliability.

REFERENCES

ALEXOPOULOS, C., AND A. F. SEILA [1998], "Output Data Analysis," Chapter 7 in *Handbook of Simulation*, J. Banks, ed., Wiley, New York.

BRATLEY, P., B. L. FOX, AND L. E. SCHRAGE [1996], *A Guide to Simulation*, 2d ed., Springer-Verlag, New York.

FISHMAN, G. S., AND L. S. YARBERRY [1997], "An Implementation of the Batch Means Method," *INFORMS Journal on Computing*, Vol. 9, pp. 296–310.

GOLDSMAN, D., L. SCHRUBEN, AND J. J. SWAIN [1994], "Tests for Transient Means in Simulated Time Series," *Naval Research Logistics*, Vol. 41, pp. 171–187.

KELTON, W. D. [1986], "Replication Splitting and Variance for Simulating Discrete-Parameter Stochastic Processes," *Operations Research Letters*, Vol. 4, pp. 275–279.

KELTON, W. D. [1989], "Random Initialization Methods in Simulation," *IIE Transactions*, Vol. 21, pp. 355–367.

KELTON, W. D., AND A. M. LAW [1983], "A New Approach for Dealing with the Startup Problem in Discrete Event Simulation," *Naval Research Logistics Quarterly*, Vol. 30, pp. 641–658.

KLEIJNEN, J. P. C. [1987], *Statistical Tools for Simulation Practitioners*, Dekker, New York.

LAW, A. M. [1977], "Confidence Intervals in Discrete Event Simulation: A Comparison of Replication and Batch Means," *Naval Research Logistics Quarterly*, Vol. 24, pp. 667–78.

LAW, A. M. [1980], "Statistical Analysis of the Output Data from Terminating Simulations," *Naval Research Logistics Quarterly*, Vol. 27, pp. 131–43.

LAW, A. M., AND J. S. CARSON [1979], "A Sequential Procedure for Determining the Length of a Steady-State Simulation," *Operations Research*, Vol. 27, pp. 1011–1025.

LAW, A. M., AND W. D. KELTON [2000], *Simulation Modeling and Analysis*, 3d ed., McGraw–Hill, New York.

NELSON, B. L. [2001], "Statistical Analysis of Simulation Results," Chapter 94 in *Handbook of Industrial Engineering*, 3d ed., G. Salvendy, ed., Wiley, New York.

SCHMEISER, B. [1982], "Batch Size Effects in the Analysis of Simulation Output," *Operations Research*, Vol. 30, pp. 556–568.

SCHRUBEN, L. [1980], "Detecting Initialization Bias in Simulation Output," *Operations Research*, Vol. 30, pp. 569–590.

SNELL, M., AND L. SCHRUBEN [1985], "Weighting Simulation Data to Reduce Initialization Effects," *IIE Transactions*, Vol. 17, pp. 354–363.

STEIGER, N. M., AND J. R. Wilson [2002], "An Improved Batch Means Procedure for Simulation Output Analysis," *Management Science*, Vol. 48, pp. 1569–1586.

WELCH, P. D. [1983], "The Statistical Analysis of Simulation Results," in *The Computer Performance Modeling Handbook*, S. Lavenberg, ed., Academic Press, New York, pp. 268–328.

EXERCISES

1. (a) Consider Example 2.1. Under what circumstances would it be appropriate to use a terminating simulation, versus a steady-state simulation, to analyze this system?

   (b) Repeat part (a) for Example 2.2 (the Able–Baker problem).

   (c) Repeat part (a) for Example 2.3.

   (d) Repeat part (a) for Example 2.4.

   (e) Repeat part (a) for Example 2.5.

   (f) Repeat part (a) for Example 2.6.

   (g) Repeat part (a) for Example 3.5 (the dump-truck problem).

2. Suppose that, in Example 11.14, the simulation analyst decided to investigate the bias by using batch means over a batching interval of 2000 minutes. By definition, a batch mean for the interval $[(j-1)2000, j(2000))$ is defined by

$$Y_j = \frac{1}{2000} \int_{(j-1)2000}^{j(2000)} L_Q(t)\, dt$$

   (a) Show algebraically that such a batch mean can be obtained from two adjacent batch means over the two halves of the interval.

   (b) Compute the seven averaged batch means for the intervals [0, 2000), [2000, 4000), ... for the $M/G/1$ simulation. Use the data $(\bar{Y}_{.j})$ in Table 11.6 (ignoring $\bar{Y}_{.15} = 8.76$).

(c) Draw plots of the type used in Figures 11.4 and 11.5. Does it still appear that deletion of the data over [0, 2000) (the first "new" batch mean) is sufficient to remove most of the point-estimator bias?

3. Suppose, in Example 11.14, that the simulation analyst could only afford to run 5 independent replications (instead of 10). Use the batch means in Table 11.5 for replications 1 to 5 to compute a 95% confidence interval for mean queue length L_Q. Investigate deletion of initial data. Compare the results from using 5 replications with those from using 10 replications.

4. In Example 11.7, suppose that management desired 95% confidence in the estimate of mean system time w and that the error allowed was $\epsilon = 0.4$ minute. Using the same initial sample of size $R_0 = 4$ (given in Table 11.1), figure out the required total sample size.

5. Simulate the dump-truck problem in Example 3.5. At first, make the run length $T_E = 40$ hours. Make four independent replications. Compute a 90% confidence interval for mean cycle time, where a cycle time for a given truck is the time between its successive arrivals to the loader. Investigate the effect of different initial conditions (all trucks initially at the loader queue, versus all at the scale, versus all traveling, versus the trucks distributed throughout the system in some manner).

6. Consider an (M, L) inventory system, in which the procurement quantity, Q, is defined by

$$Q = \begin{cases} M - I & \text{if } I < L \\ 0 & \text{if } I \geq L \end{cases}$$

where I is the level of inventory on hand plus on order at the end of a month, M is the maximum inventory level, and L is the reorder point. M and L are under management control, so the pair (M, L) is called the inventory policy. Under certain conditions, the analytical solution of such a model is possible, but the computational effort can be prohibitive. Use simulation to investigate an (M, L) inventory system with the following properties: The inventory status is checked at the end of each month. Backordering is allowed at a cost of $4 per item short per month. When an order arrives, it will first be used to relieve the backorder. The lead time is given by a uniform distribution on the interval (0.25, 1.25) months. Let the beginning inventory level stand at 50 units, with no orders outstanding. Let the holding cost be $1 per unit in inventory per month. Assume that the inventory position is reviewed each month. If an order is placed, its cost is $60 + $5Q$, where $60 is the ordering cost and $5 is the cost of each item. The time between demands is exponentially distributed with a mean of 1/15 month. The sizes of the demands follow this distribution:

Demand	Probability
1	1/2
2	1/4
3	1/8
4	1/8

(a) Make four independent replications, each of run length 100 months preceded by a 12-month initialization period, for the $(M, L) = (50, 30)$ policy. Estimate long-run mean monthly cost with a 90% confidence interval.

(b) Using the results of part (a), estimate the total number of replications needed to estimate mean monthly cost within \$5.

7. Reconsider Exercise 6, except that, if the inventory level at a monthly review is zero or negative, a rush order for Q units is placed. The cost for a rush order is $\$120 + \$12Q$, where \$120 is the ordering cost and \$12 is the cost of each item. The lead time for a rush order is given by a uniform distribution on the interval (0.10, 0.25) months.

(a) Make four independent replications for the (M, L) policy, and estimate long-run mean monthly cost with a 90% confidence interval.

(b) Using the results of part (a), estimate the total number of replications needed to estimate mean monthly cost within \$5.

8. Suppose that the items in Exercise 6 are perishable, with a selling price given by the following data:

On the Shelf (Months)	Selling Price
0-1	\$10
1-2	5
>2	0

Thus, any item that has been on the shelf greater than 2 months cannot be sold. The age is measured at the time the demand occurs. If an item is outdated, it is discarded, and the next item is brought forward. Simulate the system for 100 months.

(a) Make four independent replications for the $(M, L) = (50, 30)$ policy, and estimate long-run mean monthly cost with a 90% confidence interval.

(b) Using the results of part (a), estimate the total number of replications needed to estimate mean monthly cost within \$5.

At first, assume that all the items in the beginning inventory are fresh. Is this a good assumption? What effect does this "all-fresh" assumption have on the estimates of long- run mean monthly cost? What can be done to improve these estimates? Carry out a complete analysis.

9. Consider the following inventory system:

(a) Whenever the inventory level falls to or below 10 units, an order is placed. Only one order can be outstanding at a time.

(b) The size of each order is Q. Maintaining an inventory costs \$0.50 per day per item in inventory. Placing an order incurs a fixed cost, \$10.00.

(c) Lead time is distributed in accordance with a discrete uniform distribution between zero and 5 days.

(d) If a demand occurs during a period when the inventory level is zero, the sale is lost at a cost of $2.00 per unit.

(e) The number of customers each day is given by the following distribution:

Number of Customers per Day	Probability
1	0.23
2	0.41
3	0.22
4	0.14

(f) The demand on the part of each customer is Poisson distributed with a mean of 3 units.

(g) For simplicity, assume that all demands occur at noon and that all orders are placed immediately thereafter.

Assume further that orders are received at 5:00 P.M., or after the demand that occurred on that day. Consider the policy having $Q = 20$. Make five independent replications, each of length 100 days, and compute a 90% confidence interval for long-run mean daily cost. Investigate the effect of initial inventory level and existence of an outstanding order on the estimate of mean daily cost. Begin with an initial inventory of $Q + 10$ and no outstanding orders.

10. A store selling Mother's Day cards must decide 6 months in advance on the number of cards to stock. Reordering is not allowed. Cards cost $0.45 and sell for $1.25. Any cards not sold by Mother's Day go on sale for $0.50 for 2 weeks. However, sales of the remaining cards is probabilistic in nature according to the following distribution:

32% of the time, all cards remaining get sold.

40% of the time, 80% of all cards remaining are sold.

28% of the time, 60% of all cards remaining are sold.

Any cards left after 2 weeks are sold for $0.25. The card-shop owner is not sure how many cards can be sold, but thinks it is somewhere (i.e., uniformly distributed) between 200 and 400. Suppose that the card-shop owner decides to order 300 cards. Estimate the expected total profit with an error of at most $5.00. (*Hint*: Make three or four initial replications. Use these data to estimate the total sample size needed. Each replication consists of one Mother's Day.)

11. A very large mining operation has decided to control the inventory of high-pressure piping by a "periodic review, order up to M" policy, where M is a target level. The annual demand for this piping is normally distributed, with mean 600 and variance 800. This demand occurs fairly uniformly over the year. The lead time for resupply is Erlang distributed of order $k = 2$ with its mean at 2 months. The cost of each unit is $400. The inventory carrying charge, as a proportion of item cost on an annual basis, is expected to fluctuate normally about the mean 0.25 (simple interest), with a standard deviation of 0.01. The cost of making a review

and placing an order is \$200, and the cost of a backorder is estimated to be \$100 per unit backordered. Suppose that the inventory level is reviewed every 2 months, and let $M = 337$.

(a) Make five independent replications, each of run length 100 months, to estimate long-run mean monthly cost by means of a 90% confidence interval.
(b) Investigate the effects of initial conditions. Calculate an appropriate number of monthly observations to delete to reduce initialization bias to a negligible level.

12. Consider some number, say N, of $M/M/1$ queues in series. The $M/M/1$ queue, described in Section 6.4, has Poisson arrivals at some rate λ customers per hour, exponentially distributed service times with mean $1/\mu$, and a single server. (Recall that "Poisson arrivals" means that interarrival times are exponentially distributed.) By $M/M/1$ queues in series, it is meant that, upon completion of service at a given server, a customer joins a waiting line for the next server. The system can be shown as follows:

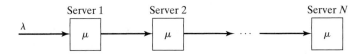

All service times are exponentially distributed with mean $1/\mu$, and the capacity of each waiting line is assumed to be unlimited. Assume that $\lambda = 8$ customers per hour and $1/\mu = 0.1$ hour. The measure of performance is response time, which is defined to be the total time a customer is in the system.

(a) By making appropriate simulation runs, compare the initialization bias for $N = 1$ (i.e., one $M/M/1$ queue) to $N = 2$ (i.e., two $M/M/1$ queues in series). Start each system with all servers idle and no customers present. The purpose of the simulation is to estimate mean response time.
(b) Investigate the initialization bias as a function of N, for $N = 1, 2, 3, 4$, and 5.
(c) Draw some general conclusions concerning initialization bias for "large" queueing systems when at time 0 the system is assumed to be empty and idle.

13. Jobs enter a job shop in random fashion according to a Poisson process at a stationary overall rate, two every 8-hour day. The jobs are of four types. They flow from work station to work station in a fixed order, depending on type, as shown next. The proportions of each type are also shown.

Type	Flow through Stations	Proportion
1	1, 2, 3, 4	0.4
2	1, 3, 4	0.3
3	2, 4, 3	0.2
4	1, 4	0.1

Processing times per job at each station depend on type, but all times are (approximately) normally distributed with mean and s.d. (in hours) as follows:

Type	Station			
	1	2	3	4
1	(20, 3)	(30, 5)	(75, 4)	(20, 3)
2	(18, 2)		(60, 5)	(10, 1)
3		(20, 2)	(50, 8)	(10, 1)
4	(30, 5)			(15, 2)

Station i will have c_i workers ($i = 1, 2, 3, 4$). Each job occupies one worker at a station for the duration of a processing time. All jobs are processed on a first-in–first-out basis, and all queues for waiting jobs are assumed to have unlimited capacity. Simulate the system for 800 hours, preceded by a 200-hour initialization period. Assume that $c_1 = 8$, $c_2 = 8$, $c_3 = 20$, $c_4 = 7$. Based on $R = 5$ replications, compute a 97.5% confidence interval for average worker utilization at each of the four stations. Also, compute a 95% confidence interval for mean total response time for each job type, where a total response time is the total time that a job spends in the shop.

14. Change Exercise 13 to give priority at each station to the jobs by type. Type 1 jobs have priority over type 2, type 2 over type 3, and type 3 over type 4. Use 800 hours as run length, 200 hours as initialization period, and $R = 5$ replications. Compute four 97.5% confidence intervals for mean total response time by type. Also, run the model without priorities and compute the same confidence intervals. Discuss the trade-offs when using *first in, first out* versus a priority system.

15. Consider a single-server queue with Poisson arrivals at rate $\lambda = 10.82$ per minute and normally distributed service times with mean 5.1 seconds and variance 0.98 seconds^2. It is desired to estimate the mean time in the system for a customer who, upon arrival, finds i other customers in the system—that is, to estimate

$$w_i = E(W | N = i) \quad \text{for } i = 0, 1, 2, \ldots$$

where W is a typical system time and N is the number of customers found by an arrival. For example, w_0 is the mean system time for those customers who find the system empty, w_1 is the mean system time for those customers who find one other customer present upon arrival, and so on. The estimate $\widehat{w}_i$ of w_i will be a sample mean of system times taken over all arrivals who find i in the system. Plot $\widehat{w}_i$ vs i. Hypothesize and attempt to verify a relation between w_i and i.

(a) Simulate for a 10-hour period with empty and idle initial conditions.

(b) Simulate for a 10-hour period after an initialization of one hour. Are there observable differences in the results of (a) and (b)?

(c) Repeat parts (a) and (b) with service times exponentially distributed with mean 5.1 seconds.

(d) Repeat parts (a) and (b) with deterministic service times equal to 5.1 seconds.

(e) Find the number of replications needed to estimate $w_0, w_1, \ldots, w_6$ with a standard error for each of at most 3 seconds. Repeat parts (a)–(d), but using this number of replications.

16. At Smalltown U., there is one specialized graphics workstation for student use located across campus from the computer center. At 2:00 A.M. one day, six students arrive at the workstation to complete an assignment. A student uses the workstation for 10 ± 8 minutes, then leaves to go to the computer center to pick up graphics output. There is a 25% chance that the run will be OK and the student will go to sleep. If it is not OK, the student returns to the workstation and waits until it becomes free. The roundtrip from workstation to computer center and back takes 30 ± 5 minutes. The computer becomes inaccessible at 5:00 A.M. Estimate the probability, p, that at least five of the six students will finish their assignment in the 3-hour period. First, make $R = 10$ replications, and compute a 95% confidence interval for p. Next, work out the number of replications needed to estimate p within $\pm.02$, and make this number of replications. Recompute the 95% confidence interval for p.

17. Four workers are spaced evenly along a conveyor belt. Items needing processing arrive according to a Poisson process at the rate 2 per minute. Processing time is exponentially distributed, with mean 1.6 minutes. If a worker becomes idle, then he or she takes the first item to come by on the conveyor. If a worker is busy when an item comes by, that item moves down the conveyor to the next worker, taking 20 seconds between two successive workers. When a worker finishes processing an item, the item leaves the system. If an item passes by the last worker, it is recirculated on a loop conveyor and will return to the first worker after 5 minutes.

Management is interested in having a balanced workload—that is, management would like worker utilizations to be equal. Let ρ_i be the long-run utilization of worker i, and let ρ be the average utilization of all workers. Thus, $\rho = (\rho_1 + \rho_2 + \rho_3 + \rho_4)/4$. According to queueing theory, ρ can be estimated by $\rho = \lambda/c\mu$, where $\lambda = 2$ arrivals per minute, $c = 4$ servers, and $1/\mu = 1.6$ minutes is the mean service time. Thus, $\rho = \lambda/c\mu = (2/4)1.6 = 0.8$; so, on the average, a worker will be busy 80% of the time.

(a) Make 5 independent replications, each of run length 40 hours preceded by a one hour initialization period. Compute 95% confidence intervals for ρ_1 and ρ_4. Draw conclusions concerning workload balance.

(b) Based on the same 5 replications, test the hypothesis $H_0 : \rho_1 = 0.8$ at a level of significance $\alpha = 0.05$. If a difference of $\pm.05$ is important to detect, determine the probability that such a deviation is detected. In addition, if it is desired to detect such a deviation with probability at least 0.9, figure out the sample size needed to do so. (See any basic statistics textbook for guidance on hypothesis testing.)

(c) Repeat (b) for $H_0 : \rho_4 = 0.8$.

(d) From the results of (a)–(c), draw conclusions for management about the balancing of workloads.

18. At a small rock quarry, a single power shovel dumps a scoop full of rocks at the loading area approximately every 10 minutes, with the actual time between scoops modeled well as being exponentially distributed, with mean 10 minutes. Three scoops of rocks make a pile; whenever one pile of rocks is completed, the shovel starts a new pile.

    The quarry has a single truck that can carry one pile (3 scoops) at a time. It takes approximately 27 minutes for a pile of rocks to be loaded into the truck and for the truck to be driven to the processing plant, unloaded, and return to the loading area. The actual time to do these things (altogether) is modeled well as being normally distributed, with mean 27 minutes and standard deviation 12 minutes.

    When the truck returns to the loading area, it will load and transport another pile if one is waiting to be loaded; otherwise, it stays idle until another pile is ready. For safety reasons, no loading of the truck occurs until a complete pile (all three scoops) is waiting.

    The quarry operates in this manner for an 8-hour day. We are interested in estimating the utilization of the trucks and the expected number of piles waiting to be transported if an additional truck is purchased.

19. Big Bruin, Inc. plans to open a small grocery store in Juneberry, NC. They expect to have two check-out lanes, with one lane being reserved for customers paying with cash. The question they want to answer is: how many grocery carts do they need?

    During business hours (6 A.M.–8 P.M.), cash-paying customers are expected to arrive at 8 per hour. All other customers are expected to arrive at 9 per hour. The time between arrivals of each type can be modeled as exponentially distributed random variables.

    The time spent shopping is modeled as normally distributed, with mean 40 minutes and standard deviation 10 minutes. The time required to check out after shopping can be modeled as lognormally distributed, with (a) mean 4 minutes and standard deviation 1 minute for cash-paying customers; (b) mean 6 minutes and standard deviation 1 minute for all other customers.

    We will assume that every customer uses a shopping cart and that a customer who finishes shopping leaves the cart in the store so that it is available immediately for another customer. We will also assume that any customer who cannot obtain a cart immediately leaves the store, disgusted.

    The primary performance measures of interest to Big Bruin are the expected number of shopping carts in use and the expected number of customers lost per day. Recommend a number of carts for the store, remembering that carts are expensive, but so are lost customers.

20. Develop a simulation model of the total time in the system for an $M/M/1$ queue with service rate $\mu = 1$; therefore, the traffic intensity is $\rho = \lambda/\mu = \lambda$, the arrival rate. Use the simulation, in conjunction with the technique of plotting ensemble averages, to study the effect of traffic intensity on initialization bias when the queue starts empty. Specifically, see how the initialization phase T_0 changes for $\rho = 0.5, 0.7, 0.8, 0.9, 0.95$.

21. Many simulation-software tools come with built-in support for output analysis. Use the web to research the features of one of these products.

12

Comparison and Evaluation of Alternative System Designs

Chapter 11 dealt with the precise estimation of a measure of performance for one system. This chapter discusses a few of the many statistical methods that can be used to compare two or more system designs on the basis of some performance measure. One of the most important uses of simulation is the comparison of alternative system designs. Because the observations of the response variables contain random variation, statistical analysis is needed to discover whether any observed differences are due to differences in design or merely to the random fluctuation inherent in the models.

The comparison of two system designs is computationally easier than the simultaneous comparison of multiple (more than two) system designs. Section 12.1 discusses the case of two system designs, using two possible statistical techniques: *independent sampling* and *correlated sampling*. Correlated sampling is also known as the *common random numbers* (CRN) technique; simply put, the same random numbers are used to simulate both alternative system designs. If implemented correctly, CRN usually reduces the variance of the estimated difference of the performance measures and thus can provide, for a given sample size, more precise estimates of the mean difference than can independent sampling. Section 12.2 extends the statistical techniques of Section 12.1 to the comparison of multiple (more than two) system designs, using the Bonferroni approach to confidence-interval estimation, screening, and selecting the best. The Bonferroni approach is limited to twenty or fewer system designs, but Section 12.3 describes how a large number of complex system designs can sometimes be represented by a simpler metamodel. Finally, for comparison and evaluation of a very large number of system designs that are related in a less structured way, Section 12.4 presents optimization via simulation.

12.1 COMPARISON OF TWO SYSTEM DESIGNS

Suppose that a simulation analyst desires to compare two possible configurations of a system. In a queueing system, perhaps two possible queue disciplines, or two possible sets of servers, are to be compared. In a supply-chain inventory system, perhaps two possible ordering policies will be compared. A job shop could have two possible scheduling rules; a production system could have in-process inventory buffers of various capacities. Many other examples of alternative system designs can be provided.

The method of replications will be used to analyze the output data. The mean performance measure for system i will be denoted by $\theta_i (i = 1, 2)$. If it is a steady-state simulation, it will be assumed that deletion of data, or other appropriate techniques, have been used to ensure that the point estimators are approximately unbiased estimators of the mean performance measures, θ_i. The goal of the simulation experiments is to obtain point and interval estimates of the difference in mean performance, namely $\theta_1 - \theta_2$. Three methods of computing a confidence interval for $\theta_1 - \theta_2$ will be discussed, but first an example and a general framework will be given.

Example 12.1

A vehicle-safety inspection station performs three jobs: (1) brake check, (2) headlight check, and (3) steering check. The present system has three stalls in parallel; that is, a vehicle enters a stall, where one attendant makes all three inspections. The current system is illustrated in Figure 12.1(a). Using data from the existing system, it has been assumed that arrivals occur completely at random (i.e., according to a Poisson process) at an average rate of 9.5 per hour and that the times for a brake check, a headlight check, and a steering check are normally distributed with means of 6.5, 6, and 5.5 minutes, respectively, all having standard deviations of approximately 0.5 minute. There is no limit on the queue of waiting vehicles.

An alternative system design is shown in Figure 12.1(b). Each attendant will specialize in a single task, and each vehicle will pass through three work stations in series. No space is allowed for vehicles between the brake and headlight check, or between the headlight and steering check. Therefore, a vehicle in the brake or headlight check must move to the next attendant, and a vehicle in the steering check must exit before the next vehicle can move ahead. The increased specialization of the inspectors suggests that mean inspection times for each type of check will decrease by 10%: to 5.85, 5.4, and 4.95 minutes, respectively, for the brake, headlight, and steering inspections. The Safety Inspection Council has decided to compare the two systems on the basis of mean response time per vehicle, where a response time is defined as the total time from a vehicle arrival until its departure from the system.

When comparing two systems, such as those in Example 12.1, the simulation analyst must decide on a run length $T_E^{(i)}$ for each model ($i = 1, 2$), and a number of replications R_i to be made of each model. From replication r of system i, the simulation analyst obtains an estimate Y_{ri} of the mean performance measure, θ_i. In Example 12.1, Y_{ri} would be the average response

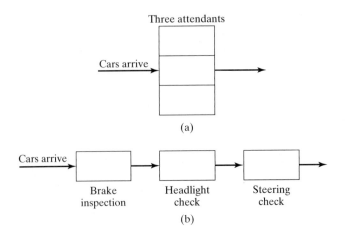

Figure 12.1   Vehicle safety inspection station and a possible alternative design.

Table 12.1   Simulation Output Data and Summary Measures for Comparing Two Systems

| System | Replication | | | | Sample Mean | Sample Variance |
	1	2	...	R_i		
1	Y_{11}	Y_{21}	...	$Y_{R_1 1}$	$\bar{Y}_{\cdot 1}$	S_1^2
2	Y_{12}	Y_{22}	...	$Y_{R_2 2}$	$\bar{Y}_{\cdot 2}$	S_2^2

time observed during replication r for system i ($r = 1, \ldots, R_i; i = 1, 2$). The data, together with the two summary measures, the sample means $\bar{Y}_{\cdot i}$, and the sample variances S_i^2, are exhibited in Table 12.1. Assuming that the estimators Y_{ri} are (at least approximately) unbiased, it follows that

$$\theta_1 = E(Y_{r1}), \quad r = 1, \ldots, R_1; \quad \theta_2 = E(Y_{r2}), r = 1, \ldots, R_2$$

In Example 12.1, the Safety Inspection Council is interested in a comparison of the two system designs, so the simulation analyst decides to compute a confidence interval for $\theta_1 - \theta_2$, the difference between the two mean performance measures. The confidence interval is used to answer two questions: (1) How large is the mean difference, and how precise is the estimator of mean difference? (2) Is there a significant difference between the two systems? This second question will lead to one of three possible conclusions:

1. If the confidence interval (c.i.) for $\theta_1 - \theta_2$ is totally to the left of zero, as shown in Figure 12.2(a), then there is strong evidence for the hypothesis that $\theta_1 - \theta_2 < 0$, or equivalently $\theta_1 < \theta_2$.

In Example 12.1, $\theta_1 < \theta_2$ implies that the mean response time for system 1 (the original system) is smaller than for system 2 (the alternative system).

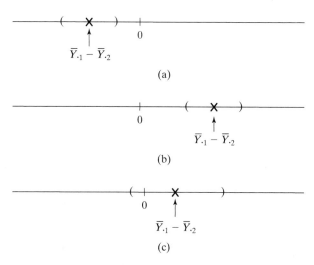

Figure 12.2   Three confidence intervals that can occur in the comparing of two systems.

2. If the c.i. for $\theta_1 - \theta_2$ is totally to the right of zero, as shown in Figure 12.2(b), then there is strong evidence that $\theta_1 - \theta_2 > 0$, or equivalently, $\theta_1 > \theta_2$.

In Example 12.1, $\theta_1 > \theta_2$ can be interpreted as system 2 being better than system 1, in the sense that system 2 has smaller mean response time.

3. If the c.i. for $\theta_1 - \theta_2$ contains zero, then, in the data at hand, there is no strong statistical evidence that one system design is better than the other.

Some statistics textbooks say that the weak conclusion $\theta_1 = \theta_2$ can be drawn, but such statements can be misleading. A "weak" conclusion is often no conclusion at all. Most likely, if enough additional data were collected (i.e., R_i increased), the c.i. would shift, and definitely shrink in length, until conclusion 1 or 2 would be drawn. In addition to one of these three conclusions, the confidence interval provides a measure of the precision of the estimator of $\theta_1 - \theta_2$.

In this chapter, a two-sided $100(1 - \alpha)\%$ c.i. for $\theta_1 - \theta_2$ will always be of the form

$$\bar{Y}_{\cdot 1} - \bar{Y}_{\cdot 2} \pm t_{\alpha/2, \nu} \text{s.e.}(\bar{Y}_{\cdot 1} - \bar{Y}_{\cdot 2}) \tag{12.1}$$

where $\bar{Y}_{\cdot i}$ is the sample mean performance measure for system i over all replications

$$\bar{Y}_{\cdot i} = \frac{1}{R_i} \sum_{r=1}^{R_i} Y_{ri} \tag{12.2}$$

and ν is the degrees of freedom associated with the variance estimator, $t_{\alpha/2, \nu}$ is the $100(1 - \alpha/2)$ percentage point of a t distribution with ν degrees of freedom, and s.e.$(\cdot)$ represents the standard error of the specified point estimator. To obtain the standard error and the degrees of freedom, the analyst uses one of three statistical techniques. All three techniques assume that the basic data, Y_{ri} of Table 12.1, are approximately normally distributed. This assumption is reasonable provided that each Y_{ri} is itself a sample mean of observations from replication r (which is indeed the situation in Example 12.1).

By design of the simulation experiment, $Y_{r1}(r = 1, \ldots, R_1)$ are independently and identically distributed (i.i.d.) with mean θ_1 and variance σ_1^2 (say). Similarly, $Y_{r2}(r = 1, \ldots, R_2)$ are i.i.d. with mean θ_2 and variance σ_2^2 (say). The three techniques for computing the confidence interval in (12.1), which are based on three different sets of assumptions, are discussed in the following subsections.

There is an important distinction between *statistically significant* differences and *practically significant* differences in systems performance. Statistical significance answers the following question: Is the observed difference $\bar{Y}_{.1} - \bar{Y}_{.2}$ larger than the variability in $\bar{Y}_{.1} - \bar{Y}_{.2}$? This question can be restated as: Have we collected enough data to be confident that the difference we observed is real, or just chance? Conclusions 1 and 2 imply a statistically significant difference, while Conclusion 3 implies that the observed difference is not statistically significant (even though the systems may indeed be different). Statistical significance is a function of the simulation experiment and the output data.

Practical significance answers the following question: Is the true difference $\theta_1 - \theta_2$ large enough to matter for the decision we need to make? In Example 12.1, we may reach the conclusion that $\theta_1 > \theta_2$ and decide that system 2 is better (smaller expected response time). However, if the actual difference $\theta_1 - \theta_2$ is very small—say, small enough that a customer would not notice the improvement—then it might not be worth the cost to replace system 1 with system 2. Practical significance is a function of the actual difference between the systems and is independent of the simulation experiment.

Confidence intervals do not answer the question of practical significance directly. Instead, they bound (with probability $1 - \alpha$) the true difference $\theta_1 - \theta_2$ within the range

$$\bar{Y}_{.1} - \bar{Y}_{.2} - t_{\alpha/2, \nu}\text{s.e.}(\bar{Y}_{.1} - \bar{Y}_{.2}) \le \theta_1 - \theta_2 \le \bar{Y}_{.1} - \bar{Y}_{.2} + t_{\alpha/2, \nu}\text{s.e.}(\bar{Y}_{.1} - \bar{Y}_{.2})$$

Whether a difference within these bounds is practically significant depends on the particular problem.

12.1.1 Independent Sampling with Equal Variances

Independent sampling means that different and independent random number streams will be used to simulate the two systems. This implies that all the observations of simulated system 1, namely $\{Y_{r1}, r = 1, \ldots, R_1\}$, are statistically independent of all the observations of simulated system 2, namely $\{Y_{r2}, r = 1, \ldots, R_2\}$. By Equation (12.2) and the independence of the replications, the variance of the sample mean, $\bar{Y}_{.i}$, is given by

$$V(\bar{Y}_{.i}) = \frac{V(Y_{ri})}{R_i} = \frac{\sigma_i^2}{R_i}, \qquad i = 1, 2$$

For independent sampling, $\bar{Y}_{.1}$ and $\bar{Y}_{.2}$ are statistically independent; hence,

$$V(\bar{Y}_{.1} - \bar{Y}_{.2}) = V(\bar{Y}_{.1}) + V(\bar{Y}_{.2})$$

$$= \frac{\sigma_1^2}{R_1} + \frac{\sigma_2^2}{R_2} \tag{12.3}$$

In some cases, it is reasonable to assume that the two variances are equal (but unknown in value); that is, $\sigma_1^2 = \sigma_2^2$. The data can be used to test the hypothesis of equal variances; if rejected, the method of Section 12.1.2 must be used. In a steady-state simulation, the variance σ_i^2 decreases as the run length $T_E^{(i)}$ increases; therefore, it might be possible to adjust the two run lengths, $T_E^{(1)}$ and $T_E^{(2)}$, to achieve at least approximate equality of σ_1^2 and σ_2^2.

If it is reasonable to assume that $\sigma_1^2 = \sigma_2^2$ (approximately), a two-sample-t confidence-interval approach can be used. The point estimate of the mean performance difference is

$$\bar{Y}_{.1} - \bar{Y}_{.2} \tag{12.4}$$

with $\bar{Y}_{.i}$ given by Equation (12.2). Next, compute the sample variance for system i by

$$S_i^2 = \frac{1}{R_i - 1} \sum_{r=1}^{R_i} (Y_{ri} - \bar{Y}_{.i})^2$$

$$= \frac{1}{R_i - 1} \left(\sum_{r=1}^{R_i} Y_{ri}^2 - R_i \bar{Y}_{.i}^2 \right) \tag{12.5}$$

Note that S_i^2 is an unbiased estimator of the variance σ_i^2. By assumption, $\sigma_1^2 = \sigma_2^2 = \sigma^2$ (say), so a pooled estimate of σ^2 is obtained by

$$S_p^2 = \frac{(R_1 - 1)S_1^2 + (R_2 - 1)S_2^2}{R_1 + R_2 - 2}$$

which has $\nu = R_1 + R_2 - 2$ degrees of freedom. The c.i. for $\theta_1 - \theta_2$ is then given by Expression (12.1) with the standard error computed by

$$\text{s.e.}(\bar{Y}_{.1} - \bar{Y}_{.2}) = S_p \sqrt{\frac{1}{R_1} + \frac{1}{R_2}} \tag{12.6}$$

This standard error is an estimate of the standard deviation of the point estimate, which, by Equation (12.3), is given by $\sigma \sqrt{1/R_1 + 1/R_2}$.

In some cases, the simulation analyst could have $R_1 = R_2$, in which case it is safe to use the c.i. in Expression (12.1) with the standard error taken from Equation (12.6), even if the variances (σ_1^2 and σ_2^2) are not equal. However, if the variances are unequal and the sample sizes differ, it has been shown that use of the two-sample-t c.i. could yield invalid confidence intervals whose true probability of containing $\theta_1 - \theta_2$ is much less than $1 - \alpha$. Thus, if there is no evidence that $\sigma_1^2 = \sigma_2^2$, and if $R_1 \neq R_2$, the approximate procedure in the next subsection is recommended.

12.1.2 Independent Sampling with Unequal Variances

If the assumption of equal variances cannot safely be made, an approximate $100(1 - \alpha)\%$ c.i. for $\theta_1 - \theta_2$ can be computed as follows. The point estimate and sample variances are computed by Equations (12.4) and (12.5). The standard error of the point estimate is given by

$$\text{s.e.}(\bar{Y}_{.1} - \bar{Y}_{.2}) = \sqrt{\frac{S_1^2}{R_1} + \frac{S_2^2}{R_2}} \tag{12.7}$$

with degrees of freedom, ν, approximated by the expression

$$\nu = \frac{(S_1^2/R_1 + S_2^2/R_2)^2}{[(S_1^2/R_1)^2/(R_1 - 1)] + [(S_2^2/R_2)^2/(R_2 - 1)]} \tag{12.8}$$

rounded to an integer. The confidence interval is then given by Expression (12.1), using the standard error of Equation (12.7). A minimum number of replications $R_1 \geq 6$ and $R_2 \geq 6$ is recommended for this procedure.

12.1.3 Common Random Numbers (CRN)

CRN means that, for each replication, the same random numbers are used to simulate both systems. Therefore, R_1 and R_2 must be equal, say $R_1 = R_2 = R$. Thus, for each replication r, the two estimates, Y_{r1} and Y_{r2}, are no longer independent, but rather are correlated. However, independent streams of random numbers are used on different replications, so the pairs (Y_{r1}, Y_{s2}) are mutually independent when $r \neq s$. (For example, in Table 12.1, the observation Y_{11} is correlated with Y_{12}, but Y_{11} is independent of all other observations.) The purpose of using CRN is to induce a positive correlation between Y_{r1} and Y_{r2} (for each r) and thus to achieve a variance reduction in the point estimator of mean difference, $\bar{Y}_{.1} - \bar{Y}_{.2}$. In general, this variance is given by

$$V(\bar{Y}_{.1} - \bar{Y}_{.2}) = V(\bar{Y}_{.1}) + V(\bar{Y}_{.2}) - 2\text{cov}(\bar{Y}_{.1}, \bar{Y}_{.2})$$

$$= \frac{\sigma_1^2}{R} + \frac{\sigma_2^2}{R} - \frac{2\rho_{12}\sigma_1\sigma_2}{R} \tag{12.9}$$

where ρ_{12} is the correlation between Y_{r1} and Y_{r2}. [By definition, $\rho_{12} = \text{cov}(Y_{r1}, Y_{r2})/\sigma_1\sigma_2$, which does not depend on r.]

Now compare the variance of $\bar{Y}_{.1} - \bar{Y}_{.2}$ arising from the use of CRN [Equation (12.9), call it V_{CRN}] to the variance arising from the use of independent sampling with equal sample sizes [Equation (12.3) with $R_1 = R_2 = R$, call it V_{IND}]. Notice that

$$V_{CRN} = V_{IND} - \frac{2\rho_{12}\sigma_1\sigma_2}{R} \tag{12.10}$$

If CRN works as intended, the correlation ρ_{12} will be positive; hence, the second term on the right side of Equation (12.10) will be positive, and, therefore,

$$V_{CRN} < V_{IND}$$

That is, the variance of the point estimator will be smaller with CRN than with independent sampling. A smaller variance (for the same sample size R) implies that the estimator based on CRN is more precise, leading to a shorter confidence interval on the difference, which implies that smaller differences in performance can be detected.

To compute a $100(1 - \alpha)\%$ c.i. with correlated data, first compute the differences

$$D_r = Y_{r1} - Y_{r2} \tag{12.11}$$

which, by the definition of CRN, are i.i.d.; then compute the sample mean difference as

$$\bar{D} = \frac{1}{R} \sum_{r=1}^{R} D_r \tag{12.12}$$

(Thus, $\bar{D} = \bar{Y}_{.1} - \bar{Y}_{.2}$.) The sample variance of the differences $\{D_r\}$ is computed as

$$S_D^2 = \frac{1}{R - 1} \sum_{r=1}^{R} (D_r - \bar{D})^2$$

$$= \frac{1}{R - 1} \left(\sum_{r=1}^{R} D_r^2 - R\bar{D}^2 \right) \tag{12.13}$$

which has degrees of freedom $\nu = R - 1$. The $100(1 - \alpha)\%$ c.i. for $\theta_1 - \theta_2$ is given by Expression (12.1), with the standard error of $\bar{Y}_{.1} - \bar{Y}_{.2} = \bar{D}$ estimated by

$$\text{s.e.}(\bar{D}) = \text{s.e.}(\bar{Y}_{.1} - \bar{Y}_{.2}) = \frac{S_D}{\sqrt{R}} \tag{12.14}$$

Because $S_D/\sqrt{R}$ of Equation (12.14) is an estimate of $\sqrt{V_{CRN}}$ and Expression (12.6) or (12.7) is an estimate of $\sqrt{V_{IND}}$, CRN typically will produce a c.i. that is shorter for a given sample size than the c.i. produced by independent sampling if $\rho_{12} > 0$. In fact, the expected length of the c.i. will be shorter with use of CRN if $\rho_{12} > 0.1$, provided $R > 10$. The larger R is, the smaller ρ_{12} can be and still yield a shorter expected length [Nelson 1987].

For any problem, there are many ways of implementing common random numbers. It is never enough to simply use the same seed on the random-number generator(s). Each random number used in one model for some purpose should be used for the same purpose in the second model—that is, the use of the random numbers must be synchronized. For example, if the ith random number is used to generate a service time at work station 2 for the 5th arrival in model 1, the ith random number should be used for the very same purpose in model 2. For queueing systems or service facilities, synchronization of the common random numbers guarantees that the two systems face identical work loads: both systems face arrivals at the same instants of time, and these arrivals demand equal amounts of service. (The actual service times of a given arrival in the two models may not be equal; they could be proportional if the server in one model were faster than the server in the other model.) For an inventory system, in comparing of different ordering policies, synchronization guarantees that the two systems face identical demand for a given product. For production or reliability systems, synchronization guarantees that downtimes for a given machine will occur at exactly the same times and will have identical durations, in the two models. On the other hand, if some aspect of one of the systems is totally different from the other system, synchronization could be inappropriate—or even impossible to achieve. In summary, those aspects of the two system designs that are sufficiently similar should be simulated with common random numbers in such a way that the two models "behave" similarly; but those aspects that are totally different should be simulated with independent random numbers.

Implementation of common random numbers is model dependent, but certain guidelines can be given that will make CRN more likely to yield a positive correlation. The purpose of the guidelines is to ensure that synchronization occurs:

1. Dedicate a random-number stream to a specific purpose, and use as many different streams as needed. (Different random-number generators, or widely spaced seeds on the same generator, can be used to get two different, nonoverlapping streams.) In addition, assign independently chosen seeds to each stream at the beginning of each replication. It is not sufficient to assign seeds at the beginning of the first replication and then let the random-number generator merely continue for the second and subsequent replications. If simulation is conducted in this manner, the first replication will be synchronized, but subsequent replications might not be.

2. For systems (or subsystems) with external arrivals: As each entity enters the system, the next interarrival time is generated, and then immediately all random variables (such as service times, order sizes, etc.) needed by the arriving entity and identical in both models are generated in a fixed order and stored as attributes of the entity, to be used later as needed. Apply guideline 1: Dedicate one random-number stream to these external arrivals and all their attributes.

3. For systems having an entity performing given activities in a cyclic or repeating fashion, assign a random-number stream to this entity. (Example: a machine that cycles between two states: up–down–up–down–... . Use a dedicated random-number stream to generate the uptimes and downtimes.)

4. If synchronization is not possible, or if it is inappropriate for some part of the two models, use independent streams of random numbers for this subset of random variates.

Unfortunately, there is no guarantee that CRN will always induce a positive correlation between comparable runs of the two models. It is known that if, for each input random variate X, the estimators Y_{r1} and Y_{r2} are increasing functions of the random variate X (or both are decreasing functions of X), then ρ_{12} will be positive. The intuitive idea is that both models (i.e., both Y_{r1} and Y_{r2}) respond in the same direction to each input random variate, and this results in positive correlation. This increasing or decreasing nature of the response variables (called *monotonicity*) with respect to the input random variables is known to hold for certain queueing systems (such as the $GI/G/c$ queues), when the response variable is customer delay, so some evidence exists that common random numbers is a worthwhile technique for queueing simulations. (For simple queues, customer delay is an increasing function of service times and a decreasing function of interarrival times.) Wright and Ramsay [1979] reported a negative correlation for certain inventory simulations, however. In summary, the guidelines recently described should be followed, and some reasonable notion that the response variable of interest is a monotonic function of the random input variables should be evident.

Example 12.1: Continued

The two inspection systems shown in Figure 12.1 will be compared by using both independent sampling and CRN, in order to illustrate the greater precision of CRN when it works.

Each vehicle arriving to be inspected has four input random variables associated with it:

A_n = interarrival time between vehicles n and $n + 1$
$S_n^{(1)}$ = brake inspection time for vehicle n in model 1
$S_n^{(2)}$ = headlight inspection time for vehicle n in model 1
$S_n^{(3)}$ = steering inspection time for vehicle n in model 1

For model 2 (of the proposed system), mean service times are decreased by 10%. When using independent sampling, different values of service (and interarrival) times would be generated for models 1 and 2 by using different random numbers. But when using CRN, the random number generator should be used in such a way that exactly the same values are generated for $A_1, A_2, A_3, \ldots$ in both models. For service times, however, we do not want the same service times in both models, because the mean service time for model 2 is 10% smaller, but we do want strongly correlated service times. There are at least two ways to do this:

1. Let $S_n^{(i)}(i = 1, 2, 3; n = 1, 2, \ldots)$ be the service times generated for model 1; then use $S_n^{(i)} - 0.1E(S_n^{(i)})$ as the service times in model 2. In words, we take each service time from model 1 and subtract 10% of its true mean.
2. Recall that normal random variates are usually produced by first generating a standard normal variate and then using Equation (8.29) to obtain the correct mean and variance. Therefore, the service times for, say, a brake inspection could be generated by

$$E(S_n^{(1)}) + \sigma Z_n^{(1)} \tag{12.15}$$

where $Z_n^{(1)}$ is a standard normal variate, $\sigma = 0.5$ minute, but $E(S_n^{(1)}) = 6.5$ minutes for model 1 and $E(S_n^{(1)}) = 5.85$ minutes (10% less) for model 2. The other two inspection times would be generated in a similar fashion. To implement (synchronized) common random numbers, the simulation analyst would generate identical $Z_n^{(i)}$ sequences ($i = 1, 2, 3; n = 1, 2, \ldots$) in both models and then use the appropriate version of Equation (12.15) to generate the inspection times.

For the synchronized runs, the service times for a vehicle were generated at the instant of arrival (by guideline 2) and stored as an attribute of the vehicle, to be used as needed. Runs were also made with nonsynchronized common random numbers, in which case one random number stream was used as needed.

Table 12.2 gives the average response time for each of $R = 10$ replications, each of run length $T_E = 16$ hours. It was assumed that two cars were present at time 0, waiting to be inspected. Column 1 gives the outputs from model 1. Model 2 was run with independent random numbers (column 2I) and with common random numbers without synchronization (column 2C*) and with

Table 12.2   Comparison of System Designs for the Vehicle-Safety Inspection System

Replication	Average Response Time for Model *1*	*2I*	*2C**	*2C*	Observed Differences $D_{1,2C^*}$	$D_{1,2C}$
1	29.59	51.62	56.47	29.55	−26.88	0.04
2	23.49	51.91	33.34	24.26	−9.85	−0.77
3	25.68	45.27	35.82	26.03	−10.14	−0.35
4	41.09	30.85	34.29	42.64	6.80	−1.55
5	33.84	56.15	39.07	32.45	−5.23	1.39
6	39.57	28.82	32.07	37.91	7.50	1.66
7	37.04	41.30	51.64	36.48	−14.60	0.56
8	40.20	73.06	41.41	41.24	−1.21	−1.04
9	61.82	23.00	48.29	60.59	13.53	1.23
10	44.00	28.44	22.44	41.49	21.56	2.51
Sample mean	37.63	43.04			−1.85	0.37
Sample variance	118.90	244.33			208.94	1.74
Standard error	6.03				4.57	0.42

synchronization (column 2C). The purpose of the simulation is to estimate mean difference in response times for the two systems.

For the two independent runs (1 and 2I), it was assumed that the variances were not necessarily equal, so the method of Section 12.1.2 was applied. Sample variances and the standard error were computed by Equations (12.5) and (12.7), yielding

$$S_1^2 = 118.9, \qquad S_{2I}^2 = 244.3$$

and

$$\text{s.e.}(\bar{Y}_{\cdot 1} - \bar{Y}_{\cdot 2I}) = \sqrt{\frac{118.9}{10} + \frac{244.3}{10}} = 6.03$$

with degrees of freedom, ν, equal to 17, as given by Equation (12.8). The point estimate is $\bar{Y}_{\cdot 1} - \bar{Y}_{\cdot 2I} = -5.4$ minutes, and a 95% c.i. [Expression (12.1)] is given by

$$-5.4 \pm 2.11(6.03)$$

or

$$-18.1 \le \theta_1 - \theta_2 \le 7.3 \qquad\qquad (12.16)$$

The 95% confidence interval in Inequality (12.16) contains zero, which indicates that there is no strong evidence that the observed difference, -5.4 minutes, is due to anything other than random variation in the output data. In other words, it is not statistically significant. Thus, if the simulation analyst had decided to use independent sampling, no strong conclusion would be possible, because the estimate of $\theta_1 - \theta_2$ is quite imprecise.

For the two sets of correlated runs (1 and 2C*, and 1 and 2C), the observations are paired and analyzed as given in Equations (12.11) through (12.14). The point estimate when not synchronizing the random numbers is given by Equation (12.12) as

$$\bar{D} = -1.9 \text{ minutes}$$

the sample variance by $S_D^2 = 208.9$ (with $\nu = 9$ degrees of freedom), and the standard error by s.e.$(\bar{D}) = 4.6$. Thus, a 95% c.i. for the true mean difference in response times, as given by expression (12.1), is

$$-1.9 \pm 2.26(4.6)$$

or

$$-12.3 < \theta_1 - \theta_2 < 8.5 \qquad (12.17)$$

Again, no strong conclusion is possible, because the confidence interval contains zero. Notice, however, that the estimate of $\theta_1 - \theta_2$ is slightly more precise than that in Inequality (12.16), because the length of the interval is smaller.

When complete synchronization of the random numbers was used, in run 2C, the point estimate of the mean difference in response times was

$$\bar{D} = 0.4 \text{ minute}$$

the sample variance was $S_D^2 = 1.7$ (with $\nu = 9$ degrees of freedom), and the standard error was s.e.$(\bar{D}) = 0.4$. A 95% c.i. for the true mean difference is given by

$$-0.50 < \theta_1 - \theta_2 < 1.30 \qquad (12.18)$$

The confidence interval in Inequality (12.18) again contains zero, but it is considerably shorter than the previous two intervals. This greater precision in the estimation of $\theta_1 - \theta_2$ is due to the use of synchronized common random numbers. The short length of the interval in Inequality (12.18) suggests that the true difference, $\theta_1 - \theta_2$, is close to zero. In fact, the upper bound, 1.30, indicates that system 2 is at most 1.30 minutes faster, in expectation. If such a small difference is not practically significant, then there is no need to look further into which system is truly better.

As is seen by comparing the confidence intervals in inequalities (12.16), (12.17), and (12.18), the width of the confidence interval is reduced by 18% when using nonsynchronized common random numbers, by 93% when using common random numbers with full synchronization. Comparing the estimated variance of $\bar{D}$ when using synchronized common random numbers with the variance of $\bar{Y}_{\cdot 1} - \bar{Y}_{\cdot 2}$ when using independent sampling shows a variance reduction of 99.5%, which means that, to achieve precision comparable to that achieved by CRN, a total of approximately $R = 209$ independent replications would have to be made.

The next few examples show how common random numbers can be implemented in other contexts.

Example 12.2:   The Dump-Truck Problem, Revisited

Consider Example 3.4 (the dump-truck problem), shown in Figure 3.7. Each of the trucks repeatedly goes through three activities: loading, weighing, and traveling. Assume that there are eight trucks and that, at time 0, all eight are at the loaders. Weighing time per truck on the single scale is uniformly distributed between 1 and 9 minutes, and travel time per truck is exponentially distributed, with mean 85 minutes. An unlimited queue is allowed before the loader(s) and before the scale. All trucks can be traveling at the same time. Management desires to compare one fast loader against the two slower loaders currently being used. Each of the slow loaders can fill a truck in from 1 to 27 minutes, uniformly distributed. The new fast loader can fill a truck in from 1 to 19 minutes, uniformly distributed. The basis for comparison is mean system response time, where a response time is defined as the duration of time from a truck arrival at the loader queue to that truck's departure from the scale.

To implement synchronized common random numbers, a separate and distinct random number stream was assigned to each of the eight trucks. At the beginning of each replication (i.e., at time 0), a new and independently chosen set of eight seeds was specified, one seed for each random number stream. Thus, weighing times and travel times for each truck were identical in both models, and the loading time for a given truck's ith visit to the fast loader was proportional to the loading time in the original system (with two slow loaders). Implementation of common random numbers without synchronization (e.g., using one random number stream to generate all loading, weighing, and travel times as needed) would likely lead to a given random number being used to generate a loading time in model 1 but a travel time in model 2, or vice versa, and from that point on the use of a random number would most likely be different in the two models.

Six replications of each model were run, each of run length $T_E = 40$ hours. The results are shown in Table 12.3. Both independent sampling and CRN were used, to illustrate the advantage of CRN. The first column (labeled model 1) contains the observed average system response time for the existing system with two loaders. The columns labeled 2I and 2C are for the alternative design having one loader; the independent sampling results are in 2I, and the CRN results are in the column labeled 2C. The rightmost column, labeled $D_{1,2C}$, lists the observed differences between the runs of model 1 and model 2C.

For independent sampling assuming unequal variances, the following summary statistics were computed by using Equations (12.2), (12.5), (12.7), (12.8), and (12.1) and the data (in columns 1 and 2I) in Table 12.3:

Point Estimate: $\bar{Y}_{\cdot 1} - \bar{Y}_{\cdot 2I} = 22.44 - 26.52 = -4.08$ minutes

Sample variances: $S_1^2 = 1.28$, $S_{2I}^2 = 2.86$

Standard Error: s.e.$(\bar{Y}_{\cdot 1} - \bar{Y}_{\cdot 2}) = (S_1^2/R_1 + S_{2I}^2/R_2)^{1/2} = 0.831$

Degrees of freedom: $\nu = 8.73 \approx 9$

95% c.i. for $\theta_1 - \theta_2$: $-4.08 \pm 2.26(0.831)$ or -4.08 ± 1.878

$$-5.96 \leq \theta_1 - \theta_2 \leq -2.20$$

Table 12.3 Comparison of System Designs for the Dump Truck Problem

Replication	Average System Response Time for Model			Differences, $D_{1,2C}$
	1 (2 Loaders)	2I (1 Loader)	2C (1 Loader)	
1	21.38	29.01	24.30	−2.92
2	24.06	24.70	27.13	−3.07
3	21.39	26.85	23.04	−1.65
4	21.90	24.49	23.15	−1.25
5	23.55	27.18	26.75	−3.20
6	22.36	26.91	25.62	−3.26
Sample mean	22.44	26.52		−2.56
Sample variance	1.28	2.86		0.767
Sample standard deviation	1.13	1.69		0.876

For CRN, implemented by the use of synchronized common random numbers, the following summary statistics were computed by using Equations (12.12), (12.13), (12.14), and (12.1) plus the data (in columns 1 and 2C) in Table 12.3:

Point Estimate: $\bar{D} = \bar{Y}_{.1} - \bar{Y}_{.2C} = -2.56$ minutes

Sample variance: $S_D^2 = 0.767$

Standard Error: s.e.$(\bar{D}) = S_D/\sqrt{R} = 0.876/\sqrt{6} = 0.358$

Degrees of freedom: $\nu = R - 1 = 5$

95% c.i. for $\theta_1 - \theta_2$: $-2.56 \pm 2.57(0.358)$ or -2.56 ± 0.919

$$-3.48 \le \theta_1 - \theta_2 \le -1.641$$

By comparing the c.i. widths, we see that the use of CRN with synchronization reduced c.i. width by 50%. This reduction could be important if a difference of as much as, say, 5.96 is considered practically significant, but a difference of at most 3.48 is not. Equivalently, if equal precision were desired, independent sampling would require approximately four times as many observations as would CRN: approximately 24 replications of each model instead of six.

To illustrate how CRN can fail when not implemented correctly, consider the dump-truck model again. There were eight trucks, and each was assigned its own random number stream. For each of the six replications, eight seeds were randomly chosen, one seed for each random number stream. Therefore, a total of 48 (6 times 8) seeds were specified for the correct implementation of common random numbers. When the authors first developed and ran this example, eight seeds were specified at the beginning of the first replication only; on the remaining five replications the random numbers were generated by continuing down the eight original streams. Since comparable replications with one and two loaders required different numbers of random variables, only the first replications of the two models were synchronized. The remaining five were not synchronized. The resulting confidence interval for $\theta_1 - \theta_2$ under CRN was approximately the same length as, or only slightly shorter than, the confidence interval under independent sampling. Therefore, CRN is quite likely to fail in reducing the standard error of the estimated difference unless proper care is taken to guarantee synchronization of the random number streams on all replications.

Example 12.3 ———————————————————————————————————

In Example 2.5, two policies for replacing bearings in a milling machine were compared. The bearing-life distribution, assumed discrete in Example 2.5 (Table 2.22), is now more realistically assumed to be continuous on the range from 950 to 1950 hours, with the first column of Table 2.22 giving the midpoint of 10 intervals of width 100 hours. The repairperson delay-time distribution of Table 2.23 is also assumed continuous, in the range from 2.5 to 17.5 minutes, with interval midpoints as given in the first column. The probabilities of each interval are given in the second columns of Tables 2.22 and 2.23.

The two models were run by using CRN and, for illustrative purposes, by using independent sampling, each for $R = 10$ replications. The purpose was to estimate the difference in mean total costs per 10,000 bearing hours, with the cost data given in Example 2.5. The estimated total cost for the two policies is given in Table 12.4.

Policy 1 was to replace each bearing as it failed. Policy 2 was to replace all three bearings whenever one bearing failed. Policy 2 was run first, and then policy 1 was run, using independent sampling (column 1I), and using CRN (column 1C). The 95% confidence intervals for mean cost difference are as follows:

Independent sampling: 4426 ± 439

CRN: 4625 ± 211

(The computation of these confidence intervals is left as an exercise for the reader.)

Notice that the confidence interval for mean cost difference when using CRN is approximately 50% of the length of the confidence interval based on independent sampling. Therefore, for the same computer costs, (i.e., for $R = 10$ replications), CRN produces estimates that are twice as precise in this example. If CRN were used, the simulation analyst could conclude with 95% confidence that the mean cost difference between the two policies is between $4414 and $4836.

12.1.4 Confidence Intervals with Specified Precision

Section 11.4.2 described a procedure for obtaining confidence intervals with specified precision. Confidence intervals for the *difference* between two systems' performance can be obtained in an analogous manner.

Suppose that we want the error in our estimate of $\theta_1 - \theta_2$ to be less than $\pm\epsilon$ (the quantity ϵ might be a practically significant difference). Therefore, our goal is to find a number of replications R such that

$$H = t_{\alpha/2,\nu}\text{s.e.}(\bar{Y}_{\cdot 1} - \bar{Y}_{\cdot 2}) \leq \epsilon \tag{12.19}$$

Table 12.4 Total Costs for Alternative Designs of Bearing Replacement Problem

Replication r	Total Cost for Policy			Difference in Total Cost
	2	II	IC	$D_{IC,2}$
1	13,340	17,010	17,556	4,216
2	12,760	17,528	17,160	4,400
3	13,002	17,956	17,808	4,806
4	13,524	17,920	18,012	4,488
5	13,754	18,880	18,200	4,446
6	13,318	17,528	17,936	4,618
7	13,432	17,574	18,350	4,918
8	14,208	17,954	19,398	5,190
9	13,224	18,290	17,612	4,388
10	13,178	17,360	17,956	4,778
Sample mean	13,374	17,800		4,624
Sample variance	160,712	276,188		87,353

As in Section 11.4.2, we begin by making $R_0 \geq 2$ replications of each system to obtain an initial estimate of s.e.$(\bar{Y}_{.1} - \bar{Y}_{.2})$. We then solve for the total number of replications $R \geq R_0$ needed to achieve the half-length criterion (12.19). Finally, we make an additional $R - R_0$ replications (or a fresh R replications) of each system, compute the confidence interval, and check that the half-length criterion has been attained.

Example 12.1: Continued _____

Recall that $R_0 = 10$ replications and complete synchronization of the random numbers yielded the 95% confidence interval for the difference in expected response time of the two vehicle-inspection stations in Inequality (12.18); this interval can be rewritten as 0.4 ± 0.90 minutes. Although system 2 appears to have the smaller expected response time, the difference is not statistically significant, since the confidence interval contains 0. Suppose that a difference larger than ± 0.5 minute is considered to be practically significant. We therefore want to make enough replications to obtain a $H \leq \epsilon = 0.5$.

The confidence interval used in Example 12.1 was $\bar{D} \pm t_{\alpha/2,R_0-1}S_D/\sqrt{R_0}$, with the specific values $\bar{D} = 0.4$, $R_0 = 10$, $t_{0.025,9} = 2.26$, and $S_D^2 = 1.7$. To obtain the desired precision, we need to find R such that

$$\frac{t_{\alpha/2,R-1}S_D}{\sqrt{R}} \leq \epsilon$$

Therefore, R is the smallest integer satisfying $R \geq R_0$ and

$$R \geq \left(\frac{t_{\alpha/2,R-1}S_D}{\epsilon} \right)^2$$

Since $t_{\alpha/2,R-1} \leq t_{\alpha/2,R_0-1}$, a conservative estimate for R is given by

$$R \geq \left(\frac{t_{\alpha/2,R_0-1}S_D}{\epsilon} \right)^2$$

Substituting $t_{0.025,9} = 2.26$ and $S_D^2 = 1.7$, we obtain

$$R \geq \frac{(2.26)^2(1.7)}{(0.5)^2} = 34.73$$

implying that 35 replications are needed, 25 more than in the initial experiment.

12.2 COMPARISON OF SEVERAL SYSTEM DESIGNS

Suppose that a simulation analyst desires to compare K alternative system designs. The comparison will be made on the basis of some specified performance measure, θ_i, of system i, for $i = 1, 2, \ldots, K$. Many different statistical procedures have been developed that can be used to analyze simulation data and draw statistically sound inferences concerning the parameters θ_i. These procedures can be classified as being either fixed-sample-size procedures or sequential-sampling (or *multistage*) procedures. In the first type, a predetermined sample size (i.e., run length and number of replications) is used to draw inferences via hypothesis tests or confidence intervals. Examples of fixed-sample-size procedures include the interval estimation of a mean performance measure (Section 11.3) and the interval estimation of the difference between mean performance measures of two systems [as by Expression (12.1) in Section 12.1]. Advantages of fixed-sample-size procedures include a known or easily estimated cost in terms of computer time before running the experiments. When computer time is limited, or when a pilot study is being conducted, a fixed-sample-size procedure might be appropriate. In some cases, clearly inferior system designs may be ruled out at this early stage. A major disadvantage is that a strong conclusion could be impossible. For example, the confidence interval could be too wide for practical use, since the width is an indication of the precision of the point estimator. A hypothesis test may lead to a failure to reject the null hypothesis, a weak conclusion in general, meaning that there is no strong evidence one way or the other about the truth or falsity of the null hypothesis.

A sequential sampling scheme is one in which more and more data are collected until an estimator with a prespecified precision is achieved or until one of several alternative hypotheses is selected, with the probability of correct selection being larger than a prespecified value. A two-stage (or multistage) procedure is one in which an initial sample is used to estimate how many additional observations are needed to draw conclusions with a specified precision. An example of a two-stage procedure for estimating the performance measure of a single system was given in Sections 11.4.2 and 12.1.4.

The proper procedure to use depends on the goal of the simulation analyst. Some possible goals are the following:

1. estimation of each parameter, θ_i;
2. comparison of each performance measure, θ_i, to a control, θ_1 (where θ_1 could represent the mean performance of an existing system);
3. all pairwise comparisons, $\theta_i - \theta_j$, for $i \neq j$;
4. selection of the best θ_i (largest or smallest).

The first three goals will be achieved by the construction of confidence intervals. The number of such confidence intervals is $C = K$, $C = K - 1$, and $C = K(K - 1)/2$, respectively. Hochberg and Tamhane [1987] and Hsu [1996] are comprehensive references for such multiple-comparison procedures. The fourth goal requires the use of a type of statistical procedure known as a multiple ranking and selection procedure. Procedures to achieve these and other goals are discussed by Kleijnen [1975, Chapters II and V], who also discusses their relative merit and disadvantages. Goldsman and Nelson [1998] and Law and Kelton [2000] discuss those selection procedures most relevant to simulation. A comprehensive reference is Bechhofer, Santner, and Goldsman [1995]. The next subsection presents a fixed-sample-size procedure that can be used to meet goals 1, 2, and 3 and is applicable in a wide range of circumstances. Subsections 12.2.2–12.2.3 present related procedures to achieve goal 4.

12.2.1 Bonferroni Approach to Multiple Comparisons

Suppose that C confidence intervals are computed and that the ith interval has confidence coefficient $1 - \alpha_i$. Let S_i be the statement that the ith confidence interval contains the parameter (or difference of two parameters) being estimated. This statement might be true or false for a given set of data, but the procedure leading to the interval is designed so that statement S_i will be true with probability $1 - \alpha_i$. When it is desired to make statements about several parameters simultaneously, as in goals 1, 2 and 3, the analyst would like to have high confidence that *all* statements are true simultaneously. The Bonferroni inequality states that

$$P(\text{all statements } S_i \text{ are true}, i = 1, \ldots, C) \geq 1 - \sum_{j=1}^{C} \alpha_j = 1 - \alpha_E \tag{12.20}$$

where $\alpha_E = \sum_{j=1}^{C} \alpha_j$ is called the overall error probability. Expression (12.20) can be restated as

$$P(\text{one or more statements } S_i \text{ is false}, i = 1, \ldots, C) \leq \alpha_E$$

or equivalently,

$$P(\text{one or more of the } C \text{ confidence intervals does not}$$
$$\text{contain the parameter being estimated}) \leq \alpha_E$$

Thus, α_E provides an upper bound on the probability of a false conclusion. To conduct an experiment that involves making C comparisons, first select the overall error probability, say $\alpha_E =$

0.05 or 0.10. The individual α_j may be chosen to be equal ($\alpha_j = \alpha_E/C$), or unequal, as desired. The smaller the value of α_j, the wider the jth confidence interval will be. For example, if two 95% c.i.'s ($\alpha_1 = \alpha_2 = 0.05$) are constructed, the overall confidence level will be 90% or greater ($\alpha_E = \alpha_1 + \alpha_2 = 0.10$). If ten 95% c.i.'s are constructed ($\alpha_i = 0.05, i = 1, \ldots, 10$), the resulting overall confidence level could be as low as 50% ($\alpha_E = \sum_{i=1}^{10} \alpha_i = 0.50$), which is far too low for practical use. To guarantee an overall confidence level of 95%, when 10 comparisons are being made, one approach is to construct ten 99.5% confidence intervals for the parameters (or differences) of interest.

The Bonferroni approach to multiple confidence intervals is based on expression (12.20). A major advantage is that it holds whether the models for the alternative designs are run with independent sampling or with common random numbers.

The major disadvantage of the Bonferroni approach in making a large number of comparisons is the increased width of each individual interval. For example, for a given set of data and a large sample size, a 99.5% c.i. will be $z_{0.0025}/z_{0.025} = 2.807/1.96 = 1.43$ times longer than a 95% c.i. For small sample sizes—say, for a sample of size 5—a 99.5% c.i. will be $t_{0.0025,4}/t_{0.025,4} = 5.598/2.776 = 1.99$ times longer than an individual 95% c.i. The width of a c.i. is a measure of the precision of the estimate. For these reasons, it is recommended that the Bonferroni approach be used only when a small number of comparisons are being made. Twenty or so comparisons appears to be the practical upper limit.

Corresponding to goals 1, 2, and 3, there are at least three possible ways of using the Bonferroni Inequality (12.20) when comparing K alternative system designs:

1. (*Individual c.i.'s*): Construct a $100(1 - \alpha_i)$% c.i. for parameter θ_i by using Expression (11.12), in which case the number of intervals is $C = K$. If independent sampling were used, the K c.i.'s would be mutually independent, and thus the overall confidence level would be $(1 - \alpha_1) \times (1 - \alpha_2) \times \cdots \times (1 - \alpha_C)$, which is larger (but not much larger) than the right side of Expression (12.20). This type of procedure is most often used to estimate multiple parameters of a single system, rather than to compare systems—and, because multiple parameter estimates from the same system are likely to be dependent, the Bonferroni inequality typically is needed.

2. (*Comparison to an existing system*): Compare all designs to one specific design—usually, to an existing system: that is, construct a $100(1 - \alpha_i)$% c.i. for $\theta_i - \theta_1 (i = 2, 3, \ldots, K)$, using Expression (12.1). (System 1 with performance measure θ_1 is assumed to be the existing system). In this case, the number of intervals is $C = K - 1$. This type of procedure is most often used to compare several competitors to the present system in order to learn which are better.

3. (*All pairwise comparisons*): Compare all designs to each other—that is, for any two system designs $i \neq j$, construct a $100(1 - \alpha_{ij})$% c.i. for $\theta_i - \theta_j$. With K designs, the number of confidence intervals computed is $C = K(K - 1)/2$. The overall confidence coefficient would be bounded below by $1 - \alpha_E = 1 - \sum\sum_{i \neq j} \alpha_{ij}$ (which follows by Expression (12.20)). It is generally believed that CRN will make the true overall confidence level larger than the right side of Expression (12.20), and usually larger than will independent sampling. The right side of Expression (12.20) can be thought of as giving the worst case (i.e., the lowest possible overall confidence level).

Example 12.4
Reconsider the vehicle-inspection station of Example 12.1. Suppose that the construction of addi-tional space to hold one waiting car is being considered. The alternative system designs are the following:

1. existing system (parallel stations);
2. no space between stations in series;
3. one space between brake and headlight inspection only;
4. one space between headlight and steering inspection only.

Design 2 was compared to the existing setup in Example 12.1. Designs 2, 3, and 4 are series queues, as shown in Figure 12.1(b), the only difference being the number or location of a waiting space between two successive inspections. The arrival process and the inspection times are as given in Example 12.1. The basis for comparison will be mean response time, θ_i, for system i, where a response time is the total time it takes for a car to get through the system. Confidence intervals for $\theta_2 - \theta_1, \theta_3 - \theta_1$, and $\theta_4 - \theta_1$ will be constructed, each having an overall confidence level of 95%. The run length T_E has now been set at 40 hours (instead of the 16 hours used in Example 12.1), and the number of replications R of each model is 10. Common random numbers will be used in all models, but this does not affect the overall confidence level, because, as mentioned, the Bonferroni Inequality (12.20) holds regardless of the statistical independence or dependence of the data.

Since the overall error probability is $\alpha_E = 0.05$ and $C = 3$ confidence intervals are to be constructed, let $\alpha_i = 0.05/3 = 0.0167$ for $i = 2, 3, 4$. Then use Expression (12.1) (with proper modifications) to construct $C = 3$ confidence intervals with $\alpha = \alpha_i = 0.0167$ and degrees of freedom $\nu = 10 - 1 = 9$. The standard error is computed by Equation (12.14), because common random numbers are being used. The output data Y_{ri} are displayed in Table 12.5; Y_{ri} is the sample mean response time for replication r on system i ($r = 1, \ldots, 10; i = 1, 2, 3, 4$). The differences $D_{ri} = Y_{r1} - Y_{ri}$ are also shown, together with the sample mean differences, $\bar{D}_{\cdot i}$, averaged over all replications as in Equation (12.12), the sample variances $S_{D_i}^2$, and the standard error. By Expression (12.1), the three confidence intervals, with overall confidence coefficient at least $1 - \alpha_E$, are given by

$$\bar{D}_{\cdot i} - t_{\alpha_i/2, R-1}\text{s.e.}(\bar{D}_{\cdot i}) \le \theta_1 - \theta_i \le \bar{D}_{\cdot i} + t_{\alpha_i/2, R-1}\text{s.e.}(\bar{D}_{\cdot i}), \qquad i = 2, 3, 4$$

The value of $t_{\alpha_i/2, R-1} = t_{0.0083, 9} = 2.97$ is obtained from Table A.5 by interpolation. For these data, with 95% confidence, it is stated that

$$-1.19 \le \theta_1 - \theta_2 \le 2.79$$
$$0.67 \le \theta_1 - \theta_3 \le 10.71$$
$$-0.94 \le \theta_1 - \theta_4 \le 3.46$$

The simulation analyst has high confidence (at least 95%) that all three confidence statements are correct. Notice that the c.i. for $\theta_1 - \theta_2$ again contains zero; thus, there is no statistically significant difference between design 1 and design 2, a conclusion that supports the previous

Table 12.5 Analysis of Output Data for Vehicle Inspection System When Using CRN

Replication,	Average Response Time for System Design				Observed Difference with System Design 1		
	1,	2,	3,	4,			
r	Y_{r1}	Y_{r2}	Y_{r3}	Y_{r4}	D_{r2}	D_{r3}	D_{r4}
1	63.72	63.06	57.74	62.63	0.66	5.98	1.09
2	32.24	31.78	29.65	31.56	0.46	2.59	0.68
3	40.28	40.32	36.52	39.87	−0.04	3.76	0.41
4	36.94	37.71	35.71	37.35	−0.77	1.23	−0.41
5	36.29	36.79	33.81	36.65	−0.50	2.48	−0.36
6	56.94	57.93	51.54	57.15	−0.99	5.40	−0.21
7	34.10	33.39	31.39	33.30	0.71	2.71	0.80
8	63.36	62.92	57.24	62.21	0.44	6.12	1.15
9	49.29	47.67	42.63	47.46	1.62	6.66	1.83
10	87.20	80.79	67.27	79.60	6.41	19.93	7.60
Sample mean, $\bar{D}_{\cdot i}$					0.80	5.686	1.258
Sample standard deviation, S_{D_i}					2.12	5.338	2.340
Sample variance, $S_{D_i}^2$					4.498	28.498	5.489
Standard error, $S_{D_i}/\sqrt{R}$					0.671	1.688	0.741

results in Example 12.1. The c.i. for $\theta_1 - \theta_3$ lies completely above zero and so provides strong evidence that $\theta_1 - \theta_3 > 0$—that is, that design 3 is better than design 1 because its mean response time is smaller. The c.i. for $\theta_1 - \theta_4$ contains zero, so there is no statistically significant difference between designs 1 and 4.

If the simulation analyst now decides that it would be desirable to compare designs 3 and 4, more simulation runs would be necessary, because it is not formally correct to decide which confidence intervals to compute after the data have been examined. On the other hand, if the simulation analyst had decided to compute all possible confidence intervals (and had made this decision before collecting the data, Y_{ri}), the number of confidence intervals would have been $C = 6$ and the three c.i.'s would have been $t_{0.0042,9}/t_{0.0083,9} \approx 3.32/2.97 = 1.12$ times (or 12%) longer. There is always a trade-off between the number of intervals (C) and the width of each interval. The simulation analyst should carefully consider the possible conclusions before running the simulation experiments and choose those runs and analyses that will provide the most useful information. In particular, the number of confidence intervals computed should be as small as possible—preferably, 20 or less.

Table 12.6 Analysis of Output Data for the Vehicle-Inspection System that Uses Independent Sampling

Replication, r	Average Response Time for System Design			
	Y_{r1}^1	Y_{r2}^2	Y_{r3}^3	Y_{r4}^4
1	63.72	59.37	52.00	59.03
2	32.24	50.06	47.04	49.97
3	40.28	60.63	53.21	60.18
4	36.94	46.36	40.88	45.44
5	36.29	68.87	50.85	66.65
6	56.94	66.44	60.42	66.03
7	34.10	27.51	26.70	27.45
8	63.36	47.98	40.12	47.50
9	49.29	29.92	28.59	29.84
10	87.20	47.14	41.62	46.44
Sample mean $\bar{Y}_{.i}$	50.04	50.43	44.14	49.85
S_i	17.70	13.98	10.76	13.64
S_i^2	313.38	195.54	115.74	185.98
$\bar{Y}_{.1} - \bar{Y}_{.i}$		−0.39	5.89	0.18
s.e.$(\bar{Y}_{.1} - \bar{Y}_{.i})$		7.13	6.55	7.07

For purposes of illustration, 10 replications of each of the four designs were run, using independent sampling (i.e., different random numbers for all runs). The results are presented in Table 12.6, together with sample means ($\bar{Y}_{.i}$), sample standard deviations (S_i), and sample variances (S_i^2), plus the observed difference of sample means ($\bar{Y}_{.1} - \bar{Y}_{.i}$) and the standard error (s.e.) of the observed difference. It is observed that all three confidence intervals for $\theta_1 - \theta_i (i = 2, 3, 4)$ contain zero. Therefore, no strong conclusion is possible from these data and this sample size. By contrast, a sample size of ten was sufficient, when using CRN, to provide strong evidence that design 3 is superior to design 1.

Notice the large increase in standard error of the estimated difference with independent sampling versus with common random numbers. These standard errors are compared in Table 12.7. In addition, a careful examination of Tables 12.5 and 12.6 illustrates the superiority of CRN. In Table 12.5, in all 10 replications, system design 3 has a smaller average response time than does system design 1. By comparing replications 1 and 2 in Table 12.5, it can be seen that a random-number stream that leads to high congestion and large response times in system design 1, as in the first replication, produces results of similar magnitude across all four system designs. Similarly, when system design 1 exhibits relatively low congestion and low response times, as in the second replication, all system designs produce relatively low average response times. This similarity of results on each replication is due, of course, to the use of common random numbers across systems. By contrast, for independent sampling, Table 12.6 shows no such similarity across system designs. In only 5 of the 10 replications is the average response time for system design 3 smaller than that for system design 1, although the average difference in response times across

Table 12.7 Comparison of Standard Errors Arising from CRN with those from Independent Sampling, for the Vehicle-Inspection Problem

	Standard Error When Using		
Difference in Sample Means	CRN Sampling	Independent Sampling	Percentage Increase
$\bar{Y}_{.1} - \bar{Y}_{.2}$	0.67	7.13	1064%
$\bar{Y}_{.1} - \bar{Y}_{.3}$	1.69	6.55	388%
$\bar{Y}_{.1} - \bar{Y}_{.4}$	0.74	7.07	955%

all 10 replications is approximately the same magnitude in each case: 5.69 minutes when using CRN, and 5.89 minutes when using independent sampling. The greater variability of independent sampling is reflected also in the standard errors of the point estimates: ± 1.69 minutes for CRN versus ± 6.55 minutes for independent sampling, an increase of 388%, as seen in Table 12.7. This example illustrates again the advantage of CRN.

As stated previously, CRN does not yield a variance reduction in all simulation models. It is recommended that a pilot study be undertaken and variances estimated to confirm (or possibly deny) the assumption that CRN will reduce the variance (or standard error) of an estimated difference. The reader is referred to the discussion in Section 12.1.3.

Some of the exercises at the end of this chapter provide an opportunity to compare CRN and independent sampling and to compute simultaneous confidence intervals under the Bonferroni approach.

12.2.2 Bonferroni Approach to Selecting the Best

Suppose that there are K system designs, and the ith design has expected performance θ_i. At a gross level, we are interested in which system is best, where "best" is defined to be having maximum expected performance.[1] At a more refined level, we could also be interested in how much better the best is relative to each alternative, because secondary criteria that are not reflected in the performance measure θ_i (such as ease of installation, cost to maintain, etc.) could tempt us to choose an inferior system if it is not deficient by much.

If system design i is the best, then $\theta_i - \max_{j \neq i} \theta_j$ is equal to the difference in performance between the best and the second best. If system design i is not the best, then $\theta_i - \max_{j \neq i} \theta_j$ is equal to the difference between system i and the best. The selection procedure we describe in this section focuses on the parameters $\theta_i - \max_{j \neq i} \theta_j$ for $i = 1, 2, \ldots, K$.

Let i^* denote the (unknown) index of the best system. As a general rule, the smaller the true difference $\theta_{i^*} - \max_{j \neq i^*} \theta_j$ is, and the more certain we want to be that we find the best system, the more replications are required to achieve our goal. Therefore, instead of demanding that we find

---

[1] If "best" is defined to be having minimum expected performance, then the procedure in this section is easily modified, as we illustrate in the example.

i^*, we can compromise and ask to find i^* with high probability whenever the difference between system i^* and the others is at least some practically significant amount. More precisely, we want the probability that we select the best system to be at least $1 - \alpha$ whenever $\theta_{i^*} - \max_{j \neq i^*} \theta_j \geq \epsilon$. If there are one or more systems that are within ϵ of the best, then we will be satisfied to select either the best or any one of the near best. Both the probability of correct selection, $1 - \alpha$, and the practically significant difference, ϵ, will be under our control.

The following procedure achieves the desired probability of correct selection (Nelson and Matejcik [1995]). And because we are also interested in how much each system differs from the best, it also forms $100(1 - \alpha)\%$ confidence intervals for $\theta_i - \max_{j \neq i} \theta_j$ for $i = 1, 2, \ldots, K$. The procedure is valid for normally distributed data when either CRN or independent sampling is being used.

Two-Stage Bonferroni Procedure

1. Specify the practically significant difference ϵ, the probability of correct selection $1 - \alpha$, and the first-stage sample size $R_0 \geq 10$. Let $t = t_{\alpha/(K-1), R_0-1}$.
2. Make R_0 replications of system i to obtain $Y_{1i}, Y_{2i}, \ldots, Y_{R_0,i}$, for systems $i = 1, 2, \ldots, K$.
3. Calculate the first-stage sample means $\bar{Y}_{\cdot i}$, $i = 1, 2, \ldots, K$. For all $i \neq j$, calculate the sample variance of the difference,[2]

$$S_{ij}^2 = \frac{1}{R_0 - 1} \sum_{r=1}^{R_0} \left(Y_{ri} - Y_{rj} - (\bar{Y}_{\cdot i} - \bar{Y}_{\cdot j}) \right)^2$$

   Let $\widehat{S}^2 = \max_{i \neq j} S_{ij}^2$, the largest sample variance.
4. Calculate the second-stage sample size,

$$R = \max \left\{ R_0, \left\lceil \frac{t^2 \widehat{S}^2}{\epsilon^2} \right\rceil \right\}$$

   where $\lceil \cdot \rceil$ means to round up.
5. Make $R - R_0$ additional replications of system i to obtain the output data $Y_{R_0+1,i}, Y_{R_0+2,i}, \ldots, Y_{R,i}$, for $i = 1, 2, \ldots, K$.[3]
6. Calculate the overall sample means

$$\bar{\bar{Y}}_i = \frac{1}{R} \sum_{r=1}^{R} Y_{ri}$$

   for $i = 1, 2, \ldots, K$.
7. Select the system with largest $\bar{\bar{Y}}_i$ as the best.

---

[2] Notice that S_{ij}^2 is algebraically equivalent to S_D^2, the sample variance of $D_r = Y_{ri} - Y_{rj}$, for $r = 1, 2, \ldots, R_0$.
[3] If it is more convenient, a total of R replications can be generated from system i by restarting the entire experiment.

Also form the confidence intervals

$$\min\{0, \bar{\bar{Y}}_i - \max_{j \neq i} \bar{\bar{Y}}_j - \epsilon\} \leq \theta_i - \max_{j \neq i} \theta_j \leq \max\{0, \bar{\bar{Y}}_i - \max_{j \neq i} \bar{\bar{Y}}_j + \epsilon\}$$

for $i = 1, 2, \ldots, K$.

The confidence intervals in Step 7 are not like the usual $\pm$ intervals presented elsewhere in this chapter. Perhaps the most useful interpretation of them is as follows. Let $\widehat{i}$ be the index of the system selected as best. Then, for each of the other systems i, we make one of the declarations:

- If $\bar{\bar{Y}}_i - \bar{\bar{Y}}_{\widehat{i}} + \epsilon \leq 0$, then declare system i to be inferior to the best.
- If $\bar{\bar{Y}}_i - \bar{\bar{Y}}_{\widehat{i}} + \epsilon > 0$, then declare system i to be statistically indistinguishable from the best (and, therefore, system i might be the best).

Example 12.4: Continued _____

Recall that, in Example 12.4, we considered $K = 4$ different designs for the vehicle-inspection station. Suppose that we would like 0.95% confidence of selecting the best (smallest expected response time) system design when the best differs from the second best by at least two minutes. This is a minimization problem, so we focus on the differences $\theta_i - \min_{j \neq i} \theta_j$ for $i = 1, 2, 3, 4$. Then we can apply the Two-Stage Bonferroni Procedure as follows:

1. $\epsilon = 2$ minutes, $1 - \alpha = 0.95$, $R_0 = 10$, and $t = t_{0.0167, 9} = 2.508$.
2. The data in Table 12.5, which was obtained by using CRN, is employed.
3. From Table 12.5, we get $S_{12}^2 = S_{D_2}^2 = 4.498$, $S_{13}^2 = S_{D_3}^2 = 28.498$, and $S_{14}^2 = S_{D_4}^2 = 5.489$. By similar calculations, we obtain $S_{23}^2 = 11.857$, $S_{24}^2 = 0.119$, and $S_{34}^2 = 9.849$.
4. Since $\widehat{S}^2 = S_{13}^2 = 28.498$ is the largest sample variance,

$$R = \max\left\{10, \left\lceil \frac{(2.508)^2(28.498)}{2^2} \right\rceil\right\} = \max\{10, \lceil 44.8 \rceil\} = 45$$

5. Make $45 - 10 = 35$ additional replications of each system.
6. Calculate the overall sample means

$$\bar{\bar{Y}}_i = \frac{1}{45} \sum_{r=1}^{45} Y_{ri}$$

for $i = 1, 2, 3, 4$.
7. Select the system with smallest $\bar{\bar{Y}}_i$ as the best. Also, form the confidence intervals

$$\min\{0, \bar{\bar{Y}}_i - \min_{j \neq i} \bar{\bar{Y}}_j - 2\} \leq \theta_i - \min_{j \neq i} \theta_j \leq \max\{0, \bar{\bar{Y}}_i - \min_{j \neq i} \bar{\bar{Y}}_j + 2\}$$

for $i = 1, 2, 3, 4$.

12.2.3 Bonferroni Approach to Screening

When a two-stage procedure is not possible, or when there are many systems, it could be useful to divide the set of systems into those that could be the best and those that can be eliminated from further consideration. For this purpose, a screening or *subset selection* procedure is useful. The following procedure, due to Nelson *et al.* [2001], guarantees that the retained subset contains the true best system with probability $\geq 1 - \alpha$ when the data are normally distributed and either independent sampling or CRN is used. The subset may contain all K of the systems, only one system, or some number in between, depending on the number of replications and the sample means and sample variances.

Screening Procedure

1. Specify the probability of correct selection $1 - \alpha$ and common sample size from each system, $R \geq 2$. Let $t = t_{\alpha/(K-1), R-1}$.
2. Make R replications of system i to obtain $Y_{1i}, Y_{2i}, \ldots, Y_{Ri}$ for systems $i = 1, 2, \ldots, K$.
3. Calculate the sample means $\bar{Y}_{\cdot i}$ for $i = 1, 2, \ldots, K$. For all $i \neq j$, calculate the sample variance of the difference,

$$
S_{ij}^2 = \frac{1}{R-1} \sum_{r=1}^{R} \left(Y_{ri} - Y_{rj} - (\bar{Y}_{\cdot i} - \bar{Y}_{\cdot j}) \right)^2
$$

4. If bigger is better, then retain system i in the selected subset if

$$
\bar{Y}_{\cdot i} \geq \bar{Y}_{\cdot j} - t \frac{S_{ij}}{\sqrt{R}} \quad \text{for all } j \neq i
$$

If smaller is better, then retain system i in the selected subset if

$$
\bar{Y}_{\cdot i} \leq \bar{Y}_{\cdot j} + t \frac{S_{ij}}{\sqrt{R}} \quad \text{for all } j \neq i
$$

All system designs that are not retained can be eliminated from further consideration.

Example 12.4: Continued

Suppose we want to see whether any of the designs for the vehicle-inspection station can be eliminated on the basis of only the 10 replications in Table 12.5. Summaries of the sample means and variances of the differences are as follows:

$\bar{Y}_{\cdot i}$	1	2	3	4
	50.04	49.24	44.35	48.78

S_{ij}^2	1	2	3	4
1		4.498	28.498	5.489
2			11.857	0.119
3				9.84

The appropriate critical value to obtain 95% confidence that the selected subset contains the true best is $t = t_{0.0167,9} = 2.508$. Recall that smaller response time is better. Applying the Subset Selection Procedure, system designs 1, 2, and 4 can all be eliminated, because

$$\bar{Y}_{.1} = 50.04 \nleq \bar{Y}_{.3} + t\sqrt{\frac{S_{13}^2}{R}} = 44.35 + 2.508\sqrt{\frac{28.498}{10}} = 48.58$$

$$\bar{Y}_{.2} = 49.24 \nleq \bar{Y}_{.3} + t\sqrt{\frac{S_{23}^2}{R}} = 44.35 + 2.508\sqrt{\frac{11.857}{10}} = 47.08$$

$$\bar{Y}_{.4} = 48.78 \nleq \bar{Y}_{.3} + t\sqrt{\frac{S_{43}^2}{R}} = 44.35 + 2.508\sqrt{\frac{9.84}{10}} = 46.84$$

Thus, in this case there was adequate data to select the best, system design 3, with 95% confidence. Had more than one system survived the subset selection, then we could perform additional analysis on that subset, perhaps using the Two-Stage Bonferroni Procedure.

12.3 METAMODELING

Suppose that there is a simulation output response variable, Y, that is related to k independent variables, say $x_1, x_2, \dots, x_k$. The dependent variable, Y, is a random variable, while the independent variables $x_1, x_2, \dots, x_k$ are called design variables and are usually subject to control. The true relationship between the variables Y and x is represented by the (often complex) simulation model. Our goal is to approximate this relationship by a simpler mathematical function called a metamodel. In some cases, the analyst will know the exact form of the functional relationship between Y and $x_1, x_2, \dots, x_k$, say $Y = f(x_1, x_2, \dots, x_k)$. However, in most cases, the functional relationship is unknown, and the analyst must select an appropriate f containing unknown parameters, and then estimate those parameters from a set of data (Y, x). Regression analysis is one method for estimating the parameters.

Example 12.5

An insurance company promises to process all claims it receives each day by the end of the next day. It has developed a simulation model of its proposed claims-processing system to evaluate how hard it will be to meet this promise. The actual number and types of claims that will need to be processed each day will vary, and the number may grow over time. Therefore, the company would like to have a model that predicts the total processing time as a function of the number of claims received.

The primary value of a metamodel is to make it easy to answer "what if" questions, such as, what the processing time will be if there are x claims. Evaluating a function f, or perhaps its derivatives, at a number of values of x is typically much easier than running a simulation experiment for each value.

12.3.1 Simple Linear Regression

Suppose that it is desired to estimate the relationship between a single independent variable x and a dependent variable Y, and suppose that the true relationship between Y and x is suspected to be linear. Mathematically, the expected value of Y for a given value of x is assumed to be

$$E(Y|x) = \beta_0 + \beta_1 x \qquad (12.21)$$

where β_0 is the intercept on the Y axis, an unknown constant; and β_1 is the slope, or change in Y for a unit change in x, also an unknown constant. It is further assumed that each observation of Y can be described by the model

$$Y = \beta_0 + \beta_1 x + \epsilon \qquad (12.22)$$

where ϵ is a random error with mean zero and constant variance σ^2. The regression model given by Equation (12.22) involves a single variable x and is commonly called a simple linear regression model.

Suppose that there are n pairs of observations $(Y_1, x_1), (Y_2, x_2), \ldots, (Y_n, x_n)$. These observations can be used to estimate β_0 and β_1 in Equation (12.22). The method of least squares is commonly used to form the estimates. In the method of least squares, β_0 and β_1 are estimated in such a way that the sum of the squares of the deviations between the observations and the regression line is minimized. The individual observations in Equation (12.22) can be written as

$$Y_i = \beta_0 + \beta_1 x_i + \epsilon_i, \qquad i = 1, 2, \ldots, n \qquad (12.23)$$

where $\epsilon_1, \epsilon_2, \ldots$ are assumed to be uncorrelated random variables.

Each ϵ_i in Equation (12.23) is given by

$$\epsilon_i = Y_i - \beta_0 - \beta_1 x_i \qquad (12.24)$$

and represents the difference between the observed response, Y_i, and the expected response, $\beta_0 + \beta_1 x_i$, predicted by the model in Equation (12.21). Figure 12.3 shows how ϵ_i is related to x_i, Y_i, and $E(Y_i|x_i)$.

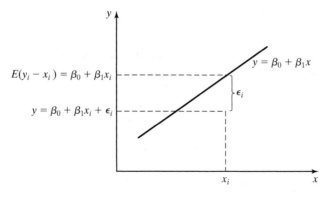

Figure 12.3 Relationship of ϵ_i to x_i, Y_i, and $E(Y_i|x_i)$.

The sum of squares of the deviations given in Equation (12.24) is given by

$$L = \sum_{i=1}^{n} \epsilon_i^2 = \sum_{i=1}^{n} (Y_i - \beta_0 - \beta_1 x_i)^2 \tag{12.25}$$

and L is called the least-squares function. It is convenient to rewrite Y_i as follows:

$$Y_i = \beta_0' + \beta_1(x_i - \bar{x}) + \epsilon_i \tag{12.26}$$

where $\beta_0' = \beta_0 + \beta_1 \bar{x}$ and $\bar{x} = \sum_{i=1}^{n} x_i/n$. Equation (12.26) is often called the transformed linear regression model. Using Equation (12.26), Equation (12.25) becomes

$$L = \sum_{i=1}^{n} [Y_i - \beta_0' - \beta_1(x_i - \bar{x})]^2$$

To minimize L, find $\partial L/\partial \beta_0'$ and $\partial L/\partial \beta_1$, set each to zero, and solve for $\widehat{\beta_0'}$ and $\widehat{\beta}_1$. Taking the partial derivatives and setting each to zero yields

$$n\widehat{\beta_0'} = \sum_{i=1}^{n} Y_i$$

$$\widehat{\beta}_1 \sum_{i=1}^{n} (x_i - \bar{x})^2 = \sum_{i=1}^{n} Y_i(x_i - \bar{x}) \tag{12.27}$$

Equations (12.27) are often called the "normal equations," which have the solutions

$$\widehat{\beta_0'} = \bar{Y} = \sum_{i=1}^{n} \frac{Y_i}{n} \tag{12.28}$$

and

$$\widehat{\beta}_1 = \frac{\sum_{i=1}^{n} Y_i(x_i - \bar{x})}{\sum_{i=1}^{n} (x_i - \bar{x})^2} \tag{12.29}$$

The numerator in Equation (12.29) is rewritten for computational purposes as

$$S_{xy} = \sum_{i=1}^{n} Y_i(x_i - \bar{x}) = \sum_{i=1}^{n} x_i Y_i - \frac{\left(\sum_{i=1}^{n} x_i\right)\left(\sum_{i=1}^{n} Y_i\right)}{n} \tag{12.30}$$

where S_{xy} denotes the corrected sum of cross products of x and Y. The denominator of Equation (12.29) is rewritten for computational purposes as

$$S_{xx} = \sum_{i=1}^{n} (x_i - \bar{x})^2 = \sum_{i=1}^{n} x_i^2 - \frac{\left(\sum_{i=1}^{n} x_i\right)^2}{n} \tag{12.31}$$

where S_{xx} denotes the corrected sum of squares of x. The value of $\widehat{\beta}_0$ can be retrieved easily as

$$\widehat{\beta}_0 = \widehat{\beta}_0' - \widehat{\beta}_1 \bar{x} \tag{12.32}$$

Example 12.6:   Calculating $\widehat{\beta}_0$ and $\widehat{\beta}_1$ _____

The simulation model of the claims-processing system in Example 12.5 was executed with initial conditions $x = 100, 150, 200, 250,$ and 300 claims received the previous day. Three replications were obtained at each setting. The response Y is the number of hours required to process x claims. The results are shown in Table 12.8. The graphical relationship between the number of claims received and total processing time is shown in Figure 12.4. Such a display is called a scatter diagram. Examination of this scatter diagram indicates that there is a strong relationship between number of claims and processing time. The tentative assumption of the linear model given by Equation (12.22) appears to be reasonable.

With the processing times as the Y_i values (the dependent variables) and the number of claims as the x_i values (the independent variables), $\widehat{\beta}_0$ and $\widehat{\beta}_1$ can be found by the following computations: $n = 15$, $\sum_{i=1}^{15} x_i = 3000$, $\sum_{i=1}^{15} Y_i = 178$, $\sum_{i=1}^{15} x_i^2 = 675{,}000$, $\sum_{i=1}^{15} x_i Y_i = 39080$, and $\bar{x} = 3000/15 = 200$.

Table 12.8 Simulation Results for Processing Time Given x Claims

Number of Claims x	Hours of Processing Time Y
100	8.1
100	7.8
100	7.0
150	9.6
150	8.5
150	9.0
200	10.9
200	13.3
200	11.6
250	12.7
250	14.5
250	14.7
300	16.5
300	17.5
300	16.3

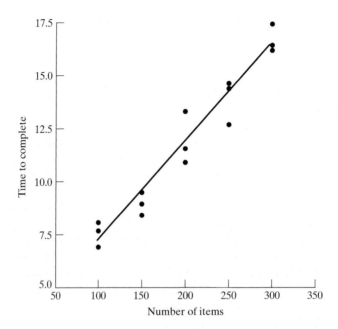

Figure 12.4   Relationship between number of claims and hours of processing time.

From Equation (12.30) S_{xy} is calculated as

$$S_{xy} = 39,080 - \frac{(3000)(178)}{15} = 3480$$

From Equation (12.31), S_{xx} is calculated as

$$S_{xx} = 675,000 - \frac{(3000)^2}{15} = 75,000$$

Then, $\widehat{\beta}_1$ is calculated from Equation (12.29) as

$$\widehat{\beta}_1 = \frac{S_{xy}}{S_{xx}} = \frac{3480}{75,000} = 0.0464$$

As shown in Equation (12.28), $\widehat{\beta}_0'$ is just $\bar{Y}$, or

$$\widehat{\beta}_0' = \frac{178}{15} \approx 11.8667$$

To express the model in the original terms, compute $\widehat{\beta}_0$ from Equation (12.32) as

$$\widehat{\beta}_0 = 11.8667 - 0.0464(200) = 2.5867$$

Then an estimate of the mean of Y given x, $E(Y|x)$, is given by

$$\widehat{y} = \widehat{\beta}_0 + \widehat{\beta}_1 x = 2.5867 + 0.0464x \tag{12.33}$$

For a given number of claims, x, this model can be used to predict the number of hours required to process them. The coefficient $\widehat{\beta}_1$ has the interpretation that each additional claim received adds an expected 0.0464 hours, or 2.8 minutes, to the expected total processing time.

Regression analysis is widely used and frequently misused. Several of the common abuses are briefly mentioned here. Relationships derived in the manner of Equation (12.33) are valid for values of the independent variable within the range of the original data. The linear relationship that has been tentatively assumed may not be valid outside the original range. In fact, we know from queueing theory that mean processing time may increase rapidly as the number of claims approaches the capacity of the system. Therefore, Equation (12.33) can be considered valid only for $100 \leq x \leq 300$. Regression models are not advised for extrapolation purposes.

Care should be taken in selecting variables that have a plausible causal relationship with each other. It is quite possible to develop statistical relationships that are unrelated in a practical sense. For example, an attempt might be made to relate monthly output of a steel mill to the weight of reports appearing on a manager's desk during the month. A straight line may appear to provide a good model for the data, but the relationship between the two variables is tenuous. A strong observed relationship does not imply that a causal relationship exists between the variables. Causality can be inferred only when analysis uncovers some plausible reasons for its existence. In Example 12.5 it is reasonable that starting with more claims implies that more time is needed to process them. Therefore, a relationship of the form of Equation (12.33) is at least plausible.

12.3.2 Testing for Significance of Regression

In Section 12.3.1, it was assumed that a linear relationship existed between Y and x. In Example 12.5, a scatter diagram, shown in Figure 12.4, relating number of claims and processing time was prepared to evaluate whether a linear model was a reasonable tentative assumption prior to the calculation of $\widehat{\beta}_0$ and $\widehat{\beta}_1$. However, the adequacy of the simple linear relationship should be tested prior to using the model for predicting the response, Y_i, given an independent variable, x_i. There are several tests which may be conducted to aid in determining model adequacy. Testing whether the order of the model tentatively assumed is correct, commonly called the "lack-of-fit test," is suggested. The procedure is explained by Box and Draper [1987], Hines, Montgomery, Goldsman, and Borror [2002], and Montgomery [2000].

Testing for the significance of regression provides another means for assessing the adequacy of the model. The hypothesis test described next requires the additional assumption that the error component ϵ_i is normally distributed. Thus, the complete assumptions are that the errors are NID$(0, \sigma^2)$—that is, normally and independently distributed with mean zero and constant variance σ^2. The adequacy of the assumptions can and should be checked by residual analysis, discussed by Box and Draper [1987], Hines, Montgomery, Goldsman, and Borror [2002], and Montgomery [2000].

Testing for significance of regression is one of many hypothesis tests that can be developed from the variance properties of $\widehat{\beta}_0$ and $\widehat{\beta}_1$. The interested reader is referred to the references just

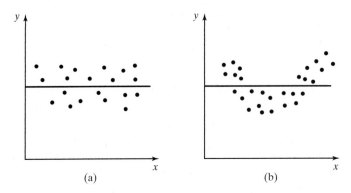

Figure 12.5   Failure to reject $H_0 : \beta_1 = 0$.

cited for extensive discussion of hypothesis testing in regression. Just the highlights of testing for significance of regression are given in this section.

Suppose that the alternative hypotheses are

$H_0 : \beta_1 = 0$

$H_1 : \beta_1 \neq 0$

Failure to reject H_0 indicates that there is no linear relationship between x and Y. This situation is illustrated in Figure 12.5. Notice that two possibilities exist. In Figure 12.5(a), the implication is that x is of little value in explaining the variability in Y, and that $\widehat{y} = \bar{Y}$ is the best estimator. In Figure 12.5(b), the implication is that the true relationship is not linear.

Alternatively, if H_0 is rejected, the implication is that x is of value in explaining the variability in Y. This situation is illustrated in Figure 12.6. Here, also, two possibilities exist. In Figure 12.6(a), the straight-line model is adequate. However, in Figure 12.6(b), even though there is a linear effect of x, a model with higher-order terms (such as $x^2, x^3, \dots$) is necessary. Thus, even though there may be significance of regression, testing of the residuals and testing for lack of fit are needed to confirm the adequacy of the model.

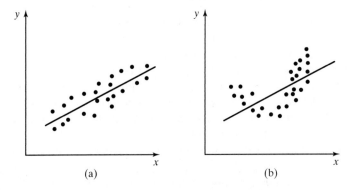

Figure 12.6 $H_0 : \beta_1 = 0$ is rejected.

The appropriate test statistic for significance of regression is given by

$$t_0 = \frac{\widehat{\beta}_1}{\sqrt{MS_E/S_{xx}}} \tag{12.34}$$

where MS_E is the mean squared error. The error is the difference between the observed value, Y_i, and the predicted value, $\widehat{y}_i$, at x_i, or $e_i = Y_i - \widehat{y}_i$. The squared error is given by $\sum_{i=1}^{n} e_i^2$, and the mean squared error, given by

$$MS_E = \sum_{i=1}^{n} \frac{e_i^2}{n-2} \tag{12.35}$$

is an unbiased estimator of $\sigma^2 = V(\epsilon_i)$. The direct method can be used to calculate $\sum_{i=1}^{n} e_i^2$: Calculate each $\widehat{y}_i$, compute e_i^2, and sum all the e_i^2 values, $i = 1, 2, \ldots, n$. However, it can be shown that

$$\sum_{i=1}^{n} e_i^2 = S_{yy} - \widehat{\beta}_1 S_{xy} \tag{12.36}$$

where S_{yy}, the corrected sum of squares of Y, is given by

$$S_{yy} = \sum_{i=1}^{n} Y_i^2 - \frac{\left(\sum_{i=1}^{n} Y_i\right)^2}{n} \tag{12.37}$$

and S_{xy} is given by Equation (12.30). Equation (12.36) could be easier to use than the direct method.

The statistic defined by Equation (12.34) has the t distribution with $n-2$ degrees of freedom. The null hypothesis H_0 is rejected if $|t_0| > t_{\alpha/2, n-2}$.

Example 12.7: Testing for Significance of Regression _____

Given the results in Example 12.6, the test for the significance of regression is conducted. One more computation is needed prior to conducting the test. That is, $\sum_{i=1}^{n} Y_i^2 = 2282.94$. Using Equation (12.37) yields

$$S_{yy} = 2282.94 - \frac{(178)^2}{15} = 170.6734$$

Then $\sum_{i=1}^{15} e_i^2$ is computed according to Equation (12.36) as

$$\sum_{i=1}^{15} e_i^2 = 170.6734 - 0.0464(3480) = 9.2014$$

Now, the value of MS_E is calculated from Equation (12.35):

$$MS_E = \frac{9.2014}{13} = 0.7078$$

The value of t_0 can be calculated by using Equation (12.34) as

$$t_0 = \frac{0.0464}{\sqrt{0.7078/75000}} = 15.13$$

Since $t_{0.025,13} = 2.16$ from Table A.5, we reject the hypothesis that $\beta_1 = 0$. Thus, there is significant evidence that x and Y are related.

12.3.3 Multiple Linear Regression

If the simple linear regression model of Section 12.3.1 is inadequate, several other possibilities exist. There could be several independent variables, so that the relationship is of the form

$$Y = \beta_0 + \beta_1 x_1 + \beta_2 x_2 + \cdots + \beta_m x_m + \epsilon \tag{12.38}$$

Notice that this model is still linear, but has more than one independent variable. Regression models having the form shown in Equation (12.38) are called multiple linear regression models. Another possibility is that the model is of a quadratic form such as

$$Y = \beta_0 + \beta_1 x + \beta_2 x^2 + \epsilon \tag{12.39}$$

Equation (12.39) is also a linear model which may be transformed to the form of Equation (12.38) by letting $x_1 = x$ and $x_2 = x^2$.

Yet another possibility is a model of the form such as

$$Y = \beta_0 + \beta_1 x_1 + \beta_2 x_2 + \beta_3 x_1 x_2 + \epsilon$$

which is also a linear model. The analysis of these three models with the forms just shown, and related models, can be found in Box and Draper [1987], Hines, Montgomery, Goldsman, and Borror [2002], Montgomery [2000], and other applied statistics texts; and also in Kleijnen [1987, 1998], which is concerned primarily with the application of these models in simulation.

12.3.4 Random-Number Assignment for Regression

The assignment of random-number seeds or streams is part of the design of a simulation experiment.[4] Assigning a different seed or stream to different design points (settings for $x_1, x_2, \ldots, x_m$ in a multiple linear regression) guarantees that the responses Y from different design points will be statistically independent. Similarly, assigning the same seed or stream to different design points induces dependence among the corresponding responses, by virtue of their all having the same source of randomness.

Many textbook experiment designs assume independent responses across design points. To conform to this assumptions, we must assign different seeds or streams to each design point. However, it is often useful to assign the same random number seeds or streams to all of the design points—in other words, to use common random numbers.

---

[4]This section is based on Nelson [1992].

The intuition behind common random numbers for metamodels is that a fairer comparison among design points is achieved if the design points are subjected to the same experimental conditions, specifically the same source of randomness. The mathematical justification is as follows: Suppose we fit the simple linear regression $Y_i = \beta_0 + \beta_1 x_i + \epsilon_i$ and obtain least squares estimates $\widehat{\beta}_0$ and $\widehat{\beta}_1$. Then an estimator of the expected difference in performance between design points i and j is

$$\widehat{\beta}_0 + \widehat{\beta}_1 x_i - (\widehat{\beta}_0 + \widehat{\beta}_1 x_j) = \widehat{\beta}_1 (x_i - x_j)$$

when x_i and x_j are fixed design points, $\widehat{\beta}_1$ determines the estimated difference between design points i and j, or for that matter between any other two values of x. Therefore, common random numbers can be expected to reduce the variance of $\widehat{\beta}_1$ and, more generally, reduce the variance of all of the slope terms in a multiple linear regression. Common random numbers typically do not reduce the variance of the intercept term, $\widehat{\beta}_0$.

The least-squares estimators $\widehat{\beta}_0$ and $\widehat{\beta}_1$ are appropriate regardless of whether we use common random numbers, but the associated statistical analysis is affected by that choice. For statistical analysis of a metamodel under common random numbers, see Kleijnen [1988] and Nelson [1992].

12.4 OPTIMIZATION VIA SIMULATION

Consider the following examples.[5]

Example 12.8:   Materials Handling System (MHS) _____
Engineers need to design a MHS consisting of a large automated storage and retrieval device, automated guided vehicles (AGVs), AGV stations, lifters, and conveyors. Among the design variables they can control are the number of AGVs, the load per AGV, and the routing algorithm used to dispatch the AGVs. Alternative designs will be evaluated according to AGV utilization, transportation delay for material that needs to be moved, and overall investment and operation costs.

Example 12.9:   Liquified Natural Gas (LNG) Transportation _____
A LNG transportation system will consist of LNG tankers and of loading, unloading, and storage facilities. In order to minimize cost, designers can control tanker size, number of tankers in use, number of jetties at the loading and unloading facilities, and capacity of the storage tanks.

Example 12.10:   Automobile Engine Assembly _____
In an assembly line, a large buffer (queue) between workstations could increase station utilization—because there will tend to be something waiting to be processed—but drive up space requirements and work-in-process inventory. An allocation of buffer capacity that minimizes the sum of these competing costs is desired.

---

[5]Some of these descriptions are based on Boesel, Nelson, and Ishii [2003].

Example 12.11:   Traffic Signal Sequencing ——————————————————

Civil engineers want to sequence the traffic signals along a busy section of road to reduce driver delay and the congestion occurring along narrow cross streets. For each traffic signal, the length of the red, green, and green-turn-arrow cycles can be set individually.

Example 12.12:   On-Line Services ——————————————————

A company offering on-line information services over the Internet is changing its computer architecture from central mainframe computers to distributed workstation computing. The numbers and types of CPUs, the network structure, and the allocation of processing tasks all need to be chosen. Response time to customer queries is the key performance measure.

What do these design problems have in common? Clearly, a simulation model could be useful in each, and all have an implied goal of finding the best design relative to some performance measures (cost, delay, etc.). In each example, there are potentially a very large number of alternative designs, ranging from tens to thousands, and certainly more than the 2 to 10 we considered in Section 12.2.2. Some of the examples contain a diverse collection of decision variables: discrete (number of AGVs, number of CPUs), continuous (tanker size, red-cycle length) and qualitative (routing strategy, algorithm for allocating processing tasks). This makes developing a metamodel, as described in Section 12.3, difficult.

All of these problems fall under the general topic of "optimization via simulation," where the goal is to minimize or maximize some measures of system performance and system performance can be evaluated only by running a computer simulation. Optimization via simulation is a relatively new, but already vast, topic, and commercial software has become widely available. In this section, we describe the key issues that should be considered in undertaking optimization via simulation, provide some pointers to the available literature, and give one example algorithm.

12.4.1 What Does 'Optimization via Simulation' Mean?

Optimization is a key tool used by operations researchers and management scientists, and there are well-developed algorithms for many classes of problems, the most famous being linear programming. Much of the work on optimization deals with problems in which all aspects of the system are treated as being known with certainty; most critically, the performance of any design (cost, profit, makespan, etc.) can be evaluated exactly.

In stochastic, discrete-event simulation, the result of any simulation run is a random variable. For notation, let $x_1, x_2, \ldots, x_m$ be the m controllable design variables and let $Y(x_1, x_2, \ldots, x_m)$ be the observed simulation output performance on one run. To be concrete, x_1, x_2, x_3 might denote the number of AGVs, the load per AGV, and the routing algorithm used to dispatch the AGVs, respectively, in Example 12.8, while $Y(x_1, x_2, x_3)$ could be total MHS acquisition and operation cost.

What does it mean to "optimize" $Y(x_1, x_2, \ldots, x_m)$ with respect to $x_1, x_2, \ldots, x_m$? Y is a random variable, so we cannot optimize the *actual* value of Y. The most common definition of optimization is

$$\text{maximize or minimize } E(Y(x_1, x_2, \ldots, x_m)) \tag{12.40}$$

In other words, the mathematical expectation, or long-run average, of performance is maximized or minimized. *This is the default definition of optimization used in all commercial packages of which we are aware.* In our example, $E(Y(x_1, x_2, x_3))$ is the expected, or long-run average cost of operating the MHS with x_1 AGVs, x_2 load per AGV, and routing algorithm x_3.

It is important to note that (12.40) is not the only possible definition, however. For instance, we might want to select the MHS design that has the best chance of costing less than $\$D$ to purchase and operate, changing the objective to

$$\text{maximize } \Pr(Y(x_1, x_2, x_3) \leq D)$$

We can fit this objective into formulation (12.40) by defining a new performance measure

$$Y'(x_1, x_2, x_3) = \begin{cases} 1, \text{ if } Y(x_1, x_2, x_3) \leq D \\ 0, \text{ otherwise} \end{cases}$$

and maximizing $E(Y'(x_1, x_2, x_3))$ instead.

A more complex optimization problem occurs when we want to select the system design that is *most likely* to be the best. Such an objective is relevant when one-shot, rather than long-run average, performance matters. Examples include a Space Shuttle launch, or the delivery of a unique, large order of products. Bechhofer, Santner, and Goldsman [1995] address this problem under the topic of "multinomial selection."

We have been assuming that a system design $x_1, x_2, \ldots, x_m$ can be evaluated in terms of a single performance measure, Y, such as cost. Obviously, this may not always be the case. In the MHS example, we might also be interested in some measure of system productivity, such as throughput or cycle time. At present, multiple objective optimization via simulation is not well developed. Therefore, one of three strategies is typically employed:

1. Combine all of the performance measures into a single measure, the most common being cost. For instance, the revenue generated by each completed product in the MHS could represent productivity and be included as a negative cost.
2. Optimize with respect to one key performance measure, but then evaluate the top solutions with respect to secondary performance measures. For instance, the MHS could be optimized with respect to expected cost, and then the cycle time could be compared for the top 5 designs. *This approach requires that information on more than just the best solution be maintained.*
3. Optimize with respect to one key performance measure, but consider only those alternatives that meet certain constraints on the other performance measures. For instance, the MHS could be optimized with respect to expected cost for those alternatives whose expected cycle time is less than a given threshold.

12.4.2 Why is Optimization via Simulation Difficult?

Even when there is no uncertainty, optimization can be very difficult if the number of design variables is large, the problem contains a diverse collection of design variable types, and little is known about the structure of the performance function. Optimization via simulation adds an additional complication: The performance of a particular design cannot be evaluated exactly, but instead must be estimated. Because we have estimates, it is not possible to conclude with assurance that one design is better than another, and this uncertainty frustrates optimization algorithms that try to move in improving directions. In principle, one can eliminate this complication by making so many replications, or such long runs, at each design point that the performance estimate has essentially no variance. In practice, this could mean that very few alternative designs will be explored, because of the time required to simulate each one.

The existence of sampling variability forces optimization via simulation to make compromises. The following are the standard ones:

- **Guarantee a prespecified probability of correct selection.** The Two-Stage Bonferroni Procedure in Section 12.2.2 is an example of this approach, which allows the analyst to specify the desired chance of being right. Such algorithms typically require either that every possible design be simulated or that a strong functional relationship among the designs (such as a metamodel) apply. Other algorithms can be found in Goldsman and Nelson [1998].

- **Guarantee asymptotic convergence.** There are many algorithms that guarantee convergence to the global optimal solution as the simulation effort (number of replications, length of replications) becomes infinite. These guarantees are useful because they indicate that the algorithm tends to get to where the analyst wants it to go. However, convergence can be slow, and there is often no guarantee as to how good the reported solution is when the algorithm is terminated in finite time (as it must be in practice). See Andradóttir [1998] for specific algorithms that apply to discrete- or continuous-variable problems.

- **Optimal for deterministic counterpart.** The idea here is to use an algorithm that would find the optimal solution *if the performance of each design could be evaluated with certainty*. An example might be applying a standard nonlinear programming algorithm to the simulation optimization problem. It is typically up to the analyst to make sure that enough simulation effort is expended (replications or run length) to insure that such an algorithm is not misled by sampling variability. Direct application of an algorithm that assumes deterministic evaluation to a stochastic simulation is not recommended.

- **Robust heuristics.** Many heuristics have been developed for deterministic optimization problems that do not guarantee finding the optimal solution, but nevertheless been shown to be very effective on difficult, practical problems. Some of these heuristics use randomness as part of their search strategy, so one might argue that they are less sensitive to sampling variability than other types of algorithms. Nevertheless, it is still important to make sure that enough simulation effort is expended (replications or run length) to insure that such an algorithm is not misled by sampling variability.

Robust heuristics are the most common algorithms found in commercial optimization via simulation software. We provide some guidance on their use in the next section. See Fu [2002] for a comprehensive discussion of optimization theory versus practice.

12.4.3 Using Robust Heuristics

By a "robust heuristic" we mean a procedure that does not depend on strong problem structure—such as continuity or convexity of $E\left(Y(x_1, \ldots, x_m)\right)$—to be effective, can be applied to problems with mixed types of decision variables, and—ideally—is tolerant of some sampling variability. Genetic algorithms (GA) and tabu search (TS) are two prominent examples, but there are many others and many variations of them. Such heuristics form the core of most commercial implementations. To give a sense of these heuristics, we describe GA and TS next. We caution the reader that only a high-level description of the simplest version of each procedure is provided. The commercial implementations are much more sophisticated.

Suppose that there are k possible solutions to the optimization via simulation problem. Let $\mathbf{X} = \{\mathbf{x}_1, \mathbf{x}_2, \dots, \mathbf{x}_k\}$ denote the solutions, where the ith solution $\mathbf{x}_i = (x_{i1}, x_{i2}, \dots, x_{im})$ provides specific settings for the m decision variables. The simulation output at solution $\mathbf{x}_i$ is denoted $Y(\mathbf{x}_i)$; this could be the output of a single replication, or the average of several replications. Our goal is to find the solution $\mathbf{x}^*$ that minimizes $E(Y(\mathbf{x}))$.

On each iteration (known as a "generation"), a GA operates on a "population" of p solutions. Denote the population of solutions on the jth iteration as $\mathbf{P}(j) = \{\mathbf{x}_1(j), \mathbf{x}_2(j), \dots, \mathbf{x}_p(j)\}$. There may be multiple copies of the same solution in $\mathbf{P}(j)$, and $\mathbf{P}(j)$ may contain solutions that were discovered on previous iterations. From iteration to iteration, this population evolves in such a way that good solutions tend to survive and give birth to new, and hopefully better, solutions, while inferior solutions tend to be removed from the population. The basic GA is given here:

Basic GA

Step 1. Set the iteration counter $j = 0$, and select (perhaps randomly) an initial population of p solutions $\mathbf{P}(0) = \{\mathbf{x}_1(0), \dots, \mathbf{x}_p(0)\}$.

Step 2. Run simulation experiments to obtain performance estimates $Y(\mathbf{x})$ for all p solutions $\mathbf{x}(j)$ in $\mathbf{P}(j)$.

Step 3. Select a population of p solutions from those in $\mathbf{P}(j)$ in such a way that those with smaller $Y(\mathbf{x})$ values are more likely, but not certain, to be selected. Denote this population of solutions as $\mathbf{P}(j + 1)$.

Step 4. Recombine the solutions in $\mathbf{P}(j + 1)$ via crossover (which joins parts of two solutions $\mathbf{x}_i(j + 1)$ and $\mathbf{x}_\ell(j + 1)$ to form a new solution) and mutation (which randomly changes a part of a solution $\mathbf{x}_i(j + 1)$).

Step 5. Set $j = j + 1$ and go to Step 2.

The GA can be terminated after a specified number of iterations, when little or no improvement is noted in the population, or when the population contains p copies of the same solution. At termination, the solution $\mathbf{x}^*$ that has the smallest $Y(\mathbf{x})$ value in the last population is chosen as best (or alternatively, the solution with the smallest $Y(\mathbf{x})$ over all iterations could be chosen).

GAs are applicable to almost any optimization problem, because the operations of selection, crossover, and mutation can be defined in a very generic way that does not depend on specifics of the problem. However, when these operations are not tuned to the specific problem, a GA's progress can be very slow. Commercial versions are often self-tuning, meaning that they update selection, crossover, and mutation parameters during the course of the search. There is some evidence that GAs are tolerant of sampling variability in $Y(\mathbf{x})$ because they maintain a population of solutions rather than focusing on improving a current-best solution. In other words, it is not critical that the GA rank the solutions in a population of solutions perfectly, because the next iteration depends on the entire population, not on a single solution.

TS, on the other hand, identifies a current best solution on each iteration and then tries to improve it. Improvements occur by changing the solution via "moves." For example, the solution (x_1, x_2, x_3) could be changed to the solution $(x_1 + 1, x_2, x_3)$ by the move of adding 1 to the first decision variable (perhaps x_1 represents the number of AGVs in Example 12.8, so the move would add one more AGV). The "neighbors" of solution $\mathbf{x}$ are all of those solutions that can be reached by legal moves. TS finds the best neighbor solution and moves to it. However, to avoid making moves that return the search to a previously visited solution, moves may become "tabu" (not usable) for some number of iterations. Conceptually, think about how you would find your way through a maze: If you took a path that lead to a dead end, then you would avoid taking that path again (it would be tabu).

The basic TS algorithm is given next. The description is based on Glover [1989].

Basic TS

Step 1. Set the iteration counter $j = 0$ and the list of tabu moves to empty. Select an initial solution $\mathbf{x}^*$ in $\mathbf{X}$ (perhaps randomly).

Step 2. Find the solution $\mathbf{x}'$ that minimizes $Y(\mathbf{x})$ over all of the neighbors of $\mathbf{x}^*$ that are not reached by tabu moves, running whatever simulations are needed to do the optimization.

Step 3. If $Y(\mathbf{x}') < Y(\mathbf{x}^*)$, then $x^* = x'$ (move the current best solution to x').

Step 4. Update the list of tabu moves and go to Step 2.

The TS can be terminated when a specified number of iterations have been completed, when some number of iterations has passed without changing $\mathbf{x}^*$, or when there are no more feasible moves. At termination, the solution $\mathbf{x}^*$ is chosen as best.

TS is fundamentally a discrete-decision-variable optimizer, but continuous decision variables can be discretized, as described in Section 12.4.4. TS aggressively pursues improving solutions, and therefore tends to make rapid progress. However, it is more sensitive to random variability in $Y(\mathbf{x})$, because $\mathbf{x}^*$ is taken to be the *true* best solution so far and attempts are made to improve it. There are probabilistic versions of TS that should be less sensitive, however. An important feature of commercial implementations of TS, which is not present in the Basic TS, is a mechanism for overiding the tabu list when doing so is advantageous.

Next, we offer two suggestions for using commercial products that employ a GA, TS, or other robust heuristic controlling sampling variability, and *restarting*.

Control Sampling Variability

In many cases, it will up to the user to determine how much sampling (replications or run length) will be undertaken at each potential solution. This is a difficult problem in general. Ideally, sampling should increase as the heuristic closes in on the better solutions, simply because it is much more difficult to distinguish solutions that are close in expected performance from those that differ widely. Early in the search, it may be easy for the heuristic to identify good solutions and search directions, because clearly inferior solutions are being compared to much better ones, but late in the search this might not be the case.

If the analyst must specify a fixed number of replications per solution that will be used through the search, then a preliminary experiment should be conducted. Simulate several designs, some at the extremes of the solution space and some nearer the center. Compare the apparent best and apparent worst of these designs, using the approaches in Section 12.1. Using the technique described in Section 12.1.4, find the minimum for the number of replications required to declare these designs to be statistically significantly different. This is the minimum number of replications that should be used.

After the optimization run has completed, perform a second set of experiments on the top 5 to 10 designs identified by the heuristic. Use the comparison techniques in Section 12.2–12.2.3 to rigorously evaluate which are the best or near-best of these designs.

Restarting

Because robust heuristics provide no guarantees that they converge to the optimal solution for optimization via simulation, it makes sense to run the optimization two or more times to see which run yields the best solution. Each optimization run should use different random number seeds or streams and, ideally, should start from different initial solutions. Try starting the optimization at solutions on the extremes of the solution space, in the center of the space, and at randomly generated solutions. If people familiar with the system suspect that certain designs will be good, be sure to include them as possible starting solutions for the heuristic.

12.4.4 An Illustration: Random Search

In this section, we present an algorithm for optimization via simulation known as random search. The specific implementation is based on Algorithm 2 in Andradóttir [1998], which provides guaranteed asymptotic convergence under certain conditions. Thus, it will find the true optimal solution if permitted to run long enough. However, in practice, convergence can be slow, and the memory requirements of this particular version of random search can be quite large. Even though random search is not a "robust heuristic," we will also use it to demonstrate some strategies we would employ in conjunction with such heuristics and to demonstrate why optimization via simulation is tricky even with what appears to be an uncomplicated algorithm.

The random-search algorithm that we present requires that there be a finite number of possible system designs (although that number may be quite large). This might seem to rule out problems with continuous decision variables, such as conveyor speed. In practice, however, apparently continuous decision variables can often be discretized in a reasonable way. For instance, if conveyor speed can be anything from 60 to 120 feet per minute, little may be lost by treating the possible conveyor speeds as $60, 61, 62, \ldots, 120$ feet per minute (61 possible values). Note, however, that there are algorithms designed specifically for continuous-variable problems (Andradóttir [1998]).

Again, let the k possible solutions to the optimization via simulation problem be denoted $\{\mathbf{x}_1, \mathbf{x}_2, \ldots, \mathbf{x}_k\}$, where the ith solution $\mathbf{x}_i = (x_{i1}, x_{i2}, \ldots, x_{im})$ provides specific settings for the m decision variables. The simulation output at solution $\mathbf{x}_i$ is denoted $Y(\mathbf{x}_i)$; this could be the output of a single replication or the average of several replications. Our goal is to find the solution $\mathbf{x}^*$ that minimizes $E(Y(\mathbf{x}))$.

On each iteration of the random-search algorithm, we compare a current good solution to a randomly chosen competitor. If the competitor is better, then it becomes the current good solution. When we terminate the search, the solution we choose is the one that has been visited most often (which means that we expect to revisit solutions many times).

Random-Search Algorithm

Step 1. Initialize counter variables $C(i) = 0$ for $i = 1, 2, \ldots, k$. Select an initial solution i^0, and set $C(i^0) = 1$. ($C(i)$ counts the number of times we visit solution i.)

Step 2. Choose another solution i' from the set of all solutions *except* i^0 in such a way that each solution has an equal chance of being selected.

Step 3. Run simulation experiments at the two solutions i^0 and i' to obtain outputs $Y(i^0)$ and $Y(i')$. If $Y(i') < Y(i^0)$, then set $i^0 = i'$. (See note following Step 4.)

Step 4. Set $C(i^0) = C(i^0) + 1$. If not done, then go to Step 2. If done, then select as the estimated optimal solution x_{i^*} such that $C(i^*)$ is the largest count.

Note that, if the problem is a maximization problem, then replace Step 3 with

Step 3. Run simulation experiments at the two solutions i^0 and i' to obtain outputs $Y(i^0)$ and $Y(i')$. If $Y(i') > Y(i^0)$, then set $i^0 = i'$.

One of the difficult problems with many optimization-via-simulation algorithms is knowing when to stop. (Exceptions include algorithms that guarantee a probability of correct selection.) Typical rules might be to stop after a certain number of iterations, stop when the best solution has not changed much in several iterations, or stop when all time available to solve the problem has been exhausted. Whatever rule is used, we recommend applying a statistical selection procedure, such as the Two-Stage Bonferroni Procedure in Section 12.2.2, to the 5 to 10 apparently best solutions. This is done to evaluate which among them is the true best with guaranteed confidence. If the raw data from the search have been saved, then these data can be used as the first-stage sample for a two-stage selection procedure (Boesel, Nelson, and Ishii [2003]).

Example 12.13: Implementing Random Search

Suppose that a manufacturing system consists of 4 stations in series. The zeroth station always has raw material available. When the zeroth station completes work on a part, it passes the part along to the first station, then the first passes the part to the second, and so on. Buffer space between stations 0 and 1, 1 and 2, and 2 and 3 is limited to 50 parts total. If, say, station 2 finishes a part but there is no buffer space available in front of station 3, then station 2 is blocked, meaning that it cannot do any further work. The question is how to allocate these 50 spaces to minimize the expected cycle time per part over one shift.

Let x_i be the number of buffer spaces in front of station i. Then the decision variables are x_1, x_2, x_3 with the constraint that $x_1 + x_2 + x_3 = 50$ (it makes no sense to allocate fewer buffer spaces than we have available). This implies a total of 1326 possible designs (can you figure out how this number is computed?).

To simplify the presentation of the random-search algorithm, let the counter for solution (x_1, x_2, x_3) be denoted as $C(x_1, x_2, x_3)$.

Random Search Algorithm

Step 1. Initialize 1326 counter variables $C(x_1, x_2, x_3) = 0$, one for each of the possible solutions (x_1, x_2, x_3). Select an initial solution, say $(x_1 = 20, x_2 = 15, x_3 = 15)$ and set $C(20, 15, 15) = 1$.

Step 2. Choose another solution from the set of all solutions *except* $(20, 15, 15)$ in such a way that each solution has an equal chance of being selected. Suppose $(11, 35, 4)$ is chosen.

Step 3. Run simulation experiments at the two solutions to obtain estimates of the expected cycle time $Y(20, 15, 15)$ and $Y(11, 35, 4)$. Suppose that $Y(20, 15, 15) < Y(11, 35, 4)$. Then $(20, 15, 15)$ remains as the current good solution.

Step 4. Set $C(20, 15, 15) = C(20, 15, 15) + 1$.

Step 2. Choose another solution from the set of all solutions *except* $(20, 15, 15)$ in such a way that each solution has an equal chance of being selected. Suppose $(28, 12, 10)$ is chosen.

Step 3. Run simulation experiments at the two solutions to obtain estimates of the expected cycle time $Y(20, 15, 15)$ and $Y(28, 12, 10)$. Suppose that $Y(28, 12, 10) < Y(20, 15, 15)$. Then $(28, 12, 10)$ becomes the current good solution.

Step 4. Set $C(28, 12, 10) = C(28, 12, 10) + 1$.

Step 2. Choose another solution from the set of all solutions *except* $(28, 12, 10)$ in such a way that each solution has an equal chance of being selected. Suppose $(0, 14, 36)$ is chosen.

Step 3. Continue...

When the search is terminated, we select the solution (x_1, x_2, x_3) that gives the largest $C(x_1, x_2, x_3)$ count. As we discussed earlier, the top 5 to 10 solutions should then be subjected to a separate statistical analysis to determine which among them is the true best (with high confidence). In this case, the solutions with the largest counts would receive the second analysis.

Despite the apparent simplicity of the Random-Search Algorithm, we have glossed over a subtle issue that often arises in algorithms with provable performance. In Step 2, the algorithm must randomly choose a solution such that all are equally likely to be selected (except the current one). How can this be accomplished in Example 12.13? The constraint that $x_1 + x_2 + x_3 = 50$ means that x_1, x_2 and x_3 cannot be sampled independently. One might be tempted to sample x_1 as a discrete uniform random variable on 0 to 50, then sample x_2 as a discrete uniform on 0 to $50 - x_1$, and finally set $x_3 = 50 - x_1 - x_2$. But this method does not make all solutions equally likely, as the following illustration shows: Suppose that x_1 is randomly sampled to be 50. Then the trial solution must be $(50, 0, 0)$; there is only one choice. But if $x_1 = 49$, then both $(49, 1, 0)$ and $(49, 0, 1)$ are possible. Thus, $x_1 = 49$ should be more likely than $x_1 = 50$ if all solutions with $x_1 + x_2 + x_3 = 50$ are to be equally likely.

12.5 SUMMARY

This chapter provided a basic introduction to the comparative evaluation of alternative system designs based on data collected from simulation runs. It was assumed that a fixed set of alternative system designs had been selected for consideration. Comparisons based on confidence intervals and the use of common random numbers were emphasized. A brief introduction to metamodels—whose purpose is to describe the relationship between design variables and the output response—and to optimization via simulation—whose purpose is to select the best from among a large and diverse collection of system designs—was also provided. There are many additional topics of potential interest (beyond the scope of this text) in the realm of statistical analysis techniques relevant to simulation. Some of these topics are

1. experimental design models, whose purpose is to discover which factors have a significant impact on the performance of system alternatives;
2. output-analysis methods other than the methods of replication and batch means;
3. variance-reduction techniques, which are methods to improve the statistical efficiency of simulation experiments (common random numbers being an important example).

The reader is referred to Banks [1998] and Law and Kelton [2000] for discussions of these topics and of others relevant to simulation.

The most important idea in Chapters 11 and 12 is that simulation output data require a statistical analysis in order to be interpreted correctly. In particular, a statistical analysis can provide a measure of the precision of the results produced by a simulation and can provide techniques for achieving a specified precision.

REFERENCES

ANDRADÓTTIR, S. [1998], "Simulation Optimization," Chapter 9 in *Handbook of Simulation*, J. Banks, ed., Wiley, New York.

BANKS, J., ed. [1998], *Handbook of Simulation*, Wiley, New York.

BECHHOFER, R. E., T. J. SANTNER, AND D. GOLDSMAN [1995], *Design and Analysis for Statistical Selection, Screening and Multiple Comparisons*, Wiley, New York.

BOESEL, J., B. L. NELSON, AND N. ISHII [2003], "A Framework for Simulation-Optimization Software," *IIE Transactions*, Vol. 35, pp. 221–229.

BOX, G. E. P., AND N. R. DRAPER [1987], *Empirical Model-Building and Response Surfaces*, Wiley, New York.

FU, M. C. [2002], "Optimization for Simulation: Theory vs. Practice," *INFORMS Journal on Computing*, Vol. 14, pp. 192–215.

GLOVER, F. [1989], "Tabu Search—Part I," *ORSA Journal on Computing*, Vol. 1, pp. 190–206.

GOLDSMAN, D., AND B. L. NELSON [1998], "Comparing Systems via Simulation," Chapter 8 in *Handbook of Simulation*, J. Banks, ed., Wiley, New York.

HINES, W. W., D. C. MONTGOMERY, D. M. GOLDSMAN, AND C. M. BORROR [2002], *Probability and Statistics in Engineering*, 4th ed., Wiley, New York.

HOCHBERG Y., AND A. C. TAMHANE [1987], *Multiple Comparison Procedures*, Wiley, New York.

HSU, J. C. [1996], *Multiple Comparisons: Theory and Methods*, Chapman & Hall, New York.

KLEIJNEN, J. P. C. [1975], *Statistical Techniques in Simulation, Parts I and II*, Dekker, New York.

KLEIJNEN, J. P. C. [1987], *Statistical Tools for Simulation Practitioners*, Dekker, New York.

KLEIJNEN, J. P. C. [1988], "Analyzing Simulation Experiments with Common Random Numbers," *Management Science*, Vol. 34, pp. 65–74.

KLEIJNEN, J. P. C. [1998], "Experimental Design for Sensitivity Analysis, Optimization, and Validation of Simulation Models," Chapter 6 in *Handbook of Simulation*, J. Banks, ed., Wiley, New York.

LAW, A. M., AND W. D. KELTON [2000], *Simulation Modeling and Analysis*, 3d ed., McGraw–Hill, New York.

MONTGOMERY, D. C. [2000], *Design and Analysis of Experiments*, 5th ed., Wiley, New York.

NELSON, B. L., AND F. J. MATEJCIK [1995], "Using Common Random Numbers for Indifference-Zone Selection and Multiple Comparisons in Simulation," *Management Science*, Vol. 41, pp. 1935–1945.

NELSON, B. L., J. SWANN, D. GOLDSMAN, AND W.-M. T. SONG [2001], "Simple Procedures for Selecting the Best System when the Number of Alternatives is Large," *Operations Research*, Vol. 49, pp. 950–963.

NELSON, B. L. [1987], "Some Properties of Simulation Interval Estimators Under Dependence Induction," *Operations Research Letters*, Vol. 6, pp. 169–176.

NELSON, B. L. [1992], "Designing Efficient Simulation Experiments," *1992 Winter Simulation Conference Proceedings*, pp. 126–132.

WRIGHT, R. D., AND T. E. RAMSAY, JR. [1979], "On the Effectiveness of Common Random Numbers," *Management Science*, Vol. 25, pp. 649–656.

EXERCISES

1. Reconsider the dump-truck problem of Example 3.5, which was also analyzed in Example 12.2. As business expands, the company buys new trucks, making the total number of trucks now equal to 16. The company desires to have a sufficient number of loaders and scales so that the average number of trucks waiting at the loader queue plus the average number at the weigh queue is no more than three. Investigate the following combinations of number of loaders and number of scales:

Number of	Number of Loaders		
Scales	2	3	4
1	–	–	–
2	–	–	–

The loaders being considered are the "slow" loaders in Example 12.2. Loading time, weighing time, and travel time for each truck are as previously defined in Example 12.2. Use common random numbers to the greatest extent possible when comparing alternative systems designs. The goal is to find the smallest number of loaders and scales to meet the company's objective of an average total queue length of no more than three trucks. In your solution, take into account the initialization conditions, run length, and number of replications needed to achieve a reasonable likelihood of valid conclusions.

2. In Exercise 11.6, consider the following alternative (M, L) policies:

			L	
			Low 30	High 40
M	Low	50	(50,30)	(50, 40)
	High	100	(100,30)	(100, 40)

Investigate the relative costs of these policies, using suitable modifications of the simulation model developed in Exercise 11.6. Compare the four system designs on the basis of long-run mean monthly cost. First make four replications of each (M, L) policy, using common random numbers to the greatest extent possible. Each replication should have a 12-month initialization phase followed by a 100-month data-collection phase. Compute confidence intervals having an overall confidence level of 90% for mean monthly cost for each policy. Then estimate the additional replications needed to achieve confidence intervals that do not overlap. Draw conclusions as to which is the best policy.

3. Reconsider Exercise 11.7. Compare the four inventory policies studied in Exercise 2, taking the cost of rush orders into account when computing monthly cost.

4. Reconsider Exercise 11.8. Compare the four monthly inventory policies studied in Exercise 2, taking into account the selling price of the perishable items.

5. In Exercise 11.9, investigate the effect of the order quantity on long-run mean daily cost. Each order arrives on a pallet on a delivery truck, so the permissible order quantities, Q, are multiples of 10 (i.e., Q may equal 10, or 20, or 30, ...). In Exercise 11.9, the policy $Q = 20$ was investigated.

   (a) First, investigate the two policies $Q = 10$ and $Q = 50$. Use the run lengths, and so on, suggested in Exercise 11.9. On the basis of these runs, decide whether the optimal Q, say Q^*, is between 10 and 50 or is greater than 50. (The cost curve as a function of Q should have what kind of shape?)

   (b) Using the results in part (a), suggest two additional values for Q and simulate the two policies. Draw conclusions. Include an analysis of the strength of your conclusions.

6. In Exercise 11.10, find the number of cards Q that the card-shop owner should purchase to maximize the profit within an error (for total profit) of approximately $5.00 at most. First, make runs for the policy of ordering $Q = 250$ and $Q = 350$ cards. With these results, plus the results of Exercise 11.10 for ordering $Q = 300$ cards, decide on a range of Q worth considering further (e.g., $200 \leq Q \leq 250$, or $250 \leq Q \leq 300$, etc.). Then investigate this restricted range for two additional policies (e.g., $Q = 200$ and $Q = 225$, or $Q = 265$ and $Q = 285$, etc.).

7. In Exercise 11.11, investigate the effect of target level M and review period N on mean monthly cost. Consider two target levels, M, determined by ± 10 from the target level used in Exercise 11.11, and consider review periods N of 1 month and 3 months. Which (N, M) pair is best, according to these simulations?

8. Reconsider Exercises 11.13 and 11.14, which involved the scheduling rules (or queue disciplines) first-in–first-out (FIFO) and priority-by-type (PR) in a job shop. In addition to these two rules, consider a shortest imminent operation (SIO) scheduling rule. For a given station, all jobs of the type with the smallest mean processing time are given highest priority. For example, when using an SIO rule at station 1, jobs are processed in the following order: type 2 first, then type 1, and type 3 last. Two jobs of the same type are processed on a FIFO basis. Develop a simulation experiment to compare the FIFO, PR, and SIO rules on the basis of mean total response time over all jobs.

9. In Exercise 11.13 (the job shop with FIFO rule), find the minimum number of workers needed at each station to avoid bottlenecks. A bottleneck occurs when average queue lengths at a station increase steadily over time. (Do not confuse increasing average queue length due to an inadequate number of servers with increasing average queue length due to initialization bias. In the former case, average queue length continues to increase indefinitely and server utilization is 1.0. In the latter case, average queue length eventually levels off and server utilization is less than 1.) Report on utilization of workers and total time it takes for a job to get through the job shop, by type and over all types. (*Hint*: If server utilization at a work station is 1.0, and if average queue length tends to increase linearly as simulation run length increases, it is a good possibility that the work station is unstable and therefore is a bottleneck. In this case, at least one additional worker is needed at the work station. Use queueing theory, namely $\lambda / c_i \mu < 1$, to suggest the minimum number of workers needed at station 1. Recall that λ is the arrival rate, $1/\mu$ is the overall mean service time for one job with one worker, and c_i is the number of workers at station i. Attempt to use the same basic condition, $\lambda / c_i \mu < 1$, to suggest an initial number of servers at station i for $i = 2, 3, 4$.)

10. (a) Repeat Exercise 9 for the PR scheduling rule (see Exercise 11.14).
    (b) Repeat Exercise 9 for the SIO scheduling rule (see Exercise 12.8).
    (c) Compare the minimum required number of workers for each scheduling rule: FIFO, versus PR, versus SIO.

11. With the minimum number of workers found in Exercises 9 and 10 for the job shop of Exercise 11.13, consider adding one worker to the entire shop. This worker can be trained to handle the processing at only one station. At which station should this worker be placed? How does this additional worker affect mean total response time over all jobs? Over type 1 jobs? Investigate the job shop with and without the additional worker for each scheduling rule: FIFO, PR, SIO.

12. In Exercise 11.17, suppose that a buffer of capacity one item is constructed in front of each worker. Design an experiment to investigate whether this change in system design has a significant impact upon individual worker utilizations (ρ_1, ρ_2, ρ_3, and ρ_4). At the very least, compute confidence intervals for $\rho_1^0 - \rho_1^1$ and $\rho_4^0 - \rho_4^1$, where ρ_i^s, is utilization for worker i when the buffer has capacity s.

13. A clerk in the admissions office at Small State University processes requests for admissions materials. The time to process requests depends on the program of interest (e.g., industrial engineering, management science, computer science, etc.) and on the level of the program (Bachelors, Masters, Ph.D.). Suppose that the processing time is modeled well as normally distributed, with mean 7 minutes and standard deviation 2 minutes. At the beginning of the day it takes the clerk some time to get set to begin working on requests; suppose that this time is modeled well as exponentially distributed, with mean 20 minutes. The admissions office typically receives between 40 and 60 requests per day.

    Let x be the number of applications received on a day, and let Y be the time required to process them (including the set-up time). Fit a metamodel for $E(Y|x)$ by making n replications at the design points $x = 40, 50, 60$. Notice that, in this case, we know that the correct model is

$$E(Y|x) = \beta_0 + \beta_1 x = 20 + 7x$$

    (Why?) Begin with $n = 2$ replications at each design point and estimate β_0 and β_1. Gradually increase the number of replications and observe how many are required for the estimates to be close to the true values.

14. Repeat the previous exercise using CRN. How do the results change?

15. The usual statistical analysis used to test for $\beta_1 \neq 0$ does not hold if we use CRN. Where does it break down?

16. Riches and Associates retains its cash reserves primarily in the form of certificates of deposit (CDs), which earn interest at an annual rate of 8%. Periodically, however, withdrawals must be made from these CDs in order to pay suppliers, etc. These cash outflows are made through a checking account that earns no interest. The need for cash cannot be predicted with certainty. Transfers from CDs to checking can be made instantaneously, but there is a "substantial penalty" for early withdrawal from CDs. Therefore, it might make sense for R&A to make use of the overdraft protection on their checking account, which charges interest at a rate of $0.00033 per dollar per day (i.e., 12% per year) for overdrafts.

    R&A likes simple policies in which it transfers a fixed amount, a fixed number of times, per year. Currently, it makes 6 transfers per year, of $18,250 each time. Your job is to find a policy that reduces its long-run cost per day.

    Judging from historical patterns, demands for cash arrive a rate of about 1 per day, with the arrivals being modeled well as a Poisson process. The amount of cash needed to satisfy each demand is reasonably represented by a lognormally distributed random variable with mean $300 and standard deviation $150.

    The penalty for early withdrawal is different for different CDs. It averages $150 for each withdrawal (regardless of size), but the actual penalty can be modeled as a uniformly distributed random variable with range $100 to $200.

Use cash level in checking to determine the length of the initialization phase. Make enough replications that your confidence interval for the difference in long-run cost per day does not contain zero. Be sure to use CRN in your experiment design.

17. If you have access to commercial optimization-via-simulation software, test how well it works as the variability of the simulation outputs increases. Use a simple model, such as $Y = x^2 + \varepsilon$, where ε is a random variable with a $N(0, \sigma^2)$ distribution, and for which the optimal solution is known ($x = 0$ for minimization, in this case). See how quickly, or whether, the software can find the true optimal solution as σ^2 increases. Next, try more complex models with more than one design variable.

18. For Example 12.13, show why there are 1326 solutions. Then derive a way to sample x_1, x_2, and x_3 such that $x_1 + x_2 + x_3 = 50$ and all outcomes are equally likely.

19. A critical electronic component with mean time to failure of x years can be purchased for $2x$ thousand dollars (thus, the more reliable the component, the more expensive it is). The value of x is restricted to being between 1 to 10 years, and the actual time to failure is modeled as exponentially distributed. The mission for which the component is to be used lasts one year; if the component fails in less than one year, then there is a cost of $20,000 for early failure. What value of x should be chosen to minimize the expected total cost (purchase plus early failure)?

To solve this problem, develop a simulation that generates a total cost for a component with mean time to failure of x years. This requires sampling an exponentially distributed random variable with mean x, and then computing the total cost as $2000x$ plus $20,000$ if the failure time is less than 1. Fit a quadratic metamodel in x and use it to find the value of x that minimizes the fitted model. [Hints: Select several values of x between 1 and 10 as design points. At each value of x, let the response variable $Y(x)$ be the average of at least 30 observations of total cost.]

20. Use optimization-via-simulation software to solve Exercise 19. If you do not have access to such software, use the random-search algorithm, and let the possible values of x be $\{1.00, 1.25, 1.50, \ldots, 10.00\}$.

21. Suppose that demand for a certain product has a Poisson distribution with mean 10 units. Use optimization-via-simulation software (or the random-search algorithm, if you do not have access to such software) to find the order size x that maximizes the probability that demand equals x units, assuming that $0 \le x \le 100$, and x is an integer.[6] [Hint: On each replication of the simulation for a trial value of x, a random demand is generated from the Poisson distribution with mean 10. If this random demand equals x, then the response is $Y(x) = 1$; otherwise it is $Y(x) = 0$.]

22. In Exercise 16, consider the following 5 policies: 6 transfers of $18,250; 5 transfers of $21,900; 4 transfers of $27,375; 3 transfers of $36,500; and 2 transfers of $54,750. Make

---

[6]This problem is based on Example 5.1 in Andradóttir [1999], "Accelerating the Convergence of Random Search Methods for Discrete Stochastic Optimization," *ACM TOMACS*, Vol. 9, pp. 349–380.

10 replications of each policy and apply the Subset Selection Procedure to choose a subset containing the policy with the least expected cost per day.

23. In Exercise 16, consider the following 5 policies: 6 transfers of $18,250; 5 transfers of $21,900; 4 transfers of $27,375; 3 transfers of $36,500; and 2 transfers of $54,750. Make 10 replications of each policy and then apply the Two-Stage Bonferroni Procedure to select the policy with the least expected cost per day. Use $\epsilon = \$2$ per day.

Part V

Applications

13

Simulation of Manufacturing and Material-Handling Systems

Manufacturing and material-handling systems provide one of the most important applications of simulation. Simulation has been used successfully as an aid in the design of new production facilities, warehouses, and distribution centers. It has also been used to evaluate suggested improvements to existing systems. Engineers and analysts using simulation have found it valuable for evaluating the impact of capital investments in equipment and physical facility and of proposed changes to material handling and layout. They have also found it useful to evaluate staffing and operating rules and proposed rules and algorithms to be incorporated into production control systems, warehouse-management control software, and material-handling controls. Managers have found simulation useful in providing a "test drive" before making capital investments, without disrupting the existing system with untried changes.

Section 13.1 provides an introduction and discusses some of the features of simulation models of manufacturing and material-handling systems. Section 13.2 discussed the goals of manufacturing simulation and the most common measures of system performance. Section 13.3 discusses a number of the issues common to many manufacturing and material-handling simulations, including the treatment of downtimes and failure, and trace-driven simulations using actual historical data or historical order files. Section 13.4 provides brief abstracts of a number of reported simulation projects, with references for additional reading. Section 13.5 gives an extended example of a simulation of a small production line, emphasizing the experimentation and analysis of system performance to achieve a desired throughput. For an overview of simulation software for manufacturing and material-handling applications, see Section 4.7.

13.1 MANUFACTURING AND MATERIAL-HANDLING SIMULATIONS

As do all modeling projects, manufacturing and material-handling simulation projects need to address the issues of scope and level of detail. Consider scope as analagous to breadth and level of detail as analagous to depth. Scope describes the boundaries of the project: what's in the model, and what's not. For a subsystem, process, machine, or other component, the project scope determines whether the object is in the model. Then, once a component or subsystem is treated as part of a model, often it can be simulated at many different levels of detail.

The proper scope and level of detail should be determined by the objectives of the study and the questions being asked. On the other hand, level of detail could be constrained by the availability of input data and the knowledge of how system components work. For new, nonexistent systems, data availability might be limited, and system knowledge might be based on assumptions.

Some general guidelines can be provided, but the judgment of experienced simulation analysts working with the customer to define, early in the project, the questions the model is being designed to address provides the most effective basis for selecting a proper scope and a proper level of detail.

Should the model simulate each conveyor section or vehicle movement, or can some be replaced by a simple time delay? Should the model simulate auxiliary parts, or the handling of purchased parts, or can the model assume that such parts are always available at the right location when needed for assembly?

At what level of detail does the control system need to be simulated? Many modern manufacturing facilities, distribution centers, baggage-handling systems, and other material-handling systems are computer controlled by a management-control software system. The algorithms built into such control software play a key role in system performance. Simulation is often used to evaluate and compare the effectiveness of competing control schemes and to evaluate suggested improvements. It can be used to debug and fine-tune the logic of a control system before it is installed.

These questions are representative of the issues that need to be addressed in choosing the correct level of model detail and scope of a project. In turn, the scope and level of model detail limit the type of questions that can be addressed by the model. In addition, models can be developed in an iterative fashion, adding detail for peripheral operations at later stages if such operations are later judged to affect the main operation significantly. It is good advice to start as simple as possible and add detail only as needed.

13.1.1 Models of Manufacturing Systems

Models of manufacturing systems might have to take into account a number of characteristics of such systems, some of which are the following:

Physical layout
Labor
     Shift schedules
     Job duties and certification
Equipment
     Rates and capacities

Breakdowns
> Time to failure
> Time to repair
> Resources needed for repair

Maintenance
> PM schedule
> Time and resources required
> Tooling and fixtures

Workcenters
> Processing
> Assembly
> Disassembly

Product
> Product flow, routing, and resources needed
> Bill of materials

Production schedules
> Made-to-stock
> Made-to-order
>> Customer orders
>> Line items and quantities

Production control
> Assignment of jobs to work areas
> Task selection at workcenters
> Routing decisions

Supplies
> Ordering
> Receipt and storage
> Delivery to workcenters

Storage
> Supplies
> Spare parts
> Work-in-process (WIP)
> Finished goods

Packing and shipping
> Order consolidation
> Paperwork
> Loading of trailers

13.1.2 Models of Material Handling Systems

In manufacturing systems, it is not unusual for 80 to 85% of an item's total time in system to be expended on material handling or on waiting for material handling to occur. This work-in-process (WIP) represents a vast investment, and reductions in WIP and associated delays can result in large cost savings. Therefore, for some studies, detailed material-handling simulations are cost effective.

In some production lines, the material-handling system is an essential component. For example, automotive paint shops typically consist of a power-and-free conveyor system that transports automobile bodies or body parts through the paint booths.

In warehouses, distribution centers, and flow-through and cross-docking operations, material handling is clearly a key component of any material-flow model. Manual warehouses typically use manual fork trucks to move pallets from receiving dock to storage and from storage to shipping dock. More automated distribution centers might use extensive conveyor systems to support putaway, order picking, order sortation, and consolidation.

Models of material-handling systems often have to contain some of the following types of subsystems:

Conveyors
> Accumulating
> Nonaccumulating
> Indexing and other special purpose
> Fixed window or random spacing
> Power and free

Transporters
> Unconstrained vehicles (e.g., manually guided fork trucks)
> Guided vehicles (automated or operator controlled, wire guided
>> chemical paths, rail guided)
> Bridge cranes and other overhead lifts

Storage systems
> Pallet storage
> Case storage
> Small-part storage (totes)
> Oversize items
> Rack storage or block stacked
> Automated storage and retrieval systems (AS/RS) with storage–retrieval
>> machines (SRM)

13.1.3 Some Common Material-Handling Equipment

There are numerous types of material-handling devices common to manufacturing, warehousing, and distribution operations. They include unconstrained transporters, such as carts, manually driven fork-lift trucks, and pallet jacks; guided path transporters, such as AGVs (automated guided vehicles); and fixed-path devices, such as various types of conveyor.

The class of unconstrained transporters, sometimes called free-path transporters, includes carts, fork-lift trucks, pallet jacks, and other manually driven vehicles that are free to travel throughout a facility unconstrained by a guide path of any kind. Unconstrained transporters are not constrained to a network of paths and may choose an alternate path or move around an obstruction. In contrast, the guided-path transporters move along a fixed path, such as chemical trails on the floor, wires imbedded in the floor, or infrared lights placed strategically, or by self-guidance, using radio communications, laser guidance and dead reckoning, and rail. Guided-path transporters sometimes contend with each other for space along their paths and usually have limited options upon meeting obstacles and congestion. Examples of guided-path transporters include the automated guided vehicle (AGV); a rail-guided turret truck for storage and retrievals of pallets in rack storage; and a crane in an AS/RS (automated storage and retrieval system).

The conveyor is a fixed-path device for moving entities from point to point, following a fixed path with specific load, stopping or processing points, and unload points. A conveyor system can consist of numerous connected sections with merges and diverts. Each section can be of one of a number of different types. Examples of conveyor types include belt, powered and gravity roller, bucket, chain, tilt tray, and power-and-free, each with its own characteristics that must be modeled accurately.

Most conveyor sections can be classified as either accumulating or nonaccumulating. An accumulating conveyor section runs continuously. If the forward progress of an item is halted while on the accumulating conveyor, slippage occurs, allowing the item to remain stationary and items behind it to continue moving until they reach the stationary item. Some belt and most roller conveyors operate in this manner. Only items that will not be damaged by bumping into each other can be placed on an accumulating conveyor.

In contrast, after an item is on a nonaccumulating conveyor section, its spacing relative to other items does not change. If one item stops moving, the entire section stops moving, and hence all items on the section stop. For example, nonaccumulating conveyor is used for moving televisions not yet in cartons, for they must be kept at a safe distance from each other while moving from one assembly or testing station to the next. Bucket conveyors, tilt-tray conveyors, some belt conveyors, and conveyors designed to carry heavy loads (usually, pallets) are nonaccumulating conveyors.

Conveyors can also be classified as fixed-window or random spacing. In fixed-window spacing, items on the conveyor must always be within zones of equal length, which can be pictured as lines drawn on a belt conveyor or trays pulled by a chain. For example, in a tilt-tray conveyor, continuously moving trays of fixed size are used to move items. The control system is designed to induct items in such a way that each item is in a separate tray; thus it is a nonaccumulating fixed-window conveyor. In contrast, with random spacing, items can be anywhere on the conveyor section relative to other items. To be inducted, they simply require sufficient space.

Besides these basic types, there are innumerable types of specialized conveyors for special purposes. For example, a specialized indexing conveyor may move forward in increments, always maintaining a fixed distance between the trailing edge of the load ahead and the leading edge of the load behind. Its purpose is to form a "slug" of items, equally spaced apart, to be inducted all together onto a transport conveyor. For the local behavior of some systems—that is, the performance at a particular workstation or induction point—a detailed understanding and accurate model of the physical workings and the control logic are essential for accurate results.

13.2 GOALS AND PERFORMANCE MEASURES

The purpose of simulation is insight, not numbers. Those who purchase and use simulation software and services want to gain insight and understanding into how a new or modified system will work. Will it meet throughput expectations? What happens to response time at peak periods?

Is the system resilient to short-term surges? What is the recovery time when short-term surges cause congestion and queueing? What are the staffing requirements? What problems occur? If problems occur, what is their cause and how do they arise? What is the system capacity? What conditions and loads cause a system to reach its capacity?

Simulations are expected to provide numeric measures of performance, such as throughput under a given set of conditions, but the major benefit of simulation comes from the insight and understanding gained regarding system operations. Visualization through animation and graphics provides major assistance in the communication of model assumptions, system operations, and model results. Often, visualization is the major contributor to a model's credibility, which in turn leads to acceptance of the model's numeric outputs. Of course, a proper experimental design that includes the right range of experimental conditions plus a rigorous analysis and, for stochastic simulation models, a proper statistical analysis is of utmost importance for the simulation analyst to draw correct conclusions from simulation outputs.

The major goals of manufacturing-simulation models are to identify problem areas and quantify system performance. Common measures of system performance include the following:

- throughput under average and peak loads;
- system cycle time (how long it takes to produce one part);
- utilization of resources, labor, and machines;
- bottlenecks and choke points;
- queueing at work locations;
- queueing and delays caused by material-handling devices and systems;
- WIP storage needs;
- staffing requirements;
- effectiveness of scheduling systems;
- effectiveness of control systems.

Often, material handling is an important part of a manufacturing system and its performance. Non-manufacturing material-handling systems include warehouses, distribution centers, cross-docking operations, baggage-handling systems at airports and container terminals. The major goals of these nonmanufacturing material-handling systems are similar to those identified for manufacturing systems. Some additional considerations are the following:

- how long it takes to process one day of customer orders;
- effect of changes in order profiles (for distribution centers);
- truck/trailer queueing and delays at receiving and shipping docks;
- effectiveness of material-handling systems at peak loads;
- recovery time from short-term surges (for example, with baggage-handling).

13.3 ISSUES IN MANUFACTURING AND MATERIAL-HANDLING SIMULATIONS

There are a number of modeling issues especially important for the achievement of accurate and valid simulation models of manufacturing and material-handling systems. Two of these issues are the proper modeling of downtimes and whether, for some inputs, to use actual system data or a statistical model of those inputs.

13.3.1 Modeling Downtimes and Failures

Unscheduled random downtimes can have a major effect on the performance of manufacturing systems. Many authors have discussed the proper modeling of downtime data (Williams [1994]; Clark [1994]; Law and Kelton [2000]). This section discusses the problems that can arise when downtime is modeled incorrectly and suggests a number of ways to model machine and system downtimes correctly.

Scheduled downtime, such as for preventive maintenance, or periodic downtime, such as for tool replacement, also can have a major effect on system performance. But these downtimes are usually (or should be) predictable and can be scheduled to minimize disruptions. In addition, engineering efforts or new technology might be able to reduce their duration.

There are a number of alternatives for modeling random unscheduled downtime, some better than others:

1. Ignore it.
2. Do not model it explicitly, but increase processing times in appropriate proportion.
3. Use constant values for time to failure and time to repair.
4. Use statistical distributions for time to failure and time to repair.

Of course, alternative (1) generally is not the suggested approach. This is certainly an irresponsible modeling technique if downtimes have an impact on the results, as they do in almost all situations. One situation in which ignoring downtimes could be appropriate, with the full knowledge of the customer, is to leave out those catastrophic downtimes that occur rarely and leave a production line or plant down for a long period of time. In other words, the model would incorporate normal downtimes but ignore those catastrophic downtimes, such as general power failures, snow storms, cyclones, and hurricanes, that occur rarely but stop all production when they do occur. The documented scope of the project should clearly state the assumed operating conditions and those conditions that are not included in the model. If it is generally known that a plant will be closed for some number of snow days per year, then the simulation need not take these downtimes into account, for the effect of any given number of days can easily be factored into the simulation results when making annual projections.

The second possibility, to factor into the model the effect of downtimes by adjusting processing times applied to each job or part, might be an acceptable approximation under limited circumstances. If each job or part is subject to a large number of small delays associated with downtime of equipment or tools, then the total of such delays may be added to the pure processing time to arrive at an adjusted processing time. If total delay time and pure processing time are random in nature, then an appropriate statistical distribution should be used for the total adjusted processing time. If the pure processing time is constant while the total delay time in one cycle is random and variable, it is almost never accurate to adjust the processing time by a constant factor. For example, if processing time is usually 10 minutes but the equipment is subject to downtimes that cause about a 10% loss in capacity, it is not appropriate to merely change the processing time to a constant 11 minutes. Such a deterministic adjustment might provide reasonably accurate estimates of overall system throughput, but will not provide accurate estimates of such local behavior as queue and buffer space needed at peak times. Queueing and short-term congestion are strongly influenced by randomness and variability.

The third possibility, using constant durations for time to failure and time to repair, might be appropriate when, for example, the downtime is actually due to preventive maintenance that is on a fixed schedule. In almost all other circumstances, the fourth possibility, modeling time to failure and time to repair by appropriate statistical distributions, is the appropriate technique. This requires either actual data for choosing a statistical distribution based on the techniques in Chapter 11, or, when data is lacking, a reasonable assumption based on the physical nature of the causes of downtimes.

The nature of time to failure is also important. Are times to failure completely random in nature, a situation due typically to a large number of possible causes of failure? In this case, exponential distribution might provide a good statistical model. Or are times to failure, rather, more regular—typically, due to some major component—say, a tool—wearing out? In this case, a uniform or (truncated) normal distribution could be more nearly appropriate. In the latter case, the mean of the distribution represents the average time to failure, and the distribution places a plus or minus around the mean.

Time to failure can be measured in a number of different ways:

1. by wall-clock time;
2. by machine or equipment busy time;
3. by number of cycle times;
4. by number of items produced.

Breakdowns or failures can be based on clock time, actual usage, or cycles. Note that the word breakdown or failure is used, even though preventive maintenance could be the reason for a downtime. As mentioned, breakdowns or failures can be probabilistic or deterministic in duration.

Actual usage breakdowns are based on the time during which the resource is used. For example, wear on a machine tool occurs only when the machine is in use. Time to failure is measured against machine-busy time and not against wall-clock time. If the time to failure is 90 hours, then the model keeps track of total busy time since the last downtime ended, and, when 90 hours is reached, processing is interrupted and a downtime occurs.

Clock-time breakdowns might be associated with scheduled maintenance—for example, changes of fluids every three months when a complete lubrication is required. Downtimes based on wall-clock time may also be used for equipment that is always busy or equipment that "runs" even when it is not processing parts.

Cycle breakdowns or failures are based on the number of times the resource is used. For example, after every 50 uses of a tool, it needs to be sharpened. Downtimes based on number of cycle times or number of items produced are implemented by generating the number of times or items and, in the model, simply counting until this number is reached. Typical uses of downtimes based on busy time or cycle times may be for maintenance or tool replacement.

Another issue is what happens to a part at a machine when the breakdown or failure occurs. Possibilities include scrapping the part, rework, or simply continuing processing after repair. In some cases—for example, when preventive maintenance is due—the part in the machine may complete processing before the repair (or maintenance activity) begins.

Time to repair can also be modeled in two fundamentally different ways:

1. as a pure time delay (no resources required);
2. as a wait time for a resource (e.g., maintenance person) plus a time delay for actual repair.

Of course, there are many variations on these methods in actual modeling situations. When a repair or maintenance person is a limited resource, the second approach will be a more accurate model and provide more information.

The next example illustrates the importance of using the proper approach for modeling downtimes and of the consequences and inaccurate results that sometimes result from inaccurate assumptions.

Example 13.1: Effect of Downtime on Queuing _____

Consider a single machine that processes a wide variety of parts that arrive in random mixes at random times. Data analysis has shown that an exponentially distributed processing time with a mean of 7.5 minutes provides a fairly accurate representation. Parts arrive at random, time between arrivals being exponentially distributed with mean 10 minutes. The machine fails at random times. Downtime studies have shown that time-to-failure can be reasonably approximated by an exponential distribution with mean time 1000 minutes. The time to repair the resource is also exponentially distributed, with mean time 50 minutes. When a failure occurs, the current part in the machine is removed from the machine; when the repair has been completed, the part resumes its processing.

When a part arrives, it queues and waits its turn at the machine. It is desired to estimate the size of this queue. An experiment was designed to estimate the average number of parts in the queue. To illustrate the effect of an accurate treatment of downtimes, the model was run under a number of different assumptions. For each case and replication, the simulation run length was 100,000 minutes.

Table 13.1 shows the average number of parts in the queue for six different treatments of the time between breakdowns. For each treatment that involves randomness, five replications of those treatments and the average for those five replications are shown.

Case A ignores the breakdowns. The average number in the queue is 2.31 parts. Across the 5 independent replications, the averages range from 2.05 to 2.70 parts. This treatment of breakdowns is not recommended.

Case B increases the average service time from 7.5 minutes to 8.0 minutes in an attempt to approximate the effect of downtimes. On average, each downtime and repair cycle is 1050 minutes, with the machine down for 50 minutes. Thus the machine is down, on the average in the long run, $50/1050 = 4.8\%$ of total time. Thus, some have argued that downtime has approximately the same effect as increasing the processing time of each part by 4.8%, which is about 7.86 minutes. Therefore, an assumed constant 8 minutes per part should be (it is argued) a conservative approach. For this treatment of downtimes, the average number of parts in the queue, over the five replications, is about 3.26 parts. Across the 5 replications, the range is from 2.81 to 4.03 parts. (Note that the variability as shown in the range of values is very small compared to the

Table 13.1   Average Number of Parts in Queue for Machines with Breakdowns

Case	1st Rep	2nd Rep	3rd Rep	4th Rep	5th Rep	Avg Rep
A. Ignore the breakdowns	2.36	2.05	2.38	2.05	2.70	2.31
B. Increase service time to 8.0	3.32	2.82	3.32	2.81	4.03	3.26
C. All random	4.05	3.77	4.36	3.95	4.43	4.11
D. Random processing, deterministic breakdowns	3.24	2.85	3.28	3.05	3.79	3.24
E. All deterministic						0.52
F. Deterministic processing, random breakdowns	1.06	1.04	1.10	1.32	1.16	1.13

other cases.) The treatment in Case B might be appropriate under some limited circumstances, but, as was discussed in a previous section, it is not appropriate under the assumptions of this example.

The proper treatment, shown as Case C, treats the randomness in processing and breakdowns properly, with the assumed correct exponential distributions. The average value is about 4.11 parts waiting for the machine. Across the 5 replications, the average queue length ranges from 3.77 to 4.43 parts. The average number waiting differs from that of Case B by almost one part.

Case D is a simplification that treats the processing randomly, but treats the breakdowns as deterministic. The results average about 3.24 parts in the queue. The range of averages is from 2.85 to 3.79 parts, quite a reduction in variability from Case C.

Case E treats all of the times as deterministic. Only one replication is needed, because additional replications (using the same seed) will reproduce the result. The average value in the queue is 0.52 parts, well below the value in Case C, or any other case for that matter. The conclusion: Ignoring randomness is dangerous and leads to totally unrealistic results.

Case F treats arrivals and processing as deterministic, but breakdowns are random. The average number of parts in the queue at the machine is about 1.13. The range is from 1.04 to 1.32 parts. For some machines and processing in manufacturing environments, Case F is the realistic situation: Processing times are constant, and arrivals are regulated—that is, are also constant. The reader is left to consider the inaccuracies that would result from making faulty assumptions regarding the nature of time to failure and time to repair.

In conclusion, there can be significant differences between the estimated average numbers in a queue, based on the treatment of randomness. The results using the correct treatment of randomness can be far different from those using alternatives. Often, one is tempted by the unavailability of detailed data and the availability of averages to want to use average time to failure as if it were a constant. Example 13.1 illustrates the dangers of inappropriate assumptions. Both the appropriate technique to use and the appropriate statistical distribution depend on the available data and on the situation at hand.

As discussed by Williams [1994], the accurate treatment of downtimes is essential for achieving valid models of manufacturing systems. Some of the essential ingredients are the following:

- avoidance of oversimplified and inaccurate assumptions;
- careful collection of downtime data;

- accurate representation of time to failure and time to repair by statistical distributions;
- accurate modeling of system logic when a downtime occurs, in terms both of the repair-time logic and of what happens to the part currently processing.

13.3.2 Trace-driven Models

Consider a model of a distribution center that receives customer orders that must be processed and shipped in one day. One modeling question is how to represent the day's set of orders. A typical order will contain one or more line items, and each line item can have a quantity of one or more pieces. For example, when you buy a new stereo, you might purchase an amplifier, a tuner, and a CD player (all separate line items, each having a quantity of one piece), and 4 identical speakers (another line item with a quantity of 4 pieces). The overall order profile can have a major impact on the performance of a particular system design. A system designed to handle large orders going to a small number of customers might not perform well if order profiles shift toward a larger number of customers (or larger number of separate shipments) with one or two items per order.

One approach is to characterize the order profile by using a discrete statistical distribution for each variable in an order:

1. the number of line items
2. for each line item, the number of pieces.

If these two variables are statistically independent, then this approach might provide a valid model of the order profile. For many applications, however, these two variables may be highly correlated in ways that could be difficult to characterize statistically. For example, an apparel and shoe company has six large customers (the large department stores and discount chains), representing 50% of sales volume, which typically order dozens or hundreds of line items and large quantities of many of the items. At the opposite pole, on any given day approximately 50% of the orders are for one or two pairs of shoes (just-in-time with a vengeance!). For this company, the number of line items in an order is highly positively correlated with the quantity ordered; that is, large orders with a large number of line items also usually have large quantities of many of the line items. And small orders with only a few line items typically order small quantities of each item.

What would happen if the two variables, number of line items and quantity per line item, were modeled by independent statistical distributions? When an order began processing, the model would make two random uncorrelated draws, which could result in order profiles quite different from those found in practice. Such an erroneous assumption could result, for example, in far too large a proportion of orders having one or two line items with large unrealistic quantities.

Another common but more serious error is to assume that there is an average order and to simulate only the number of orders in a day with each being the typical order. In the author's experience, analyses of many order profiles has shown (1) that there is no such thing as a typical order and (2) that there is no such thing as a typical order profile.

An alternative approach, and one that has proven successful in many studies, is for the company to provide the actual orders for a sample of days over the previous year. Usually, it is desirable to simulate peak days. A model driven by actual historical data is called a trace-driven model.

A trace-driven model eliminates all possibility of error due to ignoring or misestimating correlations in the data. One apparent limitation could be a customer's desire, at times, to be able to simulate hypothesized changes to the order profile, such as a higher proportion of smaller orders in terms of both line items and quantities. In practice, this limitation can be removed by adding "dials" to the order-profile portion of the model, so that a simulation analyst can "dial up" more or less of certain characteristics, as desired. One approach is to treat the day's orders as a statistical population from which the model draws samples in a random fashion. This approach makes it easy to change overall order volume without modifying the profile. A second related approach would be to subdivide a day's orders into subgroups based on number of line items, quantities or other numeric parameters, and then sample in a specified proportion from each subgroup. By changing the proportion of each subgroup, different order profiles can be "dialed up" and fed into the model. A third approach is to use factors to adjust the number of daily orders, the number of line items, and/or the quantities. In practice one of these approaches might be as accurate as can be expected for hypothesized future order profiles and might provide a cost effective and reasonably accurate model, especially for testing the robustness of a system design for assumed changes in order characteristics.

Other examples of trace-driven models include the following:

- orders to a custom job shop, using actual historical orders;
- product mix and quantities, and production sequencing, for an assembly line making 100 styles and sizes of hot-water heaters;
- time to failure and downtime, using actual maintenance records;
- Truck arrival times to a warehouse, using gate records.

Whether to make an input variable trace-driven or to characterize it as a statistical distribution depends on a number of issues, including the nature of the variable itself, whether it is correlated with or independent of other variables, the availability of accurate data, and the questions being addressed.

13.4 CASE STUDIES OF THE SIMULATION OF MANUFACTURING AND MATERIAL HANDLING SYSTEMS

The *Winter Simulation Conference Proceedings*, IIE Magazine, Modern Material Handling and other periodicals are excellent sources of information for short cases in the simulation of manufacturing and material-handling systems.

An abstract of some of the papers from past *Winter Simulation Conference Proceedings* will provide some insight into the types of problems that can be addressed by simulation. These abstracts have been paraphrased and shortened where appropriate; our goal is to provide an indication of the breadth of real-world applications of simulation.

Session: Semiconductor Wafer Manufacturing
Paper: Modeling and Simulation of Material Handling for Semiconductor Wafer Manufacturing

Authors: Neal G. Pierce and Richard Stafford

Abstract: This paper presents the results of a design study to analyze the interbay material-handling systems for semiconductor wafer manufacturing. The authors developed discrete-event simulation models of the performance of conventional cleanroom material handling including manual and automated systems. The components of a conventional cleanroom material-handling system include an overhead monorail system for interbay (bay-to-bay) transport, work-in-process stockers for lot storage, and manual systems for intrabay movement. The authors constructed models and experiments that assisted with analyzing cleanroom material-handling issues such as designing conventional automated material-handling systems and specifying requirements for transport vehicles.

Session: Simulation in Aerospace Manufacturing
Paper: Modeling Aircraft Assembly Operations
Authors: Harold A. Scott

Abstract: A simulation model is used to aid in the understanding of complex interactions of aircraft assembly operations. Simulation helps to identify the effects of resource constraints on dynamic process capacity and cycle time. To analyze these effects, the model must capture job and crew interactions at the control code level. This paper explores five aspects of developing simulation models to analyze crew operations on aircraft assembly lines:

Representing job precedence relationships
Simulating crew members with different skill and job
    proficiency levels
Reallocating crew members to assist ongoing jobs
Depicting shifts and overtime
Modeling spatial constraints and crew movements in the
    production area.

Session: Control of Manufacturing Systems
Paper: Discrete Event Simulation for Shop Floor Control
Authors: J. S. Smith, R. A. Wysk, D. T. Sturrock, S. E. Ramaswamy,
           G. D. Smith, S. B. Joshi

Abstract: This paper describes an application of simulation to shop floor control of a flexible manufacturing system. The simulation is used not only as an analysis and evaluation tool, but also as a "task generator" for the specification of shop floor control tasks. Using this approach, the effort applied to the development of control logic in the simulation is not duplicated in the development of the control system. Instead the same control logic is used for the control system as was used for the simulation. Additionally, since the simulation implements the control, it provides very high fidelity performance predictions. The paper describes implementation experience in two flexible manufacturing laboratories.

Session:	Flexible Manufacturing
Paper:	Developing and Analyzing Flexible Cell Systems Using Simulation
Authors:	Edward F. Watson and Randall P. Sadowski
Abstract:	This paper develops and evaluates flexible cell alternatives to support an agile production environment at a mid-sized manufacturer of industrial equipment. Three work cell alternatives were developed based on traditional flow analysis studies, past experience, and common sense. The simulation model allowed the analyst to evaluate each cell alternative under current conditions as well as anticipated future conditions that included changes to product demand, product mix, and process technology.

Session:	Modeling of Production Systems
Paper:	Inventory Cost Model for Just-in-Time Production
Authors:	Mahesh Mathur
Abstract:	This paper presents a simulation model used to compare setup and inventory carrying costs with varying lot sizes. While reduction of lot sizes is a necessary step towards implementation of Just-in-Time (JIT) in a job shop environment, a careful cost study is required to determine the optimum lot size under the present set-up conditions. The simulation model graphically displays the fluctuation of carrying costs and accumulation of set-up costs on a time scale in a dynamic manner. The decision of the optimum lot size can then be based on realistic cost figures.

Session:	Analysis of Manufacturing Systems
Paper:	Modeling Strain of Manual Work in Manufacturing Systems
Authors:	I. Ehrhardt, H. Herper, and H. Gebhardt
Abstract:	This paper describes a simulation model that considers manual operations for increasing the effectiveness of planning logistic systems. Even though there is ever increasing automation, there are vital tasks in production and logistics that are still assigned to humans. Present simulation modeling efforts rarely concentrate on the manual activities assigned to humans.

Session:	Manufacturing Case Studies
Paper:	Simulation Modeling for Quality and Productivity in Steel Cord Manufacturing

Authors: C. H. Turkseven and G. Ertek
Abstract: The paper describes the application of simulation modeling to esti-
            mate and improve quality and productivity performance of a steel
            cord manufacturing system. It focuses on wire fractures, which can
            be an important source of system disruption.

Session: Manufacturing Analysis and Control
Paper: Shared Resource Capacity Analysis in Biotech Manufacturing
Author: P. V. Saraph
Abstract: This paper discusses an application of simulation in analyzing the
            capacity needs of a shared resource, the Blast Freezer, at one of the
            Bayer Corporation's manufacturing facilities. The simulation model
            was used to analyze the workload patterns, run different workload
            scenarios, taking into consideration uncertainty and variability, and
            provide recommendations on a capacity increase plan. This anal-
            ysis also demonstrated the benefits of certain operational scheduling
            policies. The analysis outcome was used to determine capital invest-
            ments for 2002.

Session: Manufacturing Analysis and Control
Paper: Behavior of an Order Release Mechanism in a Make-to-Order
            Manufacturing System with Selected Order Acceptance
Authors: A. Nandi and P. Rogers
Abstract: The authors used a simulation model to evaluate a controver-
            sial policy, namely, holding orders in a pre-shop pool prior to
            their release to the factory floor. In a make-to-order manufacturing
            system, if capacity is fixed and exogenous due dates are inflexible,
            having orders wait in a pre-shop pool may cause the overall due
            date performance of the system to deteriorate. The model was used
            to evaluate an alternative approach, the selective rejection of orders
            for dealing with surges in demand while maintaining acceptable due
            date performance.

13.5 MANUFACTURING EXAMPLE: AN ASSEMBLY-LINE SIMULATION

This section describes a model of a production line for the final assembly of "gizmos". It then
focuses on how simulation can be used to analyze system performance.

13.5.1 System Description and Model Assumptions

At a manufacturing facility, an engineering team has designed a new production line for the
final assembly of gizmos. Before making the investment to install the new system, some team
members propose using simulation to analyze the system's performance, specifically to predict
system throughput (gizmos per 8-hour shift, on the average). In addition, the engineers desire

to evaluate potential improvements to the designed system. One such potential improvement is adding buffer space for holding work-in-process (WIP) between adjacent workstations.

The team decides to develop a simulation model and conduct an analysis. The team's primary objective is to predict throughput (completed gizmos per shift on the average) for the given system design and to evaluate whether it meets the desired throughput. In addition, should throughput be less than expected, the team wants to use the model to help in identifying bottlenecks, gaining insight into the system's dynamic behavior and evaluating potential design improvements.

The proposed production line has six workstations and a special rack for WIP storage between adjacent stations. There are four manual stations, each having its own operator, and two automated stations, which share a single operator. The six stations perform production tasks in the following sequence:

> Station 1: initial manual station begins final assembly of a new gizmo
> Station 2: manual assembly station
> Station 3: manual assembly station
> Station 4: automatic assembly station
> Station 5: automatic testing station
> Station 6: manual packing station

At each manual station, an operator loads a gizmo onto a workbench, performs some tasks, and on completion unloads the gizmo and places it into the WIP storage for the next workstation. The operator takes 10 seconds and 5 seconds for the loading and unloading tasks, respectively.

The WIP storage racks between each pair of adjacent stations have limited capacity. If a station completes its tasks on a gizmo but the downstream rack is full, the gizmo must remain in the station, blocking any further work. In the initial design, the WIP storage racks have the capacities shown in Table 13.2. (By assumption, the WIP storage preceding Station 1 is always kept full at 4 units; since it is assumed to always be full, its specific capacity plays no role.) The system design with capacities given in Table 13.2 is called the Baseline configuration.

From time to time, a tool will fail, causing unscheduled downtime or unexpected extra work at a manual or automated station. In addition, all operators are scheduled to take a 30-minute lunch break at the same time. Work is interrupted and resumes where it left off after lunch. This interrupt/resume rule applies to operator tasks including assembly work, parts resupply, and repairs during a downtime.

At the automatic stations, a machine performs an assembly or testing task. The automatic stations might have unscheduled (random) downtimes, but they continue to operate during the operator's lunch break. One operator services both machines to load and unload gizmos (10 seconds and 5 seconds, respectively). After being loaded, a machine processes the gizmo without

Table 13.2   Capacity of WIP Storage Buffers for Baseline
Configuration

Rack Before Station	1	2	3	4	5	6
Buffer Capacity	4	2	2	2	1	2

Table 13.3   Assembly and Parts Resupply Times

Station	Assembly per Gizmo (Seconds)	Part Number	Parts Resupply Time (seconds per Batch)	No. of Parts per Batch
1	40	A	10	15
		B	15	10
2	38	C	20	8
		D	15	14
3	38	E	30	25
4	35			
5	35			
6	40	F[a]	30	32

[a]: At station 6, the part number (F) represents the shipping containers.

Table 13.4   Assumptions and Data for Unscheduled Downtimes

Station	TTF	MTTF (Minutes)	TTR	MTTR (Minutes)	+/−	Expected Availability
1	Exponential	36.0	Uniform	4.0	1.0	90%
2	Exponential	4.5	Uniform	0.5	0.1	90%
3	Exponential	27.0	Uniform	3.0	1.0	90%
4	Exponential	9.0	Uniform	1.0	0.5	90%
5	Exponential	18.0	Uniform	2.0	1.0	90%

further operator intervention unless a downtime occurs. At all stations, the operator performs repairs as needed whenever the station experiences a downtime.

Table 13.3 gives the total assembly time and parts resupply times for each station, plus the number of parts in a batch. The assembly time for the manual stations is assumed to vary by plus/minus 2 seconds (uniformly distributed) from the times given in Table 13.3. Parts resupply time does not occur for each gizmo, but rather after a batch of parts has been assembled onto the gizmo. The machines at stations 4 and 5 do not consume parts.

Each station is subject to unscheduled (random) downtime. Manual stations 1–3 have tool failures or other unexpected problems. The automatic stations occasionally jam or have some other problem that requires the assigned operator to fix it. Station 6 (packing) is not subject to these downtimes. Table 13.4 shows time to failure (TTF) and time to repair (TTR) distributional assumptions and the assumed mean time to failure (MTTF), mean time to repair (MTTR) and spread (+/−) of repair times. For example, at Station 1, repair time is uniformly distributed with mean 4.0 minutes plus or minus 1.0 minutes—that is, uniformly distributed between 3.0 and 5.0 minutes. Failure can only occur when an operator or machine is working; hence, TTF is modeled by measuring only busy or processing time until a failure occurs.

The primary model output or response is average throughput during the assumed 7.5 working hours per 8-hour shift. The model also measures detailed station utilization, including busy or processing time, idle or starved time (no parts ready for processing), blocked time (part cannot leave station, because downstream WIP buffer is full), unscheduled downtime, and time waiting for an operator.

Station starvation occurs when the operator and station are ready to work on the next gizmo, the just-completed gizmo leaves the station, but upstream conditions cause no gizmo to be ready for this production step. In short, the upstream WIP buffer is empty.

Station blockage occurs when a station completes all tasks on a gizmo, but cannot release the part because the downstream WIP buffer is full. For both starvation and blockage, production time is lost at the given station and cannot be made up.

When an operator services more than one station, as does the operator servicing Stations 4 and 5, it is possible for both stations to need the operator at the same time. This could cause additional delay at the station and is measured by a "wait for operator" state. Blockage, starvation, and wait-for-operator at each station will be measured in order to help explain any throughput shortfall, should it occur, and to assist in identifying potential system improvements.

13.5.2 Presimulation Analysis

A presimulation analysis, taking into account the average station cycle time as well as expected station availability (90%), indicates that each station, if unhindered, can achieve the desired throughput. This initial analysis is carried out as described in this section.

From the assumed downtime data, the team was able to estimate expected station availability, under the (ideal) assumption of no interaction between stations. The expected availability shows each station's individual availability during working (nonlunch, nonbreak) hours, assuming that the operator can always place a completed gizmo into the downstream rack storage and the next gizmo is ready to begin work at the station. Expected availability is computed by MTTF/(MTTF+ MTTR), or expected busy time during a downtime "cycle" divided by the length of a downtime cycle (a busy cycle plus a repair cycle) and is given in Table 13.4. This calculation ignores certain aspects of the problem, including the parts resupply times and any delay caused by having only one operator to service both Stations 4 and 5.

The design goal for the modeled system is 390 finished gizmos per 8-hour shift. After taking lunch into account, each shift has up to 7.5 hours of available work time. With unscheduled (random) downtime expected to be 10% of available time, this further reduces working time to 0.90×7.5 hours $= 6.75$ hours. This implies that the station with the slowest total cycle time must be able to produce 390 gizmos in the available 6.75 hours. Therefore the total cycle time per gizmo at each station must not exceed 6.75 hours/390 $= 62.3$ seconds.

Now, total cycle time consists of assembly, testing or packing time, and parts resupply time (as given in Table 13.3), plus gizmo loading time of 10 seconds and unload time of 5 seconds. Parts resupply is not taken on every gizmo, but rather after a given number of gizmos corresponding to using all parts in a given batch of parts. For example, using the values in Table 13.3 for Station 1, parts resupply will take 10 seconds every 15 gizmos for Part A, plus 15 seconds every 10 gizmos for Part B, for a total time on the average of $10/15 + 15/10$ seconds per gizmo.

Using this information, the (minimum) total cycle time for each station is estimated in Table 13.5. These presimulation estimates indicate, first, that each theoretical cycle time is well below

Table 13.5 Estimated Total Cycle Time at Each Station

Station	Formula to Estimate Cycle Time (Seconds)	Estimate (Seconds)
1	$10 + 40 + 5 + 10/15 + 15/10$	57.2
2	$10 + 38 + 5 + 20/8 + 15/14$	56.6
3	$10 + 38 + 5 + 30/25$	54.2
4	$10 + 35 + 5$	50.0
5	$10 + 35 + 5$	50.0
6	$10 + 40 + 5 + 30/32$	55.9

the requirement of 62.3 seconds. Secondly, they indicate that Stations 1 and 2 are potential bottlenecks, if there are any.

As the simulation analysis will later show, Station 1 experiences blockage due to Station 2 downtime, and Station 2 occasionally experiences starvation due to downtime at Station 1 and blockage due to downtime at Station 3. These blockage and starvation conditions reduce the available work time below the calculated 90%; hence, for the Baseline Configuration, they reduce the design throughput well below the desired value, 390 gizmos per shift. In summary, a presimulation analysis, although valuable, at best can provide a rough estimate of system performance. As the simulation will show, ignoring blockage and starvation gives an overly optimistic estimate of system throughput.

13.5.3 Simulation Model and Analysis of the Designed System

Using the simulation model, the first experiment was conducted to estimate system performance of the system as designed. The simulation analyst on the team made 10 replications of the model, each having a 2-hour warm-up or initialization followed by a 5-day simulation (each day being 24 hours). A 95% confidence interval was computed for mean throughput per shift:

$$95\% \text{ CI for mean throughput: } (364.5, 366.8), or\, 365.7 \pm 1.14.$$

With 95% confidence, the model predicts that mean (or long-run average) throughput will be between 364.5 and 366.8 gizmos per 8-hour shift with the system as designed. This is well below the design throughput, 390 gizmos per shift.

The team decided to conduct further analyses to identify possible bottlenecks and potential areas of improvement.

13.5.4 Analysis of Station Utilization

At this point, the team desired to have some explanation of the shortfall in throughput. They suspected that perhaps it had to do with the small WIP buffer capacity and the resulting blockage and starvation. The same model was used to estimate detailed workstation utilization in hopes that it would provide an explanation of throughput shortfall. Table 13.6 contains 95% confidence-interval estimates for the first five workstations for percent of time down, blocked, starved, and waiting for an operator. (Waiting for operator affects only stations 4 and 5, as these two stations share one operator. The other stations have a dedicated operator. In addition to the utilization statistics in Table 13.6, the operators have a 30-minute lunch per 8-hour shift, representing 6.25% of available time.)

Table 13.6   Detailed Station Utilization for Baseline Configuration

Station	% Down	% Blocked	% Starved	% Wait for Operator
1	(8.8,9.6)	(11.4,12.5)	(0.0,0.0)	(0.0,0.0)
2	(8.2,8.4)	(8.0,8.8)	(4.9,5.6)	(0.0,0.0)
3	(7.9,8.6)	(9.9,10.4)	(6.1,6.9)	(0.0,0.0)
4	(8.9,9.6)	(2.0,2.8)	(7.5,8.2)	(13.1,14.4)
5	(8.3,9.0)	(0.0,0.2)	(19.4,20.4)	(3.9,4.7)

From the results in Table 13.6, it appears that blockage and starvation explain some portion of the shortfall in throughput. In addition, another possible explanation surfaces: Station 4 experiences a significant time waiting for the single operator that services stations 4 and 5. This delay at Station 4 could result in a full WIP buffer, which in turn would help explain the blockage at Station 3 preceding it. Percent of time blocked is higher than percent starved for Stations 1 to 3, so it appears that downstream delays could be a significant bottleneck.

The team proposed some possible system improvements:

1. having two operators to service Stations 4 and 5 (instead of the currently proposed one operator);
2. increasing the capacity of some of the WIP buffers;
3. a combination of both.

The expense of additional WIP storage space induced the team to desire to keep total buffer space as small as possible, and to require an additional operator only if absolutely necessary, while achieving the design goal of 390 gizmos per shift.

13.5.5 Analysis of Potential System Improvements

To evaluate the addition of an operator and larger WIP buffers, the model was revised appropriately to allow these changes, and a new analysis was conducted. In this analysis, the capacity of each WIP buffer for Stations 2–6 was allowed to increase by one unit above the Baseline value given in Table 13.2. In addition, the effect of a second operator at Stations 4 and 5 is considered. These possibilities result in a total of 64 scenarios or model configurations. (Why?) Making 10 replications per scenario results in a total of 640 simulation runs.

To facilitate the analysis, the team decided to use the Common Random Number technique discussed in Section 12.1.3. To implement it with proper synchronization, each source of random variability was identified and assigned a dedicated random-number stream. In this model, processing time, TTF, and TTR are modeled by statistical distributions at each of the six workstations. Therefore, a total of 18 random-number streams were defined, with 3 used at each workstation. In this way, in each set of runs, each workstation experienced the same workload and random downtimes no matter which configuration was being simulated. For a given number of replications the CRN technique, also known as correlated sampling, is expected to give shorter confidence intervals for differences in system performance.

The model configurations with the most improvement in system throughput, compared with the Baseline configuration, are shown in Table 13.7. These configurations were chosen for further

Table 13.7   Improvement in System Throughput for Alternative Configurations

Number of Operators Stations 4 & 5	Buffer Capacities						Increase in Mean Throughput per Shift (Compared to Baseline)		
	Buffer 2	Buffer 3	Buffer 4	Buffer 5	Buffer 6	Total	Ave. Diff.	CI Low	CI High
2	3	3	3	2	2	13	31.7	30.3	33.1
2	3	3	3	2	3	14	31.7	30.4	33.0
2	3	3	2	2	3	13	30.0	28.6	31.3
2	3	3	3	1	3	13	29.8	28.6	31.0
2	3	3	2	2	2	12	29.7	28.1	31.3
2	3	3	3	1	2	12	29.5	28.1	31.0
2	3	3	2	1	3	12	26.6	25.4	27.9
2	2	3	3	2	2	12	26.6	25.1	28.1
2	2	3	3	2	3	13	26.6	25.0	28.1
2	3	2	3	2	3	13	26.5	25.0	28.0
2	3	2	3	2	2	12	26.4	25.3	27.5
2	3	3	2	1	2	11	26.3	25.1	27.5

evaluation because each shows a potential improvement in throughput of approximately 25 units or more—that is, the lower end of the 95% confidence interval is 25 or higher. The values shown for "Ave Diff" represent the increase in throughput compared to the Baseline configuration. Recall that the Baseline throughput was previously estimated, with 95% confidence, to be in the interval (364.5, 366.8). Being conservative, the engineering team would like to see an improvement of $390 - 364.5 = 25.5$ gizmos per shift. The top six configurations in Table 13.7 have a lower confidence interval larger than 25.5 and hence are likely candidates for achieving the desired throughput. Interpreted statistically: The lower end of the confidence interval is larger than 25.5, so the results yield a 95% confidence that mean throughput will increase by 25.5 or more in the top six configurations listed in Table 13.7.

Note that all the most improved configurations include two operators at Stations 4 and 5. The simulation results for configurations with one operator (not shown here) indicate that a 390 throughput cannot be achieved with one operator, at least not with the buffer sizes considered.

Some configurations can be ruled out because a less expensive option achieves a similar throughput. Consider, for example, the first two configurations in Table 13.7. They are identical except for Buffer 6 capacity. Since WIP buffer capacity is expensive, the smaller total buffer capacity will be the less expensive option. Clearly, there is no need to expand from 2 to 3 units at Buffer 6. The "Total" column can assist in quickly ruling out configurations that do no better than a similar one with smaller total buffer capacity.

The model configuration that increases throughput by 25.5 or better and has the smallest total buffer capacity is the fifth one in Table 13.7, with capacities of (3,3,2,2,2) for Buffers 2 to 6, respectively. On these considerations, this system design becomes the team's top candidate for further evaluation. The next step (not included here) would be to conduct a financial analysis of each alternative configuration.

13.5.6 Concluding Words: The Gizmo Assembly-Line Simulation

Real-life examples similar to this example model include assembly lines for automotive parts and automobile bodies, automotive pollution-control assemblies, consumer items such as washing machines, ranges, and dishwashers, and any number of other assembly operations with a straight flow and limited buffer space between workstations. Similar models and analyses may also apply to a job shop with multiple products, variable routing, and limited work-in-process storage.

13.6 SUMMARY

This chapter introduced some of the ideas and concepts most relevant to manufacturing and material handling simulation. Some of the key points are the importance of modeling downtimes accurately, the advantages of trace-driven simulations with respect to some of the inputs, and the need in some models for accurate modeling of material-handling equipment and the control software.

REFERENCES

BANKS, J. [1994], "Software for Simulation," in *1994 Winter Simulation Conference Proceedings*, eds. J. D. Tew, S. Manivannan, D. A. Sadowski, and A. F. Seila, Association for Computing Machinery, New York, pp. 26–33.

CLARK, G. M. [1994], "Introduction to Manufacturing Applications," in *1994 Winter Simulation Conference Proceedings*, eds. J. D. Tew, S. Manivannan, D. A. Sadowski, and A. F. Seila, Association for Computing Machinery, New York, pp. 15–21.

EHRHARDT, I., H. HERPER, AND H. GEBHARDT [1994], "Modelling Strain of Manual Work in Manufacturing Systems," in *1994 Winter Simulation Conference Proceedings*, eds. J. D. Tew, S. Manivannan, D. A. Sadowski, and A. F. Seila, Association for Computing Machinery, New York, pp. 1044–1049.

LAW, A. M. AND W. D. KELTON [2000], *Simulation Modeling and Analysis*, 3d ed., McGraw–Hill, New York.

NANDI, A., AND P. ROGERS [2003], "Behavior of an Order Release Mechanism in a Make-to-Order Manufacturing System with Selected Order Acceptance," in *2003 Winter Simulation Conference Proceedings*, eds. S. E. Chick, P. J. Sánchez, D. Ferrin, and D. J. Morrice, Association for Computing Machinery, New York, pp. 1251–1259.

PIERCE, N. G., AND R. STAFFORD [1994], "Modeling and Simulation of Material Handling for Semiconductor Wafer Fabrication," in *1994 Winter Simulation Conference Proceedings*, eds. J. D. Tew, S. Manivannan, D. A. Sadowski, and A. F. Seila, Association for Computing Machinery, New York, pp. 900–906.

SARAPH, P. V. [2003], "Shared Resource Capacity Analysis in Biotech Manufacturing," in *2003 Winter Simulation Conference Proceedings*, eds. S. E. Chick, P. J. Sánchez, D. Ferrin, and D. J. Morrice, Association for Computing Machinery, New York, pp. 1247–1250.

SCOTT, H. A. [1994], "Modeling Aircraft Assembly Operations," in *1994 Winter Simulation Conference Proceedings*, eds. J. D. Tew, S. Manivannan, D. A. Sadowski, and A. F. Seila, Association for Computing Machinery, New York, pp. 920–927.

SMITH, J. S., *et al.* [1994], "Discrete Event Simulation for Shop Floor Control," in *1994 Winter Simulation Conference Proceedings*, eds. J. D. Tew, S. Manivannan, D. A. Sadowski, and A. F. Seila, Association for Computing Machinery, New York, pp. 962–969.

TURKSEVEN, C. H., AND G. ERTEK [2003], "Simulation Modeling for Quality and Productivity in Steel Cord Manufacturing," in *2003 Winter Simulation Conference Proceedings*, eds. S. E. Chick, P. J. Sánchez, D. Ferrin, and D. J. Morrice, Association for Computing Machinery, New York, pp. 1225–1229.

WATSON, E. F., AND R. P. SADOWSKI [1994], "Developing and Analyzing Flexible Cell Systems Using Simulation," in *1994 Winter Simulation Conference Proceedings*, eds. J. D. Tew, S. Manivannan, D. A. Sadowski, and A. F. Seila, Association for Computing Machinery, New York, pp. 978–985.

WILLIAMS, E. J. [1994], "Downtime Data—Its Collection, Analysis, and Importance," in *1994 Winter Simulation Conference Proceedings*, eds. J. D. Tew, S. Manivannan, D. A. Sadowski, and A. F. Seila, Association for Computing Machinery, New York, pp. 1040–1043.

EXERCISES

Instructions to the student: Many of the following exercises contain material-handling equipment such as conveyors and vehicles. The student is expected to use any simulation language or simulator that supports modeling conveyors and vehicles at a high level.

Some of the following exercises use the uniform, exponential, normal, or triangular distributions. Virtually all simulation languages and simulators support these, plus other distributions. The use of the first three distributions was explained in the note to the exercises in Chapter 4; the use of the triangular is explained in the exercise that requires it. For reference, the properties of these distributions, plus others used in simulation, are given in Chapter 5, and random-variate generation is covered in Chapter 8.

1. A case sortation system consists of one infeed conveyor and 12 sortation lanes, as shown in the following schematic (not to scale):

Cases enter the system from the left at a rate of 50 per minute at random times. All cases are 18 by 12 inches and travel along the 18 inch dimension. The incoming mainline conveyor is 20 inches wide and 60 feet in length (as shown). The sortation lanes are numbered 1 to 12 from left to right, and are 18 inches wide and 15 feet in length, with 2 feet of spacing between adjacent lanes. (Estimate any other dimensions that are needed.) The infeed conveyor runs at 180 feet/minute, the sortation lanes at 90 feet/minute. All conveyor sections are accumulating, but, upon entrance at the left, incoming cases are at least 2 feet apart from leading edge to leading edge. On the sortation lanes, the cases accumulate with no gap between them.

Incoming cases are distributed to the 12 lanes in the following proportions:

1	6%	7	11%
2	6%	8	6%
3	5%	9	5%
4	24%	10	5%
5	15%	11	3%
6	14%	12	0%

The 12th lane is an overflow lane; it is used only if one of the other lanes fill and a divert is not possible.

At the end of the sortation lanes, there is a group of operators who scan each case with a bar-code scanner, apply a label and then place it on a pallet. Operators move from lane to lane as necessary to avoid allowing a lane to fill. There is one pallet per lane, each holding 40 cases. When a pallet is full, assume a new empty one is immediately available. If a lane fills to 10 cases and another case arrives at the divert point, this last case continues to move down the 60-foot mainline conveyor and is diverted into lane 12, the overflow lane.

Assume that one operator can handle 8.5 cases per minute, on the average. Ignore walking time and assignment of an operator to a particular lane; in other words, assume the operators work as a group uniformly spread over all 12 lanes.

(a) Set up an experiment that varies the number of operators and addresses the question: How many operators are needed? The objective is to have the minimum number of operators but also to avoid overflow.

(b) For each experiment in part (a), report the following output statistics:

> Operator utilization
> Total number of cases palletized
> Number of cases palletized by lane
> Number of cases to the overflow lane

(c) For each experiment in part a, verify that all cases are being palletized. In other words, verify that the system can handle 50 cases per minute, or explain why it cannot.

2. Redo Exercise 1 to a greater level of detail by modeling operator walking time and operator assignment to lanes. Assume that operators walk at 200 feet per minute and that the walking distance from one lane to the next is 5 feet. Handling time per case is now assumed to be 7.5 cases per minute. Devise a set of rules that can be used by operators for lane changing. (For example, change lanes to that lane with the greatest number of cases only when the current lane is empty or the other lane reaches a certain level.) Assume that each operator is assigned to a certain number of adjacent lanes and handles only those lanes. However, if necessary, two operators (but no more) may be assigned to one lane—that is, operator assignments may overlap.

(a) If your lane-changing rule has any numeric parameters, experiment to find the best settings. Under these circumstances, how many operators are needed? What is the average operator utilization?

(b) Does a model that has more detail, as does Exercise 2a when compared to Exercise 1, always have greater accuracy? How about this particular model? Compare the results of Exercise 2a to the results for Exercise 1. Are the same or different conclusions drawn?

(c) Devise a second lane-changing rule. Compare results between the two rules. Compare total walking time or percent of time spent walking between the two rules.

Suggestion: A lane-changing rule could have one or two "triggers". A one-trigger rule might state that, if a lane reached a certain level, the operator moved to that lane. (Without modification, such a rule could lead to excessive operator movement, if two lanes had about the same number of cases near the trigger level.) A two-trigger rule might state that, if a lane reached a certain level and the operator's current lane became empty, then change to the new lane; but if a lane reaches a specified higher "critical" level, then the operator immediately changes lanes.

(d) Compare your results with those of other students who may have used a different lane-changing rule.

3. Redo Exercise 2 with a different operator-assignment rule. Basically, operators can be assigned to any lane as conditions warrant, but, as before, no more than two operators can be assigned to a lane at the same time. Address the same questions (a)–(d) as in Exercise 2.

4. A package-sortation system consists of one in-feed conveyor, 12 sortation lanes (or chutes), and a take-away conveyor, as shown in the following schematic (not to scale):

Packages enter the system from the left (at point E) at a rate of 100 per minute at random times. A package is diverted to a sortation lane on the basis of its shipping destination; in the model, a given percentage of the packages chosen randomly, according to the proportions in the next table, are assigned to each lane. The percentages for each lane/destination is as follows:

1	6%	7	10%
2	6%	8	6%
3	5%	9	4%
4	24%	10	4%
5	15%	11	3%
6	14%	12	3%

After a slug of 8 packages has accumulated, they are all released together onto a 50-foot section of takeaway conveyor and head toward the point labeled T; they transfer there (at conveyor speed) onto a second section of takeaway conveyor at the end of which is a bar-code scanner (at the point labeled B). A slug cannot be released until the previous slug has cleared the point T. In this model, packages disappear at point B.

All packages are 18 by 12 inches and travel along their 18 inch dimension. The incoming mainline conveyor is 20 inches wide and 60 feet in length (as shown). The 12 sortation lanes are 18 inches wide and 15 feet in length with 2 feet of spacing between adjacent lanes. (Estimate any other dimensions that are needed.) The infeed conveyor runs at 240 feet/minute, the sortation lanes at 90 feet/minute. The infeed and sortation conveyor sections are accumulating, but, upon entrance at the left, incoming packages are at least 2 feet apart from leading edge to leading edge. On the sortation lanes, the packages accumulate with no gap between them. The first 50-foot section of takeaway conveyor is a nonaccumulating belt-type conveyor; the second, 30-foot section is accumulating.

When a lane contains 8 packages, no more diverts are allowed until the lane is released and the slug completely clears the bottom of the chute. When a lane is full or before the slug has cleared, if another package arrives to divert, the affected package does not divert but rather moves to the end of the infeed conveyor and then recirculates to the start of the infeed conveyor on a conveyor that is not shown in the schematic. Simulate recirculation as a time delay of 45 seconds without explicitly modeling the recirculation conveyor itself.

(a) Set up an experiment to discover the minimum speed required of the takeaway conveyor to minimize or eliminate recirculation. Conveyors available from vendors have speeds starting at 60 feet/minute and increasing in increments of 30 feet/minute. (Faster conveyor is more expensive, so minimizing the speed will reduce costs, provided that recirculation is not a problem.)

(b) Could destination assignment to available lane be modified in such a way as to improve system performance? Devise a better assignment rule and test it, using the simulation model.

5. A fleet of AGVs (Automated Guided Vehicles) services 5 workstations in series by transporting parts on fixtures from one workstation to the next, remaining with a part from the pickup point through all workstations to the offload point. The incoming staging conveyor, the AGV guidepath, the five workstations (A–E), and the offload conveyor are shown in the following schematic (not to scale):

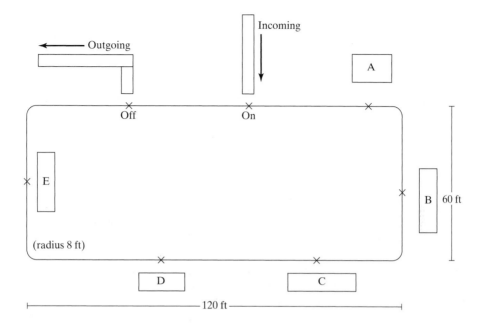

AGVs travel in a clockwise direction. Initially, all are empty and queued at the incoming conveyor at the "On" point. The workstations are labeled A–E. Parts on fixtures are staged on the incoming conveyor and picked up by an AGV at the point labeled "On". The AGV then travels to machine A and stops at the point marked on the guidepath. The part and fixture remain on the AGV during processing, then travel together to the next workstation. After processing is completed at station E, the AGV travels to the point labeled "Off", the part with fixture is automatically unloaded onto the offload conveyor, and the empty AGV queues at the point labeled "On" for the next part.

Assume that the incoming conveyor remains full. It is 15 feet long and can hold 3 parts with fixtures, each taking 5 feet of conveyor space. Therefore, when one part with fixture is picked up by an AGV, another part and fixture are placed on the end of the conveyor. Pickup time is 30 seconds. The incoming conveyor runs at 30 feet per minute.

When a part has completed processing at all stations and has been transported to the offload point labeled "Off", the offload operation takes 45 seconds. The offload conveyor itself runs at 30 feet per minute. The parts and fixtures, although taking up a space of 3 feet by 3 feet, require 5 feet on the conveyor.

Initial workstation placements are shown in approximate fashion, with no exact locations given. Constraints are that each workstation must be placed along the straight section of guidepath on which they are shown on the schematic, with the AGV stopping point no closer than 5 feet from the entrance to or exit from a curve. The incoming and offload conveyor must be placed as shown, with the offload point 20 feet from the end of the preceding curve and the pickup point 40 additional feet to the right. Otherwise, AGV stopping points at workstations can be placed at any desired location on the straight section of guidepath.

AGVs travel at 15 feet per second and accelerate and decelerate at 3 ft/sec/sec. They are 7 feet in length and queue 8 feet apart (1 foot gap between two AGVs) when waiting for the processing position at a workstation and when in the pickup/offload position at a conveyor.

At each of the workstations, processing-time-per-part, mean-time-to-failure (MTTF), and mean-time-to-repair (MTTR) are as follows:

Workstation	Processing Time (sec.)	MTTF (min.)	MTTR (min.)
A	120	90	3.1
B	100	82	4.0
C	140	110	4.5
D	45	20	5.0
E	100	240	9.0

Detailed data is not available, so certain assumptions regarding time to failure and time to repair are made: Assume that actual time to failure is exponentially distributed, with the specified mean (MTTF). Assume that repair time requires a maintenance person and that there is only one available. Actual repair times are uniformly distributed, with the mean as shown and a half-width of one-half the mean. Time to failure is measured against wall-clock time, not against station-busy time. After a repair, any remaining processing time is taken, and the part continues.

> There are three basic questions:
> What is maximum system throughput?
> How many AGVs are needed?
> What is the best workstation placement (within the stated constraints)?
> For each set of experiments in parts (a) and (b), following, consider
>   the following issues:

Should there be a "warmup" period for loading the system during which no statistics are collected? How long should the simulation run? (One suggestion: The warmup period should last at least until several parts have completed processing. The run-length should allow for at least 10 downtimes on each machine. Simulate with as many AGVs as will reasonably fit on the guidepath and then try to reduce the number without reducing throughput.)

Your experiments should vary the number of AGVs and the workstation placement. Before simulating, make your best estimate regarding workstation placement, and justify that choice. Also, make a reasonable estimate regarding the likely range for number of AGVs.

(a) In the first set of experiments, use a rough-cut model that does not model the AGVs explicitly. Compute the approximate travel time between adjacent workstations and use this value as a time delay in the model. Assume there are sufficient AGVs. (Use this model to address the question of maximum throughput, but not the number of AGVs needed.) What are the advantages and limitations of this approach?

(b) In the second set of experiments, model the AGVs and guidepath explicitly. Design an experiment to find out the maximum system throughput (in parts per hour) with the smallest number of AGVs.

6. More accurate information has been obtained on failure times for the workstations in Exercise 5. Redo Exercise 5 with the new information. Failures appear to be related more closely to the number of parts processed than to the amount of time passed. The new data is as follows:

Workstation	MTTF (parts)	MTTR (min.)
A	45	3.1
B	49	4.0
C	47	4.5
D	27	5.0
E	144	9.0

Note that MTTF (mean-time-to-failure) is measured in terms of number of parts, but assumptions about time to repair are unchanged from Exercise 5. For example, for workstation A, if time to failure happened to be 45 parts, then after a failure, the 45th part would experience the next failure. In reality, failure occurs during processing, but for modeling purposes, assume the failure occurs just as processing begins. (Does this make any difference?)

For the distribution of time to failure, assume a discrete uniform distribution with mean (MTTF) over the range from 1 to 2*MTTF − 1. For example, for workstation A, the actual times to failure are equally likely to be any value from 1 to 89 parts.

Address the same questions as in Exercise 5. Do these new assumptions make any difference? How many failures are there per 8-hour shift for the first set of assumptions as compared with those for the revised assumptions?

7. For the problem in Exercises 5 and 6, a study was done to evaluate the repair times more accurately for each workstation. It was found that a triangular distribution provided a closer fit to the actual repair times. Assume the following parameters for the triangular distribution:

Workstation	Repair Time (Triangular Distribution)			
	Minimum (min.)	Most Likely (min.)	Maximum (min.)	Average (min.)
A	1.0	2.3	6.0	3.1
B	2.0	3.0	7.0	4.0
C	2.5	3.5	7.5	4.5
D	2.0	4.0	9.0	5.0
E	4.0	7.0	16.0	9.0

Note that the average repair times are identical to the assumed averages (MTTR) in Exercises 5 and 6, but the distribution is triangular instead of uniform. (Most simulation languages or simulators have a built-in triangular distribution that requires specification of the minimum value, most likely value or mode, and maximum value. The average value is given for informational purposes only; it is the sum of the other three parameters divided by 3. The triangular distribution is discussed in Chapter 5.)

(a) Redo Exercise 5, but with the new assumptions for repair times.
(b) Redo Exercise 6, but with the new assumptions for repair times.

8. One approach to modeling downtimes discussed in this chapter, but not recommended, was to replace the processing time with a larger adjusted processing time and to leave out the downtimes. The adjusted processing time is meant to account for downtimes, at least in the long run and on the average. How would this method have affected the results for Exercise 5?

(a) For Exercise 5, compute the adjusted processing times for each workstation.
(b) For Exercise 5, use the adjusted processing times and leave out all explicit downtimes due to failure. How does this change the results? Is this a good approach for this model?

9. Suppose that improved maintenance and better machines could totally eliminate downtime due to failure for the system in Exercise 5. By how much would throughput improve?

10. Develop a model for Example 13.1 and attempt to reproduce qualitatively the results found in the text regarding different assumptions for simulating downtimes. Do not attempt to get exactly the same numerical results, but rather to show the same qualitative results.

(a) Do your models support the conclusions discussed in the text? Provide a discussion and conclusions.
(b) Make a plot of the number of entities in the queue versus time. Can you tell when failures occurred? After a repair, about how long does it take for the queue to get back to "normal"?

11. In Example 13.1, the failures occurred at low frequency compared with the processing time of an entity. Time to failure was 1000 minutes, and interarrival time was 10 minutes, implying that few entities would experience a failure. But, when an entity did experience

a failure (of 50 minutes, on average), it was several times larger than the processing time of 7.5 minutes.

Redo the model for Example 13.1, assuming high-frequency failures. Specifically, assume that the time to failure is exponentially distributed, with mean 2 minutes, and the time to repair is exponentially distributed, with mean 0.1 minute or 6 seconds. As compared with the low-frequency case, entities will tend to experience a number of short downtimes.

For low-frequency versus high-frequency downtimes, compare the average number of downtimes experienced per entity, the average duration of downtime experienced, the average time to complete service (including downtime, if any), and the percent of time down.

Note that the percentage of time the machine is down for repair should be the same in both cases:

$$50/(1000 + 50) = 4.76\%$$

$$6 \sec /(2 \min + 6 \sec) = 4.76\%$$

Verify percentage downtime from the simulation results. Are the results identical? ... close? Should they be identical, or just close? As the simulation run-length increases, what should happen to percentage of time down?

With high-frequency failures, do you come to the same conclusions as were drawn in the text regarding the different ways to simulate downtimes? Make recommendations regarding how to model low-frequency versus high-frequency failures.

12. Redo Exercise 11 (based on Example 13.1), but with one change: When an entity experiences a downtime, it must be reprocessed from the beginning. If service time is random, take a new draw from the assumed distribution. If service time is constant, it starts over again. How does this assumption affect the results?

13. Redo Exercise 11 (based on Example 13.1), but with one change: When an entity experiences a downtime, it is scrapped. How does scrapping entities on failure affect the results in the low-frequency and in the high-frequency situations? What are your recommendations regarding the handling of low- versus high-frequency downtimes when parts are scrapped?

14. Sheets of metal pass sequentially through 4 presses: shear, punch, form, and bend. Each machine is subject to downtime and die change. The parameters for each machine are as follows:

Press	Process Rate (per min.)	Time to Failure (min.)	Time to Repair (min.)	No. of Sheets to a Die Change (no. sheets)	Time to Change Die (min.)
Shear	4.5	100	8	500	25
Punch	5.5	90	10	400	25
Form	3.8	180	9	750	25
Bend	3.2	240	20	600	25

Note that processing time is given as a rate—for example, the shear press works at a rate of 4.5 sheets per minute. Assume that processing time is constant. The automated equipment makes the time to change a die fairly constant, so it is assumed to be always 25 minutes. Die changes occur between stamping of two sheets after the number shown in the table have gone through a machine. Time to failure is assumed to be exponentially distributed, with the mean given in the table. Time to repair is assumed to be uniformly distributed, with the mean taken from the table and a half-width of 5 minutes. When a failure occurs, 20% of the sheets are scrapped. The remaining 80% are reprocessed at the failed machine after the repair.

Assume that an unlimited supply of material is available in front of the shear press, which processes one sheet after the next as long as there is space available between itself and the next machine, the punch press. In general, one machine processes one sheet after another continuously, stopping only for a downtime, for a die change, or because the available buffer space between itself and the next machine becomes full. Assume that sheets are taken away after bending at the bend press. Buffer space is divided into 3 separate areas, one between the shear and the punch presses, the second between the punch and form presses, and the last between form and bend.

(a) Assume that there is an unlimited amount of space between machines. Run the simulation for 480 hours (about 1 month with 24 hour days, 5 days per week). Where do backups occur? If the total buffer space for all three buffers is limited to 15 sheets (not counting before shear or after bend), how would you recommend dividing this space among the three adjacent pairs of machines? Does this simulation provide enough information to make a reasonable decision?

(b) Modify the model so that there is a finite buffer between adjacent machines. When the buffer becomes full and the machine feeding the buffer completes a sheet, the sheet is not able to exit the machine. It remains in the machine blocking additional work. Assume that total buffer space is 15 sheets for the 3 buffers.

Use the recommendation from part (a) as a starting point for each buffer size. Attempt to minimize the number of runs. You are allowed to experiment with a maximum of 3 buffer sizes for each buffer. (How many runs does this make?) Run a set of experiments to determine the allocation of buffer space that maximizes production. Simulate each alternative for at least 1000 hours.

Report total production per hour on the average, press utilization (broken down by percentage of time busy, down, changing dies, and idle), and average number of sheets in each buffer.

14

Simulation of Computer Systems

It is only natural that simulation is used extensively to simulate computer systems, because of their great importance to the everyday operations of business, industry, government, and universities. In this chapter, we look at the motivations for simulating computer systems, the different types of approaches used, and the interplay between characteristics of the model and implementation strategies. We begin the discussion by looking at general characteristics of computer-system simulations. Next, we lay the groundwork for investigating simulation of computer systems by looking at various types of simulation tools used to perform those simulations. In section 14.3, we describe different ways that input is presented or generated for these simulations. We next work through an example of a high-level computer system one might simulate, paying attention to problems of model construction and output analysis. In section 14.5, we turn to the central processing unit (CPU) and point out what is generally simulated and how. Following this, we consider simulation of memory systems, in section 14.6.

14.1 INTRODUCTION

Computer systems are incredibly complex. A computer system exhibits complicated behavior at time scales from the time to "flip" a transistor's state (on the order of 10^{-11} seconds) to the time it takes a human to interact with it (on the order of seconds or minutes). Computer systems are designed hierarchically, in an effort to manage this complexity. Figure 14.1 illustrates the point. At a high level of abstraction (the system level), one might view computational activity in terms of

517

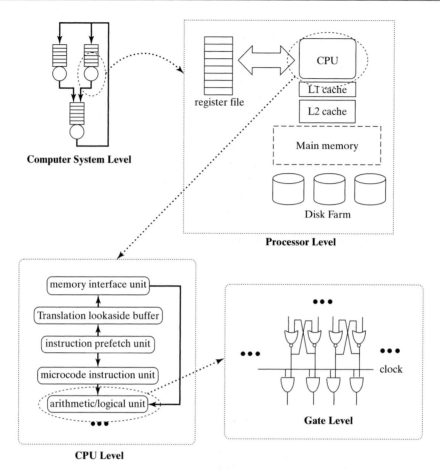

Figure 14.1   Different levels of abstraction in computer systems.

tasks circulating among servers, queueing for service when a server is busy. A lower level in the hierarchy can view the activity as being among components of a given processor (its registers, its memory hierarchy). At a lower level still, one views the activity of functional units that together make up a central processing unit, and, at an even lower level, one can view the logical circuitry that makes it all happen.

Simulation is used extensively at every level of this hierarchy, with some results from one level being used at another. For instance, engineers working on designing a new chip will begin by partitioning the chip functionally (e.g., the subsystem that does arithmetic, the subsystem that interacts with memory, and so on), establish interfaces between the subsystems, then design and test the subsystems individually. Given a subsystem design, the electrical properties of the circuit are first studied by using a circuit simulator that solves differential equations describing electrical behavior. At this level, engineers work to ensure the correctness of signals' timing throughout the circuit and to ensure that the electrical properties fall within the parameters intended by the design.

Once this level of validation has been achieved, the electrical behavior is abstracted into logical behavior (e.g., signals formerly thought of as electrical waveforms are now thought of as logical 1's and 0's). A different type of simulator is next used to test the correctness of the circuit's logical behavior. A common testing technique is to present the design with many different sets of logical inputs ("test vectors") for which the desired logical outputs are known. Discrete-event simulation is used to evaluate the logical response of the circuit to each test vector and is also used to evaluate timing (e.g., the time required to load a register with a datum from the main memory). Once a chip's subsystems are designed and tested, the designs are integrated, and then the whole system is subjected to testing, again by simulation.

At a higher level, one simulates by using functional abstractions. For instance, a memory chip could be modeled simply as an array of numbers, and a reference to memory as just an indexing operation. A special type of description language exists for this level, called "register-transfer-language" (see, for instance, Mano 1993). This is like a programming language, with reassigned names for registers and other hardware specific entities and with assignment statements used to indicate data transfer between hardware entities. For example, the following sequence loads into register r3 the data whose memory address is in register r6, subtracts one from it, and writes the result into the memory location that is word adjacent (a word in this example is 4 bytes in size) to the location first read:

```
r3 = M[r6];
r3 = r3-1;
r6 = r6+4;
M[r6] = r3;
```

A simulator of such a language might ascribe deterministic time constants to the execution of each of these statements. This is a useful level of abstraction to use when one needs to express sequencing of data transfers at a low level, but not so low as the gates themselves. The abstraction makes sense when one is content to assume that the memory works and that the time to put a datum in or out is a known constant. The "known constant" is a value resulting from analysis at a lower level of abstraction. Functional abstraction is also commonly used to simulate subsystems of a central processing unit (CPU), in the study of how an executing program exercises special architectural features of the CPU.

At a higher level still, one might study how an Input–Output (I/O) system behaves in response to execution of a computer program. The program's behavior may be abstracted to the point of being *modeled*, but with some detailed description of I/O demands (e.g., with a Markov chain that with some specificity describes an I/O operation as the Markov chain transitions). The behavior of the I/O devices may be abstracted to the point that all that is considered is how long it takes to complete a specified I/O operation. Because of these abstractions, one can simulate larger systems, and simulate them more quickly. Continuing in this vein, at a higher level of abstraction still, one dispenses with specificity altogether. The execution of a program is modeled with a randomly sampled CPU service interval; its I/O demand is modeled as a randomly sampled service time on a randomly sampled I/O device.

Different levels of abstraction serve to answer different sorts of questions about a computer system, and different simulation tools exist for each level. Highly abstract models rely on stochastically modeled behavior to estimate high-level system performance, such as throughput (average number of "jobs" processed per unit time) and mean response time (per job). Such models can also incorporate system failure and repair and can estimate metrics such as mean time to failure and availability. Less abstract models are used to evaluate specific systems components. A study of an advanced CPU design might be aimed at estimating the throughput (instructions executed per unit time); a study of a hierarchical memory system might seek to estimate the fraction of time that a sought memory reference was found immediately in the examined memory. As we have already seen, more detailed models are used to evaluate functional correctness of circuit design.

14.2 SIMULATION TOOLS

Hand in hand with different abstraction levels, one finds different tools used to perform and evaluate simulations. We next examine different types of tools and identify important characteristics about their function and their use.

An important characteristic of a tool is how it supports model building. In many tools, one constructs networks of components whose local behavior is already known and already programmed into the tool. This is a powerful paradigm for complex model construction. At the low end of the abstraction hierarchy, electrical circuit simulators and gate-level simulators are driven by network descriptions. Likewise, at the high end of the abstraction hierarchy, tools that simulate queueing networks and Petri nets are driven by network descriptions, as are sophisticated commercial communication-system simulators that have extensive libraries of preprogrammed protocol behaviors. Some of these tools allow one to incorporate user-programmed behavior, but it appears this is not the norm as a usage pattern.

A very significant player in computer-systems design at lower levels of abstraction is the VHDL language (e.g., see Ashenden [2001]). VHDL is the result of a U.S. effort in the 1980's to standardize the languages used to build electronic systems for the government. It has since undergone the IEEE standardization process and is widely used throughout the industry. As a language for describing digital electronic systems, VHDL serves both as a design specification and as a simulation specification. VHDL is a rich language, full both of constructs specific to digital systems and the constructs one expects to find in a procedural programming language. It achieves its dual role by imposing a clear separation between system topology and system behavior. Design specification is a matter of topology; simulation specification is a matter of behavior. Libraries of predefined subsystems and behaviors are widely available, but the language itself very much promotes user-defined programmed behavior. VHDL is also innovative in its use of abstract interfaces (e.g., to a functional unit) to which different "architectures" at different levels of abstraction may be attached. For instance, the interface to the Arithmetic Logical Unit (ALU) would be VHDL "signals" that identify the input operands, the operation to be applied to them, and the output. One could attach to this interface an architecture that in a few lines of code just performs the operation—if an addition is specified, just one VHDL statement assigns the output signal to be the sum (using the VHDL addition operator) of the two input signals. An alternative architecture could completely specify the gate-level logical design of the ALU. Models that interact with the ALU interface cannot tell how the semantics of the interface are implemented. This separation of interface from architecture supports modular construction of models and allows one to validate a new submodel architecture by comparing the results it returns to the interface with those returned by a different architecture given the same inputs. A substantive treatment of

VHDL is well beyond the scope of this book. VHDL is widely used in the electrical and computer engineering community, but is hardly used outside of it.

One drawback to VHDL is that it is a big language, requires a substantial VHDL compiler, and vendors typically target the commercial market at prices that exclude academic research. Of course, other simulation languages exist, and this text describes several in Chapter 4. Such languages are good for modeling certain types of computer systems at a high level, but are not designed or suited for expression of computer-systems modeling at lower levels of the abstraction hierarchy. As a result, when computer scientists need to simulate specialized model behavior, they will often write a simulation (or a simulator) from scratch. For example, if a new policy for moving data between memories in a hierarchy is to be considered, an existing language will not have that policy preprogrammed; when a new architectural feature in a CPU is designed, the modeler will have to describe that feature and its interaction with the rest of the CPU, using a general programming language. A class of tools exists that use a general programming language to express simulation-model behavior, among them SimPack (Fishwick [1992]), C++SIM (Little and McCue [1994]), CSIM (Schwetman [1986]), Awesime (Grunwald [1995]), and SSF (Cowie *et al.* [1999]). This type of tool defines objects and libraries for use with such languages as C, C++, Java. Model behavior is expressed as a computer program that manipulates these predefined objects. The technique is especially powerful when used with object-oriented languages, because the tool can define base-class objects whose behavior is extended by the modeler.

Some commercial simulation languages do support interaction with general programming languages; however, simulation languages are not frequently used in the academic computer-science world. Cost is a partial explanation. Commercial packages are developed with commercial needs and commercial budgets in mind, yet computer scientists can usually develop what they need relatively quickly, themselves. Another explanation is a matter of emphasis: Simulation languages tend to include a rich number of predefined simulation objects and actions and allow access to a programming language to express object behavior; a simulation model is expressed primarily in the constructs of the simulation language, and the model is evaluated either by compiling the model (using a simulation-language-specific compiler) and running it or by using a simulation-language-specific interpreter.

One of the many advantages to such an approach is that the relative rigidity of the programming model makes possible graphical model building, thereby raising the whole model-building endeavor to a higher level of abstraction. Some tools have so much preprogrammed functionality that it is possible to design and run a model without writing a single line of computer code.

By contrast, programming languages with simulation constructs tend to define a few elemental simulation objects; a simulation model is expressed principally via the notions and control flow of the general programming language, with references to simulation objects interspersed. To evaluate the model, one compiles or interprets the program, using a compiler or interpreter associated with the general programming language, as opposed to one associated with the simulation language. The former approach supports more rapid model development in contexts where the language is tuned to the application; the latter approach supports much greater generality in the sorts of models that can be expressed.

Among tools supporting user-programmed behavior, a fundamental characteristic is the worldview that is supported. In the following two subsections, we look closely at process orientation as it is expressed in SSF, then at an event-oriented approach using a Java base framework.

14.2.1 Process Orientation

A process-oriented view (see Chapter 3) implies that the tool must support separately schedulable threads of control. Threading is a fundamental concept in programming, and a discussion of its capabilities and implementation serves to highlight important issues in simulation modeling. Fundamentally, a "thread" is a separately schedulable unit of execution control, implemented as part of a single executing process (as seen by the operating system; see Nutt [2004]). An operating system has the notion of separate processes (which might interact), which typically have their own separate and independent memory spaces. A group of threads operate in the same process memory space, with each thread having allocated to it a relatively small portion of that space for its own use. That space is used to contain the thread's *state*, which is the full set of all information needed to restart the thread after it is suspended. State would include register values and the thread's runtime stack, which holds variables that are local to the procedures called by the thread. Once a thread is given control, it runs until it yields up control, either via an explicit statement that serves simply to relinquish control or by blocking until signaled by another thread to continue.

These ideas are made more concrete by discussing them in the context of a Java implementation of SSF. Java defines the `Thread` class; a subclass of `Thread` defines the `execute` method, which is defined in the thread body. Threads coordinate with each other through "locks," which provide mutually exclusive access to code segments. Every instance of a Java object has an associated lock (and almost every variable in Java is an object). A thread tries to execute a code fragment protected by the lock for object `obj` via the Java statement

```
synchronized(obj) { /* code fragment */ }
```

A thread must acquire the lock before executing the code fragment, and only one thread has the lock at a time. A thread that executes a `synchronized` statement at an instant at which another thread holds the lock blocks—which could mean suspension, depending on the thread scheduler. Java threads can also coordinate through `wait` and `notify` method calls, also associated with an object's lock. A thread that executes `obj.wait()` suspends. Actually, multiple threads can execute `obj.wait()`, and each will suspend. Eventually some thread executes `obj.notify()`, and the thread scheduler releases one of the suspended threads to continue.

These notions can be used to implement process orientation in a Java simulator. Each simulation process derives from the Java Thread class. One additional thread will maintain an event list; processing for that thread involves removing the least-time event from the event list, reanimating the simulation process thread (or threads) associated with that event, and blocking until those threads have completed. While a process thread is executing, it may cause additional events to be inserted into the scheduler thread's event list. When a process thread completes, it needs to block and to signal the scheduler thread that it is finished. We accomplish all this by using two locks per simulation process. One of these locks is the one Java provides automatically for every object (and a simulation process thread is an object). The other lock is a variable each simulation object defines, which we'll call `lock`. A suspended process thread blocks on a call `lock.wait()`; it remains blocked there until the scheduling thread executes `notify()` on that same object variable. After the scheduler does this, it blocks by calling `wait()` on the simulation process object's own built-in lock. So the simulation process thread notifies the scheduler that it is finished by calling `notify()` on its own built-in lock.

SSF code we discussed earlier in Chapter 4 (Figures 4.14 and 4.15) illustrates some of these points. Recall that this code models a single server with exponentially distributed interarrival times and positive normal service times. A cursory glance shows the model to be legitimate Java code that uses SSF base classes.

SSF defines five base classes around which simulation frameworks are built (discussed in Chapter 4). The key one for discussing process orientation is the **process** class; derived classes **Arrivals** in Figure 4.14 and **Server** in Figure 4.15 are examples of it. The base class specifies that method **action** be the thread body; each derived class overrides the base-class definition to specify its own thread's behavior. Every object of a given class derived from **process** defines a separate thread of control, but all execute the same thread code body.

The `waitFor` statement used in `Arrival`'s thread body suspends the thread; its argument specifies how long in simulation time the thread suspends. The Java thread-based scheduling mechanism we described earlier enables implementation of `waitFor` to cause a "wake-up" event to be inserted into the scheduling thread's event list, time stamped with the current time plus the `waitFor` argument. Here variable `time` is the future-event time; method `insertProcess` puts the process into the event queue. A non-Simple process (e.g., one implemented with a Java thread) goes through a sequence of synchronization steps to reach the `notify()` method. (We will say more about Simple processes in 14.2.2.) The scheduler thread has blocked on the process's native lock; this `notify()` releases it. The process then immediately calls `wait()` on its `lock` variable, which suspends the thread until the scheduler executes `notify()` on that same variable. From the point of view of the code fragment executing `waitFor`, the statement following the `waitFor` call executes precisely at the time implied by the `waitFor` argument. The code in Figure 14.2 (taken from an SSF implementation) illustrates this.

The call to `waitOn` in the **Server**'s `action` has a slightly different implementation. The code implementing `waitOn` first attaches the process to the `inChannel`'s list of processes that are blocked on it, then engages in the same lock synchronization sequence as `waitFor` to block itself and release the scheduler thread. The semantics of releasing a blocked process are defined in terms of SSF `Events`. An `outChannel` object to which an `Event` object is written has almost always been "mapped" to an `inChannel` object. When an `Event` is written to an `outChannel` at time t, the `outChannel`'s `write` method computes the time $t + d$ at which the `Event` is available

```
public void waitFor(long timeinterval){
  time = owner.owner.clock + timeinterval;
  owner.owner.insertProcess(this);
  if (!isSimple()){
    synchronized(lock){
     synchronized(this){
        notify();
     }
     try{lock.wait();}
     catch(InterruptedException e){}
    }
  }
}
```

Figure 14.2 SSF implementation of waitFor statement.

on the associated `inChannel` (*d* is a function of delays declared when the `outChannel` is created, the `mapTo` method is called, and the `write` method is called), and an internal event is put on the scheduler's event list, with time stamp $t + d$. The scheduler executes this event (no SSF process does) and releases all processes blocked on the `inChannel` to which the `Event` arrives. Each of these is able to get a copy of the `Event` so delivered, by calling the `inChannel`'s `activeEvents` method.

From these descriptions, we see that, normally, each event has a thread overhead cost: 2 thread reanimations, and 2 thread suspensions. Depending on how thread context switching is implemented, this cost ranges from heavy to very heavy, as compared with a purely event-oriented view. These costs can be avoided in SSF by designing processes to be *simple*, as is described next.

14.2.2 Event Orientation

From a methodological point of view, the process-oriented view is distinguished from the event-oriented view in terms of the focus of the model description. Process orientation allows for a continuous description, with pauses or suspensions. Event orientation does not. From an *implementation* point of view, the key distinguishing feature of process-oriented simulation is the need to support suspension and reanimation, which leads us to threads, as we have seen. In SSF, though, we see that the difference between process and event orientation is not very large: The SSF world encompasses both. The only difference is that, for SSF to be event oriented, its processes need to be *simple*, a technical term for the case when every statement in `action` that might suspend the process would be the last statement executed under normal execution semantics.

The implementation of `waitFor` in Figure 14.2 computes the time when the suspension is lifted and puts a reanimation event in the event list. Synchronization by threads through locks is used only if the process is not *simple*. An implementation of `waitOn` would be entirely similar. If every SSF process in a model is *simple*, there is no true code suspension, and the model is essentially event oriented. The `action` body for a *simple* process is just executed from its normal entry point when the condition that releases that process from "suspension" is satisfied. The only way an **Event** that is written into an `outChannel` is delivered is if the recipient had called `waitOn` for the corresponding `inChannel` at a time prior to that at which the `Event` was written. Thus, we see that some of the "events" implicit in an SSF model with event orientation are kernel events, which decide whether model events ought to be executed as a result. Writing to an `outChannel` schedules a kernel event at the `Event`'s receive time, but the kernel's processing of that event determines whether an `action` body is called. Nevertheless, execution of `action` bodies constitutes the essential "event processing" when SSF is used in a purely event-oriented view. It is interesting that, from a conceptual point of view, there is very little difference between process-oriented and event-oriented SSF.

To conclude this discussion on tools, we remark that *flexibility* is the key requirement in computer-systems simulation. Flexibility in most contexts means the ability to use the full power of a general programming language. This requires a level of programming expertise that is not needed by users of commercial graphically oriented modeling packages. The implementation requirements of an object-oriented event-oriented approach are much less delicate than those of

Table 14.1   Decreasing Abstraction and Model Results

Typical System	Model Results	Tools
CPU Network	job throughput,	queueing network,
	job response time	Petri net simulators, scratch
Processor	instruction throughput,	VHDL, scratch
	time/instruction	
Memory System	miss rates, response time	VHDL, scratch
ALU	timing, correctness	VHDL, scratch
Logic Network	timing, correctness	VHDL, scratch

a threaded simulator, and the amount of simulator overhead involved in delivering an event to an object is considerably less than the cost of a context switch in a threaded system. For these reasons, most of the simulators written from scratch take the event-oriented view. However, the underlying simulation framework necessarily provides a lower level of abstraction and so forces a modeler to design and implement more model-management logic. The choice between using a process-oriented or an event-oriented simulator—or writing one's own—is a function of the level of modeling ease, versus execution speed.

To summarize this section, we present a table that lists different levels of abstraction in computer-systems simulation, the sorts of questions whose answers are sought from the models, and the sorts of tools typically used for modeling. The level of abstraction decreases as one descends through Table 14.1.

14.3 MODEL INPUT

Just as there are different levels of abstraction in computer-systems simulation, there are different means of providing input to a model. The model might be driven by stochastically generated input, or it might be given trace input, measured from actual systems. Simulations at the high end of the abstraction hierarchy most typically use stochastic input; simulations at lower levels of abstraction commonly employ trace input. Stochastic input models are particularly useful when one wishes to study system behavior over a range of scenarios; it could be that all that is required is to adjust an input model parameter and rerun the simulation. Of course, using randomly generated input raises the question of how real or representative the input is; that doubt frequently induces systems people to prefer trace data on lower level simulations. Using a trace means one cannot explore different input scenarios, but traces are useful when directly comparing two different implementations of some policy or some mechanism on the same input. The realism of the input gives the simulation added authority.

In all cases, the data used to drive the simulation is intended to exercise whatever facet of the computer system is of interest. High-level systems simulations accept a stream of job descriptions; CPU simulations accept a stream of instruction descriptions; memory simulations accept a stream of memory references; and gate-level simulations accept a stream of logical signals.

Computer systems modeled as queueing networks (recall Chapter 7) typically interpret "customers" as computer programs; servers typically represent services such as attention by the CPU or an Input–Output (I/O) system. Random sampling generates customer interarrival times; it may also be used to govern routing and time in service. However, it is common in computer-systems contexts to have routing and service times be state dependent (e.g., the next server visited is already specified in the customer's description, or could be the attached server with least queue length).

Interarrival processes have historically been modeled as Poisson processes (where times between successive arrivals have an exponential distribution). However, this assumption has fallen from favor as a result of empirical observations that significantly contradict Poisson assumptions in current computer and communication systems. The real value of Poisson assumptions lies in tractability for mathematical analysis, so, as simulationists, we can discard them with little loss.

In the subsections to follow, we look at the mathematical formulation of common input models, stochastic input models for virtual memory, and direct-execution techniques.

14.3.1 Modulated Poisson Process

Stochastic input models ought to reflect the real-life phenomenon called *burstiness*—that is, brief periods when traffic intensity is much higher than normal. An input model sometimes used to support this, retaining a useful level of mathematical tractability, is a *Modulated Poisson Process*, or MPP. (See Fischer and Meier-Hellstern, [1993].) The underlying framework is a continuous-time Markov chain (CTMC), whose details we sketch so as to employ the concept later. A CTMC is always in some *state*; for descriptive purposes, states are named by the integers: $1, 2, \ldots$. The CTMC remains in a state for a random period of time, transitions randomly to another state, stays there for a random period of time, transitions again, and so on. The CTMC behavior is completely described by its *generator matrix*, $\mathcal{Q} = \{q_{i,j}\}$. For states $i \neq j$, entry $q_{i,j}$ describes the rate at which the chain transitions from state i into state j (this is the total transition rate out of state i, times the probability that it transitions then into state j). The rate describes how quickly the transition is made; its units are transitions per unit simulation time. Diagonal element $q_{i,i}$ is the negated sum of all rates out of state i : $q_{i,i} = -\sum_{j \neq i} q_{i,j}$. An operational view of the CTMC is that, upon entering a state i, it remains in that state for an exponentially distributed period of time, the exponential having rate $-q_{i,i}$. When making the transition, it chooses state j with probability $-q_{i,j}/q_{i,i}$. Many CTMCs are *ergodic*, meaning that, if it is left to run forever, every state is visited infinitely often. In an ergodic chain, π_i denotes state i's *stationary probability*, which we can interpret as the long-term average fraction of time the CTMC is in state i. A critical relationship exists between stationary probabilities and transition rates : For every state i,

$$\pi_i \sum_{j \neq i} q_{i,j} = \sum_{j \neq i} \pi_j q_{j,i}$$

If we think of $q_{i,j}$ as describing a probability "flow" that is enabled when the CTMC is in state i, then these equations say that, in the long term, the sum of all flows out of state i is the same as the sum of all flows into the state. We will see in the example that follows that we can use

the balance equations to build a stochastic input with desired characteristics. To complete the
definition of a MPP, it remains only to associate a customer arrival rate λ_i with state i. When the
CTMC is in state i, customers are generated as a Poisson process with rate λ_i.

To illustrate, let us consider an input process that is either OFF, ON, or BURSTY (the output
rate is much higher in the BURSTY state than in the ON state). We wish for the process to be
OFF half of the time—on average, for 1 second—and, when it is not OFF, we wish for it to be
BURSTY for 10% of the time. We will assume that the CTMC transitions into BURSTY only
from the ON state and transitions out of BURSTY only into the ON state. We will say that state 0
corresponds to OFF, 1 to ON, and 2 to BURSTY. Our problem statement implies that $\pi_0 = 0.5$,
$\pi_1 = 0.45$, and $\pi_2 = 0.05$. The only transition from OFF is to ON, and the mean OFF time is 1,
so we infer that $q_{0,1} = 1$. The balance equation for state 0 can be rewritten as

$$0.5 = 0.45q_{1,0}$$

and hence $q_{1,0} = (0.5/0.45)$. The balance equation for state 1 can be rewritten as

$$0.45((0.5/0.45) + q_{1,2}) = 0.5 + 0.05q_{2,1}$$

and the balance equation for state 2 is

$$0.05q_{2,1} = 0.45q_{1,2}$$

The equations for states 1 and 2 are identical; mathematically, we don't have enough conditions
to force a unique solution. If we add the constraint that a BURSTY period lasts, on average, $1/10$
of a second, we thereby define that $q_{2,1} = 10$ and, hence, that $q_{1,2} = (0.5/0.45)$. Operationally,
the simulation of this CTMC is straightforward. In state 0, one samples an exponential with mean
1 to determine the state's holding time. Following this period, the CTMC transitions into state 1
and samples a holding time from an exponential with mean 0.45, after which it transitions to OFF
or BURSTY with equal probability. In the BURSTY state, it samples an exponential holding time
with mean 0.1. Now all that is left is for us to define the state-dependent customer arrival rates.
Obviously, $\lambda_0 = 0$; for illustration, we choose $\lambda_1 = 10$ and $\lambda_2 = 500$.

Figure 14.3 presents a snippet of code used to generate times of arrivals in this process.
Transitions between states are sampled by using the inverse-transform technique, described in
Chapter 9. (The variable `acc` computes the cumulative probability function in the distribution
described by the row vector `P[state]`.) Figure 14.4 plots total customers generated as a function
of time—for a short period of a sample run, and for a longer period. In the shorter run, we see
regions where the graph increases sharply; they correspond to periods in the BURSTY state.
While the CTMC is not in this state, a mixture of OFF and ON periods moves the accumulated
packet count up at a much more gradual rate. The MPP model can describe burstiness, but the
burstiness is limited in time scale. The longer run views the data at a time scale that is two orders
of magnitude larger, and we see that the irregularities are largely smoothed.

In contrast to the Markovian essence of the MPP model, consider a traffic source that remains
OFF for an exponentially distributed period of time with mean 1.0, but, when it comes ON, remains
on for a period of time sampled from a Pareto distribution. While it is ON, packets arrive as a
Poisson process. As we will see in the chapter on simulation of computer networks, the Pareto
distribution is of particular interest because it gives rise to "self-similarity," which informally
means preservation of irregularities at multiple time scales. Figure 14.5 parallels the MPP data,
displaying accumulated packet counts as a function of time; it presents behavior for the first 1000

```
class mpp {

  public static double Finish;          // sim termination
  public static double time = 0.0;      // current clock
  public static double htime, etime;    // transition times
  public static int state = 0;          // current state id
  public static int total = 0;          // total pkts emitted
  public static Random stream;
  ...

public static void main(String argv[]) {
  ...
while( time < Finish ) {

  // generate exponential holding time, state-dependent mean
  htime = time+exponential( stream, hold[state] );

  // emit packets until state transition time. State dependent
  // rate. Note assignment made to etime in while condition test
  while( (etime = time+exponential( stream, 1.0/rate[state]))
                   < min( htime, Finish) )  {
        System.out.println( etime + `` '' + total);
     total++;
     time = etime;    // advance to packet issue time
  }
  time = htime;

  // select next state
  double trans = stream.nextDouble();
  double acc = P[state][0];
  int i = 0;

  while( acc < trans ) acc += P[state][++i];
  state = i;
   }
  }
  ...
}
```

Figure 14.3   Java code generating MPP trace.

units of time and for the first 100,000 units of time. Here, despite two orders of magnitude of difference in run length, the visual impression of behavior is much the same between the two traces. This sort of behavior is frequently seen in computer and communication systems; the long lengths reflect burstiness of packets, file lengths, and demand on a server.

14.3.2 Virtual-Memory Referencing

Randomness can also be used to drive models in the middle levels of abstraction. An example is a model of program-execution behavior in a computer with virtual memory. (See Nutt [2004].) In such a system, the data and instructions used by the program are organized in units called *pages*. All pages are the same size, typically 2^{10} to 2^{12} bytes in size. The physical memory of a

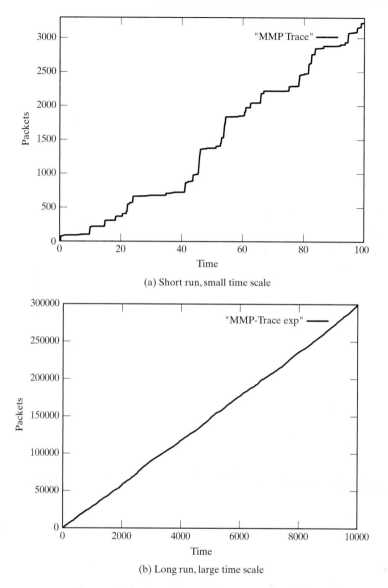

(a) Short run, small time scale

(b) Long run, large time scale

Figure 14.4 Sample runs from MPP model.

computer is divided into *page frames*, each capable of holding exactly one page. The decision of which page to map to which frame is made by the operating system. As the program executes, it makes memory references to the "virtual memory," as if it occupied a very large memory starting at address 0 and were the only occupant of the memory. On every memory reference made by the program, the hardware looks up the identity of the page frame containing the reference and translates the virtual address into a physical address. The hardware might discover that the

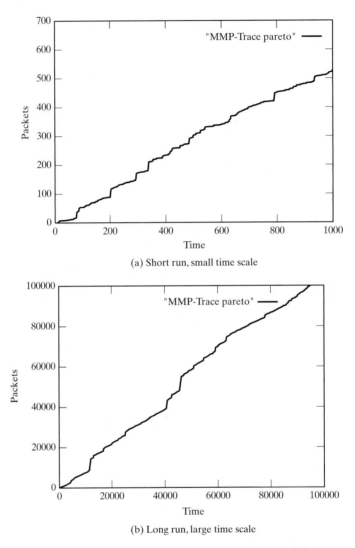

(a) Short run, small time scale

(b) Long run, large time scale

Figure 14.5   Sample runs from self-similar model.

referenced page is not present in the main memory; this situation is called a *page fault*. When a page fault occurs, the hardware alerts the operating system, which then takes over to bring in the referenced page from a disk and decides which page frame should contain it. The operating system could need to evict a page from a page frame to make room for the new one. The policy the operating system uses to decide which page to evict is called the "replacement policy." The quality of a replacement policy is often measured in terms of the *hit ratio*—the fraction of references made whose page frames are found immediately.

Virtual-memory systems are used in computers that support concurrent execution of multiple programs. In order to study different replacement policies, one could simulate the memory-referencing behavior of several different programs, simulate the replacement policy, and count the number of references that page fault. For this simulation to be meaningful, it is necessary that the stochastically generated references capture essential characteristics of program behavior. Virtual memory works well precisely because programs do tend to exhibit a certain type of behavior; this behavior is called locality of reference. What this means intuitively is that program references tend to cluster in time and space and that, when a reference to a new page is made and the page is brought in from the disk, it is likely that the other data or instructions on the page will also soon be referenced. In this way, the overhead of bringing in the page is amortized over all the references made to that page before it is eventually evicted. A program's referencing behavior can usually be separated into a sequence of "phases"; during each phase, the program makes references to a relatively small collection of pages, called its *working set*. Phase transitions essentially change the program's working set. The challenge for the operating system is to recognize when the pages used by a program are no longer in its working set, for these are the pages it can safely evict to make room for pages that *are* in some program's working set.

Figure 14.6 illustrates a stream of memory references taken from an execution of the commonly used gcc compiler. One graph gives a global picture; the other cuts out references to pages over number 100 and shows more fine detail. Each graph depicts points of the form (i, p_i) where p_i is the page number of the ith reference made by the program (arithmetically shifted so that the smallest page number referenced is 10). The phases are clearly seen; each member of the working set of a phase is seen as lines (which are really just a concatenation of many points). One striking facet of this graph is how certain pages remain in almost all working sets. However, other kinds of programs exhibit other behaviors. A common characteristic of scientific programs is that the execution is dominated by an inner loop that sweeps over arrays of data; the pages containing the instructions are in the working set throughout the loop, but data pages migrate in and out.

Despite various differences, a near-invariant among program executions is the presence of phase-like behavior and of working sets. In the building of a stochastic reference generator, it therefore makes sense to focus modeling effort on phase and working-set definition. As a starting point, we might, with every reference generated, randomly choose (with some small probability) whether to start a new phase (by changing the working set). Given a working set, we would choose to reference some page in the working set with high probability and, if choosing to stay in the set, choose with high probability the same page as the one last referenced in the working set. The inner loop of a program that generates references in this fashion appears in Figure 14.7. Details of working-set definition are hidden inside of routine new_wrkset and might vary with the type of program being modeled. For the purposes of illustration here, we wrote a version that defined a working set by randomly choosing a working-set size between 2 and 8 and a maximum page number of 100. A working set of size n is constructed by randomly choosing a "center" page c from among all pages, randomly choosing an integer dispersion factor d from 2 to 6, and then randomly selecting a working set from among all pages within distance $d \times n$ from center page c (with appropriate wraparound of page numbers at the endpoints 0 and 100). In order to model the referencing pattern of a scientific program's instruction stream, we manipulated the logic illustrated above to "lock down" a working set for a long time in the middle of the program execution. Figure 14.8 illustrates the result. As designed, phases and working sets are precisely defined.

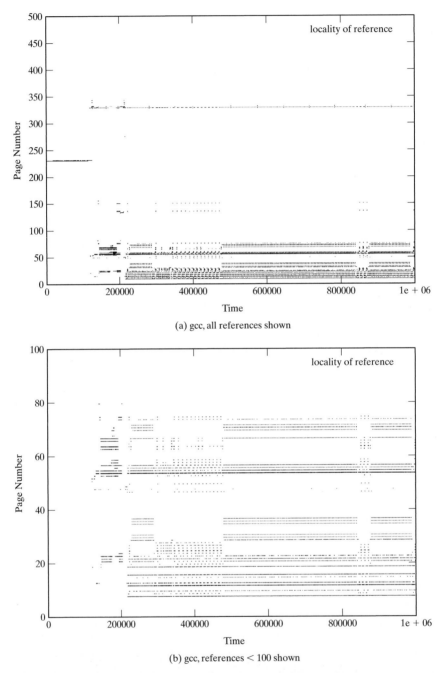

(a) gcc, all references shown

(b) gcc, references < 100 shown

Figure 14.6   Scatter-plotted referencing pattern of gcc compiler. Referenced page number is plotted as a function of reference number ("time"). Horizontal sequences indicate frequent rereferences to the same page number.

```
double ppt = 0.0001;   // Pr{phase transition}
double psw = 0.999;    // Pr{ref in WS}
double psp = 0.9;      // Pr{reference same page}

// method new_wrkset() creates a new working set
// method from_wrkset() samples from the working set
// method not_from_wrkset() samples from outside the working set

int ref;               // last page referenced
int sv_ref;            // save ref
Random stream;         // random number stream

...

for(int i=0; i<length; i++) {
   if( stream.nextDouble() <  ppt ) new_wrkset();   // phase transition
   if( stream.nextDouble() < psw ) {                // stay in working set?
    if( psp < stream.nextDouble() )                 // change page, in wrkset
       ref = sv_ref = from_wrkset();
   } else ref = not_from_wrkset();   // step outside of wrkset

 System.out.println(i + `` '' + ref);
 ref = sv_ref;
 }
```

Figure 14.7   Java pseudocode for generating a reference trace.

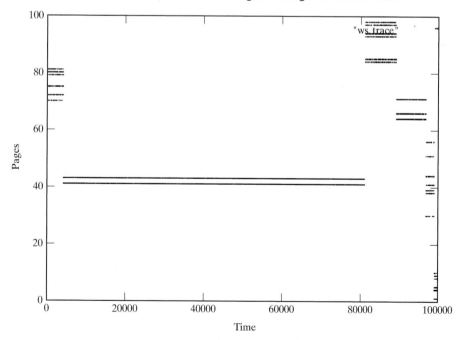

Figure 14.8   A synthetic trace modeling a scientific-program instruction stream.

The preceding example illustrates how one can in principle generate an execution path stochastically, but simulations at the middle level of abstraction also commonly use traces. Studies of CPU design will use a measured trace of instructions executed by a running program; studies of memory systems will use a measured trace of the addresses referenced by an executing program. Such traces get to be lengthy. A small piece of a typical trace of memory references is shown here:

```
2       430d70
2       430d74
2       415130
0     1000acac
2       414134
1     7fff00ac
2       414138
```

The first number is a code describing the type of access; 2 represents an instruction fetch, 0 a data read, 1 a data write. The second number represents a memory address, in hexadecimal. If the trace were also to describe the instruction stream, a hexadecimal word giving the machine code of the instruction fetched could follow the memory address on every instruction fetch line. Two or three words of memory are needed to represent one reference, even when the information is efficiently packed (not as characters, as shown, which take much more space!). Consider also the amount of computation needed to simulate a CPU or memory for the execution of a significantly long run of a nontrivial program. These observations help us understand the motivation for techniques that compress the address trace and for techniques that allow one to infer information about multiple systems from a single pass through a long trace. We will say more about these techniques later in this chapter.

Another method of generating input is called "direct execution" simulation. (For examples, see Covington *et al.* [1991], Lebeck and Wood [1997], Dickens *et al.* [1996]). One approach to it is illustrated in Figure 14.9. Direct execution is like generating a trace and driving the simulation with that trace, all at once. Computer programs are "instrumented" with additional code that observes the instructions the program executes and the memory and I/O references the program makes as it executes. The instrumented program is compiled and linked with a simulation kernel library. Execution control rests with the simulation kernel, which calls the instrumented program to provide the next instruction or reference that the program generates. The simulation kernel uses the returned information to drive the model for the next step. The simulation model driven by the program's execution can be of an entirely different CPU design, or a memory system, or even (given multiple instrumented programs) the internals of a communications network. Direct-execution simulation solves the problem of storing very large traces—the trace is consumed as it is being generated. However, it is tricky to modify computer programs to get at the trace information and to coordinate the trace generator with discrete-event simulator. The only practical way an ordinary simulator practitioner can use such methods is when the system has a software tool for making such modifications, but this feature is not common.

14.4 HIGH-LEVEL COMPUTER-SYSTEM SIMULATION

In this section, we illustrate concepts typical of high-level computer simulations by sketching a simulation model of a computer system that services requests from the World Wide Web.

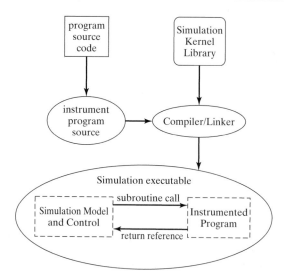

Figure 14.9   Direct-execution simulation.

Example 14.1

A company that provides a major website for searching and links to sites for travel, commerce, entertainment, and the like wishes to conduct a capacity-planning study. The overall architecture of its system is shown in Figure 14.10. At the back end, one finds data servers responsible for all aspects of handling specific queries and updating databases. Data servers receive requests for service from application servers—machines dedicated to running specific applications (e.g., a search engine) supported by the site. In front of the applications are Web servers, which manage the interaction of applications with the World Wide Web; the portal to the whole system is a load-balancing router that distributes requests directed to the website among the Web servers.

The goal of the study is to evaluate the site's ability to handle load at peak periods. The desired output is an empirical distribution of the access response time. Thus, the high-level simulation model should focus on the impact of timing at each level that is used, system factors that affect that timing, and the effects of timing on contention for resources. To understand where those delays occur, let us consider the processing associated with a typical query.

All entries into the system are through a dedicated router, which examines the request and forwards it to some Web server. Time is required to exercise the logic of looking at the request to discern whether it is a new request (requiring load balancing) or part of an ongoing session. It is reasonable to assume one switching time for a preexisting request and a different time for a new request. The result of the first step is selection of a Web server and the enqueueing there of a request for service. A Web server can be thought of as having one queue of threads of new requests, a second queue of threads that are suspended awaiting a response from an application server, and a third queue of threads "ready" to process responses from application servers. An accepted request from the router creates a new request thread. We may assume the Web server has adequate memory to deal with all requests. It has a queueing policy that manages access to the CPU; the distinction between new requests and responses from application servers is maintained

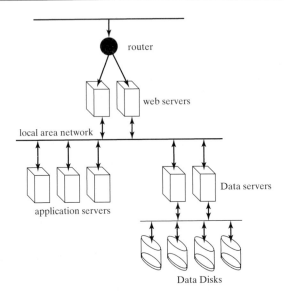

Figure 14.10 Website server system.

for the sake of scheduling and for the sake of assigning service times, the distributions of which depend on the type. The servicing of a new request amounts to identification of an application and the associated application server. A request for service is formatted and forwarded to an application server, and the requesting thread joins the suspended queue. At an application server, requests for service are organized along application types. A new request creates a thread that joins a new-request queue associated with the identified application. An application request is modeled as a sequence of sets of requests from data servers, interspersed with computational bursts—for example,

```
burst 1
request data from D1, D3, and D5
burst 2
request data from D1 and D2
burst 3
```

In this model, we assume that all data requests from a set must be satisfied before the subsequent computational burst can begin. Query search on a database is an example of an application that could generate a long sequence of bursts and data requests, with large numbers of data requests in each set. We need not assume that every execution of an application is identical in terms of data requests or execution bursts; these can be generated stochastically. An application thread's state will include description of its location in its sequence and a list of data requests still outstanding before the thread can execute again. Thus, for each application, we will maintain a list of threads that are ready to execute and a list of threads that are suspended awaiting responses from data servers. An application server will implement a scheduling policy over sets of ready application threads. A data server creates a new thread to respond to a data request and places it in a queue of ready threads. Some data server might implement memory-management policies and could require further coordination with the application server to know when to release used memory. Upon receiving service, the thread requests data from a disk, then suspends until the disk operation completes, at which point the thread is moved from the suspended list to the ready list and, when executed, again reports back to the application server associated with the request. The thread suspended at the application server responds; eventually, the application thread finishes and reports its completion back to the Web-server thread that initiated it, which in turn communicates the results back over the Internet.

Stepping back from the details, we see that a simulation model of this system must specify a number of features, listed in Table 14.2. All of these affecting timing in some way. The query-response-time distribution can be estimated by measuring, for each query, the time between at which a request first hits the router and the time at which the Web-server thread communicates the results. From the set of simulated queries, one can build up a histogram. As should be evident, a response time reflects a great many different factors related to execution bursts, scheduling policies, and disk-access times. Deeper understanding of the system is obtained by measuring behavior at each server of each type. One would look especially for evidence of bottlenecks. CPU bottlenecks would be reflected at servers with high CPU utilization; IO bottlenecks at disks with high utilization. To assess system capacity at peak loads, we would simulate to identify bottlenecks, then look to see how to reduce load at bottleneck devices by changes in scheduling policies, by binding of applications to servers, or by increasing the number of CPUs or disks in the system. Normally, one must resimulate a reconfigured system under the same load as before to assess the effects of the changes.

The website model is an excellent candidate for a threaded (process-oriented) approach to modeling. The most natural process-oriented approach is to associate processes with servers. The simulation model is expressed from an abstracted point of view of the servers' operating system. Individual queries become messages that are passed between server processes. In additional to limiting the number of processes, an advantage of this approach is that it explicitly exposes the scheduling of query processing at the user level. The modeler has both the opportunity and the responsibility to provide the logic of scheduling actions that model processing done on behalf of a query. It is a modeling viewpoint that simplifies analysis of server behavior—an overloaded server is easily identified by the (modeler-observable) length of its queue of runnable queries. However, it is a modeling viewpoint that is a bit lower in abstraction than the first one and requires more modeling and coding on the part of the user.

An event-oriented model of this system need not look a great deal different from the second of our process-oriented models. A query passed as a message between servers have an obvious event-oriented expression. A modeler would have to add to the logic, events, and event handlers that describe the way a CPU passes through simulation time. For example, consider a call to *hold(qt)*

Table 14.2 Required Specification for Web System Model

Subsystem	Specifications
Router	load-balancing policy, execution times
Web Server	server count, queueing policy, execution times
Application Server	server count, queueing policy, behavior model
Data Server	server count, disk count, queueing policy, memory policy, disk timing

in a process-oriented model to express that the CPU is allocating `qt` units of service to a query, during which time it does nothing else. In an event-oriented model, one would need to define events that reflect "starting" and "stopping" the processing of a query, with some scheduling logic interspersed. Additional events and handlers need to be defined for any "signaling" that might be done between servers in a process-oriented model—for example, when a data-server process awaits completion of modeled IO requests sent to its disks. A process-oriented approach, even one focused on servers rather than queries, lifts the level of model expression to a higher level of abstraction and reduces the amount of code that must be written. In a system as complex as the website, one must factor complexity of expression into the overall model-development process.

14.5 CPU SIMULATION

Next, we consider a lower level of abstraction and look at the simulation of a central processing unit. Whereas the high-level simulation of the previous example treated execution time of a program as a constant, at the lower level we do the simulation to discover what the execution time is. The input driving this simulation is a stream of instructions. The simulation works through the mechanics of the CPU's logical design to find out what happens in response to that stream, how long it takes to execute the program, and where bottlenecks exist in the CPU design. Our discussion illustrates some of the functionality of a modern CPU and the model characteristics that such a simulation seeks to discern. Examples of such simulations include those described in Cmelik and Keppel [1994], Bedicheck [1995], Witchel and Rosenblum [1996], Austin, Larson, and Ernst [2002], Bohrer *et al.* [2004] and Magnusson *et al.* [2002]. The view of the CPU taken in our discussion is similar to that taken by the RSIM system (Hughes *et al.* [2002]).

The main challenge to making effective use of a CPU is to avoid stalling it; stalling happens whenever the CPU commits to executing an instruction whose inputs are not all present. A leading cause of stalls is the latency delay between CPU and main memory, which can be tens of CPU cycles. One instruction might initiate a read—for example,

```
load $2, 4($3)
```

which is an assembly language statement that instructs the CPU to use the data in register 3 (after adding value 4 to it) as a memory address and to put the data found at that address into register 2. If the CPU insisted on waiting for that data to appear in register 2 before further execution, the instruction could stall the CPU for a long time if the referenced address is not found in the cache. High-performance CPUs avoid this by recognizing that additional instructions *can* be executed, up to the point where the CPU attempts to execute an instruction that reads the contents of register 2—for example,

```
add $4,$2,$5
```

This instruction adds the contents of registers 2 and 5, and places the result in register 4. If the data expected in register 2 is not yet present, the CPU will stall. So we see that, to allow the CPU to continue past a memory load, it is necessary to (1) mark the target register as being unready, (2) allow the memory system to load the target register asynchronously while the CPU continues on in the instruction stream, (3) stall the CPU if it attempts to read a register marked as unready, and (4) clear the unready status when the memory operation completes.

The sort of arrangement just described was first used in the earliest supercomputers, designed in the 1960s. Modern microprocessors add some additional capabilities to exploit *instruction level parallelism* (ILP). We outline some of the current architecture ideas in use to illustrate what a simulation model of an ILP CPU involves.

The technique of pipelining has long been recognized as a way of accelerating the execution of computer instructions. (See Patterson and Hennessy [1997].) Pipelining exploits the fact that each instruction goes sequentially through several stages in the course of being processed; separate hardware resources are dedicated to each stage, permitting multiple instructions to be in various stages of processing concurrently. A typical sequence of stages in an ILP CPU is as follows:

1. *Instruction fetch*: The instruction is fetched from the memory.
2. *Instruction decode*: The memory word holding the instruction is interpreted to discover what operation is specified; the registers involved are identified.
3. *Instruction issue*: An instruction is "issued" when there are no constraints holding it back from being executed. Constraints that keep an instruction from being issued include data not yet being ready in an input register and unavailability of a functional unit (e.g., Arithmetic Logical Unit) needed to execute the instruction.
4. *Instruction execute*: The instruction is performed.
5. *Instruction complete*: Results of the instruction are stored in the destination register.
6. *Instruction graduate*: Executed instructions are graduated in the order that they appear in the instruction stream.

Ordinary pipelines permit at most one instruction to be represented in each stage; the degree of parallelism (number of concurrent instructions) is limited to the number of stages. ILP designs allow multiple instructions to be represented in some stages. This necessarily implies the possibility of executing some stages of successively fetched instructions out of order. For example, it is entirely possible for the nth instruction, I_n, to be constrained from being issued for several clock cycles while the next instruction, I_{n+1}, is not so constrained. An ILP processor will push the evaluation of I_{n+1} along as far as it can without waiting on I_n. However, the *instruction graduate* stage will reimpose order and insist on graduating I_n before I_{n+1}.

ILP CPUs use architectural slight of hand with respect to register useage to accelerate performance. An ILP machine typically has more registers available than appear in the instruction set. Registers named in instructions need not precisely be the registers actually used in the implementation of those instructions. This is acceptable, of course, as long as the effect of the instructions is the same in the end. One factor motivating this design is the possibility of having multiple instructions involving the same logical registers (those named by the instructions themselves) actively being processed concurrently. By providing each instruction with its own "copy" of a register, we eliminate one source of stalls. Another factor involves branches—that is, instructions that interrupt the sequential flow of control. An ILP, encountering a branch instruction, will predict whether the branch is taken or not and possibly alter the instruction stream as a result. Various methods exist to predict branching, but any of them will occasionally predict incorrectly. When an incorrect prediction is made, the register state computed as a result of speculating on branch outcome needs to be discarded and execution resumed at the branch point. Thus, another use of additional registers is to store the "speculative register state." With dedicated hardware resources to track register useage following speculative branch decision, speculative state can be discarded in a single cycle and control resumed at the mispredicted branch point. In all of these cases, the hardware implements techniques for renaming the logical registers that appear in the instructions to physical registers, for maintaining the mapping of logical to physical registers, and for managing physical register useage.

A simulation model of an ILP CPU will model the logic of each stage and coordinate the movement of instructions from stage to stage. We consider each stage in turn.

An instruction-fetch stage could interact with the simulated memory system, if that is present. However, if the CPU simulation is driven by a direct-execution simulation or by a trace file, there is little for a model of this stage to do but get the next instruction in the stream. If a memory system is present, this stage could look into an instruction cache for the next referenced instruction, stalling if a miss is suffered.

Following an instruction fetch, an instruction will be in the CPU's list of active instructions until it exits altogether from the pipeline. The instruction-decode stage places an instruction in this list; a logical register that appears as the target of an operation is assigned a physical register—registers used as operand sources will have been assigned physical registers in instructions that defined their values. (Sequencing issues associated with having multiple representations of the same register are dealt with at a later stage in the pipeline.) Branch instructions are identified in this stage, predictions of branch outcomes are made, and resources for tracking speculative execution are committed here.

Decoded instructions pass into the instruction-issue stage. The logic here is complex and very much timing dependent. An instruction cannot be issued until values in its input registers are available and a functional unit needed to perform the instruction is available. An input value might be not yet in a register, for instance, if that value is loaded from memory by a previous instruction and has not yet appeared. A functional unit could be unavailable because all appropriate ones are busy with multicycle operations initiated by other instructions. Implementation of the issue-stage model (and hardware) depends on marking registers and functional units as busy or pending and on making sure that, when the state of a register or functional unit changes, any instruction that cannot yet issue because of that register or functional unit is reconsidered for issue.

Simulation of the instruction-execute stage is a matter of computing the result specified by the instruction (e.g., an addition). At this point, the action of depositing the result into a register or memory is scheduled for the instruction-complete stage. This latter stage also cleans up the status bits associated with registers and functional units involved in the instruction and resolves the final outcome of a predicted branch. If a branch was mispredicted, the speculatively fetched and processed instructions that follow it are removed from other pipeline stages, the hardware that tracks speculative instruction is released, and the instruction stream is reset to follow the branch's other decision direction.

Between the instruction-issue and instruction-complete stages, instructions could get processed in an order that does not correspond to the original instruction stream. The last stage, graduation, reorders them. Architecturally, this permits an ILP CPU to associate an exception (e.g., a page fault or a division by zero) with the precise instruction that caused it. Simulation of this stage is a matter of knowing the sequence number of the next instruction to be graduated, then graduating it when it appears.

Example 14.2

An example helps to show what goes on. Consider the following sequence of assembly-language instructions for a hypothetical computer:

```
load $2, 0($6) ; I1-  load $2 from memory
mult $5, 2      ; I2-  multiply $5 with constant 2
add  $4, 12     ; I3-  add constant 12 to $4
add  $5, $2     ; I4-  $5 <- $5 + $2
add  $5, $4     ; I5-  $5 <- $5 + $4
```

Let us suppose that the register load misses the first-level cache but hits in the second-level cache, resulting in a delay of 4 cycles before the register gets the value. Suppose further that separate hardware exists for addition and multiplication, that addition takes one cycle, and that multiplication takes 2 cycles to complete. Time is assumed to advance in units of a single clock tick.

Table 14.3 shows a timeline of when each instruction is in each stage. Cycles in which an instruction cannot proceed through the pipeline are marked as "stall" cycles. Processing is most easily understood by tracing individual instructions through.

I1.   After being fetched in cycle 1, the decode of I1 assigns physical register $p1 as the target of the load operation and marks $p1 as unready. No constraints prohibit I1 from being issued in cycle 3 nor executed in cycle 4. Because the memory operation takes 4 cycles to finish, I1 is stalled in cycles 5–8. Cycle 9 commits the data from memory to physical register $p1 and clears its unready flag; the instruction is graduated in cycle 10.

Table 14.3   Pipeline Stages, ILP CPU Simulation

Inst./Cycle	1	2	3	4	5	6	7
I1	fetch	decode	issue	execute	stall	stall	stall
I2		fetch	decode	issue	execute	stall	complete
I3			fetch	decode	issue	execute	complete
I4				fetch	decode	stall	stall
I5					fetch	decode	stall

Inst./Cycle	8	9	10	11	12	13	14
I1	stall	complete	graduate				
I2	stall	stall	stall	graduate			
I3	stall	stall	stall	stall	graduate		
I4	stall	stall	issue	execute	complete	graduate	
I5	stall	stall	stall	stall	stall	issue	complete

I2. Instruction I2 is fetched in cycle 2 and has physical register $p2 allocated to receive the results of the multiplication in the cycle-3 decode stage; $p2's unready flag is raised. No constraints keep I2 from being issued in cycle 4 or executed in cycle 5, but the 2-cycle delay of the multiplier means the result is not committed to register $p2 until cycle 7, at which point the $p2 unready flag is cleared. The instruction remains stalled through cycles 8–10, awaiting the graduation of I1.

I3. Instruction I3 is fetched in cycle 3 and has physical register $p3 allocated to receive the results of its addition in the cycle-4 decode stage. The $p3 unready flag is raised. There are no constraints keeping I3 from being issued in cycle 5 and executed in cycle 6, with results written into $p3 in cycle 7, at which time $p3's unready flag is cleared. I3 must stall, however, during cycles 8–11, awaiting the graduation of I2.

I4. Instruction I4 is fetched in cycle 4 and has physical register $p4 allocated to receive the results of the addition during the cycle-5 decode stage. $p4's unready flag is raised at that point. Physical registers $p1 and $p2 are operands to the addition; I4 stalls in cycles 6–9, waiting for their unready flags to clear. It then passes the remaining stages without further delay, clearing the $p4 unready flag in cycle 12.

I5. Instruction I5 is fetched in cycle 5 and has physical register $p5 allocated to receive the results of its addition in the cycle-6 decode stage, at which point the $p5 unready flag is set. Physical registers $p3 and $p4 contain the addition's operands; I5 stalls through cycles 7–12, waiting for their unready flags both to clear. From that point forward, I5 passes through the remaining stages without further delay.

The performance benefit of pipelining and ILP can be appreciated if we compare the execution time of this sequence on a nonpipelined, non-ILP machine. Assuming that each stage must be performed for each instruction but that one instruction is processed in its entirety before another one begins, 51 cycles are needed to execute I1 through I5. With the advanced architectural features, only 15 cycles are needed. The example illustrates both the parallelism that pipelining exposes and the latency tolerance that the ILP design supports. Even though I1 stalls for four cycles while awaiting a result from memory, the pipeline keeps moving other instructions through to some extent. The bottom line for someone using a model like this is the rate at which instructions are graduated, as this reflects the effectiveness of the CPU design. Secondary statistics would try to pinpoint where in the design stalls occur that might be alleviated (e.g., if many stalls occur because of waiting for the multiplier (no such stalls occur in the example), then one could consider including an additional multiplier in the CPU design).

Our explanation of the model's workings was decidedly process oriented, taking the view of an instruction. However, the computational demands of a model like this are enormous, owing to the very large number of instructions that must be simulated to assess the CPU design on, say, a single program run. The relatively high cost of context switching would deter use of a normal process-oriented language. One could implement what is essentially a process-oriented view by using events—each time an instruction passes through a stage, an event is scheduled to take that instruction through the next stage, accounting for stalls. The amount of simulation work accomplished per event is thus the amount of work done on behalf of one instruction in one stage. An alternative approach is to eschew explicit events altogether and simply use a cycle-by-cycle activity scan. At each cycle, one would examine each active instruction to see whether any activity associated with that instruction can be done. An instruction that was at one stage at cycle j will, at cycle $j + 1$, be examined for constraints that would keep it at cycle j. Finding none, that instruction would be advanced to the next stage. An activity-scanning approach has the attractiveness of eliminating event-list overhead, but the disadvantage of expending computational effort on checking the status of a stalled instruction on every cycle during which it is stalled. Implementation details and model behavior largely determine whether an activity-scanning approach is faster than an event-oriented approach (with the nod going to activity scanning when few instructions stall).

14.6 MEMORY SIMULATION

One of the great challenges of computer architecture is finding ways to deal effectively with the increasing gap in operation speed between CPUs and main memory. A factor of 100 in speed is not far from the mark. The main technique that has evolved is to build hierarchies of memories. A relatively small memory—the L1 cache—operates at CPU speed. A larger memory—the L2 cache—is larger and operates more slowly. The main memory is larger still and slower still. The smaller memories hold data that was referenced recently and nearby data that one hopes will also be referenced soon. Data moves up the hierarchy on demand and ages out as it becomes disused, to make room for the data in current use. For instance, when the CPU wishes to read memory location 100,000, hardware will look for it in the L1 cache; if it fails to find it there, it will look in the L2 cache. If it is found there, an entire block containing that reference is moved from the L2 cache into the L1 cache. If it is not found in the L2 cache, a (larger) block of data containing location 10,000 is copied from the main memory to the L2 cache, and part of that block (containing location 10,000 of course) is copied into the L1 cache. It could take 50 cycles or more to accomplish this. After this cost has been suffered, the hope and expectation is that the CPU will continue to make references to data in the block brought in, because accesses to L1 data are made at CPU speeds. Fortunately, most programs exhibit locality of reference at this scale (as well as at the paging scale discussed earlier in the chapter), so the strategy works. However, after a block ceases to be referenced for a time, it is ejected from the L1 cache. It could remain in the L2 cache for a while and later be brought back into the L1 cache if any element of the block is referenced again. Eventually a block remains unreferenced long enough so that it is ejected also from the L2 cache.

The astute reader will realize that data that is written into an L1 cache by the CPU creates a consistency problem, in that a memory address then has different values associated with it at different levels of the memory hierarchy. One way of dealing with this is to write through to all cache levels every time there is a write—the new value is asynchronously pushed from L1 through L2 to the main memory. An alternative method copies back a block from one memory level to the lower level, at the point the block is being ejected from the faster level. The write-through strategy avoids writing back blocks when they are ejected, whereas the write-back strategy requires that an entire block be written back when ejected, even if only one word of the block was modified, once. One of the roles simulation plays is to compare performance of these two write-back strategies, taking into consideration all costs and contention for the resources needed to support writing back modifications.

Like paging systems, the principle measure of the quality of a memory hierarchy is its hit ratio at each level. As with CPU models, to evaluate a memory hierarchy design, one must study the design in response to a very long string of memory references. Direct-execution simulation can provide such a reference stream, as can long traces of measured reference traffic. Nearly every caching system is a demand system, which means that a new block is not brought into a cache before a reference is made to a word in that block. Decisions left still to the designer include whether to write-through or write-back modifications, the replacement policy, and the "set associativity."

The concept of set associativity arises in response to the cost of the mechanism used to look for a match. Imagine we have an L2 cache with 2 million memory words (an actual figure from an actual machine). The CPU references location 10000—the main memory has, say, 2^{12} words, so the L2 cache holds but a minute fraction of the memory. How does the hardware find out whether location 10000 is in the L2 cache? It uses what is called an associative memory, one that associates search keys with data. One queries an associative memory by providing some search key. If the key is found in the memory, then the data associated with the key is returned; otherwise, indication of failure is given. In the caching context, the search key is derived from the reference address, and the return data is the data stored at that address. Caches must be very very fast, which means that the search process has to be abbreviated. This is accomplished by dedicating comparison hardware with every location in the associative memory. Presented with a search key, every comparator looks for a match with the key at its location. At most one comparator will see a match and return the data; it is possible that none will. A *fully associative* cache is one where any address can appear anywhere in the cache. This means building the cache to have a unique comparator associated with every address in the cache; doing so is prohibitively expensive. Tricks are played with memory addresses in order to reduce the costs greatly. The idea is to partition the address space into sets. Figure 14.11 illustrates how a 48-bit memory address might be partitioned in key, set id, and block offset. Any given memory address is mapped to the set identified by its set-id address bits. This scheme assigns the first block of 2^b addresses to set 0, the second block of 2^b addresses to set 1, and so on, wrapping around back to set 0 after 2^s blocks have been assigned. Each set is give a small portion of the cache—the set size—typically, 2 or 4 or 8 words. Only those addresses mapped to the same set compete for storage in that space. Only as many comparators are needed as there are words in the set. Given an address, the hardware uses the set-id bits to identify the set number and the key bits to identify the key. The hardware matches the keys of the blocks already in the identified set to comparator inputs and also provides the key of the sought address as input to all the comparators. Comparisons are made in parallel; in the case of a match, the block-offset bits are used to index into the identified block to select the particular address being referenced.

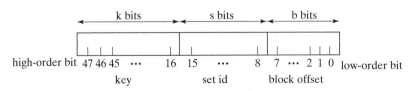

Figure 14.11   48-bit address partitioned for cache.

The overall size of this cache is seen to be the total number of sets times the set size. One role of simulation is to work out, for a given cache size, how the space ought to be partitioned into sets. This is largely a cost consideration, for increasing the set size (thereby reducing the number of sets) typically increases the hit ratio. However, if a set size of 4 yields a sufficiently large hit ratio, then there is little point to increasing the set size (and cost).

Least Recently Used (LRU) is the replacement policy most typically used. When a reference is made but is not found in a set, some block in the set is ejected to make room for the one containing the new reference. Under LRU, the block selected for ejection is the one which, among all blocks in the set, was last referenced most distantly in the past.

LRU is one of several replacement policies known as *stack* policies. (See Stone [1990].) These are characterized by the behavior that, for any reference in any reference string, if that reference misses in a cache of size n, then it also misses in every cache of size $m < n$, and that, if it hits in a cache of size m, then it hits in every cache of size $n > m$. Simulations can exploit this fact to compute the miss ratio of many different set sizes, in just one pass of the reference string! Suppose that we do not wish to consider any set size larger than 64. Now we conduct the simulation with set sizes of 64. Every block in the cached set is marked with a priority—namely, the temporal index of the last reference made to it (e.g., the block containing the first reference in the string is marked with 1, the block containing the second reference is marked with a 2 (overwriting the 1, if the same as the previous block), etc.). When a block must be replaced, the one with the smallest index is selected. Imagine that the simulation organizes and maintains the contents of a cached set in LRU order, with the most recently referenced block first in the order. The *stack distance* of a block in this list is its distance from the front; the most recently referenced block has stack distance 1, the block referenced next most recently has stack distance 2, the LRU block has stack distance 64. Presented with a reference, the simulation searches the list of cache blocks for a match. If no match is found, then, by the stack property, no match will be found in any cache of a size smaller than 64, on this reference, for this reference string. If a match is found and the block has stack distance k, then no match will be found in any cache smaller than size k, and a match will always be found in a cache of size larger than k. Rather than record a hit or miss, one increments the kth element of a 64-element array that records the number of matches at each LRU level. To find out how many hits occurred in a cache of size n, one sums up the counts of the first n elements of the array. Thus, with a little arithmetic at the end of the run, one can count (for each set cache) the number of hits for every set of every size between 1 and 64.

Figure 14.12 illustrates the evolution of an LRU list in response to a reference string. Under each reference (given as a alphabetic symbol rather than actual memory address) is the state of the LRU stack *after* the reference is processed. The horizontal direction from left to right symbolizes the trace, read from left to right. A hit is illustrated by a circle, with an arrow showing the migration of the symbol to the top of the heap. The "hits" array counts the number of hits found at each stack distance. Thus we see that a cache of size 1 will have the hit ratio 0/15, a cache of size 1 will have the hit ratio 1/15, and a cache of size 3 will have the hit ratio 6/15.

In the context of a set-associative cache simulation, each set must be managed separately, as shown in the figure. In one pass, one can get hit ratios for varying set sizes, but it is important to note that each change in set size corresponds to a change in the overall size of the entire cache. This technique alone does not let us in one pass discover the hit ratios for all the different ways reference trace   A   B   C   A   D   B   A   D   C   D   F   C   B   F   E   hits array

stack distance 1   A   B   C   A   D   B   A   D   C   D   F   C   B   F   E   0

stack distance 2   A   B   C   A   D   B   A   D   C   D   F   C   B   F   1

stack distance 3   A   B   C   A   D   B   A   A   C   D   F   C   B   5

Figure 14.12   LRU stack evolution.

one might partition a cache of a given capacity (e.g., 256 sets with set size 1 versus 128 sets with set size 2 versus 64 sets with set size 4). It actually is possible to evaluate all these possibilities in one pass, but the technique is beyond the scope of this discussion.

14.7 SUMMARY

This chapter looked at the broad area of simulating computer systems. It emphasized that computer-system simulations are performed at a number of levels of abstraction. Inevitably, it discussed a good deal of computer science along with the simulation aspects, for in computer-systems simulation the two are inseparable.

The chapter outlined fundamental implementation issues behind computer-system simulators—principally, how process orientation is implemented and how object-oriented concepts such as inheritance are fruitfully employed. Next it considered model input, ranging from stochastically generated traffic, to stochastically generated memory-referencing patterns, to measured traces and direct-execution techniques. The chapter was brought to a conclusion by looking at examples of simulation at different levels of abstraction: a WWW-site server system, an instruction-level CPU simulation, and simulation of set-associative memory systems.

The main point is that computer-system simulators are tailored to the tasks at hand. Appropriate levels of abstraction need to be chosen, as must appropriate simulation techniques.

REFERENCES

ASHENDEN, P.J. [2001], *The Designer's Guide to VHDL*, 2d ed., Morgan Kaufmann, San Fransisco.

AUSTIN T., E. LARSON, AND D. ERNST [2002], "SimpleScalar: An Infrastructure for Computer System Modeling," *IEEE Computer*, Vol. 35, No. 2, pp. 59–67.

BEDICHECK, R.C. [1995], "Talisman: Fast and Accurate Multicomputer Simulation," *Proceedings of the 1995 ACM SIGMETRICS Conference*, pp. 14–24, Ottawa, ON, May.

BOHRER P., M. ELNOZAHY, A. GHEITH, G. LEFURGY, T. NAKRA, J. PETERSON, R. RAJAMONY, R. ROCKHOLD, H. SHAFI, R. SIMPSON, E. SPEIGHT, K. SUDEEP, E. VAN HENSGERGEN, AND L. ZHANG [2004], "Mambo—A Full System Simulator for the PowerPC Architecture," *Performance Evaluation Review*, Vol. 31, No. 4, 8–12.

CMELIK, B., AND D. KEPPLE [1994], "Shade: A Fast Instruction-Set Simulator for Execution Profiling," *Proceedings of the 1994 ACM SIGMETRICS Conference*, pp. 128–137, Nashville, TN, May.

COVINGTON, R., S. DWARKADA, J. JUMP, S. MADALA, AND J. SINCLAIR [1991], "Efficient Simulation of Parallel Computer Systems," *International Journal on Computer Simulation*, vol. 1, No. 1, 1991.

COWIE, J., A. OGIELSKI, AND D. NICOL [1999], "Modeling the Global Internet," Vol. 1, No. 1, pp. 42–50.

DICKENS, P., P. HEIDELBERGER, AND D. NICOL [1996], "Parallelized Direct Execution Simulation of Message Passing Programs," *IEEE Trans. on Parallel and Distributed Systems*, Vol. 7, No. 10, pp. 1090–1105.

FISCHER, W., AND K. MEIER-HELLSTERN [1993], "The Markov-Modulated Poisson (MMPP) Cookbook," *Performance Evaluation*, Vol. 18, No. 2, pp. 149–171.

FISHWICK, P. [1992], "SIMPACK: Getting Started with Simulation Programming in C and C++," *Proceedings of the 1992 Winter Simulation Conference*, pp. 154–162, Washington, DC.

GRUNWALD, D. [1995], *User's Guide to Awesime-II*, Department of Computer Science, Univ. of Colorado, Boulder, CO.

HENNESSY, J.L., AND D.A. PATTERSON [1997], *Computer Organization and Design, The Hardware/Software Interface*, 2d ed., Morgan Kaufmann Publishers, Inc., Palo Alto, CA.

HUGHES, C., V. PAI, P. RANGANATHAN, AND S. ADVE [2002], "RSIM: Simulating Shared Memory Multiprocessors with ILP Processors," *IEEE Computer*, Vol. 35, No. 2, pp. 40–49.

LEBECK, A., AND D. WOOD [1997], "Active Memory: A New Abstraction for Memory System Simulation," *ACM Transactions on Modeling and Computer Simulation*, Vol. 7, No. 1, pp. 42–77.

LITTLE, M.C., AND D.L. McCUE [1994], "Construction and Use of A Simulation Package in C++," *C User's Journal*, Vol. 3, No. 12.

MAGNUSSON, P., M. CHRISTENSSON, J. ESKILSON, D. FORSGREN, G. HALLBER, J. HOGBERG, F. LARSSON, A. MOESTEDT, AND B. WERNER [2002], "SIMICS: A Full System Simulation Platform," *IEEE Computer*, Vol. 35, No. 2, pp. 50–58.

MANO, M. [1993], *Computer Systems Architecture*, 3d ed., Prentice Hall, Englewood Cliffs, NJ.

NUTT, G. [2004], *Operating Systems, A Modern Prespective*, 3d ed., Addison–Wesley, Reading MA.

SCHWETMAN, H. [1986], "CSIM: AC-Based, Process-Oriented Simulation Language," *Proceedings of the 1986 Winter Simulation Conference*, pp. 387–396.

SCHWETMAN, H.D. [2001], "CSIM 19: A Powerful Tool for Building Systems Models," *Proceedings of the 2001 Winter Simulation Conference*, pp. 250–255.

STONE, H. [1990], *High Performance Computer Architecture*, Addison–Wesley, Reading MA.

WITCHEL, E., AND M. ROSENBLUM [1996], "EMBRA: Fast and Flexible Machine Simulation," *Proceedings of the 1996 ACM SIGMETRICS Conference*, pp. 68–79, Philedelphia, PA, May.

EXERCISES

1. Sketch the logic of an event-oriented model of an M/M/1 queue. Estimate the number of events executed when processing the arrival of 5000 jobs. How many context switches on average does a process-oriented implementation of this queue incur if patterned after the SSF implementation of the single-server queue in Chapter 4?

2. For each of the systems listed, sketch the logic of a process-oriented model and of an event-oriented model. For both approaches, develop and simulate the model in any language:

   - a central-server queueing model: when a job leaves the CPU queue, it joins the I/O queue with shortest length.
   - a queueing model of a database system, that implements fork join: a job receives service in two parts. When it first enters the server it spends a small amount of

simulation time generating a random number of requests to disks. It then suspends (freeing the server) until such time as *all* the requests it made have finished, and then enqueues for its second phase of service, where it spends a larger amount of simulation time, before finally exiting. Disks may serve requests from various jobs concurrently, but serve them using FCFS ordering. Your model should report on the statistics of a job in service—how long (on average) it waited for phase 1, how long it waits on average for its I/O requests to complete, and how long it waits on average for service after its I/O requests complete.

3. Consider a Markov Modulated Process with three states (OFF, ON, BURSTY) and having the following characteristics:

   - The MMP is in the OFF state for 25% of the time, on average;
   - The MMP is in the BURSTY state for 5% of the time, on average;
   - The MMP transitions from OFF to BURSTY with probability 0.05, and from OFF to ON with probability 0.95;
   - The MMP transitions from ON to OFF with probability 0.9, and from ON to BURSTY with probability 0.1;
   - The MMP transitions from BURSTY to ON with probability 0.5, and from BURSTY to OFF with probability 0.5.

   If the time spent in the OFF state is an exponential with mean 0.25, what are the means of the exponential times spent in states ON and BURSTY? Simulate and plot the results.

4. Recall the pseudo-code for generating reference traces (Figure 14.7). Write routines `new_wrkset`, `from_wrkset`, and `not_from_wrkset` to model the following types of programs:

   (a) a scientific program with a large working set during initialization, a small working set for the bulk of the computation, and a different working set to complete the computation. (You will need to modify the control code in the figure slightly to force phase transitions in desired places);

   (b) a program whose working set always contains a core set of pages present in every phase, with the rest of the pages clustered elsewhere in the address space.

5. Using any simulator or language you like, model some aspect of your local computing-system environment (e.g., the print queue for a network printer). Carefully describe your workload model.

6. Using any simulator or language you like, model the router-to-Web-server logic of the system described in section 14.1. Pay special attention to the load-balancing mechanism that the router employs.

7. Using any simulator or language you like, model the interaction between application server and data server described in section 14.4. Pay special attention to the logic of requesting multiple data services and of waiting until all are completed until advancing to the next burst.

8. Consider the following language for describing CPU instructions:

   ```
   op   r1   r2
   ```

   The preceding expressions describe an operation, where

   ```
   op=1   means add, op=2 means subtract. Each require 1 cycle.
   op=3   means mult.  r1 receives the result r1 op r2. A multiplication
          requires 2 cycles.
   op=4   means a load from memory, into r1, using the value
          in r2 as the memory address.  Every 10th load requires
          4 cycles, the remaining loads require 1.
   op=5   means a store to memory, storing the data found in r1,
          using the value in r2 as the memory address. Each store
          requires 1 cycle.
   ```

   Write a CPU simulation along the lines of that described in 14.5 that accepts a stream of instructions in the format just described. Your simulator should use a logical-to-physical register mapping, use the timing information previously sketched, and use stall instructions as described in the example.

9. Write a one-pass simulator of a fully associative cache of size N that accepts as input reference traces from the trace generator in Problem 4. At the end of the simulation, report on the hit ratio for all cache sizes that are powers of 2 no greater than N.

10. Integrate the trace generator created in Problem 4 with the one-pass simulator written in the previous problem, in effect creating a pseudo "direct-execution simulator."

11. Analyze the log of WWW requests to your site's server, produce a stochastic model of the request stream, and simulate it.

15

Simulation of Computer Networks

15.1 INTRODUCTION

Computers and the networks that connect them have become part of modern working life. In this chapter, we illustrate by example some of the ways that discrete-event simulation is used to understand network systems, the software that controls them, and the traffic that they carry.

Like computer systems, network systems exhibit complexity at multiple layers. Networked systems are designed (with varying degrees of fidelity) in accordance with the so-called Open System Interconnection (OSI) Stack Mode (Zimmerman, 1980). The fundamental idea is that each layer provides certain services and guarantees to the layer above it. An application or protocol at a particular layer communicates only with protocols directly above and below it in the stack, implementing communication with a corresponding application or protocol at the same stack layer in a different device. Simulation is used to study behavior at all these layers, although not generally all in the same model. Different layers encapsulate different levels of communication abstraction.

The **Physical Layer** is concerned with the communication of a raw bit-stream, over a physical medium. The specification of a physical layer has to address all the physical aspects of the communication: voltage or radio signal strength, standards for connecting a physical device to the medium, and so on. Models of this layer describe physics.

The **Data Link Layer** implements the communication of so-called *data-frames*, which contain a limited chunk of data and some addressing information. Protocols at the Data Link Layer interact with the physical layer to send and receive frames, but also provide the service of "error-free" communication to the layer above it. Protocols at the Data Link Layer must therefore implement error-detection and retransmission when needed. A critical component of avoiding errors is access control, which ensures that at most one device is transmitting at a time on a shared medium. Techniques for access control have significant impact on how long it takes to deliver data and on the overall capacity of the network to move data. Simulation plays an important role in understanding tradeoffs between access-control techniques; in this chapter, we will look at some protocols and the characteristics that simulation reveals.

The **Network Layer** is responsible for all aspects of delivering data frames across subnetworks. A given frame may cross multiple physical mediums en-route to its final destination; the Network Layer is responsible for logical addresses across subnets, for routing across subnets, for flow control, and so on. The success of the Internet is due in no small part to widespread adaption of the *Internet Protocol*, more commonly known as IP (Comer, 2000). IP specifies a global addressing scheme that allows communication between devices across the globe. The specification of IP packets includes fields that describe the type of data being carried in the packet, the size of the packet, the protocol suitable for interpreting the packet, source/destination addressing information, and more. The Network Layer provides error-free end-to-end delivery of packets to the layer above it. Simulation is frequently used to study algorithms that manage devices (routers) that implement the Network Layer.

The **Transport Layer** accepts a message from the layer above, segments it into packets that are passed to the Network Layer for transmission, and provides the assurance that received packets are delivered to the layer above in the order in which they appear in the original message, error free, without loss, and without duplication. Thus, the Transport Layer protocol in the sending device coordinates with the Transport Layer protocol in the receiving device in such a way that the receiving device can infer packet-order information. Variants of the Transmission Control Protocol (TCP) are most commonly used at this layer of the stack (Comer, 2000). Dealing with packet loss is the responsibility of the transport layer. Packet loss is distinct from error-free transmission—a packet could be transmitted to a routing device without error, only to find that device does not have the buffering capacity to store it; the packet is received without error, but is deliberately dropped. Transport layer protocols need to detect and react to packet loss, because they're responsible for replacing the packets that are dropped. One of the ways they do this is to apply flow-control algorithms that simultaneously try to utilize the available bandwidth fully, yet avoid the loss of packets. Simulation has historically played a critical role in studying the behavior of different transport protocols, and in this chapter we will examine simulation of TCP.

The first four OSI layers are well defined and separated in actual implementation. The remaining three have not emerged so strongly in practice. Officially the **Session Layer** is responsible for the creation, maintenance, and termination of a "session" abstraction, a session being a prolonged period of interaction between two entities. Above this one finds the **Presentation Layer**, whose specification includes conversion between data formats. An increasingly important conversion function is encryption/decryption. Finally, the **Application Layer** serves as the interface between users and network services. Services typically associated with the Application Layer include email, network management tools, remote printer access, and sharing of other computational resources.

Any simulation of networking must include models of data traffic, and so we begin the discussion there. At the time of this writing, the field of traffic modeling is very active, and we bring to the discussion key elements of an exciting area of current work.

Devices with traffic they wish to transmit must somehow gain access to the networks that carry traffic. Our second area of discussion then considers the problem of how devices coordinate to use the network medium, sometimes called Media Access Control (MAC) protocols. Historically, simulation has played an important role in helping engineers to understand the performance of different MAC protocols.

Finally, we describe the Transport Control Protocol (TCP) and discuss how simulation plays an important role in its study.

15.2 TRAFFIC MODELING

Our discussion of network simulation begins with modeling of the data traffic that the networks carry. We'll consider two levels of detail for this, corresponding to two different levels of abstraction. The first is at the application level: consideration of how commonly used network applications create demand for a network. Such models are appropriate when one's interest is in the details of a relatively small network and the impact that its native applications have on it. The second level of abstraction is of aggregated application flows. This level is appropriate when one's focus is on the Internet's core infrastructure, where the global impact of global traffic needs to be represented.

One of the easiest models of traffic-load generation is that of moving files across the network. Our interest here is not in the mechanics of the protocols that accomplish the movement so much as it is in the model of the traffic load that is offered to the network. Simulation studies that model file transfer typically are focused on the impact that the traffic has on servers holding the files. A given transfer can be characterized by the size of the file and by the rate at which its bytes are presented to the network. We usually also characterize how often a user initiates an ftp transfer. A simple model of a file-transfer request process is as an on–off source, whose *off* period is randomly distributed (e.g., an exponential think time) and whose *on* period is driven by the arrival of a file. The *on* period lasts as long as needed to push or pull a file of the referenced length. File size is sampled from another probability distribution. Measurements suggest that a heavy-tailed distribution is appropriate. This is especially appropriate given the level of music-sharing activity on the Internet.

Another significant source of application traffic is the World Wide Web. Traffic associated with web pages is more complex than individual file transfers and so bears separate treatment. We describe a model expounded upon in (Barford and Crovella [1998]), called Surge. Here we model the delay between successive sessions with an intersession delay distribution. Within a session, a number of different URLs will be accessed, with another delay time between each such access; this is illustrated in Figure 15.1.

The Surge model incorporates a number of important characteristics of files, most importantly, including

- the distribution of file sizes, among all files on a web server,
- the distribution of the file sizes of those files that are actually *requested*,
- temporal locality of file-referenced file.

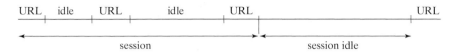

Figure 15.1   Nested on–off periods in Surge WWW traffic generation.

The first and third characteristics, coupled with a model of referencing pattern, essentially define the second characteristic. Suppose that we've selected the first k files already—call them $f_1, f_2, \ldots, f_k$—and suppose that this set of references is organized in a Least-Recently-Used stack. We select the $(k+1)^{\text{st}}$ file by sampling an integer from a stack-distance distribution. If that sample has value j, the next file selected has position j in the LRU stack (position 1 being the last file referenced). Empirical studies of reference strings of files suggest that a lognormal distribution is appropriate. This distribution places significant weight on small values; hence, it induces temporal locality of reference. When the stack-distance sample is larger than the number of files in the LRU stack, a new file is sampled from the set of files not yet in the reference stream.

This description gives a general, but simplistic idea of the structure of Surge. Its authors pay much attention to issues of identifying distributional parameters that are internally consistent and that produce traffic that can be validated against real traffic. Our goal here is to introduce the fundamental notions behind a model of web traffic.

Models of other interesting and important application types can be found in the literature. We expect that the Internet will increasingly support telephony—"voice over IP (VoIP)" (Black [2001]), and so attendant models should be developed. A sampling of the current literature suggests that a VoIP source be modeled as an on–off process, where both phases have distributions with tails somewhat heavier than exponential (e.g., an appropriate Weibull). Increasingly, the Internet will be used to stream video content. Models for video are more complex, because they must capture a number of facets of video compression, at different time-scales.

All of the application models we've considered describe the traffic workload offered to a network by individual programs. There are contexts in which a modeler needs instead to consider the impact of aggregated application flows on a network device. One *could* create the aggregate stream by piecing together many individual application streams—or one could start with an aggregated model in the first place. We next consider direct models of aggregated offered load.

Classical models of telephone traffic assume that aggregated call arrivals to the telephony network follow a Poisson distribution and that call completions likewise are Poisson. The early days of modeling and engineering *data* networks made the same assumption. However, with time, it became clear that this assumption didn't match reality well. In telephony, the increased use of faxes, and then Internet connections, radically transformed the statistical behavior of traffic. Two things emerged as being particularly different: First, data traffic exhibits a burstiness that flies in the face of the exponential's memoryless property. MMP processes described in Chapter 14 can be used to introduce burstiness explicitly into the arrival pattern of packets to a data network. However, studies indicate that the durations of burstiness aren't Markovian, as in the MMP model. Instead, traffic seems to exhibit long-term temporal dependence—correlations in the number of active sessions that extend past what, statistically, can be expected from MPP models.

Researchers noticed that there is tremendous variance in the size of files transferred within a session. It seemed that a heavy-tailed distribution like the Pareto does a good job of capturing this spread. Heavy-tailed distributions have the characteristic that, infrequently, very very large samples emerge. These large samples are large enough relative to their probability to exert a very significant influence on the moments of the distribution; in some cases, the integral defining variance diverges. It was hypothesized then that long-range dependence in session counts was due to the correlations induced by the concurrency of very long-lived sessions.

A model that appears to capture these explanations is the "Poisson Pareto Burst Process" (Zukeman et al. [2003]), in which bursts (e.g., sessions) of traffic arrive as a Poisson process. Each session length has duration sampled from a Pareto distribution. Bursts may be concurrent. More formally, let t_i be the arrival time of the ith burst, equal to $t_{i-1} + e_i$ where e_i is sampled from an exponential, and let b_i be the Pareto-sampled duration of that burst, and let $d_i = t_i + b_i$ be the finishing time of the ith burst. The state at t, $X(t)$, is the number of bursts t_j with $t_j \leq t < d_j$.

The Pareto distribution with parameters a and b has the probability distribution function

$$D(x) = 1 - \left(\frac{b}{x}\right)^a$$

for $x \geq b$. The distribution has mean $(ab)/(a-1)$ and variance $ab^2/((a-1)^2(a-2))$. One can sample a Pareto with these parameters, using the inverse transform technique:

$$x = b \times (1.0 - U)^{-1.0/a}$$

In this equation U is a uniformly distributed random variable.

It is instructive to consider how traffic is analyzed for evidence of long-range dependence and whether the style of synthetic traffic generation described here exhibits it. Let $X_1, X_2, \ldots,$ be a stationary time series, whose samples have mean μ and variance σ^2. The autocorrelation function $\rho(k)$ describes how well correlated are samples k apart in the time series:

$$\rho(k) = \frac{E[(X_t - \mu)(X_{t+k} - \mu)]}{\sigma^2}$$

The sample autocorrelation function can be constructed from an actual sample by estimating the expectation in the numerator. Long-range dependence is observed when $\rho(k)$ decays slowly as a function of k. Long-range dependence is more formally defined in terms of the autocorrelation function, if there exists a real number $\alpha \in (0, 1)$, and a constant $\beta > 0$ such that

$$\lim_{k \to \infty} \frac{\rho(k)}{\beta k^{-\alpha}} = 1$$

The denominator of this limit describes how slowly $\rho(k)$ needs to go to zero as k increases. The smaller α is, the slower is the degradation. $H = 1 - \alpha/2$ is known as the *Hurst parameter* for the sequence. Values of H with $0.5 < H < 1.0$ define long-range dependence; the larger H is, the more significant is the long-range dependence.

To see evidence that PPBP does yield long-range dependence, we ran an experiment where the mean burst interarrival time was 1 second and the Pareto parameters were $a = 1.1$ and $b = 10$. We computed the sample autocorrelation function, shown in Figure 15.2. Here we see directly that the autocorrelation decays very slowly. We also used the SELFIS tool (Karagiannis et al. [2003])

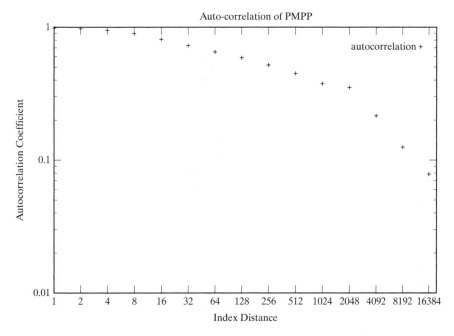

Figure 15.2   Autocorrelation function of aggregated stream of 50 sources.

to estimate the Hurst parameter; all of its estimators indicate strong long-range dependence in the sampled series.

Burstiness is not the only consideration in traffic modeling. Traffic intensity exhibits a strong diurnal characteristic—that is, source intensity varies with the source's time of day; furthermore, weekends and holidays behave differently still. To accommodate time-of-day considerations, one can allow the exponential burst interarrival distribution of the PPBP to have a parameter that is dependent on the time of day.

The PPBP describes the number of active sessions $X(t)$ as a function of time. $X(t)$ may be transformed into packet arrival rates, and hence into packets, by including a packet-rate parameter λ. The process $\lambda X(t)$ thus gives an arrival rate of packets from an aggregated set of sources to a network device that handles such.

The main point to be understood about traffic-modeling is that models of aggregated traffic ought to exhibit characteristics of aggregation, whereas application traffic ought to focus on what makes the applications distinct. Next, we look at how traffic acquires a shared medium for carrying traffic.

15.3 MEDIA ACCESS CONTROL

Computers in an office or university environment are usually integrated into a local area network (LAN). Computers access the network through cables (a.k.a. wireline), although an increasing fraction access it through radio (wireless). In either case, when a computer wishes to use the network to transmit some information, it engages in a Media Access Control (MAC) protocol.

Different MAC protocols give traffic different characteristics. Simulation is an extremely important tool for assessing the behavior of a given protocol. A MAC protocol gives traffic specific qualities of latency (average and maximum are usually interesting) and throughput. The behavior of these qualities as a function of "offered load" (traffic intensity) is of critical interest, for some protocols allow throughput to actually *decrease* as the demands on the network go up—a lose–lose situation.

15.3.1 Token-Passing Protocols

One class of MAC protocols is based on the notion of a "token," or permission to transmit. In the "polling protocol" variation, a master controller governs which device on the shared medium may transmit (Kurose and Ross [2002]). The controller selects a network device and sends it the token. If the recipient has "frames" (the basic unit of transmission) buffered up, it sends them, up to a maximum number of frames. The controller listens to the network and detects when the token holder either has selected not to transmit or has finished transmission. The controller then selects another network device and sends it the token. Devices are visited in round-robin fashion.

One drawback of the polling protocol is that the controller is a device with separate functionality from the others. A more homogeneous approach is achieved by using a *token bus* protocol. In this approach a device is programmed to transmit frames (again up to a maximum number) when it receives a token, but is programmed to pass the token directly to a different specified network device after it is finished. There is no controller; the network devices pass the token among themselves, effectively creating a decentralized round-robin polling scheme.

A drawback of both types of token-passing protocol is that a single failure can stop the network in its tracks—in the case of the polling protocol, the network stops if the controller dies; in the case of the token bus, a token passed to a dead device in effect gets lost. In the latter case, one can detect that a device failed to pass the token on and so amend the protocol to deal with like failures.

Token-passing networks are "fair," in the sense that each device is assured its turn within each round. The overhead of access control is the time that the network spends on transmitting the token (rather than data) and the time that the network is idle long enough for a device to ascertain that a transmission has ended or is not going to occur. An important characteristic of token-passing protocols is that the throughput (bits per second of useful traffic) is monotone nondecreasing as a function of the "offered load" (traffic that the network is requested to carry). To illustrate this point, Figure 15.3 plots data from a set of experiments on a modeled 10 Mbits (10 million bits per second) network, with 10 devices, evenly spaced, with a latency delay of 25.6 μsec between the most distant pair. (We use this figure in order to compare this network with one managed by using Ethernet, later.) Five different experiments are displayed on the graph; right now, we are interested only in the one labeled "token bus, Poisson." The experiments assume that the data frame is 1500 bytes long and that the token is 10 bytes long. They assume that, once a device gains the token, it may send at most one frame and then must release the token. This set of data uses a Poisson process to generate frame arrivals. The *x*-axis gives the "offered load," measured here as the total sum of bits presented to the network before the simulation end time,

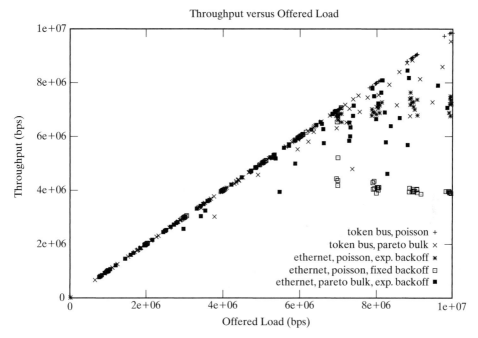

Figure 15.3   Throughput versus Offered Load, for Token Bus and Ethernet MAC protocols, Poisson and Bulk Pareto arrival processes, Exponential and Fixed Backoff (for Ethernet)

divided by the length of the simulation run. The y-axis plots the measured throughput. For each off-time rate, we run 10 independent experiments. For each experiment, we plot the observed pair (offered load, throughput). For the experiment of interest, the throughput increases linearly with the offered load, right up to network saturation. It is interesting to note, though, the impact of a change in the traffic-arrival pattern. We replaced the Poisson arrival process with the arrival process that defines an PPBP, a Poisson bulk arrival process, where the number of frames in each bulk arrival is a truncated Pareto. We use the same Pareto parameters as before ($a = 1.1$ and $b = 10$) and reduced the arrival by a factor of the inverse Pareto mean ($(a - 1)/(ab)$) to obtain the same average bit-arrival rate. The set of data points associated with the label "token bus, Pareto bulk" reflect the impact of this change. Throughput grows linearly with offered load until the bus is roughly 60% utilized. For larger loads, we begin to see some deviations from linear. For a point (x, y) off the diagonal, the difference between x and y reflects the volume of unserved frames at the end of the simulation—the frames in queue. This is no surprise; queueing theory tells us that we should expect significant queue lengths when the arrival pattern is highly variant.

Another important aspect is the average time a frame awaits transmission after arrival. Knowledge of queueing theory and the protocol's operation identifies two factors that ought to contribute to growth in the queuing length. One factor is the time required by a token to reach a new frame arrival. As the offered load increases, the amount of work that the token encounters and must serve prior to reaching the new arrival increases linearly. A second factor is from queueing theory;

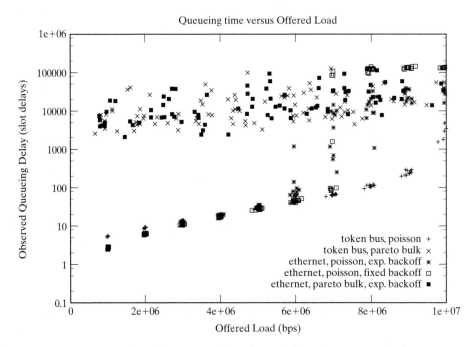

Figure 15.4 Average Queue Delay versus Offered Load: for Token Bus and Ethernet MAC protocols, for Poisson and Bulk Pareto arrival processes, and for Exponential and Fixed Backoff (for Ethernet).

the view from a station is of an M/G/1 queue. In this view, the service time incorporates the time spent waiting for a token to arrive, a mean that increases with the offered load. A job's average time in an M/G/1 queue grows with $1/(1 - \rho)$, where $\rho = \lambda/\mu$ is the ratio of arrival rate to service-completion rate. As the offered load grows, ρ increases; this fact explains the second factor of waiting-time growth. As ρ approaches unity, the asymptotic waiting time increases rapidly.

Figure 15.4 confirms this intuition, plotting the average time a frame waits in queue between the time of its arrival and the time at which it begins transmission. Again we execute 10 independent experiments for a given offered load and plot the raw pair (x, q), where x is the average number of bits presented to the network per second in that run and q is the average time a job is enqueued. Units of queueing delay are "slot times", the length of time required for a bit to traverse a cable at the limit of what is permissible for Ethernet (25.6 µsec). The extreme range of queueing delays observed for the five experiment types encourages our use of a log scale on the y-axis. Tracking data from the experiments by using Poisson arrivals, we see stability in the growth pattern, up to the point where the bus is fully saturated. We know to expect extremes there. What is very interesting, though, are the extremely high average queueing delays experienced under the "bulk Pareto" assumptions. If nothing else, these kinds of experiments point out the importance of the traffic model in the analyzing of network behavior.

A straight-forward implementation of a token-bus protocol models devices, the bus, and the explicit and continuous passing of the token among stations. However, this implementation has an undesirable characteristic. Under low traffic load, the model creates a discrete event approximately every 10.84 μsec, the time it takes to transmit a token between adjacent stations. Under low traffic load, the token could completely cycle through the network many times before reaching a point in simulation time when there is a frame available for transmission. Unless the simulation has some particular reason for pushing the token around an otherwise idle network (e.g., if, at each hop, there is a nonzero probability of the token's being lost or corrupted, forcing the protocol to detect and react), there are more efficient ways of executing the simulation, at the cost of incorporating extra logic. We may suppose that each device samples the next future time at which a batch of frames arrives. Before that time, if the device has no frames to transmit, it will make no further demands on the network. When the simulation has reached a time at which no frame is being transmitted and no device has a frame waiting for transmission, we perform a calculation to advance simulation time past an epoch during which the only activity is token passing. Because the time required to circulate the token around the ring is computable, and the next time at which a frame is available at any station is known, we can advance simulation time to the cycle in which the next frame is transmitted and save ourselves the computational effort of getting to that place by pushing a token around.

15.3.2 Ethernet

Token-based access protocols have been popular, but they have drawbacks when it comes to network management. In particular, every time a device is added to or removed from the network, configuration actions must be taken to ensure that a new device gets the token and that a removed device is never again sent the token. The Ethernet access protocol is a solution to this problem (Spurgeon [2000]). A device attached to an Ethernet cable has no specific idea of other devices on that cable; however, when it wants to use the cable, it must coordinate with such other devices. Consider the problem—a device has a frame to send; when can it send it? Ethernet is a decentralized protocol, meaning that there is no controller granting access. A device can "listen" to the Ethernet cable to see whether it is currently in use. If the cable is already in use, the device holds off until the cable is free. However, two or more devices could independently and more or less simultaneously decide to transmit, shortly after which the transmission on the cable is garbled. Both devices can detect this "collision" (e.g., by comparing what they are transmitting on the cable with what they are receiving from the cable). Collision detection and reaction to it is the one of the key components of the Ethernet protocol; it is a so-called Carrier Sense Multiple Access/Collision Detection (CSMA/CD) protocol.

The format of an Ethernet frame is illustrated in Figure 15.5. The 8-byte preamble is a special sequence of bits (alternating 1's and 0's, except for the last bit which is also a '1') that listeners on the cable recognize and use to prepare to examine the next frame field, a 6-byte Destination address that may specify one device, a group of devices, or a broadcast to all listening devices. After scanning the full Destination address, a device listening to the cable knows whether it is an intended recipient. The next 6 bytes identify the sending device; then comes a 2-byte field describing the number of data bytes. The data follow, and the frame is terminated with a 4-byte code used for error detection.

When a device decides to transmit, it begins in the knowledge that it is possible for another device to begin also, not yet having heard the new transmission. Ethernet specifications on network

size (bytes)	8	6	6	2		4
field	Preamble	Destination MAC adrs	Source MAC adrs	Length	Data	Cyclic Redundancy Check

Figure 15.5   Format of Ethernet Frame

design ensure that any transmission will be heard by another device within $\delta = 25.6$ μsec. This is called a slot time. The worst case is that the device begins to transmit at time t, yet before time $t + \delta$, a device at the other end of the cable decides to transmit and does so just before time $t + \delta$, and another δ time is needed by the first device to detect the collision.

The length of the data portion of an ethernet frame is not specified by the protocol. However, there is a lower bound on the allowable length of the data portion. The frame must be large enough so that it takes longer than 2 slot times to transmit it. This bound ensures that, if a collision does occur, the sending device will be transmitting when the effects of the collision reach it, and hence it can detect the collision. This minimum is 46 bytes of data; furthermore, a frame is not permitted to carry more than 1500 bytes of data.

Some of the complexities of Ethernet exist because of physics. An accurate simulation of Ethernet must therefore pay attention to the delicacies of signal latency. The model used to generate Ethernet performance figures specifically accounts for signal latency. It assumes that the devices are evenly spaced along a cable that requires a full slot time (25.6 μsec) for a signal to traverse. When a device listens to the cable to see whether it is free, the model really answers the question of whether the device can, at that instant, hear any transmission that might have already started. This is a matter of measuring the distance between a sending device and a listening device, computing the signal latency time between them, and working out whether the sender started longer ago than that latency. Likewise, when a device has a frame to send and is listening to the cable to find out when it is idle, its view of the cable state is one that accounts for a certain delay between when a transmission ends and when that end is seen by an observer.

A device with a frame to send listens to the cable and, if it hears nothing begins to transmit. If it successfully transmits the frame without collision and has another frame to send, it waits 2 slot times before making the attempt. If a device wanting to send a frame hears that the cable is in use, it simply waits until the cable is quiet and then begins to transmit. The most interesting part of Ethernet is its approach to collisions. If a device transmitting a frame F detects a collision, it continues to transmit—but jumbled—long enough to ensure that it transmits a full minimum frame's worth of bits. This "jamming" ensures that all devices on the cable detect the collision. Next, it backs off and waits a while before trying to send F again.

The backoff period following a collision has been a topic of some study, one in which simulation has played an important role. If the backoff time is short, there is a chance of not overly increasing the delay time of a frame, but there is also a significant chance of incurring another collision. On the other hand, if the backoff time is large, one reduces the risk of a subsequent collision, but ensures that the delay of the frame in the system will be large. Over time, the following strategy, called "exponential backoff", has become the Ethernet standard. Following the mth collision while attempting to transmit frame F, the device randomly samples an integer k from $[0, 2^m - 1]$, and waits $2k$ slot times before making another attempt. If 10 attempts are made without success, the frame is simply dropped. The term "exponential backoff" describes the doubling in length of the mean backoff time on each successive collision. Successive collisions are measures, of a sort, of the level of congestion in the network. A device strives to reduce its contribution to the congestion, and so enable other frames to get through and relieve the congestion.

Simulation is a useful tool to investigate both backoff schemes and other variants of Ethernet one might consider. We did experiments (assuming Poisson arrivals) on exponential backoff and on "fixed" backoff—where, after a collision occurs, the sender chooses $k \in [0, 4]$ slot times to wait, uniformly at random. Figure 15.3 illustrates the effects on throughput. Under exponential backoff, throughput increases linearly with offered load until after about 60% utilization. For greater load, throughput hovers in the 70% of bandwidth regime, without significant degradation. The story is quite different under fixed backoff. When offered load is 70% of the network bandwidth, the throughput plummets from 60+% and settles in at around 40% of bandwidth—under higher load, the network delivers poorer service. Queueing delays are affected too, as one would expect. Under high load, the delays under fixed backoff are an order of magnitude larger than those under exponential backoff.

A final set of experiments used the same Poisson bulk arrival process, with Pareto-based bulk arrivals, assuming exponential backoff. The results are similar to those for the token bus: large and highly variable queueing delays, and some deviation of throughput from linear at high load. This set of experiments suggests that Ethernet may be more sensitive to the Pareto's high variance than is the token-bus protocol.

15.4 DATA LINK LAYER

A network is far more complicated than the single channel seen by a MAC protocol. A frame might be sent and received many times, by many devices, before it reaches its ultimate destination. Consequently, data traveling at the physical layer contains at least two addresses. One address is a hardware address of the intended endpoint of the current hop. This address (like an Ethernet address) is recognizable by a device's network-interface hardware. The second address is the ultimate destination's network address, typically an IP address. Different types of devices make up the network. A *hub* is a device that simply copies every bit received on one interface to all its other interfaces. Hubs are useful for connecting separated networks, but have the disadvantage that the connection brings those networks into the same Ethernet collision domain.

A *bridge* makes the same sort of connection, but keeps component subnetworks in different collision domains. For every frame heard on one interface, the bridge takes the destination address and looks up in a table the interface through which that destination can be reached. The bridge has nothing to do if one reaches the destination through the same interface as that through which the frame was observed—the destination will recognize the frame for itself. However, if the destination is reached through a different interface, the bridge takes the responsibility of injecting the frame through that interface, moving it closer to its ultimate goal. In injecting the frame, the bridge acts like a source on that subnetwork, engaging in that subnetwork's MAC protocol. The bridge in effect moves a frame from one collision domain and puts it into another. It can also bridge different subdomain technologies (e.g., different types of Ethernet). Contexts where one would consider simulation study of MAC protocols on one subdomain are the sorts of contexts where one would use simulation and involve models of bridges.

A bridge involves only the physical layer and the data link layer. There is a practical limit on devices retaining the physical addresses of other devices, particularly devices that are in different administrative domains. A *router* is a device that can connect more widely dispersed networks, by making its connections at the Network Layer. A frame coming in to a router on one interface is pushed up to the IP layer, where the IP destination address is extracted; the IP address determines which interface should be used to forward the packet. The *forwarding tables* used to direct traffic flow are the result of complex *routing algorithms*, such as OSPF (Moy [1998]) and BGP (van Beijum [2002]). Simulation is frequently used to study variants and optimizations of these protocols.

We will see that network services commonly used provide users with delivery of data error free and in the order it was sent. These attributes are provided in spite of the real possibility that data will be corrupted in transmission or lost in transmission. A router is one place where a frame might be lost, for, if the router experiences a temporary burst of traffic, all to be routed through a particular interface, buffers holding frames waiting to be forwarded could become exhausted. We think of the traffic flowing through a router as being a set of flows, each flow being defined by the source–destination pair involved. When the arrivals become bursty, and the router's buffer becomes saturated, arrivals that cannot be buffered are deliberately dropped. Most flows actively involved in the burst will lose frames. Under TCP, data loss is the signal that congestion exists, and TCP reacts by significantly decreasing the rate at which it injects traffic into the network. But it takes time to detect this loss—a lot more time than it takes to route frames through the router. One idea that has been studied extensively (by using simulation) is *Random Early Detection* (RED) (Floyd and Jacobson [1993]) queue management. The idea behind RED is to have a router continuously monitor the number of frames enqueued for transmission and, when the average length exceeds a threshold, proactively attempt to throttle back arrival rates before the arrivals overwhelm the buffer and cause all of the flows to suffer. RED visits each frame and, with some probability, either preemptively discards it, or marks a "congestion bit" that is available in the TCP header, but is not much used by most TCP implementations. RED chooses a few flows to suffer for the hoped-for sake of the network as a whole. Complexity abounds in finding effective RED parameters (e.g., threshold queue length, probability of dropping a visited frame) and in assessing tradeoffs and impacts that use of RED could have. Simulation, of course, has played and will play a key role in making these assessments.

15.5 TCP

The Transport Control Protocol (TCP) (Comer [2000]) establishes a connection between two devices, both of which view the communication as a stream of bytes. TCP ensures error-free, in-order delivery of that stream. As we have seen, data frames might be discarded (in response to congestion) somewhere between the sender and receiver; TCP is responsible for recognizing when data loss occurs and for retransmitting data that have gone missing. TCP mechanics are focused on avoiding loss, detecting it, and rapidly responding to it. A number of TCP variants

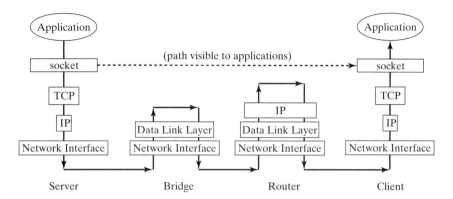

Figure 15.6 Data flow from TCP sender to TCP receiver, passing through network devices.

have been proposed and studied; all of these studies use simulation extensively to determine the protocol's behavior under different operating conditions.

Our discussion of TCP serves to illustrate further how different components of networking layers come together. Figure 15.6 illustrates data flow from a server to a client. Two applications intending to communicate establish "sockets" at each side. Sockets are viewed by the applications as buffers into which data could be written and out of which data might be read. Calls to sockets are sometimes blocking calls, in the sense that, if a socket buffer cannot accept more data on a write, or has no data to provide on a read, the calling processes blocks. On the server side, the TCP implementation is responsible for removing data from the socket's buffer and sending it down through the protocol stack to the network. Once on the network, the data pass through different devices. In this figure, we illustrate a bridge (which involves remapping of hardware addresses and does not look at the IP address) and a router (which must decode the IP address to find out the interface through which the data is passed). The client host's IP recognizes that the data ought to go up the stack to TCP, and the client side TCP is responsible for releasing the data to the socket—but only a contiguous stream of data. If the router drops a frame of this flow, the client-side TCP must somehow detect and communicate this absence to the server-side TCP.

TCP segments the data flow into *segments*. Figure 15.7 illustrates the header (in 32-bit words) that is placed around the data. First, note that the only addressing information is "port number" at the source and destination machines—IP is responsible for knowing (and remembering) the identity of the machines involved. From TCP's point of view, there is just a source and a destination. **SeqN** and **AckN** are descriptors of points in the data flow, viewed as a stream of bytes, each numbered. **SeqN** is then the "sequence number" of the first byte in the segment. At the beginning of a connection, a sender and receiver agree upon an initial sequence number (usually random); the **SeqN** value is this initial number plus the byte index within the stream of the first byte carried in the segment. Because the segment size is fixed, the receiver can infer the precise subsequence of the byte stream contained in the segment. The **AckN** field is critical for detecting lost segments. Every time a TCP receiver sends a header about the flow (e.g., in accordance with acknowledgement rules), it puts into the **AckN** field the sequence number of the next byte it needs to receive to maintain a contiguous flow. Since TCP provides a contiguous data stream to the layer above, the value in **AckN** is the initial sequence number plus the index of the next

Source port number	Destination port number
(SeqN) Sequence Number	
(AckN) Acknowledgement Number	

Header Length		U R G	A C K	P S H	R S T	S Y N	F I N	Receiver Window Size

Checksum	Pointer to urgent data
Options	
Data	

Figure 15.7   TCP header format.

byte it would provide to that layer, if it were available. The linkage of this value with packet loss is subtle. TCP requires a receiver to send an acknowledge for every segment it receives and requires a sender to detect within a certain time limit whether a segment it has sent has been acknowledged. Now imagine the effect if 3 segments are sent, and the second one is lost en route. Assume the initial sequence number is 0. The first segment is received, and the receiver sends back an acknowledgement with **AckN** equal to (say) 961 (and the ACK flag set to 1 to indicate that the **AckN** field is valid). The third segment is received, but the receiver notices that the value of **SeqN** is a segment larger than expected—it notices the hole. So it sends back an acknowledgement, but **AckN** in that header is again 961. The second segment sent is not acknowledged, of course, but interestingly, neither is the third. Eventually the TCP sender times out while waiting for these acknowledgements and resends the unacknowledged packets. The only other field in this header, that is critical to our discussion is the Receiver window size, which is included in an Acknowledge to report how many bytes of buffer are currently available to receive data from the sender.

One can visualize TCP as sliding a *send window* over the byte stream. Within the send window are bytes that have been sent, but not yet acknowledged. TCP controls the rate at which it injects segments into the network by maintaining a *congestion window size*, which at any time is the largest the send window is allowed to get. If the send-window size is smaller than the congestion-window size and there are data to send, TCP is free to send it, up until the point where the send window has the same size as the congestion window. When the TCP sender has stopped for this reason, an incoming acknowledgement can reduce the size of the send window (because bytes at the lower end of the window are now acknowledged), and so free more transmission.

TCP tries to find just how much bandwidth it can use for its connection by experimenting with the congestion-window size. When the window is too small, there is bandwidth available but it isn't being used. When the window is too large, the sender contributes to congestion in the network, and the flow could suffer data loss as a result. TCP's philosophy is to grow the congestion window aggressively until there is indication that it has overshot the (unknown) target size, then fall back and advance more slowly. This all is formally described in terms of variables *cwnd* and *ssthresh*. TCP is in *slow start* mode whenever *cwnd* < *ssthresh*, but in *congestion avoidance* mode whenever *cwnd* > *ssthresh*. Both variables change as TCP executes; *cwnd* grows with acknowledgements a certain way in slow-start, and a different way in congestion-avoidance; *ssthresh* changes when packets are lost. When a TCP connection is first established, *cwnd* is typically set to one segment size and *ssthresh* typically is initialized to a value like 2^{16}. TCP starts in *slow-start* mode, which is distinguished by the characteristic that, for every segment that is acknowledged, *cwnd* grows by a segment's worth of data.

Consider how *cwnd* behaves during slow start by thinking about TCP sending out segments in rounds. In the first round, it sends out one segment, then immediately stalls, because the send-window and congestion-window sizes are equal. When the acknowledgement eventually returns, the sender issues *two* segments as the second round—it replaces the segment that was acknowledged and sends another, because *cwnd* increased by 1. The sender stalls until acknowledgements come in. The *two* acknowledgements for the second round enable the sender to issue *four* segments: half of these due to replacing the ones acknowledged, the other half due to the one-per-acknowledgement increase of *cwnd* rule during slow start. The number of segments issued thus doubles in successive rounds.

Any one of a number of things can halt the doubling of the number of segments sent each round. One is detection of packet loss, the effects of which are to set *ssthresh* to be half the size of the send window, set the send-window size to zero, and set *cwnd* to allow retransmission of one segment (the one in the lost packet). Another way TCP ceases to double the number of segments sent each round is due to the rule that the congestion window may not be increased to exceed certain limits—an internally imposed buffer size at the sender side, or the size of the "receiver window"—the field in ACKs which reports how much space is available for new data. Finally, the doubling effect changes also if *cwnd* grows to exceed *ssthresh*, and so puts TCP into congestion-avoidance mode. Within congestion-avoidance mode, *cwnd* increases, but much more slowly. Intuitively, *cwnd* increases by one segment for every full round that is sent and acknowledged (as opposed to increasing by one segment with every segment that is acknowledged). This is sometimes described as increasing *cwnd* by *1/cwnd* with every acknowledgement.

Simulation is an excellent tool for understanding how TCP works and many of the subleties of its behavior; we now examine simple examples of that behavior. The first topology is that of a server, a client, and a 800 kbps link between them. The server is to send a 300000 byte file to the client. We attach a monitor that emits a tcpdump formatted trace (see www.tcpdump.org) of every TCP packet that passes (in either direction) through the server's network interface. Postprocessing of this trace yields information about how TCP variables of interest behave. In the first situation, we plot the values of **SeqN** in packets sent by the server and the values of **AckN** in packets sent by the receiver in response for the first six rounds, assuming an initial sequence number of 0. This is illustrated in Figure 15.8, where the Y-axis is logarithmic in order to illustrate interesting behavior at different scales. The TCP connection is requested by the client

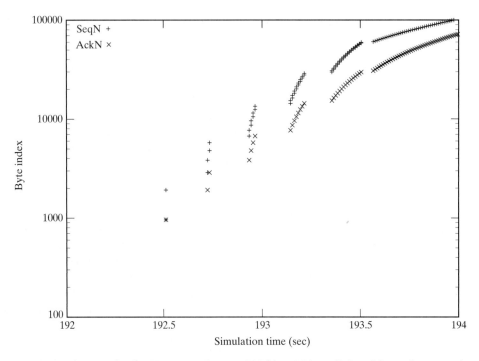

Figure 15.8 Early rounds of TCP connection on 800 kbps/100 ms link, with tcpdump probe at the server's network interface.

at time 192, the first step in TCP's three-way handshake that results in the server sending the first segment at time 192.3 (not actually shown in the graph, to allow higher resolution to later rounds). The **SeqN** in the header of that segment is 1, the index of the first byte in the segment. It takes approximately 100 ms for the segment to reach the client, and another 100 ms for the client's acknowledgment to reach the monitoring point, at time 192.5. (The exact figures are a little different, as they account for the transmission delay caused by the link bandwidth.) The **ACK** bit of that segment is set, and the **AckN** value in the header is 961—the index of the next byte the receiver expects to see. The server's send window now being empty, and *cwnd* having advanced from 1 to 2 by virtue of the received acknowledgement, the server immediately sends *two* segments, one with **SeqN** equal to 961, the next with **SeqN** equal to $961 + 960 = 1921$. The graph shows overlapping marks for byte index 961, one from the acknowledgement header, and one from the next segment the server sends. The delay between the server's sending of a segment and the ultimate acknowledgement of that segment is known as the *round-trip time*, or RTT. In this example, the network is as simple as it can be, and the RTT is just the sum of the time to send a segment across the link plus the time to send an acknowledgement back—here, a value very close to 200 ms. At times 192.3 and 192.5, the server stopped sending segments just as soon as its send window and the congestion window were the same size. After one RTT, acknowledgements from the previous round come in; they allow the server to double the number of segments sent from one round to the next. For rounds three and four (at times 192.7 and 192.9, approximately), the graph shows the slight staggering of times associated with acknowledgements coming in and new segments going out.

Figure 15.8 shows how, in slow-start mode, upon receiving a burst of acknowledgements, the server generates a burst of new segments. A moment's reflection shows that, if the acknowledgement for the first segment in that burst is received while the burst is continuing, then the burst will continue ad infinitum. For, at the instant that critical acknowledgement is received, the send window must be smaller than the congestion window, and the send window will not grow after this point, while the congestion window will. We can compute the size of the congestion window at which this phenomenon occurs—it is when the congestion window is large enough that the time needed to transmit that many bytes is precisely the RTT. Back-of-the-envelope calculations indicate that this is 20000 bytes, or just under 21 segments. In these experiments, the receiver window is limited to 32 segments, so this saturation happens before the flow is limited by that buffer. SSFNet initializes *ssthresh* to 65396 bytes, so this saturation point is reached in slow-start, before *cwnd* reaches *ssthresh* and triggers congestion-avoidance mode. Since *cwnd* starts with value 1 and doubles with every round, the server saturates its sends in the middle of the 6th round. This is observed in Figure 15.8, in the round that starts just after time 193.5.

In Figure 15.9, we illustrate this same experiment, along with another that is identical—save that the link latency is 300 ms. A larger epoch of simulation time is illustrated. There is an interesting kink in the **SeqN** data set for the 100 ms network, in the vicinity of **SeqN** = 65*K*.

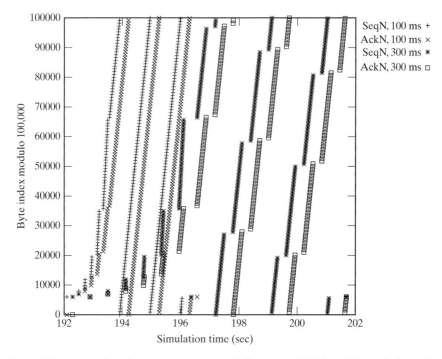

Figure 15.9   TCP connections between server and client: an 800 kbps/100-ms link, and an 800 kbps/300-ms link.

The "slope" of the data set decreases perceptibly. Up to this point, for every acknowledgement received two new segments are transmitted, and they are marked in the tcpdump trace as occurring at the same instant (SSFNet does not ascribe time advance to protocol actions, only to network transmission). At the point of the kink, the value of *cwnd* becomes equal to the receiver window, 32 segments. The sender window becomes limited by the size of the receiver window, rather than by *cwnd*, so, after the kink, there is a one-to-one correspondence between receipt of an acknowledgement and transmission of another segment. Now consider the experiments using a 300 ms latency. As we'd expect, rounds happen approximately every 600 ms. To saturate the link, the sender window has to become three times as large as in the first experiment—almost 64 segments. However, this will never happen, because the send window will be limited by the receiver window, at 32 segments. Indeed, we see that the change in slope of the **SeqN** trace happens at the same byte index as it did with the first experiment. Likewise, we see visually that there's a gap in transmission time between each successive round.

As a final example of how simulation illustrates the behavior of TCP, we consider an experiment designed to induce packet loss. The topology is that of a server, a router, and a client. Again, the server is to send 300000 bytes to the client. Both server and client connect with the router. The link between server and router has 8 Mbps of bandwidth and 5-ms latency. The link between client and router has 800 kbps of bandwidth and 100-ms latency. The router's interface with the client has a 6000-byte buffer. If a packet arrives to that interface and there is insufficient buffer space available, the packet is dropped. From earlier analysis of TCP, we can foresee, in part, what will happen. In the slow-start phase, the server begins to double the number of segments with each successive round. However, it can push packets towards the router 10 times faster than the router can push packets to the client, so a queue will form at the interface. The buffer holds at most 6 packets, so we expect that, in the round where 8 packets are sent, there will be packet loss. Figure 15.10 illustrates this experiment, adding a trace of *cwnd* behavior to that of **SeqN** and **AckN** (once again measured at the server's network interface). The effects of the packet loss are visually distinctive. Around time 193.5, the server begins to receive a sequence of acknowledgements that all carry the same **AckN** value. These acknowledgements were sent in response to packets that were sent *after* a loss. Recall that TCP rules on **AckN** specify that the receiver identify the sequence number of the next byte it needs to receive to advance the sequence of contiguously received bytes; hence, the repeated **AckN** identifies the beginning of the first lost segment. At the point at which the loss is observed, the send-window size is approximately 25000 bytes; in reaction to the loss, *ssthresh* is set to half this value, *cwnd* is set to 1, and the sender window collapses to size zero in order to cause the retransmission of all segments (from the first lost one forward). In the region between times 193 and 194, we see the impact that loss has on *cwnd* and how the slow-start doubling of *cwnd* with each round begins anew. (Notice the small periods of sharply increased growth at times 194.6, 194.8, and 195.) However, this time, congestion-avoidance mode is entered when *cwnd* reaches *ssthresh*, shortly after time 195; thereafter, it grows more or less linearly with time. This particular transfer ends just before *cwnd* reaches a size that will allow loss once again; had the transfer advanced that far, TCP's treatment of *cwnd* would look very much like the period from 193.8 on.

As these simple examples show, TCP's relatively simple rules create complex behavior. Simulation is an indispensable tool for predicting how TCP will behave in any given context and for understanding that behavior.

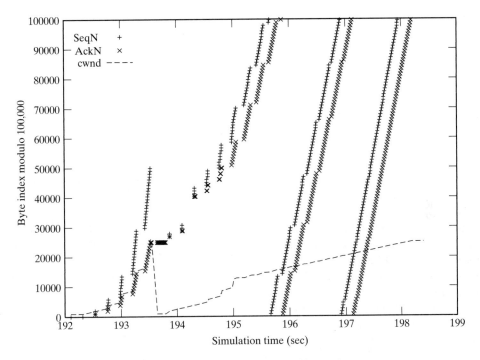

Figure 15.10   TCP connection suffering loss.

15.6 MODEL CONSTRUCTION

SSFNet is a versatile tool for building and analyzing network simulations, used in the previous section to look at how TCP behaves. Suggested homework projects encourage use of SSFNet, and so we describe the general process SSFNet uses in constructing a simulation from an input model. We then illustrate this process, in part, by describing the contents of one input file used in the last subsection. This is not a users' manual for SSFNet; very complete documentation exists at www.ssfnet.org. Our aim here to is give a sense of the approach and to encourage readers to investigate further.

15.6.1 Construction

Input to SSFNet is in the form of so-called Domain Modeling Language (DML) files. At the simplest level, a DML file contains just a recursively defined list of attribute–value pairs, where an attribute is a string and a value may be either a string or a list of attribute–value pairs. This structure naturally induces a tree, where interior nodes are attributes (labeled with the attribute string name) and leaves are values of type *string* (rather than of type *list*). To illustrate, consider this DML list:

```
Net [
     frequency 1000000
```

```
host [ id 0
        interface [ id 0 bitrate 800000]
        nhi_route [ dest 1(0) interface 0 ]
]
host [ id 1
        interface [ id 0 bitrate 800000]
        nhi_route [ dest default interface 0 ]
]
link [ attach 0(0) attach 1(0) latency 0.1 ]
]
```

This has some elements of SSFNet DML structure worth noting. Description of a network, elements within the network, and connections between them use a hierarchical naming convention known as the Network-Host-Interface convention, or just NHI. The network is defined in terms of links between interfaces, and each interface has an id number that is unique among all interfaces owned by a common host. That host has an id number that is unique among all hosts in a common net. Each net has an id, unique among all nets contained in the same parent net, and so on. The NHI address 0.1.2(4) refers to an interface named 4, within a host named 2, within a net named 1, within a net named 0. Within a net, a reference such as 2(4) is understood to mean interface 4 associated with the uniquely named host 2 within that understood net. The NHI address of an interface is derived from the nesting described within a DML file. The first interface to appear in the preceding example has NHI address 0(0); the second interface to appear has address 1(0). The `link` attribute in this example specifies two endpoints of the link, in NHI addressing (using the `attach` attribute), and a link latency of 100 ms.

The recursive structure of DML allows it be parsed easily and allows one to construct a parse-tree whose interior nodes are attributes and whose leaves are string-valued values. The parse-tree associated with the previous example is illustrated in Figure 15.11. This data structure gives a handy way of methodolically building a model from a DML description. The SSFNet engine recursively traverses the tree and configures core SSFNet objects (such as host). Attributes or values within the tree can be referenced globally by the sequence of attribute labels on nodes

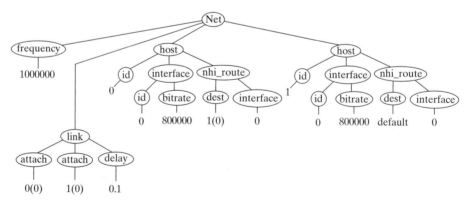

Figure 15.11   Parse tree of simple DML example.

Table 15.1 Common Attributes in SSFNet DML Models

Attribute	*Value*
Net	list describing a network
frequency	number of discrete ticks per simulation second
traffic	list of traffic patterns
pattern	description of traffic pattern, in terms of receiver (client) and server (sender).
servers	list describing a set of servers to which a client might connect—including their NHI identities and port numbers
link	list describing interfaces to be connected, and associated latency
host	list describing a host, and diverse attributes it may have
graph	list of protocols in a host's protocol stack
ProtocolSession	list specifying a protocol
interface	a list describing a connection to the network; attributes include connection bandwidth, and target file for storing monitoring information.
route	description of a forwarding table entry for IP. The *dest* attribute identifies the destination being described; the interface attribute describes which interface packets for that destination should be routed.
dictionary	a list of constants that can be referenced elsewhere within the DML file

from the root to the target. This proves to be useful: one can embed in a DML file a "library" of attribute–value pairs and reference elements of that library.

SSFNet recognizes a variety of attributes, many of which are described in Table 15.1.

15.6.2 Example

Finally, we illustrate some of these ideas by looking at the DML input file for one of our TCP examples. The file is presented in Figure 15.12 (annotated with line numbers for easier reference).

In this particular file, lines 1–8 are comments describing the architecture. Line 10 tells the SSFNet model parser where to find format descriptions of certain constructs; when the parser encounters these constructs in the DML file, it will check against the schema to ensure format correctness. Line 12 starts the overarching list: "Net" followed by a list. Line 14 specifies a clock resolution of 1 microsecond. Lines 15–20 describe the network's traffic, a single pattern that includes host 0 as client. The "servers" attribute gives a list of servers, in this case a single one at NHI address 1(0) (meaning host 1, interface), using port 10.

The "link" attribute at line 24 describes two interfaces to be connected: the one at NHI address 0(0), and the one at NHI address 1(0). The latency across this link is specified to be 0.1 seconds.

A host contains protocols and interfaces to the network. The host beginning at line 28 is given NHI id 1 and contains a "graph" of protocol sessions. Each model of a software component is described as such a session. The order of appearance in the graph is important, descending from higher to lower in the stack. Each protocol session describes its type (e.g., server, client, TCP, IP), and the Java class that describes its behavior. These classes are constructed, by using certain methologies, to be composable; builders of simulation models (in contrast to developers

```
# basic3a.dml
# ****  topology
# *       client                 server
# **      0(0) ...........          1(0)
#              800kb  ..100ms         |
#                                     tcpdump
Net [
10  _schema [ _find .schemas.Net ]
    frequency 1000000
    traffic [
15    pattern [
        client 0
        servers [ port 10 nhi 1(0) ]
      ]
20  ]
#  --------------- LINKS
25  link[attach 0(0) attach 1(0) delay 0.1]
#  --------------- SERVER
    host [
    id 1
30  graph [
        ProtocolSession [name server use SSF.OS.TCP.test.tcpServer port 10
            request_size 4
            show_report true
            debug      false
35      ]
        ProtocolSession [name socket use SSF.OS.Socket.socketMaster]
        ProtocolSession [name tcp use SSF.OS.TCP.tcpSessionMaster
            _find .dictionary.tcpinit
            debug    true
40          rttdump "basic3a.rtt"
            rexdump "basic3a.rto"
            cwnddump "basic3a.wnd"
            con_count "basic3a.con_count"
        ]
45      ProtocolSession [name ip use SSF.OS.IP]
    ]
    interface [id 0 bitrate 800000 tcpdump "basic3a.tcpdump"]
    route [dest default interface 0 ]
    ]
#  --------------- CLIENT
50  host [
    id 0
    graph [
55      ProtocolSession [
            name client use SSF.OS.TCP.test.tcpClient
            start_time 192    # earliest time to send request to server
            start_window 0    # start time jitter window size
            file_size 300000  # file size
60          request_size  4   # client request datagram size (bytes)
            show_report true  # print client-server session summary report
            debug false       # print verbose client/server diagnostics
        ]
        ProtocolSession [name socket use SSF.OS.Socket.socketMaster]
65      ProtocolSession [name tcp use SSF.OS.TCP.tcpSessionMaster
            _find .dictionary.tcpinit
        ]
        ProtocolSession [name ip use SSF.OS.IP]
70  interface [id 0 bitrate 800000]
    route [dest default interface 0 ]
    ]
75 ]


80 dictionary [
    tcpinit [
        ISS 0                    # initial sequence number
        MSS 960                  # maximum segment size
85      RcvWndSize  32           # receive buffer size
        SendWndSize 32           # maximum send window size
        SendBufferSize 32        # send buffer size
        MaxRexmitTimes 12        # maximum retransmission times before drop
        TCP_SLOW_INTERVAL 0.5    # granularity of TCP slow timer
90      TCP_FAST_INTERVAL 0.2    # granularity of TCP fast(delayack) timer
        MSL 60.0                 # maximum segment lifetime
        MaxIdleTime 600.0        # maximum idle time for drop a connection
        delayed_ack false        # delayed ack option
        fast_recovery true       # implement fast recovery algorithm
95      show_report true         # print terse TCP connection diagnostics
    ]
]
```

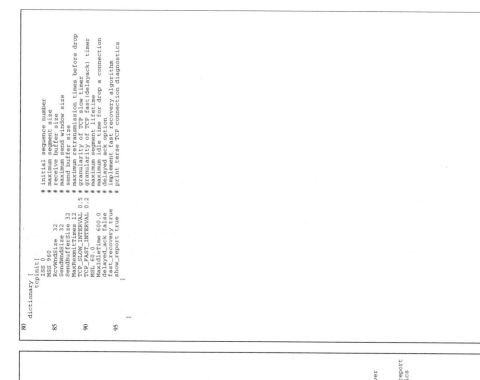

Figure 15.12 Domain Modeling Language specification of a simple network experiment.

of modeling components these builders use) need not develop new classes, but the methodology specifies how one does. A protocol session of a given type may include attributes specific to that type. For example, the tcpServer protocol beginning at line 32 specifies the port through which it is accessible (10). Line 37 begins the declaration of the tcpSessionMaster, a component that manages all TCP sessions. Characteristics of its version of TCP are described by including a list of attributes defined in a list held elsewhere in the DML file. The statement `_find .dictionary.tcpinit` causes the contents of the named list to essentially be inserted at the point of the statement. The string `.dictionary.tcpinit` names the list in terms of how to find it in the file: "." is the highest level list, "`dictionary`" is the name of an attribute in that list, "`tcpinit`" is the attribute associated with the sought list, an attribute of the value-list of `dictionary`. This list starts at line 82.

We quickly describe the meaning of each attribute not obvious from the comments, in order to illustrate the diversity of parameters in SSFNet's implementation of TCP. `RcvWnd-Size`, `SendWndSize`, and `SendBufferSize` describe units of MSS and limit buffer useage (which affects TCP behavior, as we have already seen). A missing segment will be retransmitted up to `MaxrexmitTimes` times before the TCP session is aborted. `TCP_SLOW_INTERVAL` and `TCP_FAST_INTERVAL` give timer values used to determine when enough time has gone by so that a transmitted segment has not yet been acknowledged. If a TCP session is inactive for `MaxIdle-Time` seconds, it is terminated. `delay_ack` and `fast_recovery` are Boolean flags that describe whether to use particular optimizations known for TCP.

Back within the specification of the host (at lines 40–43), we find attributes whose values are files into which the system saves descriptions of how TCP variables behaved during the simulation. Following this (at line 40) is the inclusion of the IP protocol. This, in turn, is followed by declaration of the server's single interface, given id 0 for NHI coordinates and specified to have a bandwidth of 800 Kbits per second. The last attribute for the server is an "`nhi_route`", an element in IP's forwarding table, described in NHI coordinates. The server is not a router and so needs only to direct traffic from IP to one interface. Attribute–value pair `dest default` says to route *everything* through the interface to follow, 0.

Specification of the second host is similar. In this case, the uppermost ProtocolSession is that of a client that requests data, through a socket. Attributes for the client include the simulation time at which it initiates the request. (It actually specifies a window of simulation time in which this occurs, to provide some jitter when multiple clients are to start more or less simultaneously). The length of the transfer being requested is an attribute (line 60). The rest of this host's ProtocolSessions are similar to the server's, although we don't save so much information about TCP's behavior at this host.

15.7 SUMMARY

In this chapter, we touched on some important topics related to simulation of computer networks. Traffic modeling—at different levels of abstraction—is a crucial element of simulating and modeling networks. We emphasized the importance of non-Poisson arrivals models, in some cases to better match characteristics of specific applications, in others to be sure to explain and capture long-range dependence.

Next, we focused on the Data Link layer and on the Media Access Control algorithms. We examined the token-bus and ethernet protocols, discussed subleties of their simulation, and showed by example how significant an impact traffic-model assumptions can have on network performance. Following this, we mentioned issues at the Data Link layer for which simulation has been a critical tool for investigation.

Much of the traffic on the Internet is carried by using TCP. We described TCP's basic rules and used simulation to illustrate some of the consequences of these rules. Finally, we sketched how one builds network models in the SSFNet simulator.

This chapter has barely scratched the surface of how networking uses simulation. Our hope is that what we discuss leads a student to explore more deeply any one of a number of fascinating areas of networking that can be explored only with simulation. The exercises are designed to do this and to teach the student some skill in using SSF and SSFNet.

REFERENCES

BARFORD, P., AND M. CROVELLA [1998], "An Architecture for a WWW Workload Generator," *Proceedings of the 1998 SIGMETRICS Conference*, Madison, WI, pp. 151–160.

BLACK, U. [2001], "Voice Over IP," Prentice Hall, Upper Saddle River, NJ.

COMER, D. [2000], "Networking with TCP/IP Volume 1: Principles, Protocols, and Architecture, 4th ed.," Prentice Hall, Upper Saddle River, NJ.

FLOYD, S., AND V. JACOBSON [1993], "Random Early Detection Gateways for Congestion Avoidance," *IEEE/ACM Transactions on Networking*, Vol. 1, No. 4, pp. 397–413.

KARAGIANNIS T., M. FALOUTSOS, AND M. MOLLE [2003], "A User-Friendly Self-Similarity Analysis Tool," *ACM SIGCOMM Computer Communication Review*, Vol. 33, No. 3, pp. 81–93.

KUROSE, J., AND K. ROSS [2002], *Computer Networking: A Top-Down Approach Featuring the Internet*, 2d ed., Addison–Wesley, Reading, MA.

MOY, J. [1998], *OSPF: Anatomy of an Internet Routing Protocol*, Addison–Wesley, Reading, MA.

SPURGEON, C. [2000], *Ethernet: The Definitive Guide*, O'Reilly, Cambridge MA.

VAN BEIJNUM, I. [2002], *BGP: Building Reliable Networks with the Border Gateway Protocol*, O'Reilly, Cambridge, MA.

ZIMMERMAN, H. [1980], "OSI reference model—the ISO model of architecture for open system interconnection" *IEEE Transactions on Communications*, Vol. COM-28, No. 4, pp. 425–432.

ZUKEMAN, M., D. NEAME, AND R. ADDIE [2003], "Internet Traffic Modeling and Future Technology Implications," *Proceedings of the 2003 InfoCom Conference*, San Franciso, CA.

EXERCISES

1. Survey literature in models of Voice-over-IP traffic, and build a simulator that creates traffic load corresponding to one model of particular interest.

2. Create a Markov-Modulated Poisson process (see chapter 14) and a Poisson–Pareto Burst Process that yield the same average bit-rate traffic demand. Acquire the SELFIS tool for analyzing long-range dependence (it's free), and compare traces from the MMP and PPBP models.

3. Get from www.bcnn.net the SSF models for the token-bus protocol experiments reported in this chapter. Design and perform a sensitivity analysis of mean time in system, as a function of frame size. Consider different input traffic models.

4. Get from www.bcnn.net the SSF models for the Ethernet protocol experiments reported in this chapter. Design and perform a sensitivity analysis of throughput as a function of the physical distance between ethernet ports. Likewise, design and perform a sensitivity analysis of throughput as a function of maximum frame size.

5. Acquire the SSFNet simulator from www.ssfnet.org (free for academic use) and the TCP models described in this chapter from www.bcnn.net.

- Look into how TCP behavior changes in each case by increasing the bandwidth by a factor of 10.
- Investigate how TCP behavior changes in each case by reducing the link latency by a factor of 10.
- Work out how TCP behavior changes in each case by increasing the buffer limits expressed in the DML file by a factor of 10.

Appendix

Table A.1   Random Digits

94737	08225	35614	24826	88319	05595	58701	57365	74759
87259	85982	13296	89326	74863	99986	68558	06391	50248
63856	14016	18527	11634	96908	52146	53496	51730	03500
66612	54714	46783	61934	30258	61674	07471	67566	31635
30712	58582	05704	23172	86689	94834	99057	55832	21012
69607	24145	43886	86477	05317	30445	33456	34029	09603
37792	27282	94107	41967	21425	04743	42822	28111	09757
01488	56680	73847	64930	11108	44834	45390	86043	23973
66248	97697	38244	50918	55441	51217	54786	04940	50807
51453	03462	61157	65366	61130	26204	15016	85665	97714
92168	82530	19271	86999	96499	12765	20926	25282	39119
36463	07331	54590	00546	03337	41583	46439	40173	46455
47097	78780	04210	87084	44484	75377	57753	41415	09890
80400	45972	44111	99708	45935	03694	81421	60170	58457
94554	13863	88239	91624	00022	40471	78462	96265	55360
31567	53597	08490	73544	72573	30961	12282	97033	13676
07821	24759	47266	21747	72496	77755	50391	59554	31177
09056	10709	69314	11449	40531	02917	95878	74587	60906
19922	37025	80731	26179	16039	01518	82697	73227	13160
29923	02570	80164	36108	73689	26342	35712	49137	13482
29602	29464	99219	20308	82109	03898	82072	85199	13103
94135	94661	87724	88187	62191	70607	63099	40494	49069
87926	34092	34334	55064	43152	01610	03126	47312	59578
85039	19212	59160	83537	54414	19856	90527	21756	64783
66070	38480	74636	45095	86576	79337	39578	40851	53503
78166	82521	79261	12570	10930	47564	77869	16480	43972
94672	07912	26153	10531	12715	63142	88937	94466	31388
56406	70023	27734	22254	27685	67518	63966	33203	70803
67726	57805	94264	77009	08682	18784	47554	59869	66320
07516	45979	76735	46509	17696	67177	92600	55572	17245
43070	22671	00152	81326	89428	16368	57659	79424	57604
36917	60370	80812	87225	02850	47118	23790	55043	75117
03919	82922	02312	31106	44335	05573	17470	25900	91080
46724	22558	64303	78804	05762	70650	56117	06707	90035
16108	61281	86823	20286	14025	24909	38391	12183	89393
74541	75808	89669	87680	72758	60851	55292	95663	88326
82919	31285	01850	72550	42986	57518	01159	01786	98145
31388	26809	77258	99360	92362	21979	41319	75739	98082
17190	75522	15687	07161	99745	48767	03121	20046	28013
00466	88068	68631	98745	97810	35886	14497	90230	69264

Table A.2   Random Normal Numbers

0.23	−0.17	0.43	2.18	2.13	0.49	2.72	−0.18	0.42
0.24	−1.17	0.02	0.67	−0.59	−0.13	−0.15	−0.46	1.64
−1.16	−0.17	0.36	−1.26	0.91	0.71	−1.00	−1.09	−0.02
−0.02	−0.19	−0.04	1.92	0.71	−0.90	−0.21	−1.40	−0.38
0.39	0.55	0.13	2.55	−0.33	−0.05	−0.34	−1.95	−0.44
0.64	−0.36	0.98	−0.21	−0.52	−0.02	−0.15	−0.43	0.62
−1.90	0.48	−0.54	0.60	−0.35	−1.29	−0.57	0.23	1.41
−1.04	−0.70	−1.69	1.76	0.47	−0.52	−0.73	0.94	−1.63
−.78	0.11	−0.91	−1.13	0.07	0.45	−0.94	1.42	0.75
0.68	1.77	−0.82	−1.68	−2.60	1.59	−0.72	−0.80	0.61
−0.02	0.92	1.76	−0.66	0.18	−1.32	1.26	0.61	0.83
−0.47	1.04	0.83	−2.05	1.00	−0.70	1.12	0.82	0.08
−0.40	1.40	1.20	0.00	0.21	−2.13	−0.22	1.79	0.87
−0.75	0.09	−1.50	0.14	−2.99	−0.41	−0.99	−0.70	0.51
−0.66	−1.97	0.15	−1.16	−0.60	0.50	1.36	1.94	0.11
−0.44	−0.09	−0.59	1.37	0.18	1.44	−0.80	2.11	−1.37
1.41	−2.71	−0.67	1.83	0.97	0.06	−0.28	0.04	−0.21
1.21	−0.52	−0.20	−0.88	−0.78	0.84	−1.08	−0.25	0.17
0.07	0.66	−0.51	−0.04	−0.84	0.04	1.60	−0.92	1.14
−0.08	0.79	−0.09	−1.12	−1.13	0.77	0.40	0.69	−0.12
0.53	−0.36	−2.64	0.22	−0.78	1.92	−0.26	1.04	−1.61
−1.56	1.82	−1.03	1.14	−0.12	−0.78	−0.12	1.42	−0.52
0.03	−1.29	−0.33	2.60	−0.64	1.19	−0.13	0.91	0.78
1.49	1.55	−0.79	1.37	0.97	0.17	0.58	1.43	−1.29
−1.19	1.35	0.16	1.06	−0.17	0.32	−0.28	0.68	0.54
−1.19	−1.03	−0.12	1.07	0.87	−1.40	−0.24	−0.81	0.31
0.11	−1.95	−0.44	−0.39	−0.15	−1.20	−1.98	0.32	2.91
−1.86	0.06	0.19	−1.29	0.33	1.51	−0.36	−0.80	−0.99
0.16	0.28	0.60	−0.78	0.67	0.13	−0.47	−0.18	−0.89
1.21	−1.19	−0.60	−1.22	0.07	−1.13	1.45	0.94	0.54
−0.82	0.54	−0.98	−0.13	1.52	0.77	0.95	−0.84	2.40
0.75	−0.80	−0.28	1.77	−0.16	−0.33	2.43	−1.11	1.63
0.42	0.31	1.56	0.56	0.64	−0.78	0.04	1.34	−0.01
−1.50	−1.78	−0.59	0.16	0.36	1.89	−1.19	0.53	−0.97
−0.89	0.08	0.95	−0.73	1.25	−1.04	−0.47	−0.68	−0.87
0.19	0.85	1.68	−0.57	0.37	−0.48	−0.17	2.36	−0.53
0.49	0.32	−2.08	−1.02	2.59	−0.53	0.15	0.11	0.05
−1.44	0.07	−0.22	−0.93	−1.40	0.54	−1.28	−0.15	0.67
−0.21	−0.48	1.21	0.67	−1.10	−0.75	−0.37	0.68	−0.02
−0.65	−0.12	0.94	−0.44	−1.21	−0.06	−1.28	−1.51	1.39
0.24	−0.83	1.55	0.33	−0.59	−1.24	0.70	0.01	0.15
−0.73	1.24	0.40	−0.61	0.68	0.69	0.07	−0.23	−0.66
−1.93	0.75	−0.32	0.95	1.35	1.51	−0.88	0.10	−1.19
0.08	0.16	0.38	−0.96	1.99	−0.20	0.98	0.16	0.26
−0.47	−1.25	0.32	0.51	−1.04	0.97	2.60	−0.08	1.19

Table A.3   Cumulative Normal Distribution

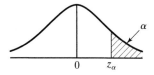

$$\phi(z_\alpha) = \int_{-\infty}^{z_\alpha} \frac{1}{\sqrt{2\pi}} e^{-u^2/2}\, du = 1 - \alpha$$

z_α	0.00	0.01	0.02	0.03	0.04	z_α
0.0	0.500 00	0.503 99	0.507 98	0.511 97	0.515 95	**0.0**
0.1	0.539 83	0.543 79	0.547 76	0.551 72	0.555 67	**0.1**
0.2	0.579 26	0.583 17	0.587 06	0.590 95	0.594 83	**0.2**
0.3	0.617 91	0.621 72	0.625 51	0.629 30	0.633 07	**0.3**
0.4	0.655 42	0.659 10	0.662 76	0.666 40	0.670 03	**0.4**
0.5	0.691 46	0.694 97	0.698 47	0.701 94	0.705 40	**0.5**
0.6	0.725 75	0.729 07	0.732 37	0.735 65	0.738 91	**0.6**
0.7	0.758 03	0.761 15	0.764 24	0.767 30	0.770 35	**0.7**
0.8	0.788 14	0.791 03	0.793 89	0.796 73	0.799 54	**0.8**
0.9	0.815 94	0.818 59	0.821 21	0.823 81	0.826 39	**0.9**
1.0	0.841 34	0.843 75	0.846 13	0.848 49	0.850 83	**1.0**
1.1	0.864 33	0.866 50	0.868 64	0.870 76	0.872 85	**1.1**
1.2	0.884 93	0.886 86	0.888 77	0.890 65	0.892 51	**1.2**
1.3	0.903 20	0.904 90	0.906 58	0.908 24	0.909 88	**1.3**
1.4	0.919 24	0.920 73	0.922 19	0.923 64	0.925 06	**1.4**
1.5	0.933 19	0.934 48	0.935 74	0.936 99	0.938 22	**1.5**
1.6	0.945 20	0.946 30	0.947 38	0.948 45	0.949 50	**1.6**
1.7	0.955 43	0.956 37	0.957 28	0.958 18	0.959 07	**1.7**
1.8	0.964 07	0.964 85	0.965 62	0.966 37	0.967 11	**1.8**
1.9	0.971 28	0.971 93	0.972 57	0.973 20	0.973 81	**1.9**
2.0	0.977 25	0.977 78	0.978 31	0.978 82	0.979 32	**2.0**
2.1	0.982 14	0.982 57	0.983 00	0.983 41	0.983 82	**2.1**
2.2	0.986 10	0.986 45	0.986 79	0.987 13	0.987 45	**2.2**
2.3	0.989 28	0.989 56	0.989 83	0.990 10	0.990 36	**2.3**
2.4	0.991 80	0.992 02	0.992 24	0.992 45	0.992 66	**2.4**
2.5	0.993 79	0.993 96	0.994 13	0.994 30	0.994 46	**2.5**
2.6	0.995 34	0.995 47	0.995 60	0.995 73	0.995 85	**2.6**
2.7	0.996 53	0.996 64	0.996 74	0.996 83	0.996 93	**2.7**
2.8	0.997 44	0.997 52	0.997 60	0.997 67	0.997 74	**2.8**
2.9	0.998 13	0.998 19	0.998 25	0.998 31	0.998 36	**2.9**
3.0	0.998 65	0.998 69	0.998 74	0.998 78	0.998 82	**3.0**
3.1	0.999 03	0.999 06	0.999 10	0.999 13	0.999 16	**3.1**
3.2	0.999 31	0.999 34	0.999 36	0.999 38	0.999 40	**3.2**
3.3	0.999 52	0.999 53	0.999 55	0.999 57	0.999 58	**3.3**
3.4	0.999 66	0.999 68	0.999 69	0.999 70	0.999 71	**3.4**
3.5	0.999 77	0.999 78	0.999 78	0.999 79	0.999 80	**3.5**
3.6	0.999 84	0.999 85	0.999 85	0.999 86	0.999 86	**3.6**
3.7	0.999 89	0.999 90	0.999 90	0.999 90	0.999 91	**3.7**
3.8	0.999 93	0.999 93	0.999 93	0.999 94	0.999 94	**3.8**
3.9	0.999 95	0.999 95	0.999 96	0.999 96	0.999 96	**3.9**

(continued overleaf)

Table A.3   (*continued*)

z_α	0.00	0.01	0.02	0.03	0.04	z_α
0.0	0.519 94	0.523 92	0.527 90	0.531 88	0.535 86	**0.0**
0.1	0.559 62	0.563 56	0.567 49	0.571 42	0.575 34	**0.1**
0.2	0.598 71	0.602 57	0.606 42	0.610 26	0.614 09	**0.2**
0.3	0.636 83	0.640 58	0.644 31	0.648 03	0.651 73	**0.3**
0.4	0.673 64	0.677 24	0.680 82	0.684 38	0.687 93	**0.4**
0.5	0.708 84	0.712 26	0.715 66	0.719 04	0.722 40	**0.5**
0.6	0.742 15	0.745 37	0.748 57	0.751 75	0.754 90	**0.6**
0.7	0.773 37	0.776 37	0.779 35	0.782 30	0.785 23	**0.7**
0.8	0.802 34	0.805 10	0.807 85	0.810 57	0.813 27	**0.8**
0.9	0.824 94	0.831 47	0.833 97	0.836 46	0.838 91	**0.9**
1.0	0.853 14	0.855 43	0.857 69	0.859 93	0.862 14	**1.0**
1.1	0.874 93	0.876 97	0.879 00	0.881 00	0.882 97	**1.1**
1.2	0.894 35	0.896 16	0.897 96	0.899 73	0.901 47	**1.2**
1.3	0.911 49	0.913 08	0.914 65	0.916 21	0.917 73	**1.3**
1.4	0.926 47	0.927 85	0.929 22	0.930 56	0.931 89	**1.4**
1.5	0.939 43	0.940 62	0.941 79	0.942 95	0.944 08	**1.5**
1.6	0.950 53	0.951 54	0.952 54	0.953 52	0.954 48	**1.6**
1.7	0.959 94	0.960 80	0.961 64	0.962 46	0.963 27	**1.7**
1.8	0.967 84	0.968 56	0.969 26	0.969 95	0.970 62	**1.8**
1.9	0.974 41	0.975 00	0.975 58	0.976 15	0.976 70	**1.9**
2.0	0.979 82	0.980 30	0.980 77	0.981 24	0.981 69	**2.0**
2.1	0.984 22	0.984 61	0.985 00	0.985 37	0.985 74	**2.1**
2.2	0.987 78	0.988 09	0.988 40	0.988 70	0.988 99	**2.2**
2.3	0.990 61	0.990 86	0.991 11	0.991 34	0.991 58	**2.3**
2.4	0.992 86	0.993 05	0.993 24	0.993 43	0.993 61	**2.4**
2.5	0.994 61	0.994 77	0.994 92	0.995 06	0.995 20	**2.5**
2.6	0.995 98	0.996 09	0.996 21	0.996 32	0.996 43	**2.6**
2.7	0.997 02	0.997 11	0.997 20	0.997 28	0.997 36	**2.7**
2.8	0.997 81	0.997 88	0.997 95	0.998 01	0.998 07	**2.8**
2.9	0.998 41	0.998 46	0.998 51	0.998 56	0.998 61	**2.9**
3.0	0.998 86	0.998 89	0.998 93	0.998 97	0.999 00	**3.0**
3.1	0.999 18	0.999 21	0.999 24	0.999 26	0.999 29	**3.1**
3.2	0.999 42	0.999 44	0.999 46	0.999 48	0.999 50	**3.2**
3.3	0.999 60	0.999 61	0.999 62	0.999 64	0.999 65	**3.3**
3.4	0.999 72	0.999 73	0.999 74	0.999 75	0.999 76	**3.4**
3.5	0.999 81	0.999 81	0.999 82	0.999 83	0.999 83	**3.5**
3.6	0.999 87	0.999 87	0.999 88	0.999 88	0.999 89	**3.6**
3.7	0.999 91	0.999 92	0.999 92	0.999 92	0.999 92	**3.7**
3.8	0.999 94	0.999 94	0.999 95	0.999 95	0.999 95	**3.8**
3.9	0.999 96	0.999 96	0.999 96	0.999 97	0.999 97	**3.9**

Source: W. W. Hines and D. C. Montgomery, *Probability and Statistics in Engineering and Management Science*, 2d ed., © 1980, pp. 592–3. Reprinted by permission of John Wiley & Sons, Inc., New York.

Table A.4 Cumulative Poisson Distribution

	$\alpha = $ Mean											
x	.01	.05	.1	.2	.3	.4	.5	.6	.7	.8	.9	x
0	.990	.951	.905	.819	.741	.670	.607	.549	.497	.449	.407	0
1	1.000	.999	.995	.982	.963	.938	.910	.878	.844	.809	.772	1
2		1.000	1.000	.999	.996	.992	.986	.977	.966	.953	.937	2
3				1.000	1.000	.999	.998	.997	.994	.991	.987	3
4						1.000	1.000	1.000	.999	.999	.998	4
5									1.000	1.000	1.000	5

	$\alpha = $ Mean											
x	1.0	1.1	1.2	1.3	1.4	1.5	1.6	1.7	1.8	1.9	2.0	x
0	.368	.333	.301	.273	.247	.223	.202	.183	.165	.150	.135	0
1	.736	.699	.663	.627	.592	.558	.525	.493	.463	.434	.406	1
2	.920	.900	.879	.857	.833	.809	.783	.757	.731	.704	.677	2
3	.981	.974	.966	.957	.946	.934	.921	.907	.891	.875	.857	3
4	.996	.995	.992	.989	.986	.981	.976	.970	.964	.956	.947	4
5	.999	.999	.998	.998	.997	.996	.994	.992	.990	.987	.983	5
6	1.000	1.000	1.000	1.000	.999	.999	.999	.998	.997	.997	.995	6
7					1.000	1.000	1.000	1.000	.999	.999	.999	7
8									1.000	1.000	1.000	8

	$\alpha = $ Mean											
x	2.2	2.4	2.6	2.8	3.0	3.5	4.0	4.5	5.0	5.5	6.0	x
0	.111	.091	.074	.061	.050	.030	.018	.011	.007	.004	.002	0
1	.355	.308	.267	.231	.199	.136	.092	.061	.040	.027	.017	1
2	.623	.570	.518	.469	.423	.321	.238	.174	.125	.088	.062	2
3	.819	.779	.736	.692	.647	.537	.433	.342	.265	.202	.151	3
4	.928	.904	.877	.848	.815	.725	.629	.532	.440	.358	.285	4
5	.975	.964	.951	.935	.916	.858	.785	.703	.616	.529	.446	5
6	.993	.988	.983	.976	.966	.935	.889	.831	.762	.686	.606	6
7	.998	.997	.995	.992	.988	.973	.949	.913	.867	.809	.744	7
8	1.000	.999	.999	.998	.996	.990	.979	.960	.932	.894	.847	8
9		1.000	1.000	.999	.999	.997	.992	.983	.968	.946	.916	9
10				1.000	1.000	.999	.997	.993	.986	.975	.957	10
11						1.000	.999	.998	.995	.989	.980	11
12							1.000	.999	.998	.996	.991	12
13								1.000	.999	.998	.996	13
14									1.000	.999	.999	14
15										1.000	.999	15
16											1.000	16

(continued overleaf)

Table A.4 (*continued*)

x	6.5	7.0	7.5	8.0	9.0	10.0	12.0	14.0	16.0	18.0	20.0	x
					α = Mean							
0	.002	.001	.001									0
1	.011	.007	.005	.003	.001							1
2	.043	.030	.020	.014	.006	.003	.001					2
3	.112	.082	.059	.042	.021	.010	.002					3
4	.224	.173	.132	.100	.055	.029	.008	.002				4
5	.369	.301	.241	.191	.116	.067	.020	.006	.001			5
6	.527	.450	.378	.313	.207	.130	.046	.014	.004	.001		6
7	.673	.599	.525	.453	.324	.220	.090	.032	.010	.003	.001	7
8	.792	.729	.662	.593	.456	.333	.155	.062	.022	.007	.002	8
9	.877	.830	.776	.717	.587	.458	.242	.109	.043	.015	.005	9
10	.933	.901	.862	.816	.706	.583	.347	.176	.077	.030	.011	10
11	.966	.947	.921	.888	.803	.697	.462	.260	.127	.055	.021	11
12	.984	.973	.957	.936	.876	.792	.576	.358	.193	.092	.039	12
13	.993	.987	.978	.966	.926	.864	.682	.464	.275	.143	.066	13
14	.997	.994	.990	.983	.959	.917	.772	.570	.368	.208	.105	14
15	.999	.998	.995	.992	.978	.951	.844	.669	.467	.287	.157	15
16	1.000	.999	.998	.996	.989	.973	.899	.756	.566	.375	.221	16
17		1.000	.999	.998	.995	.986	.937	.827	.659	.469	.297	17
18			1.000	.999	.998	.993	.963	.883	.742	.562	.381	18
19				1.000	.999	.997	.979	.923	.812	.651	.470	19
20					1.000	.998	.988	.952	.868	.731	.559	20
21						.999	.994	.971	.911	.799	.644	21
22						1.000	.997	.983	.942	.855	.721	22
23							.999	.991	.963	.899	.787	23
24							.999	.995	.978	.932	.843	24
25							1.000	.997	.987	.955	.888	25
26								.999	.993	.972	.922	26
27								.999	.996	.983	.948	27
28								1.000	.998	.990	.966	28
29									.999	.994	.978	29
30									.999	.997	.987	30
31									1.000	.998	.992	31
32										.999	.995	32
33										1.000	.997	33
34											.999	34
35											.999	35
36											1.000	36

Source: J. Banks and R. G. Heikes, *Handbook of Tables and Graphs for the Industrial Engineer and Manager,* © 1984, pp. 34–35. Reprinted by permission of John Wiley and Sons, Inc., New York.

Table A.5 Percentage Points of The Student's t Distribution with v Degrees of Freedom

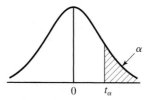

v	$t_{0.005}$	$t_{0.01}$	$t_{0.025}$	$t_{0.05}$	$t_{0.10}$
1	63.66	31.82	12.71	6.31	3.08
2	9.92	6.92	4.30	2.92	1.89
3	5.84	4.54	3.18	2.35	1.64
4	4.60	3.75	2.78	2.13	1.53
5	4.03	3.36	2.57	2.02	1.48
6	3.71	3.14	2.45	1.94	1.44
7	3.50	3.00	2.36	1.90	1.42
8	3.36	2.90	2.31	1.86	1.40
9	3.25	2.82	2.26	1.83	1.38
10	3.17	2.76	2.23	1.81	1.37
11	3.11	2.72	2.20	1.80	1.36
12	3.06	2.68	2.18	1.78	1.36
13	3.01	2.65	2.16	1.77	1.35
14	2.98	2.62	2.14	1.76	1.34
15	2.95	2.60	2.13	1.75	1.34
16	2.92	2.58	2.12	1.75	1.34
17	2.90	2.57	2.11	1.74	1.33
18	2.88	2.55	2.10	1.73	1.33
19	2.86	2.54	2.09	1.73	1.33
20	2.84	2.53	2.09	1.72	1.32
21	2.83	2.52	2.08	1.72	1.32
22	2.82	2.51	2.07	1.72	1.32
23	2.81	2.50	2.07	1.71	1.32
24	2.80	2.49	2.06	1.71	1.32
25	2.79	2.48	2.06	1.71	1.32
26	2.78	2.48	2.06	1.71	1.32
27	2.77	2.47	2.05	1.70	1.31
28	2.76	2.47	2.05	1.70	1.31
29	2.76	2.46	2.04	1.70	1.31
30	2.75	2.46	2.04	1.70	1.31
40	2.70	2.42	2.02	1.68	1.30
60	2.66	2.39	2.00	1.67	1.30
120	2.62	2.36	1.98	1.66	1.29
∞	2.58	2.33	1.96	1.645	1.28

Source: Robert E. Shannon, *Systems Simulation: The Art and Science*, © 1975, p. 375. Reprinted by permission of Prentice Hall, Upper Saddle River, NJ.

Table A.6 Percentage Points of The Chi-Square Distribution with v Degrees of Freedom

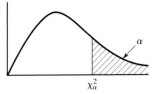

χ_α^2

v	$\chi_{0.005}^2$	$\chi_{0.01}^2$	$\chi_{0.025}^2$	$\chi_{0.05}^2$	$\chi_{0.10}^2$
1	7.88	6.63	5.02	3.84	2.71
2	10.60	9.21	7.38	5.99	4.61
3	12.84	11.34	9.35	7.81	6.25
4	14.96	13.28	11.14	9.49	7.78
5	16.7	15.1	12.8	11.1	9.2
6	18.5	16.8	14.4	12.6	10.6
7	20.3	18.5	16.0	14.1	12.0
8	22.0	20.1	17.5	15.5	13.4
9	23.6	21.7	19.0	16.9	14.7
10	25.2	23.2	20.5	18.3	16.0
11	26.8	24.7	21.9	19.7	17.3
12	28.3	26.2	23.3	21.0	18.5
13	29.8	27.7	24.7	22.4	19.8
14	31.3	29.1	26.1	23.7	21.1
15	32.8	30.6	27.5	25.0	22.3
16	34.3	32.0	28.8	26.3	23.5
17	35.7	33.4	30.2	27.6	24.8
18	37.2	34.8	31.5	28.9	26.0
19	38.6	36.2	32.9	30.1	27.2
20	40.0	37.6	34.2	31.4	28.4
21	41.4	38.9	35.5	32.7	29.6
22	42.8	40.3	36.8	33.9	30.8
23	44.2	41.6	38.1	35.2	32.0
24	45.6	43.0	39.4	36.4	33.2
25	49.6	44.3	40.6	37.7	34.4
26	48.3	45.6	41.9	38.9	35.6
27	49.6	47.0	43.2	40.1	36.7
28	51.0	48.3	44.5	41.3	37.9
29	52.3	49.6	45.7	42.6	39.1
30	53.7	50.9	47.0	43.8	40.3
40	66.8	63.7	59.3	55.8	51.8
50	79.5	76.2	71.4	67.5	63.2
60	92.0	88.4	83.3	79.1	74.4
70	104.2	100.4	95.0	90.5	85.5
80	116.3	112.3	106.6	101.9	96.6
90	128.3	124.1	118.1	113.1	107.6
100	140.2	135.8	129.6	124.3	118.5

Source: Robert E. Shannon, *Systems Simulation: The Art and Science*, © 1975, p. 372. Reprinted by permission of Prentice Hall, Upper Saddle River, NJ.

Table A.7 Percentage Points of The F Distribution with $\alpha = 0.05$

Degrees of Freedom for the Numerator (v_1)

v_2 \ v_1	1	2	3	4	5	6	7	8	9	10	12	15	20	24	30	40	60	120	∞
1	161.4	199.5	215.7	224.6	230.2	234.0	236.8	238.9	240.5	241.9	243.9	245.9	248.0	249.1	250.1	251.1	252.2	253.3	254.3
2	18.51	19.00	19.16	19.25	19.30	19.33	19.35	19.37	19.38	19.40	19.41	19.43	19.45	19.45	19.46	19.47	19.48	19.49	19.50
3	10.13	9.55	9.28	9.12	9.01	8.94	8.89	8.85	8.81	8.79	8.74	8.70	8.66	8.64	8.62	8.59	8.57	8.55	8.53
4	7.71	6.94	6.59	6.39	6.26	6.16	6.09	6.04	6.00	5.96	5.91	5.86	5.80	5.77	5.75	5.72	5.69	5.66	5.63
5	6.61	5.79	5.41	5.19	5.05	4.95	4.88	4.82	4.77	4.74	4.68	4.62	4.56	4.53	4.50	4.46	4.43	4.40	4.36
6	5.99	5.14	4.76	4.53	4.39	4.28	4.21	4.15	4.10	4.06	4.00	3.94	3.87	3.84	3.81	3.77	3.74	3.70	3.67
7	5.59	4.74	4.35	4.12	3.97	3.87	3.79	3.73	3.68	3.64	3.57	3.51	3.44	3.41	3.38	3.34	3.30	3.27	3.23
8	5.32	4.46	4.07	3.84	3.69	3.58	3.50	3.44	3.39	3.35	3.28	3.22	3.15	3.12	3.08	3.04	3.01	2.97	2.93
9	5.12	4.26	3.86	3.63	3.48	3.37	3.29	3.23	3.18	3.14	3.07	3.01	2.94	2.90	2.86	2.83	2.79	2.75	2.71
10	4.96	4.10	3.71	3.48	3.33	3.22	3.14	3.07	3.02	2.98	2.91	2.85	2.77	2.74	2.70	2.66	2.62	2.58	2.54
11	4.84	3.98	3.59	3.36	3.20	3.09	3.01	2.95	2.90	2.85	2.79	2.72	2.65	2.61	2.57	2.53	2.49	2.45	2.40
12	4.75	3.89	3.49	3.26	3.11	3.00	2.91	2.85	2.80	2.75	2.69	2.62	2.54	2.51	2.47	2.43	2.38	2.34	2.30
13	4.67	3.81	3.41	3.18	3.03	2.92	2.83	2.77	2.71	2.67	2.60	2.53	2.46	2.42	2.38	2.34	2.30	2.25	2.21
14	4.60	3.74	3.34	3.11	2.96	2.85	2.76	2.70	2.65	2.60	2.53	2.46	2.39	2.35	2.31	2.27	2.22	2.18	2.13
15	4.54	3.68	3.29	3.06	2.90	2.79	2.71	2.64	2.59	2.54	2.48	2.40	2.33	2.29	2.25	2.20	2.16	2.11	2.07
16	4.49	3.63	3.24	3.01	2.85	2.74	2.66	2.59	2.54	2.49	2.42	2.35	2.28	2.24	2.19	2.15	2.11	2.06	2.01
17	4.45	3.59	3.20	2.96	2.81	2.70	2.61	2.55	2.49	2.45	2.38	2.31	2.23	2.19	2.15	2.10	2.06	2.01	1.96
18	4.41	3.55	3.16	2.93	2.77	2.66	2.58	2.51	2.46	2.41	2.34	2.27	2.19	2.15	2.11	2.06	2.02	1.97	1.92
19	4.38	3.52	3.13	2.90	2.74	2.63	2.54	2.48	2.42	2.38	2.31	2.23	2.16	2.11	2.07	2.03	1.98	1.93	1.88
20	4.35	3.49	3.10	2.87	2.71	2.60	2.51	2.45	2.39	2.35	2.28	2.20	2.12	2.08	2.04	1.99	1.95	1.90	1.84
21	4.32	3.47	3.07	2.84	2.68	2.57	2.49	2.42	2.37	2.32	2.25	2.18	2.10	2.05	2.01	1.96	1.92	1.87	1.81
22	4.30	3.44	3.05	2.82	2.66	2.55	2.46	2.40	2.34	2.30	2.23	2.15	2.07	2.03	1.98	1.94	1.89	1.84	1.78
23	4.28	3.42	3.03	2.80	2.64	2.53	2.44	2.37	2.32	2.27	2.20	2.13	2.05	2.01	1.96	1.91	1.86	1.81	1.76
24	4.26	3.40	3.01	2.78	2.62	2.51	2.42	2.36	2.30	2.25	2.18	2.11	2.03	1.98	1.94	1.89	1.84	1.79	1.73
25	4.24	3.39	2.99	2.76	2.60	2.49	2.40	2.34	2.28	2.24	2.16	2.09	2.01	1.96	1.92	1.87	1.82	1.77	1.71
26	4.23	3.37	2.98	2.74	2.59	2.47	2.39	2.32	2.27	2.22	2.15	2.07	1.99	1.95	1.90	1.85	1.80	1.75	1.69
27	4.21	3.35	2.96	2.73	2.57	2.46	2.37	2.31	2.25	2.20	2.13	2.06	1.97	1.93	1.88	1.84	1.79	1.73	1.67
28	4.20	3.34	2.95	2.71	2.56	2.45	2.36	2.29	2.24	2.19	2.12	2.04	1.96	1.91	1.87	1.82	1.77	1.71	1.65
29	4.18	3.33	2.93	2.70	2.55	2.43	2.35	2.28	2.22	2.18	2.10	2.03	1.94	1.90	1.85	1.81	1.75	1.70	1.64
30	4.17	3.32	2.92	2.69	2.53	2.42	2.33	2.27	2.21	2.16	2.09	2.01	1.93	1.89	1.84	1.79	1.74	1.68	1.62
40	4.08	3.23	2.84	2.61	2.45	2.34	2.25	2.18	2.12	2.08	2.00	1.92	1.84	1.79	1.74	1.69	1.64	1.58	1.51
60	4.00	3.15	2.76	2.53	2.37	2.25	2.17	2.10	2.04	1.99	1.92	1.84	1.75	1.70	1.65	1.59	1.53	1.47	1.39
120	3.92	3.07	2.68	2.45	2.29	2.17	2.09	2.02	1.96	1.91	1.83	1.75	1.66	1.61	1.55	1.50	1.43	1.35	1.25
∞	3.84	3.00	2.60	2.37	2.21	2.10	2.01	1.94	1.88	1.83	1.75	1.67	1.57	1.52	1.46	1.39	1.32	1.22	1.00

Degrees of Freedom for Denominator (v_2)

Source: W. W. Hines and D. C. Montgomery, *Probability and Statistics in Engineering and Management Science*, 2d ed., © 1980, p. 599. Reprinted by permission of John Wiley & Sons, Inc., New York.

Table A.8   Kolmogorov–Smirnov Critical Values

Degrees of Freedom (N)	$D_{0.10}$	$D_{0.05}$	$D_{0.01}$
1	0.950	0.975	0.995
2	0.776	0.842	0.929
3	0.642	0.708	0.828
4	0.564	0.624	0.733
5	0.510	0.565	0.669
6	0.470	0.521	0.618
7	0.438	0.486	0.577
8	0.411	0.457	0.543
9	0.388	0.432	0.514
10	0.368	0.410	0.490
11	0.352	0.391	0.468
12	0.338	0.375	0.450
13	0.325	0.361	0.433
14	0.314	0.349	0.418
15	0.304	0.338	0.404
16	0.295	0.328	0.392
17	0.286	0.318	0.381
18	0.278	0.309	0.371
19	0.272	0.301	0.363
20	0.264	0.294	0.356
25	0.24	0.27	0.32
30	0.22	0.24	0.29
35	0.21	0.23	0.27
Over 35	$\dfrac{1.22}{\sqrt{N}}$	$\dfrac{1.36}{\sqrt{N}}$	$\dfrac{1.63}{\sqrt{N}}$

Source: F. J. Massey, "The Kolmogorov–Smirnov Test for Goodness of Fit," *The Journal of the American Statistical Association*, Vol. 46. © 1951, p. 70. Adapted with permission of the American Statistical Association.

Table A.9   Maximum Likelihood Estimates of the Gamma Distribution

$1/M$	β	$1/M$	β	$1/M$	β
0.020	0.0187	2.700	1.494	10.300	5.311
0.030	0.0275	2.800	1.545	10.600	5.461
0.040	0.0360	2.900	1.596	10.900	5.611
0.050	0.0442	3.000	1.646	11.200	5.761
0.060	0.0523	3.200	1.748	11.500	5.911
0.070	0.0602	3.400	1.849	11.800	6.061
0.080	0.0679	3.600	1.950	12.100	6.211
0.090	0.0756	3.800	2.051	12.400	6.362
0.100	0.0831	4.000	2.151	12.700	6.512
0.200	0.1532	4.200	2.252	13.000	6.662
0.300	0.2178	4.400	2.353	13.300	6.812
0.400	0.2790	4.600	2.453	13.600	6.962
0.500	0.3381	4.800	2.554	13.900	7.112
0.600	0.3955	5.000	2.654	14.200	7.262
0.700	0.4517	5.200	2.755	14.500	7.412
0.800	0.5070	5.400	2.855	14.800	7.562
0.900	0.5615	5.600	2.956	15.100	7.712
1.000	0.6155	5.800	3.056	15.400	7.862
1.100	0.6690	6.000	3.156	15.700	8.013
1.200	0.7220	6.200	3.257	16.000	8.163
1.300	0.7748	6.400	3.357	16.300	8.313
1.400	0.8272	6.600	3.457	16.600	8.463
1.500	0.8794	6.800	3.558	16.900	8.613
1.600	0.9314	7.000	3.658	17.200	8.763
1.700	0.9832	7.300	3.808	17.500	8.913
1.800	1.034	7.600	3.958	17.800	9.063
1.900	1.086	7.900	4.109	18.100	9.213
2.000	1.137	8.200	4.259	18.400	9.363
2.100	1.188	8.500	4.409	18.700	9.513
2.200	1.240	8.800	4.560	19.000	9.663
2.300	1.291	9.100	4.710	19.300	9.813
2.400	1.342	9.400	4.860	19.600	9.963
2.500	1.393	9.700	5.010	20.000	10.16
2.600	1.444	10.000	5.160		

Source: S.C. Choi and R. Wette, "Maximum Likelihood Estimates of the Gamma Distribution and Their Bias," *Technometrics*, Vol. 11, No. 4, Nov. © 1969, pp. 688–9. Adapted with permission of the American Statistical Association.

Table A.10 Operating Characteristic Curves for The Two-Sided t Test for Different Values of Sample Size n

(a) $\alpha = 0.05$

(b) $\alpha = 0.01$

Source: C. L. Ferris, F. E. Grubbs, and C. L. Weaver, "Operating Characteristics for the Common Statistical Tests of Significance," *Annals of Mathematical Statistics*, June 1946. Reproduced with permission of The Institute of Mathematical Statistics.

Table A.11 Operating Characteristic Curves for the One-Sided t Test for Different Values of Sample Size n

(a) $\alpha = 0.05$

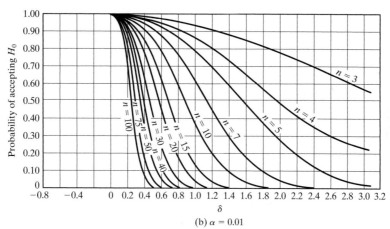

(b) $\alpha = 0.01$

Source: A. H. Bowker and G. J. Lieberman, *Engineering Statistics*, 2d ed., © 1972, p. 203. Reprinted by permission of Prentice Hall, Upper Saddle River, NJ.

Index

591